Advanced Materials in Catalysis and Adsorption

Advanced Materials in Catalysis and Adsorption

Editor

Ilya V. Mishakov

MDPI • Basel • Beijing • Wuhan • Barcelona • Belgrade • Manchester • Tokyo • Cluj • Tianjin

Editor
Ilya V. Mishakov
Department of Materials
Science and Functional
Materials
Boreskov Institute of
Catalysis
Novosibirsk
Russia

Editorial Office
MDPI
St. Alban-Anlage 66
4052 Basel, Switzerland

This is a reprint of articles from the Special Issue published online in the open access journal *Materials* (ISSN 1996-1944) (available at: www.mdpi.com/journal/materials/special_issues/Advanced_Adsorption).

For citation purposes, cite each article independently as indicated on the article page online and as indicated below:

LastName, A.A.; LastName, B.B.; LastName, C.C. Article Title. *Journal Name* **Year**, *Volume Number*, Page Range.

ISBN 978-3-0365-7429-5 (Hbk)
ISBN 978-3-0365-7428-8 (PDF)

© 2023 by the authors. Articles in this book are Open Access and distributed under the Creative Commons Attribution (CC BY) license, which allows users to download, copy and build upon published articles, as long as the author and publisher are properly credited, which ensures maximum dissemination and a wider impact of our publications.

The book as a whole is distributed by MDPI under the terms and conditions of the Creative Commons license CC BY-NC-ND.

Contents

About the Editor vii

Preface to "Advanced Materials in Catalysis and Adsorption" ix

Ilya V. Mishakov
Editorial for Special Issue "Advanced Materials in Catalysis and Adsorption"
Reprinted from: *Materials* **2023**, *16*, 2895, doi:10.3390/ma16072895 1

Anna M. Ozerova, Arina R. Potylitsyna, Yury I. Bauman, Elena S. Tayban, Inna L. Lipatnikova and Anna V. Nartova et al.
Synthesis of Chlorine- and Nitrogen-Containing Carbon Nanofibers for Water Purification from Chloroaromatic Compounds
Reprinted from: *Materials* **2022**, *15*, 8414, doi:10.3390/ma15238414 5

Alexander M. Volodin, Roman M. Kenzhin, Yury I. Bauman, Sofya D. Afonnikova, Arina R. Potylitsyna and Yury V. Shubin et al.
Comparative Study on Carbon Erosion of Nickel Alloys in the Presence of Organic Compounds under Various Reaction Conditions
Reprinted from: *Materials* **2022**, *15*, 9033, doi:10.3390/ma15249033 25

Vladimir V. Chesnokov, Igor P. Prosvirin, Evgeny Yu. Gerasimov and Aleksandra S. Chichkan
Synthesis of Boron-Doped Carbon Nanomaterial
Reprinted from: *Materials* **2023**, *16*, 1986, doi:10.3390/ma16051986 41

Arina R. Potylitsyna, Yuliya V. Rudneva, Yury I. Bauman, Pavel E. Plyusnin, Vladimir O. Stoyanovskii and Evgeny Y. Gerasimov et al.
Efficient Production of Segmented Carbon Nanofibers via Catalytic Decomposition of Trichloroethylene over Ni-W Catalyst
Reprinted from: *Materials* **2023**, *16*, 845, doi:10.3390/ma16020845 53

Sofya D. Afonnikova, Anton A. Popov, Yury I. Bauman, Pavel E. Plyusnin, Ilya V. Mishakov and Mikhail V. Trenikhin et al.
Porous Co-Pt Nanoalloys for Production of Carbon Nanofibers and Composites
Reprinted from: *Materials* **2022**, *15*, 7456, doi:10.3390/ma15217456 75

Tatyana A. Maksimova, Ilya V. Mishakov, Yury I. Bauman, Artem B. Ayupov, Maksim S. Mel'gunov and Aleksey M. Dmitrachkov et al.
Effect of Pretreatment with Acids on the N-Functionalization of Carbon Nanofibers Using Melamine
Reprinted from: *Materials* **2022**, *15*, 8239, doi:10.3390/ma15228239 91

Dmitrii German, Ekaterina Kolobova, Ekaterina Pakrieva, Sónia A. C. Carabineiro, Elizaveta Sviridova and Sergey Perevezentsev et al.
The Effect of Sibunit Carbon Surface Modification with Diazonium Tosylate Salts of Pd and Pd-Au Catalysts on Furfural Hydrogenation
Reprinted from: *Materials* **2022**, *15*, 4695, doi:10.3390/ma15134695 111

Valery Skudin, Tatiana Andreeva, Maria Myachina and Natalia Gavrilova
CVD-Synthesis of N-CNT Using Propane and Ammonia
Reprinted from: *Materials* **2022**, *15*, 2241, doi:10.3390/ma15062241 133

Adrian Walkowiak, Joanna Wolska, Anna Wojtaszek-Gurdak, Izabela Sobczak, Lukasz Wolski and Maria Ziolek
Modification of Gold Zeolitic Supports for Catalytic Oxidation of Glucose to Gluconic Acid
Reprinted from: *Materials* 2021, *14*, 5250, doi:10.3390/ma14185250 147

Vladislav Shilov, Dmitriy Potemkin, Vladimir Rogozhnikov and Pavel Snytnikov
Recent Advances in Structured Catalytic Materials Development for Conversion of Liquid Hydrocarbons into Synthesis Gas for Fuel Cell Power Generators
Reprinted from: *Materials* 2023, *16*, 599, doi:10.3390/ma16020599 169

Natalia Ruban, Vladimir Rogozhnikov, Sergey Zazhigalov, Andrey Zagoruiko, Vyacheslav Emelyanov and Pavel Snytnikov et al.
Composite Structured $M/Ce_{0.75}Zr_{0.25}O_2/Al_2O_3/FeCrAl$ (M = Pt, Rh, and Ru) Catalysts for Propane and n-Butane Reforming to Syngas
Reprinted from: *Materials* 2022, *15*, 7336, doi:10.3390/ma15207336 203

Gaoyang Liu, Faguo Hou, Xindong Wang and Baizeng Fang
Robust Porous TiN Layer for Improved Oxygen Evolution Reaction Performance
Reprinted from: *Materials* 2022, *15*, 7602, doi:10.3390/ma15217602 217

T. N. Afonasenko, D. V. Glyzdova, V. L. Yurpalov, V. P. Konovalova, V. A. Rogov and E. Yu. Gerasimov et al.
The Study of Thermal Stability of Mn-Zr-Ce, Mn-Ce and Mn-Zr Oxide Catalysts for CO Oxidation
Reprinted from: *Materials* 2022, *15*, 7553, doi:10.3390/ma15217553 229

Aleksandr S. Gorkusha, Sergey V. Tsybulya, Svetlana V. Cherepanova, Evgeny Y. Gerasimov and Svetlana N. Pavlova
Nonstoichiometry Defects in Double Oxides of the A_2BO_4-Type
Reprinted from: *Materials* 2022, *15*, 7642, doi:10.3390/ma15217642 247

Jiameng Sun, Bin Yu, Xuejiao Yan, Jianfeng Wang, Fuquan Tan and Wanfeng Yang et al.
High Throughput Preparation of Ag-Zn Alloy Thin Films for the Electrocatalytic Reduction of CO_2 to CO
Reprinted from: *Materials* 2022, *15*, 6892, doi:10.3390/ma15196892 259

Andrei T. Matveev, Liubov A. Varlamova, Anton S. Konopatsky, Denis V. Leybo, Ilia N. Volkov and Pavel B. Sorokin et al.
A New Insight into the Mechanisms Underlying the Discoloration, Sorption, and Photodegradation of Methylene Blue Solutions with and without BNO_x Nanocatalysts
Reprinted from: *Materials* 2022, *15*, 8169, doi:10.3390/ma15228169 271

Jiao Jiao, Yihua Li, Qi Song, Liujin Wang, Tianlie Luo and Changfei Gao et al.
Removal of Pharmaceuticals and Personal Care Products (PPCPs) by Free Radicals in Advanced Oxidation Processes
Reprinted from: *Materials* 2022, *15*, 8152, doi:10.3390/ma15228152 291

Diana Rakhmawaty Eddy, Dian Nursyamsiah, Muhamad Diki Permana, Solihudin, Atiek Rostika Noviyanti and Iman Rahayu
Green Production of Zero-Valent Iron (ZVI) Using Tea-Leaf Extracts for Fenton Degradation of Mixed Rhodamine B and Methyl Orange Dyes
Reprinted from: *Materials* 2022, *15*, 332, doi:10.3390/ma15010332 315

About the Editor

Ilya V. Mishakov

Dr. Ilya V. Mishakod is a leading researcher in Boreskov Instirute of Catalysis as the head of the laboratory. He is currently a lecturer at Novosibirsk State University (Catalysis) as an Associate Professor. He has co-authored around200 papers. His scientific interests are in the field of heterogeneous catalysis; nanotechnology; carbon nanofibers; functionalized carbon materials; self-organized catalysts; hierarchical composites; nanocrystalline oxides; oxidative dehydrogenation; three-way catalysts; alloys; catalytic decomposition; chemical vapor deposition; environmental catalysis; and metal dusting and corrosion.

Preface to "Advanced Materials in Catalysis and Adsorption"

The design of novel materials with improved characteristics for catalytic applications and adsorption will be always be high in researchers' interest. In this reprint, we have collected the most recent studies focused on development of advanced carbon nanostructured materials, functionalized carbons, nanostructured oxides, electrocatalysts, and catalytic composite materials. The benefits of using such novel catalytic materials and adsorbents in various applications are demonstrated.

Ilya V. Mishakov
Editor

Editorial for Special Issue "Advanced Materials in Catalysis and Adsorption"

Ilya V. Mishakov

Department of Materials Science and Functional Materials, Boreskov Institute of Catalysis, Lavrentieva Ave, 5, 630090 Novosibirsk, Russia; mishakov@catalysis.ru

This Special Issue aims to cover the latest research on the design and development of advanced materials for adsorption and catalytic applications. The synthesis of various carbon nanomaterials (CNMs), including carbon nanofibers (CNFs), carbon nanotubes (CNTs), graphene-like materials, and carriers (Sibunit), is of particular interest [1–8]. The functionalization of CNMs with heteroatoms (i.e., nitrogen [1,6–8] and boron [3]) is an effective method for boosting the applicability of carbon nanomaterials as catalytic substrates. Another group of the designed materials belongs to the oxide-based supports, including zeolite catalysts [9], composite catalytic materials [5,10–12], mixed Mn-Zr-Ce-O oxides [13], and perovskite-like materials [14]. In many cases, the active component of the developed catalysts is represented by precious metals—Pd [1,7], Pt [5,11], Rh [10,11], and Ru [11]—and other metals such as Ag [15], Au [7,9], and their alloys.

Advanced catalytic materials have been developed for diverse types of heterogeneous catalytic reactions, such as the hydrodechlorination of chloroaromatics [1]; the dehalogenation of halogenated hydrocarbons [2]; dechlorination via catalytic pyrolysis [1,2,4]; the catalytic coupling of CH_4 [14]; the catalytic processing of hydrocarbons and their mixtures into synthesis gas [10,11] or CNT and CNF materials [2,5,6,8]; the oxidation of carbon monoxide [13]; the hydrogenation of organic compounds [7]; and selective oxidation [9]. Such advanced materials can be also used for electrocatalytic applications involving the oxygen evolution reaction (OER) [12] and the electrocatalytic reduction of carbon dioxide (CO_2RR) [15]. Researchers' attention has been especially drawn to environmental protection, where advanced materials and adsorbents are highly demanded for the processing of waste components [1,16–18] and pharmaceuticals [17], the (photo)degradation of dyes [16,18], the catalytic decomposition of chlorinated hydrocarbons [1,2,4], wastewater treatment [1,18], and the abatement of CO-containing gases [13].

The functionalization of various CNMs for further application as catalytic supports and sorbents has attracted a great deal of research attention. A new method of synthesizing a boron-doped graphene material using a template concept has been proposed in [3]. The produced B-doped graphene material (composed of three to eight monolayers) is characterized by a boron concentration of about 4 wt.% and a high specific surface area (SSA) of 800 m^2/g. The post-functionalization of carbon nanofibers with nitrogen using melamine was described in [6]. Therein, the authors revealed that the preliminary acidic treatment of CNFs permits an increase in the amount of N-doping (from 1.4 to 4.3 wt.%), while preserving the structure and consistency of the initial carbon filaments. Diazonium salts can also be effectively used for the surface modification of carbonaceous supports (Sibunit) used as carriers for Pd- and Pd-Au catalysts for furfural hydrogenation [7]. The authors concluded that the modification of Sibunit with various functional groups leads to changes in the hydrophobic/hydrophilic and electrostatic properties of the surface, thus influencing its selectivity.

On the other hand, the "one-pot" synthetic approach to obtaining N-functionalized carbon nanofilaments was successfully implemented in [1,8]. Acetonitrile [1] and NH_3 [8] were employed as the N-containing co-reagents in CVD-synthesis, thus facilitating the introduction of nitrogen in one stage. It was found [8] that the number of bamboo-like and spherical structures induced by the incorporation of nitrogen into the CNTs' structure depends on the temperature (T_{opt} = 650–700 °C) and the NH_3 concentration in the reaction mixture (the optimal composition was C_3H_8/NH_3 = 50/50%).

The carbon erosion or metal dusting (MD) of bulk Ni-M and Co-M alloys can be used to produce carbon nanofibers and CNF-based composites [1,2,4–6]. The carbon erosion of nickel alloys during the catalytic pyrolysis of organic compounds with the formation of carbon nanofibers in a flow-through reactor and under reaction conditions in a close volume (Reactions under Autogenic Pressure at Elevated Temperature (RAPET)) has been extensively studied in [2]. The efficiency of using the ferromagnetic resonance (FMR) method to monitor the appearance of catalytically active nickel was demonstrated. The controllable synthesis of CNF and Co-Pt/CNF composites via ethylene decomposition over Co-Pt (0–100 at.% Pt) microdispersed alloys was described in [5]. The authors revealed the impact of the Pt content in $Co_{1-x}Pt_x$ alloys on their activity in CNFs and found that the addition of 15–25 at.% Pt to a cobalt catalyst leads to a three- to five-fold increase in productivity [5].

Furthermore, chlorinated hydrocarbons can also serve as a carbon source in catalytic pyrolysis reactions [1,4]. Thus, the research on trichloroethylene (TCE) decomposition over a microdispersed Ni-W alloy (4 wt.% W) revealed that the addition of tungsten results in an enhancement of Ni productivity (1.5–2 times) toward the generation of carbon filaments with a unique segmented structure and a high SSA of 374 m^2/g [4]. A similar approach to synthesizing CNFs via the catalytic processing of TCE (a component of organochlorine wastes) over a Ni catalyst was realized in [1]. The segmented CNFs produced were shown to be an effective adsorbent for the decontamination of water containing traces of 1,2-dichlorobenzene (1,2-DCB). The deposition of Pd nanoparticles (1.5 wt.%) on the surface of the CNFs allowed the authors to create an adsorbent with a catalytic function. It was found that the repeated use of regenerated adsorbent catalysts for the purification of aqueous solutions ensures the almost complete removal of 1,2-DCB.

Oxide-based materials with advanced features are also within the scope of the collected papers [9,13,14]. A comprehensive crystallochemical analysis of the possible structure of planar defects in Sr_2TiO_4-layered perovskite (a prospective catalyst for CH_4 oxidative coupling) was reported in [14]. The authors established a relationship between the concentration of planar defects and the non-stoichiometry of the Sr_2TiO_4 phase. The presence of defects is thought to lead to the enrichment of the surface with Sr, which might serve as an explanation for the catalytic activity of Sr_2TiO_4 in the considered reaction. In [13], researchers investigated the catalytic performance of MnO_x-CeO_2, MnO_x-ZrO_2, and MnO_x-ZrO_2-CeO_2 with a molar ratio of Mn/(Zr + Ce + Mn) = 0.3 in the CO oxidation reaction. The authors found that the thermal stability of the catalysts is determined by the decomposition temperature of the solid solution of $Mn_x(Ce,Zr)_{1-x}O_2$. The high CO oxidation activity of the samples was shown to be correlated with the presence of oxygen vacancies and the content of easily reduced fine MnO_x particles. The "metal-support" interaction and its impact on the performance of gold-supported zeolitic catalysts were investigated in [9]. Three-dimensional HBeta and layered two-dimensional MCM-36 were used as supports for gold and then studied with respect to the oxidation of glucose to gluconic acid (with O_2 and H_2O_2). The important roles of the porosity of the zeolite supports, the accumulation of a negative charge on the Au nanoparticles, and the impact of the oxidant (O_2 or H_2O_2) on the nature of the reaction's limiting step were revealed.

This Special Issue also includes studies related to the design of novel materials for electrocatalytic reactions. In [12], the researchers fabricated a high-performance and low-cost porous TiN electrocatalyst for the oxygen evolution reaction (OER). The authors applied a method of the thermal nitriding of Ti to prepare a novel TiN-Ti support, which was found to yield better dispersion of the active component (IrO_x), lower ohmic resistance, and more accessible catalytic active sites on the catalytic interface. At the same time, the study in [15] focused on searching for a selective electrocatalyst for the CO_2 reduction reaction (CO_2RR). Accordingly, the researchers fabricated a series of Ag-Zn alloy catalysts using magnetron co-sputtering and explored their electrocatalytic performance in terms of the CO_2RR process. The synergistic effects of Ag and Zn on Ag_5Zn_8 and $AgZn_3$ catalysts were established. The authors concluded that the activity and selectivity of CO_2RR are highly dependent on the element ratio and phase composition in the Ag-Zn alloyed system.

The composite structured catalysts M/$Ce_{0.75}Zr_{0.25}O_2$/Al_2O_3/FeCrAl (M = Pt, Rh, Ru) were designed in [10,11] for the catalytic processing of light hydrocarbons into synthesis gas. The catalysts are composed of a high-heat-conducting FeCrAl block covered by θ-Al_2O_3 and a layer of ceria–zirconia mixed oxide with deposited 2–3 nm sized Pt, Rh, or Ru nanoparticles. Reformates were produced via the reformation of propane steam over the designed catalysts at 600 °C; the reformates contained ~65 vol.% of H_2, which can be used as a fuel for solid oxide fuel cells [11]. In addition, the authors studied the catalytic properties of the Rh/$Ce_{0.75}Zr_{0.25}O_2$/Al_2O_3/FeCrAl composite catalyst with respect to the conversion of diesel fuel into synthesis gas and concluded that it can be used in fuel cell power generators [10].

Using advanced materials and methods for environmentally protective purposes was also within the scope of this Special Issue. For instance, waste produced by the textile-industry contains various dyes (e.g., methyl orange, methylene blue, and rhodamine B) that pose environmental risks. In the research presented in [18], the authors synthesized zerovalent iron (ZVI) using biological reducing agents isolated from tea-leaves (polyphenolic compounds) and showed the applicability of the designed ZVI system to wastewater treatment. In [16], the photodegradation of methylene blue over BNO_x catalysts (containing 4.2 and 6.5% oxygen) was studied in detail. The authors revealed the important role of surface oxygen defects in the adsorption capacity and photocatalytic activity of BNO_x.

The Editors of Special Issue are very grateful to all the authors for their excellent contributions and to the editorial team of *Materials* for their kind assistance and peer review. The readers will definitely enjoy this Special Issue and, hopefully, find new ideas and concepts for their future investigations!

Acknowledgments: I would like to thank all the Authors for submitting their papers to this Special Issue, "Advanced Materials for Catalysis and Adsorption", as well as all the Reviewers and Editors for their contributions to improving these submissions.

Conflicts of Interest: The author declares no conflict of interest.

References

1. Ozerova, A.M.; Potylitsyna, A.R.; Bauman, Y.I.; Tayban, E.S.; Lipatnikova, I.L.; Nartova, A.V.; Vedyagin, A.A.; Mishakov, I.V.; Shubin, Y.V.; Netskina, O.V. Synthesis of Chlorine- and Nitrogen-Containing Carbon Nanofibers for Water Purification from Chloroaromatic Compounds. *Materials* **2022**, *15*, 8414. [CrossRef] [PubMed]
2. Volodin, A.M.; Kenzhin, R.M.; Bauman, Y.I.; Afonnikova, S.D.; Potylitsyna, A.R.; Shubin, Y.V.; Mishakov, I.V.; Vedyagin, A.A. Comparative Study on Carbon Erosion of Nickel Alloys in the Presence of Organic Compounds under Various Reaction Conditions. *Materials* **2022**, *15*, 9033. [CrossRef] [PubMed]
3. Chesnokov, V.V.; Prosvirin, I.P.; Gerasimov, E.Y.; Chichkan, A.S. Synthesis of Boron-Doped Carbon Nanomaterial. *Materials* **2023**, *16*, 1986. [CrossRef] [PubMed]
4. Potylitsyna, A.R.; Rudneva, Y.V.; Bauman, Y.I.; Plyusnin, P.E.; Stoyanovskii, V.O.; Gerasimov, E.Y.; Vedyagin, A.A.; Shubin, Y.V.; Mishakov, I.V. Efficient Production of Segmented Carbon Nanofibers via Catalytic Decomposition of Trichloroethylene over Ni-W Catalyst. *Materials* **2023**, *16*, 845. [CrossRef] [PubMed]
5. Afonnikova, S.D.; Popov, A.A.; Bauman, Y.I.; Plyusnin, P.E.; Mishakov, I.V.; Trenikhin, M.V.; Shubin, Y.V.; Vedyagin, A.A.; Korenev, S.V. Porous Co-Pt Nanoalloys for Production of Carbon Nanofibers and Composites. *Materials* **2022**, *15*, 7456. [CrossRef] [PubMed]
6. Maksimova, T.A.; Mishakov, I.V.; Bauman, Y.I.; Ayupov, A.B.; Mel'gunov, M.S.; Dmitrachkov, A.M.; Nartova, A.V.; Stoyanovskii, V.O.; Vedyagin, A.A. Effect of Pretreatment with Acids on the N-Functionalization of Carbon Nanofibers Using Melamine. *Materials* **2022**, *15*, 8239. [CrossRef] [PubMed]
7. German, D.; Kolobova, E.; Pakrieva, E.; Carabineiro, S.A.C.; Sviridova, E.; Perevezentsev, S.; Alijani, S.; Villa, A.; Prati, L.; Postnikov, P.; et al. The Effect of Sibunit Carbon Surface Modification with Diazonium Tosylate Salts of Pd and Pd-Au Catalysts on Furfural Hydrogenation. *Materials* **2022**, *15*, 4695. [CrossRef] [PubMed]

8. Skudin, V.; Andreeva, T.; Myachina, M.; Gavrilova, N. CVD-Synthesis of N-CNT Using Propane and Ammonia. *Materials* **2022**, *15*, 2241. [CrossRef] [PubMed]
9. Walkowiak, A.; Wolska, J.; Wojtaszek-Gurdak, A.; Sobczak, I.; Wolski, L.; Ziolek, M. Modification of Gold Zeolitic Supports for Catalytic Oxidation of Glucose to Gluconic Acid. *Materials* **2021**, *14*, 5250. [CrossRef] [PubMed]
10. Shilov, V.; Potemkin, D.; Rogozhnikov, V.; Snytnikov, P. Recent Advances in Structured Catalytic Materials Development for Conversion of Liquid Hydrocarbons into Synthesis Gas for Fuel Cell Power Generators. *Materials* **2023**, *16*, 599. [CrossRef] [PubMed]
11. Ruban, N.; Rogozhnikov, V.; Zazhigalov, S.; Zagoruiko, A.; Emelyanov, V.; Snytnikov, P.; Sobyanin, V.; Potemkin, D. Composite Structured M/Ce$_{0.75}$Zr$_{0.25}$O$_2$/Al$_2$O$_3$/FeCrAl (M = Pt, Rh, and Ru) Catalysts for Propane and n-Butane Reforming to Syngas. *Materials* **2022**, *15*, 7336. [CrossRef] [PubMed]
12. Liu, G.; Hou, F.; Wang, X.; Fang, B. Robust Porous TiN Layer for Improved Oxygen Evolution Reaction Performance. *Materials* **2022**, *15*, 7602. [CrossRef] [PubMed]
13. Afonasenko, T.N.; Glyzdova, D.V.; Yurpalov, V.L.; Konovalova, V.P.; Rogov, V.A.; Gerasimov, E.Y.; Bulavchenko, O.A. The Study of Thermal Stability of Mn-Zr-Ce, Mn-Ce and Mn-Zr Oxide Catalysts for CO Oxidation. *Materials* **2022**, *15*, 7553. [CrossRef] [PubMed]
14. Gorkusha, A.S.; Tsybulya, S.V.; Cherepanova, S.V.; Gerasimov, E.Y.; Pavlova, S.N. Nonstoichiometry Defects in Double Oxides of the A$_2$BO$_4$-Type. *Materials* **2022**, *15*, 7642. [CrossRef] [PubMed]
15. Sun, J.; Yu, B.; Yan, X.; Wang, J.; Tan, F.; Yang, W.; Cheng, G.; Zhang, Z. High Throughput Preparation of Ag-Zn Alloy Thin Films for the Electrocatalytic Reduction of CO$_2$ to CO. *Materials* **2022**, *15*, 6892. [CrossRef] [PubMed]
16. Matveev, A.T.; Varlamova, L.A.; Konopatsky, A.S.; Leybo, D.V.; Volkov, I.N.; Sorokin, P.B.; Fang, X.; Shtansky, D.V. A New Insight into the Mechanisms Underlying the Discoloration, Sorption, and Photodegradation of Methylene Blue Solutions with and without BNO$_x$ Nanocatalysts. *Materials* **2022**, *15*, 8169. [CrossRef] [PubMed]
17. Jiao, J.; Li, Y.; Song, Q.; Wang, L.; Luo, T.; Gao, C.; Liu, L.; Yang, S. Removal of Pharmaceuticals and Personal Care Products (PPCPs) by Free Radicals in Advanced Oxidation Processes. *Materials* **2022**, *15*, 8152. [CrossRef] [PubMed]
18. Eddy, D.R.; Nursyamsiah, D.; Permana, M.D.; Solihudin; Noviyanti, A.R.; Rahayu, I. Green Production of Zero-Valent Iron (ZVI) Using Tea-Leaf Extracts for Fenton Degradation of Mixed Rhodamine B and Methyl Orange Dyes. *Materials* **2022**, *15*, 332. [CrossRef] [PubMed]

Disclaimer/Publisher's Note: The statements, opinions and data contained in all publications are solely those of the individual author(s) and contributor(s) and not of MDPI and/or the editor(s). MDPI and/or the editor(s) disclaim responsibility for any injury to people or property resulting from any ideas, methods, instructions or products referred to in the content.

Article

Synthesis of Chlorine- and Nitrogen-Containing Carbon Nanofibers for Water Purification from Chloroaromatic Compounds

Anna M. Ozerova [1], Arina R. Potylitsyna [1,2], Yury I. Bauman [1], Elena S. Tayban [1], Inna L. Lipatnikova [1], Anna V. Nartova [1], Aleksey A. Vedyagin [1], Ilya V. Mishakov [1], Yury V. Shubin [3] and Olga V. Netskina [1,2,*]

1. Boreskov Institute of Catalysis SB RAS, Lavrentieva Av. 5, 630090 Novosibirsk, Russia
2. Department of Natural Sciences, Novosibirsk State University, Pirogova Str. 2, 630090 Novosibirsk, Russia
3. Nikolaev Institute of Inorganic Chemistry SB RAS, Lavrentieva Av. 3, 630090 Novosibirsk, Russia
* Correspondence: netskina@catalysis.ru; Tel.: +7-383-330-7458

Abstract: Chlorine- and nitrogen-containing carbon nanofibers (CNFs) were obtained by combined catalytic pyrolysis of trichloroethylene (C_2HCl_3) and acetonitrile (CH_3CN). Their efficiency in the adsorption of 1,2-dichlorobenzene (1,2-*DCB*) from water has been studied. The synthesis of CNFs was carried out over self-dispersing nickel catalyst at 600 °C. The produced CNFs possess a well-defined segmented structure, high specific surface area (~300 m^2/g) and high porosity (0.5–0.7 cm^3/g). The addition of CH_3CN into the reaction mixture allows the introduction of nitrogen into the CNF structure and increases the volume of mesopores. As a result, the capacity of CNF towards adsorption of 1,2-*DCB* from its aqueous solution increased from 0.41 to 0.57 cm^3/g. Regardless of the presence of N, the CNF samples exhibited a degree of 1,2-*DCB* adsorption from water–organic emulsion exceeding 90%. The adsorption process was shown to be well described by the Dubinin–Astakhov equation. The regeneration of the used CNF adsorbent through liquid-phase hydrodechlorination was also investigated. For this purpose, Pd nanoparticles (1.5 wt%) were deposited on the CNF surface to form the adsorbent with catalytic function. The presence of palladium was found to have a slight effect on the adsorption capacity of CNF. Further regeneration of the adsorbent-catalyst via hydrodechlorination of adsorbed 1,2-*DCB* was completed within 1 h with 100% conversion. The repeated use of regenerated adsorbent-catalysts for purification of solutions after the first cycle of adsorption ensures almost complete removal of 1,2-*DCB*.

Keywords: chloroaromatics; adsorption; hydrodechlorination; N-containing carbon nanofibers; trichloroethylene; nickel catalyst

1. Introduction

Chlorinated organic compounds (COCs) are among the most persistent and toxic pollutants. They are widely used as pesticides, propellants, refrigerants, degreasers, solvents, as well as reagents and intermediate products in organic synthesis [1,2]. The use of COCs in various technological processes is accompanied by their release into the environment. On the other hand, the chlorination of water-containing organic substances may also result in the formation of a number of COCs which have carcinogenic and mutagenic effects [3]. COCs are known to be very stable and only slightly dissolvable. They have a high resistance to degradation. Considerable amounts of COCs can be found in industrial effluents, groundwater, rivers, seas, rainwater, and even in drinking water [1,2,4]. Therefore, the research aiming to improve the purification technologies for removing such dangerous chemicals from liquid waste and wastewater is of high demand. Various methods have been proposed for the removal of COCs from water, including catalytic oxidation [5]; biodegradation [6]; advanced oxidation processes, such as photocatalytic destruction [7] or electrochemical degradation [8]; reductive dechlorination [9]; as well as non-destructive methods, such as adsorption [10–12]. Each of these methods has its advantages and limitations [1]. For example, catalytic oxidation technology has been widely utilized because of its relative simplicity. Meanwhile, it remains costly, requires elevated temperatures, and poses a certain risk to the environment due to the potential emission of highly toxic byproducts (e.g., dioxins, furans, and phosgene) derived as a result of their incomplete combustion [13].

The catalytic pyrolysis of chlorinated hydrocarbons and related wastes is known to be one of the most powerful techniques for their processing [14,15]. This process is carried out in a reductive atmosphere, which completely eliminates the risk of the formation of side ecotoxicants. The self-dispersing Ni-based catalysts can be used for this purpose as the most active and tolerant to deactivation by chlorine systems [16]. "Self-dispersing" implies fast fragmentation (disintegration) of bulk Ni or Ni-based alloys (Ni-M, where M = Pd, Mo, W etc.) in the reaction atmosphere, which leads to the formation of disperse active particles [17]. The decomposition of aliphatic chlorinated hydrocarbons (i.e., 1,2-dichloroethane (DCE) and trichloroethylene (TCE)) over alloyed Ni-M catalysts is accompanied by accumulation of the chlorine-containing graphite-like nanomaterials represented mostly by carbon nanofibers (CNFs) [17–20]. The filamentous carbon product resulted from the catalytic decomposition of chlorinated hydrocarbons is known to possess both a unique segmental structure and high textural parameters (specific surface area of 300–400 m^2/g, pore volume of 0.5–0.8 cm^3/g), which makes this type of nanomaterial very attractive for adsorption applications [21,22]. The appearance of the segmented secondary structure of CNFs is believed to be driven by the periodic 'chlorination-dechlorination' process taking place over the surface of nickel catalyst during the H_2-assisted catalytic pyrolysis of chlorinated hydrocarbons [23]. Thus, the potential use of such carbon nanomaterial obtained via catalytic pyrolysis of COCs is of particular practical interest.

It should be noted that real industrial organochlorine wastes are usually composed of a complex mixture of polychlorinated hydrocarbons, including a portion of organics functionalized by O- and N-containing groups. As recently demonstrated, the use the nitrogen-containing compound acetonitrile (AN) as a co-reagent in the catalytic pyrolysis of DCE or TCE vapors allows the obtainment of N-containing CNF materials with a nitrogen content of up to 3 wt% [24]. The addition of AN vapors into the reaction mixture was also shown to have a boosting effect on the catalyst's productivity with respect to the CNF material. In addition, the produced N-containing CNF (N-CNF) exhibited the preserved segmental structure of fibers, along with an enhanced specific surface area (up to 550 m^2/g) and porosity [24]. The high textural parameters of such filamentous carbon product make it a very attractive nanomaterial for various fields of application, including adsorption [25,26].

Carbon materials are commonly used as adsorbents for water purification from diverse types of contamination [27,28]. As recently shown, CNFs produced via the catalytic decomposition of DCE demonstrate very high efficiency in adsorption of chloroaromatic pollutants from water [29,30]. At the same time, the possibility of reusing the carbon adsorbent for water purification should include the removal of the chloroaromatic compound adsorbed during the first purification cycle. Thus, the regeneration of the adsorbent can be carried out under mild conditions using the catalytic reaction of liquid-phase hydrodechlorination (HDC) [22,31,32]. HDC is a widespread method that is recognized as the most effective method for the degradation of polychlorinated aromatics. Various supported catalysts containing Pd, Pt, Ni and its alloys can be used as active components for this purpose [33,34]. In the case of using carbon adsorbents, palladium at a moderate concentration of 1–3% is to be deposited on their surface in order to provide the necessary catalytic function [35]. In other words, the purification cycle involves the adsorption of COCs from the aqueous solution and the reductive regeneration of the catalyst by the liquid-phase HDC of adsorbed species with molecular hydrogen [21]. Such an approach allows for both concentrating the chloroaromatic pollutants and obtaining the products useful in the chemical industry. The object of study for water purification was 1,2-dichlorobenzene (1,2-DCB), which is a high-priority pollutant. It is widely used as a high-boiling solvent in chemical industry, as well as a reagent and an intermediate for dyes and agrochemical manufacture [36].

In the present paper, a complex approach for the disposal of the chlorinated organic multi-component wastes is proposed. In the first stage, TCE, which was chosen as the model aliphatic polychlorinated hydrocarbon, was converted over Ni-catalyst into segmented CNF material. In addition, the N-containing CNF was obtained in the same way via combined pyrolysis of TCE and AN. In the second stage, the produced CNF and N-CNF samples were explored in the adsorption of 1,2-*DCB* from aqueous solutions in the static regime. The opportunity to regenerate the spent CNF adsorbent by liquid-phase catalytic hydrodechlorination was also examined. For this purpose, 1.5 wt% palladium was deposited on both of the CNF samples. These adsorbent-catalysts were then tested in the adsorption–regeneration cycle.

2. Materials and Methods

A schematic representation of the principle experimental stages is presented in Figure 1. A detailed description of each step is given below.

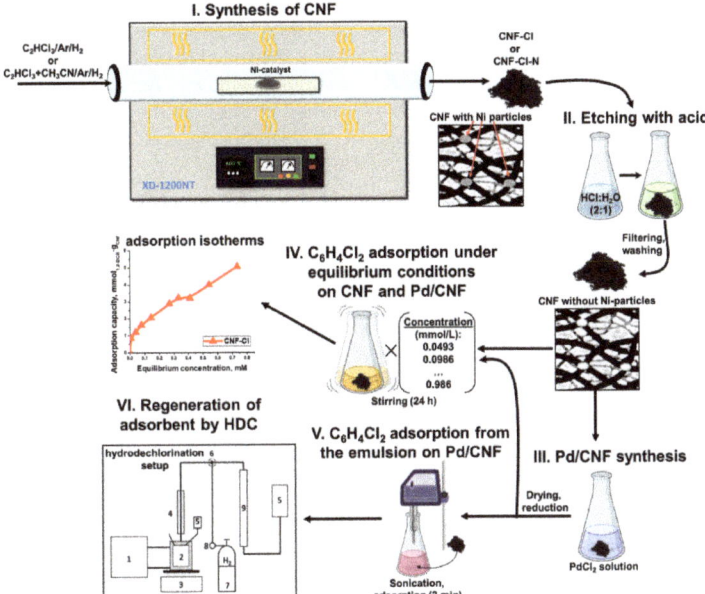

Figure 1. Schematic representation of the experimental work. The HDC setup consists of the following parts: (1) thermostat; (2) temperature-controlled reactor; (3) magnetic stirrer; (4) reflux condenser; (5) odor trap; (6) three-way valve; (7) gas bottle; (8) gas reducer; (9) measuring burette.

The following commercial reagents were used as received: trichloroethylene (C_2HCl_3, chemically pure, Component-Reactiv, Moscow, Russia); acetonitrile (CH_3CN, chemically pure, Component-Reactiv, Moscow, Russia); $PdCl_2$ (pure, Aurat, Moscow, Russia); $NaBH_4$ (purity of 98 wt%, Chemical Line, Saint Petersburg, Russia); 1,2-dichlorobenzene ($C_6H_4Cl_2$, purity of 99 wt%, Sigma-Aldrich, St. Louis, MO, USA); KOH (analytically pure, Reakhim, Moscow, Russia); and 2-propanol ($CH_3CH(OH)CH_3$, high purity grade, Baza No.1 Khim-reactivov, Staraya Kupavna, Russia). Microdispersed Ni-catalyst was synthesized as described elsewhere [37].

2.1. Synthesis of Functionalized CNFs

The H$_2$-assisted catalytic pyrolysis of C$_2$HCl$_3$ vapors (or the joint decomposition of C$_2$HCl$_3$ and CH$_3$CN) resulting in the formation of carbon nanomaterial was carried out in a flow-through horizontal quartz reactor (Zhengzhou Brother Furnace Co., Ltd., Zhengzhou, China).

A specimen of the catalyst (microdispersed powder of Ni), 300.00 ± 0.5 mg each, was spread throughout the entire quartz plate. The plate was then placed inside the reactor, after which the reactor was heated in an Ar atmosphere up to the reaction temperature of 600 °C. The samples were then treated in a hydrogen flow at 600 °C for 10 min in order to reduce the oxide film on their surfaces. After the reduction, the reactor was fed with the reaction mixture containing vapors of TCE (6 vol%) or TCE (6 vol%) + AN (25 vol%), hydrogen (37 vol%), and argon, at 600 °C. The flow rate of the reaction mixture was 54 L/h. The duration of the synthesis was 2 h. When the experiment was completed, the reactor was cooled in an argon flow to room temperature, after which the carbon product was unloaded and weighed. The measured carbon yield, Y_C (expressed in grams of CNF per 1 g of catalyst, $g_{CNF} \cdot g_{cat}^{-1}$), was found to be 9.8 $g_{CNF} \cdot g_{cat}^{-1}$ (for TCE only) and 10.6 $g_{CNF} \cdot g_{cat}^{-1}$ (in the case of TCE + AN).

The resulting carbon nanomaterials were further treated with hydrochloric acid (12%) for 12 h in order to remove the metallic particles of residual catalyst. The etched CNF material was then washed with water to neutral pH, filtered out, and dried at 120 °C. The resulting CNF material produced via the decomposition of TCE only was labeled as «CNF-Cl», whereas the CNF sample obtained by the joint decomposition of TCE and AN was labeled as «CNF-Cl-N».

2.2. Synthesis of the Pd/CNF Adsorbent-Catalysts

The adsorbent-catalysts were prepared by incipient wetness impregnation of the CNF-Cl and CNF-Cl-N samples with hydrochloric solutions of PdCl$_2$. The obtained samples were dried at 110–130 °C for 4 h. The reduction of deposited PdCl$_2$ was performed at room temperature by treatment with an aqueous solution of NaBH$_4$ (Pd:NaBH$_4$ molar ratio was 1:3). The synthesized adsorbent-catalysts were labeled as «Pd/CNF-Cl» and «Pd/CNF-Cl-N». According to X-ray fluorescence analysis data, the palladium content in all the Pd-containing samples was 1.5 ± 0.05 wt%.

2.3. Study on 1,2-DCB Adsorption from Aqueous Solution under Equilibrium Conditions

2.3.1. Measurement Procedure

The adsorption characteristics of the prepared CNF and Pd/CNF samples were studied under equilibrium conditions by a static method at 25 °C for 24 h. For this purpose, 3 mg of each sample was added to a 60 mL solution of 1,2-DCB, with concentrations in the range from 0.0493 to 0.986 mmol·L^{-1}. In order to exclude the effect of the pore-diffusion factor, the mixtures were evenly shaken with a shaker LOIP LS-110 (RNPO RusPribor, Saint-Petersburg, Russia) at an orbiting speed of 200 rpm. Changes in the concentration of 1,2-DCB during the adsorption cycle were monitored by the UV-spectroscopy technique [38–40] using a Varian Cary 100 instrument (Agilent, Santa Clara, CA, USA). To obtain reliable data, each experiment was repeated at least three times. Differences between experimental values did not exceed ±3%.

The amount of the adsorbed 1,2-DCB was calculated as follows [41]:

$$A = \frac{\Delta C \times V}{m} \quad (1)$$

where A is the adsorption capacity of the CNFs or Pd/CNF, mol·g^{-1}; ΔC is the difference in 1,2-DCB concentration before and after adsorption, mol·L^{-1}; V is the volume of 1,2-DCB solution, L; m is the loading of the adsorbent, g.

The pore filling was determined as a ratio of the volume of 1,2-DCB in CNF pores to the total pore volume:

$$F_{pores} = \frac{A^*}{V_{\Sigma pore}} \times 100\% \tag{2}$$

where F_{pores} is the pore filling of 1,2-DCB, %; A^* is the adsorption capacity, which was achieved by adsorption of 1,2-DCB from an aqueous solution, cm^3·g^{-1}; $V_{\Sigma pore}$ is the total pore volume of CNF, cm^3·g^{-1}.

2.3.2. Adsorption Isotherm Modeling

Both the Langmuir and Dubinin–Astakhov isotherm models were used to analyze the adsorption data. The applicability of the models was checked via a comparison of their correlation coefficients (R^2 values).

Langmuir Isotherm

The Langmuir isotherm is often used to study the adsorption processes of various pollutants. This model is applicable for monolayer adsorption onto a surface with a limited amount of identical sites. Uniform adsorption energies and no adsorbate transmigration in the plane of the surface are considered. The equation for the Langmuir isotherm is given below [42]:

$$A = \frac{K_L A_{max} C_{eq}}{1 + K_L C_{eq}}, \tag{3}$$

where A is the amount of 1,2-DCB adsorbed per gram, mol·g^{-1}; A_{max} is the adsorption capacity or the adsorption maximum, mol·g^{-1}; C_{eq} is the equilibrium 1,2-DCB concentration, mol·L^{-1}; K_L is the Langmuir constant, L·mol^{-1}.

The surface area of adsorbent occupied by one molecule of 1,2-DCB (ω) was calculated by the Langmuir isotherm model [43]:

$$\omega = \frac{SSA}{A_{max} N_A} \tag{4}$$

where SSA is the specific surface area of the adsorbent, m^2·g^{-1}; A_{max} is the adsorption capacity or the adsorption maximum, mol·g^{-1}; N_A is the Avogadro number (6.02×10^{23} mol^{-1}).

Dubinin-Astakhov Isotherm

As is reported [44], the Dubinin theory, which considers volume filling for vapor adsorption, can be applied to study the adsorption of organic compounds from their solutions [45,46]. The Dubinin–Astakhov equation can be expressed as follows [47]:

$$A = A_{max} \exp\left[-\left[\frac{RT \times \ln\left(\frac{C_{max}}{C_{eq}}\right)}{E_{ads}^{eff}}\right]^n\right], \tag{5}$$

where A_{max} is the adsorption maximum or adsorption capacity, mol·g^{-1}; C_{max} is the highest concentration of 1,2-DCB in water, mol·L^{-1}; C_{eq} is the equilibrium 1,2-DCB concentration, mol·L^{-1}; E_{ads}^{eff} is the characteristic energy of 1,2-DCB adsorption, kJ·mol^{-1}.

The maximum amount of 1,2-DCB (W_0) adsorbed per gram (or maximum adsorption capacity), and the average diameter (d) of the adsorbent's pores filled with 1,2-DCB were defined by the Dubinin–Astakhov isotherm model [21,43]:

$$W_0 = \frac{A_{max} \times M_{1,2-DCB}}{\rho_{1,2-DCB}}, \tag{6}$$

$$d = 2 \cdot \frac{K_{C_6H_6} \times \beta}{E_{ads}^{eff}}, \quad (7)$$

where $M_{1,2-DCB}$ is the molar mass of 1,2-DCB, g·mol^{-1}; $\rho_{1,2-DCB}$ is the density of 1,2-DCB, g·cm^{-3}; $K_{C_6H_6}$ is the coefficient for the standard (benzene), equal to 12 kJ·nm·mol^{-1}; β is the affinity coefficient (for 1,2-DCB, equal to 1.28).

2.4. Testing Pd/CNF in the Adsorptive-Catalytic Cycle

Purification of the aqueous medium from 1,2-DCB was conducted in two steps comprising an adsorptive–catalytic cycle: (1) adsorption of 1,2-DCB on Pd/CNF-Cl and Pd/CNF-Cl-N samples from the emulsion; (2) regeneration of adsorbent-catalysts by hydrodechlorination in a liquid phase. First, an emulsion of 0.15 mL of 1,2-DCB in 100 mL of water was prepared by sonication for 20 min, after which 0.5 g of catalyst was added. The adsorption time was 2 min while shaking. The catalyst was then separated by decantation (2 min) and immediately used in the liquid-phase hydrodechlorination experiment.

The calculation of the maximum amount of 1,2-DCB in the reaction medium during the regeneration of adsorbent-catalysts by hydrodechlorination was performed according to the equation:

$$C_{1,2-DCB}^{max} = \frac{A \times m \times V_0}{V}, \quad (8)$$

where $C_{1,2-DCB}^{max}$ is the concentration of 1,2-DCB adsorbed on the adsorbent-catalyst, mmol·L^{-1} (or M); A is the adsorption capacity, which was achieved by the adsorption of 1,2-DCB from an aqueous emulsion, mol·g^{-1}; m is the mass of the adsorbent-catalyst, g; V_0 is the standard volume, 1000 mL; V is the volume of 2-propanol in the reaction medium of 1,2-DCB hydrodechlorination, 11 mL.

The apparatus for 1,2-DCB hydrodechlorination consisted of two main parts: a 50 mL temperature-controlled glass reactor equipped with a magnetic stirrer and a volumetric block to maintain a constant H$_2$ pressure. The reaction medium consisted of two immiscible fluids: 11 mL of 2-propanol and 4 mL of a 50 wt% solution of KOH. Alkali was used to bind HCl formed during the hydrodechlorination, thus preventing the catalyst from deactivation [48]. 2-Propanol serves as the organic phase which is able to dissolve 1,2-DCB efficiently and facilitate the formation of active hydrogen species on the surface of the catalyst during the hydrodechlorination process [49].

The regeneration of the adsorbent-catalyst by liquid-phase hydrodechlorination of 1,2-DCB was performed at 25 °C under a constant hydrogen pressure of 0.1 MPa and continuous stirring at 1200 rpm. When the stirring rate is above 800 rpm, the external diffusion processes have a negligible influence [50]. Following the complete conversion of 1,2-DCB, the catalyst was separated from the reaction medium by filtration, washed with water, dried at 110–130 °C for 4 h, and then used in the adsorptive–catalytic cycle for the second run. It is important to note that each experiment was repeated at least 3 times. Differences between the experimental points from different repetitions do not exceed ±4%.

Analysis of the hydrodechlorination products was performed using a Chromos GH-1000 gas chromatograph (Chromos Engineering, Dzerzhinsk, Russia) with a flame-ionization detector and a column (2.5 mm × 2 m) filled with Chromaton N-AW sorbent. The instrument was operated under argon using a hydrogen flame at temperatures from 60 to 120 °C (heating rate 10 °C/min).

2.5. Characterization of the Samples

Pure CNF as Pd-containing samples were explored by means of high-resolution transmission electron microscopy (HR TEM) using a JEM-2010CX microscope (JEOL, Tokyo, Japan). The device works at an accelerating voltage of 100 kV. The spherical aberration coefficient of an objective lens is 2.8 mm. The line resolution of the microscope is 1.4 Å.

The X-ray diffraction (XRD) study of the synthesized CNF samples was performed on a Shimadzu XRD-7000 (Shimadzu, Tokyo, Japan) diffractometer (CuK$_\alpha$ radiation) at room temperature using a graphite monochromator. The average crystallite size was determined from the integral broadening of the profiles of the 002 diffraction peak described by the Pearson function VII (PVII), using the Scherrer formula in WINFIT 1.2.1 software [51].

X-ray photoelectron spectroscopy (XPS) data were collected on a SPECS (SPECS Surface Nano Analysis GmbH, Berlin, Germany) spectrometer equipped with a hemispherical PHOIBOS-150-MCD-9 analyzer. The non-monochromatic Mg K_α radiation (hν = 1253.6 eV) at 180 W was used as the primary excitation. For calibration of the spectrometer, the Au $4f_{7/2}$ (binding energy (BE) at 84.0 eV) and Cu $2p_{3/2}$ (932.7 eV) peaks from metallic gold and copper foils were used [52]. The samples were located on a holder with a 3 M double-sided adhesive copper conducting tape. The Peak 4.1 XPS software package was applied for both spectral analysis and data processing. To determine the binding energy values and the areas of XPS peaks, the Shirley background was subtracted, and an analysis of line shapes was performed. Gaussian–Lorentzian functions were used to fit the curves in each XPS area. The atomic ratios of elements were determined from the integral photoelectron peak intensities. The corresponding relative atomic sensitivity factors [52] and the transmission function of the analyzer were used for the correction.

In order to determine the specific surface area (SSA) and pore volume (V_{pore}) of the obtained CNF, nitrogen adsorption/desorption at 77 K (Brunauer–Emmett–Teller (BET) method) was used. The samples were degassed under an oil-free vacuum at 300 °C for 5 h. The isotherms of adsorption were measured using an ASAP-2400 automated instrument (Micromeritics, Norcross, GA, USA).

The loading of palladium in adsorbent-catalysts was measured by means of X-ray fluorescence analysis using a VRA-30 instrument (Carl Zeiss, Jena, Germany) with a Cr anode X-ray tube. The relative determination error was ±5%.

3. Results and Discussion

3.1. Characterization of CNF Samples

The morphology and structure of the synthesized carbon nanomaterial was studied by scanning and transmission electron microscopies. According to the SEM data, the carbon products formed as a result of the catalytic decomposition of TCE and TCE + AN mixtures over Ni catalyst are predominantly represented by rather long graphite-like filaments (Figure 2). Note that the addition of AN to the reaction mixture has no evident effect on the morphology of the produced carbon nanomaterial.

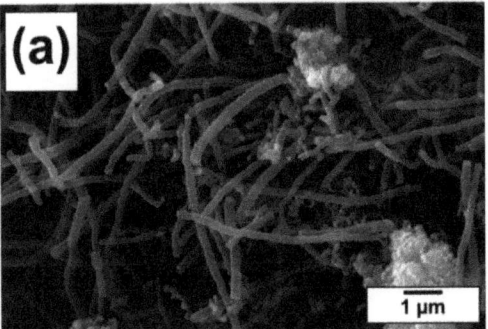

Figure 2. SEM micrographs of the CNF-Cl (**a**) and CNF-Cl-N (**b**) samples.

The detailed structure of CNF samples is represented by the selected TEM images shown in Figure 3. It is also clear from TEM data that the addition of AN into the TCE/H_2/Ar reaction mixture (Figure 3c,d) does not have any noticeable impact on the structure of segmented carbon filaments. The formation of such a structure is caused by the periodic process of chlorination–dechlorination, which takes place on the surface of active Ni particles, catalyzing the growth of CNF. The dark sections of the fibers are densely packed graphite-like "flakes", while the lighter ones are represented by graphite packs with a looser arrangement and a greater interplanar spacing [23]. The diameter of the carbon fibers varies from 100 to 300 nm.

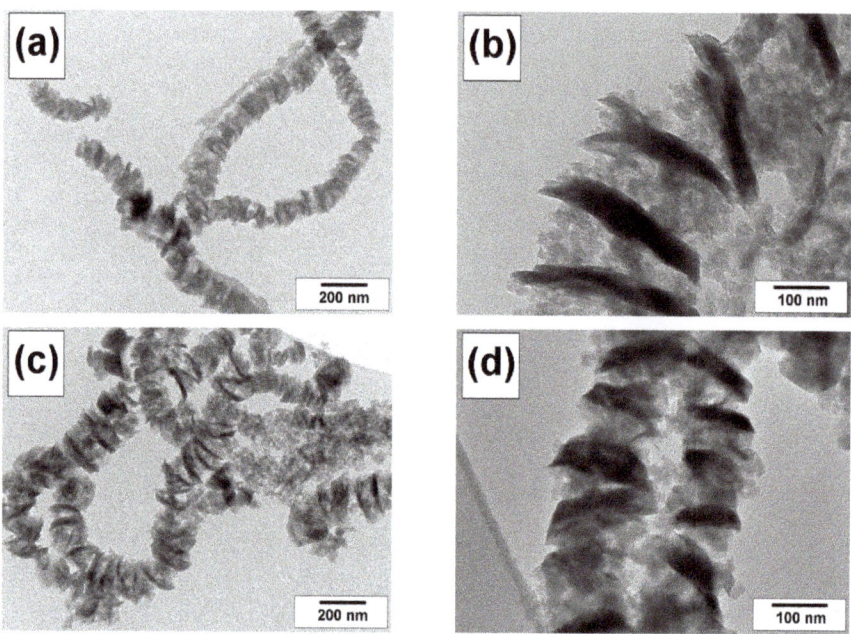

Figure 3. TEM micrograph of (**a**,**b**) CNF-Cl and (**c**,**d**) CNF-Cl-N.

XRD patterns of the synthesized CNF materials exposed to acidic treatment are presented in Figure 4. Both the CNF-Cl and CNF-Cl-N samples showed a broad reflection from the (002) graphite planes at a 2θ of 25–26° (Figure 4). No reflexes attributed to traces of the metallic component (Ni) were identified. The distance between the graphitic layers (d_{002}) determined from the position of the (002) peak was found to be 3.50 Å and 3.45 Å for CNF-Cl and CNF-Cl-N, respectively. It should be noted that the calculated average distance between the layers in CNFs (3.45–3.50 Å) is significantly larger than that corresponding to the crystalline graphite (3.35 Å).

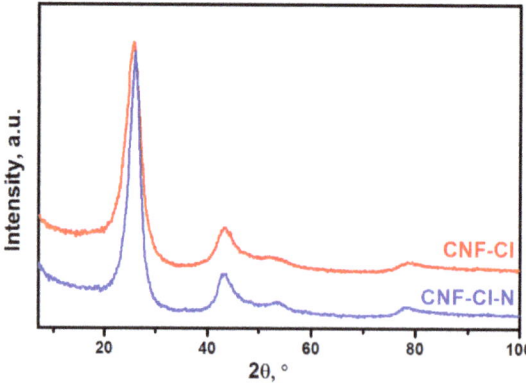

Figure 4. XRD patterns for the synthesized CNF-Cl and CNF-Cl-N samples. Both samples were washed from metallic particles by etching with HCl acid.

The functional groups present on the surface of the obtained CNF samples were studied using the XPS method (Tables S1 and S2). As seen in Figure 5, the main elements are carbon and oxygen. In addition to the corresponding lines, the detailed analysis of survey XPS spectra pointed to the presence of N1s and Cl2p low-intensity lines as well. After survey recording, the region spectra for the main lines of elements were collected to obtain a sufficient 'signal/noise' ratio for peak deconvolution and to define the positions of the peaks and analyze their area (Figure 6). The calculated contents of the elements (in atomic percent) are presented in Table 1. The results of the detailed quantitative XPS data analysis are presented in Table 2 and Figure 6.

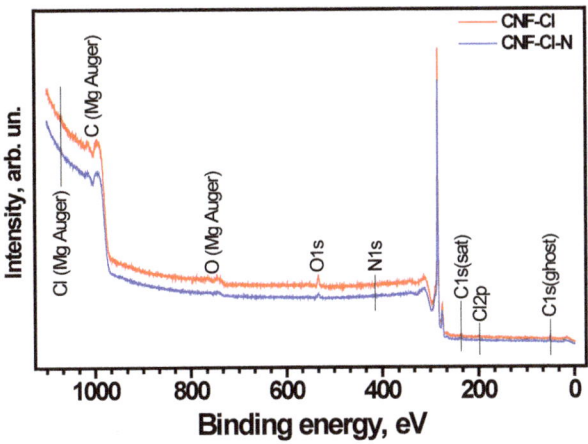

Figure 5. Survey XPS spectra of the CNF-Cl and CNF-Cl-N samples.

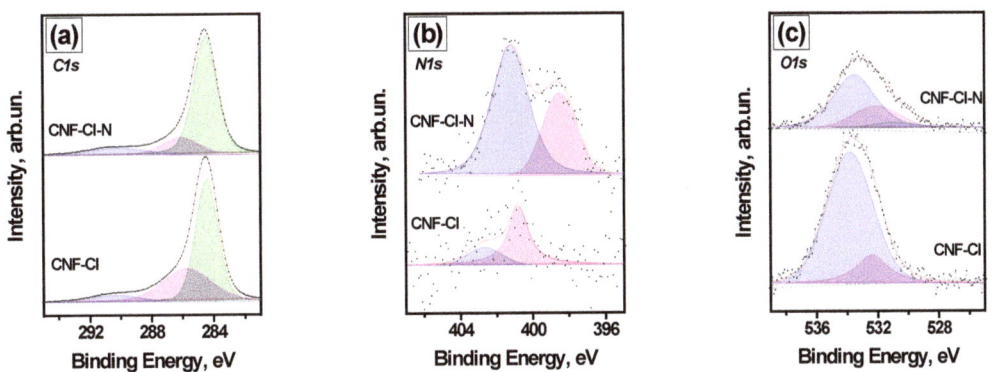

Figure 6. (a) C1s, (b) N1s, (c) O1s XPS spectra of the CNF-Cl and CNF-Cl-N samples.

Table 1. XPS data of the surface composition of the CNF-Cl and CNF-Cl-N samples.

CNF Sample	Content, wt%		
	N	O	Cl
CNF-Cl	0.07	2.65	0.09
CNF-Cl-N	0.39	1.47	0.17

Table 2. Results of the quantitative XPS data analysis.

CNF Sample	Element	BE, eV	Atomic Ratio to Carbon	State Portion	Total Atomic Ratio to Carbon
CNF-Cl	O1s	532.4	0.0049	0.18	0.027
		533.8	0.022	0.82	
	N1s	400.8	0.00057	0.75	0.00077
		402.7	0.00020	0.25	
CNF-Cl-N	O1s	531.0	0.00088	0.059	0.015
		532.1	0.0039	0.26	
		533.6	0.010	0.68	
	N1s	398.5	0.0013	0.33	0.0040
		401.2	0.0027	0.67	

The oxygen-containing groups appear in the samples as a result of their contact with atmospheric air after their removal from the reactor (see Table 2). The CNF-Cl-N sample contains the greatest amount of functional groups (0.39 wt% of N and 0.17 wt% of Cl). It was found that the introduction of nitrogen into the CNF structure promotes an almost 2-fold increase in the chlorine concentration (from 0.09 to 0.17 wt%). Presumably, the higher content of Cl in the CNF-Cl-N sample might prevent the adsorption of oxygen, thus leading to a drop in O concentration (from 2.65 to 1.47 wt%). It should be also noted that an insignificant amount of nitrogen was also detected in the composition of the CNF-Cl sample (Table 1).

The C1s spectra of the CNF samples are shown in Figure 6a. Since the amounts of O, N, and Cl are low, the presence of the components of C1s spectra attributed to the carbon bound with these elements is not expected. For both CNF-Cl and CNF-Cl-N samples, the C1s line consists of three components: C-C sp^2 (binding energy at 284.5 eV), C-C (C-H) sp^3 (at 286 eV), and $\pi \to \pi^*$ shake-up satellite (at 290 eV) [53–55]. The comparison of the C1s line shapes shows a drop in the portion of sp^3 carbon for the CNF-Cl-N sample (C sp^3 portion of 0.17) when compared with the CNF-Cl sample (C sp^3 portion of 0.36).

The N1s spectra of the studied samples are presented in Figure 6b. Analysis of the N1s region for the CNF-Cl sample reveals the presence of two components at 400.8 eV, described as N-O bonds [56], and at 402.7 eV, assigned to quaternary N [55], trapped N_2 [56] or O=N-C groups [57]. For the CNF-Cl-N sample, the components of the N1s line at 398.5 eV attributed to N≡C [56] and at 401.2 eV related to C-(NC)-C graphitic nitrogen in aromatics [57] were found in a ratio of 1:2.

In Figure 6c, the O1s spectra of the studied samples are demonstrated. Peaks were observed in the spectra of the CNF-Cl sample at 532.4 eV (C=O) and 533.8 eV (C-OH), and at 531.0 eV (C(O)O), 532.1 eV (C=O), and 533.6 eV (C-OH) for the CNF-Cl-N sample [58,59]. It should be noted that the total amount of oxygen in the CNF-Cl-N sample is lower if compared with the CNF-Cl sample.

The textural parameters of the obtained carbon nanomaterials were studied by low-temperature adsorption of nitrogen. According to the data obtained (Table 3), the specific surface area of the materials is almost the same: 297 $m^2 \cdot g^{-1}$ for CNF-Cl and 292 $m^2 \cdot g^{-1}$ for CNF-Cl-N. This observation testifies once again to the fact that the addition of AN to the reaction mixture (TCE/H_2/Ar) does not alter the structural and textural properties of the resulting carbon product.

Table 3. Textural characteristics of the CNF-Cl and CNF-Cl-N samples. BET analysis data.

Sample	SSA, $m^2 \cdot g^{-1}$	$V_{\Sigma pore}$, $cm^3 \cdot g^{-1}$	$V_{micropore}$, $cm^3 \cdot g^{-1}$	$\bar{d}$, nm
CNF-Cl	297	0.55	0.017	7.4
CNF-Cl-N	292	0.69	0.019	9.4

The pore structure of the produced CNF samples was also found to be very similar. Both samples show a bimodal pore size distribution with peaks in micro- and mesoporous regions (Figure 7). The volume of micropores is almost the same and does not exceed 3% of the total pore volume (Table 3). However, the total pore volume and the average pore diameter for the N-containing CNF-Cl-N sample are greater than those for CNF-Cl. The measured values are 0.69 $cm^3 \cdot g^{-1}$ and 9.4 nm for CNF-Cl-N and 0.55 $cm^3 \cdot g^{-1}$ and 7.4 nm for CNF-Cl, respectively (Table 3). The observed difference is due to a larger contribution of mesopores in the case of N-containing CNF (Figure 7).

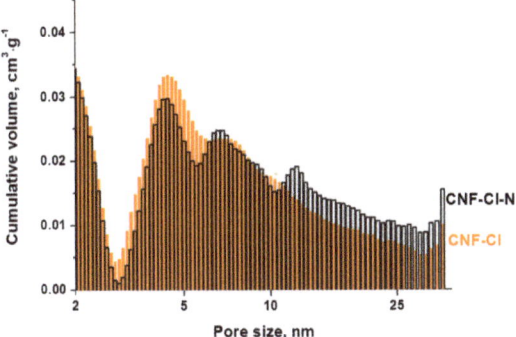

Figure 7. Pore size distribution for the CNF-Cl and CNF-Cl-N samples.

Thus, the catalytic H_2-assisted decomposition of TCE (or the mixture of TCE and AN) over Ni-catalyst allowed for the filamentous carbon material to be obtained. The produced carbon nanofibers have a diameter of 100–300 nm and a well-defined segmented structure of densely and loosely packed Cl-containing graphite layers. They are characterized by a comparatively high value of *SSA* (~300 $m^2 \cdot g^{-1}$) and a bimodal pore size distribution. The use of AN as a co-reagent for CNF synthesis results in the incorporation of nitrogen atoms into the structure of carbon product (0.39 wt%), along with an enhancement of total porosity (from 0.55 to 0.69 $cm^3 \cdot g^{-1}$), owing to the increased value of mesopores.

Using the obtained CNFs, the palladium adsorbent-catalysts with a Pd content of 1.50 ± 0.03 wt% were prepared. When palladium is supported on CNF, the specific surface area decreases from 297 to 293 $m^2 \cdot g^{-1}$ and from 292 to 284 $m^2 \cdot g^{-1}$ for CNF-Cl and CNF-Cl-N, respectively. Such insignificant changes are due to the localization of palladium nanoparticles on the outer surface of carbon fibers (Figure 8). Pd does not enter their pores, as in the case of microporous carbon black [60].

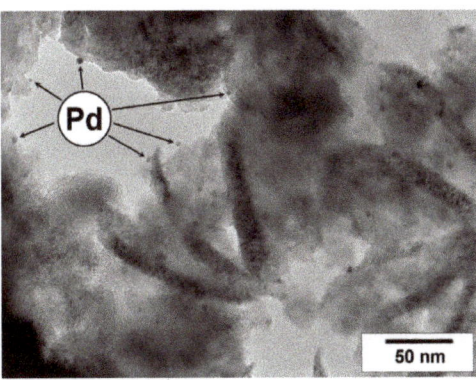

Figure 8. TEM micrograph of the Pd/CNF-Cl-N sample with spherical Pd particles.

3.2. Adsorbtion of 1,2-DCB from Aqueous Solutions

To characterize the adsorption properties of the synthesized CNF samples and related Pd/CNF materials, the 1,2-DCB adsorption from aqueous solutions under equilibrium conditions at 25 °C was investigated. The experimental adsorption isotherms (Figure 9) were analyzed using both the Langmuir and Dubinin–Astakhov models (Table 4).

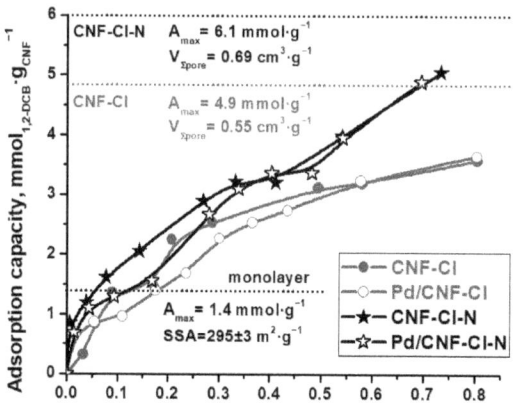

Figure 9. Experimental adsorption isotherms of 1,2-DCB at 25 °C on the CNF-Cl, Pd/CNF-Cl, CNF-Cl-N, and Pd/CNF-Cl-N samples.

Table 4. Values of parameters calculated from the Langmuir and Dubinin–Astakhov adsorption isotherms for 1,2-DCB on the CNF-Cl, Pd/CNF-Cl, CNF-Cl-N, and Pd/CNF-Cl-N samples.

Isotherm	Linear Form	Parameter		Sample			
				CNF-Cl	Pd/CNF-Cl	CNF-Cl-N	Pd/CNF-Cl-N
Langmuir	$\frac{1}{A} = \frac{1}{A_{max}} + \frac{1}{K_L A_{max}} \times \frac{1}{C_{eq}}$ $\omega = \frac{SSA}{A_{max} N_A}$	A_{max} (mol·g^{-1}) ω (nm^2) R^2		0.0022 0.22 0.710	0.0034 0.14 0.965	0.0027 0.18 0.714	0.0030 0.16 0.840
Dubinin–Astakhov	$\ln A = \ln A_{max} - \left[\frac{RT}{E_{ads}^{eff}} \times \ln \frac{C_{max}}{C_{eq}}\right]^n$ $d = 2 \times \frac{K_{C_6H_6} \times \beta}{E_{ads}^{eff}}$	A_{max} (mol·g^{-1}) W_0 (cm^3·g^{-1}) E_{ads}^{eff} (kJ·mol^{-1}) * d (nm) * R^2	$n = 1$	0.0038 0.43 5.0 6.2 0.924	0.0045 0.51 4.3 7.1 0.969	0.0048 0.54 6.4 8.5 0.950	0.0053 0.60 4.7 6.6 0.935
		A_{max} (mol·g^{-1}) W_0 (cm^3·g^{-1}) E_{ads}^{eff} (kJ·mol^{-1}) * d (nm) * R^2	$n = 2$	0.0025 0.29 7.7 4.0 0.733	0.0030 0.34 6.5 4.7 0.927	0.0033 0.38 9.6 3.2 0.765	0.0035 0.39 7.5 4.1 0.783

* The average value.

The obtained adsorption isotherms are presented in Figure 9. They are characterized by an inflection in a low range of 1,2-DCB concentration (Figure 9), which corresponds to a monolayer capacity of 0.0014 mol·g^{-1}, calculated considering the SSA value of these carbon materials (Table 4), and a 1,2-DCB molecule area of 0.35 nm^2 [61]. As the 1,2-DCB concentration increases, the adsorption capacity of CNF increases as well. This is most probably due to the volumetric filing of pores, as reported recently [21,33]. Indeed, the experimental isotherms are not linearized by the Langmuir Equation (2), but well described by the Dubinin–Astakhov Equation (4) (Table 4), which has been proposed for the mathematical description of the volumetric filling of pores [62]. The empirical parameter n, varying from 1 to 6, is linked to the degree of heterogeneity of the pore system. In our case, n is equal to 1. It is consistent with the literature that n approaches one for adsorbents with a wide pore distribution, including those with a bimodal structure [63]. As can be seen from the fitting results presented in Figure 9, the Dubinin–Astakhov model with $n = 1$ describes the adsorption of 1,2-*DCB* on CNF samples well, which are the bimodal adsorbents with two types of pores (Figure 7). The average diameter of the pores filled with 1,2-*DCB* calculated by this approximation (Figure 10a,b) was found to be close to the average diameter of the pores of CNFs obtained by the BET method (Table 3).

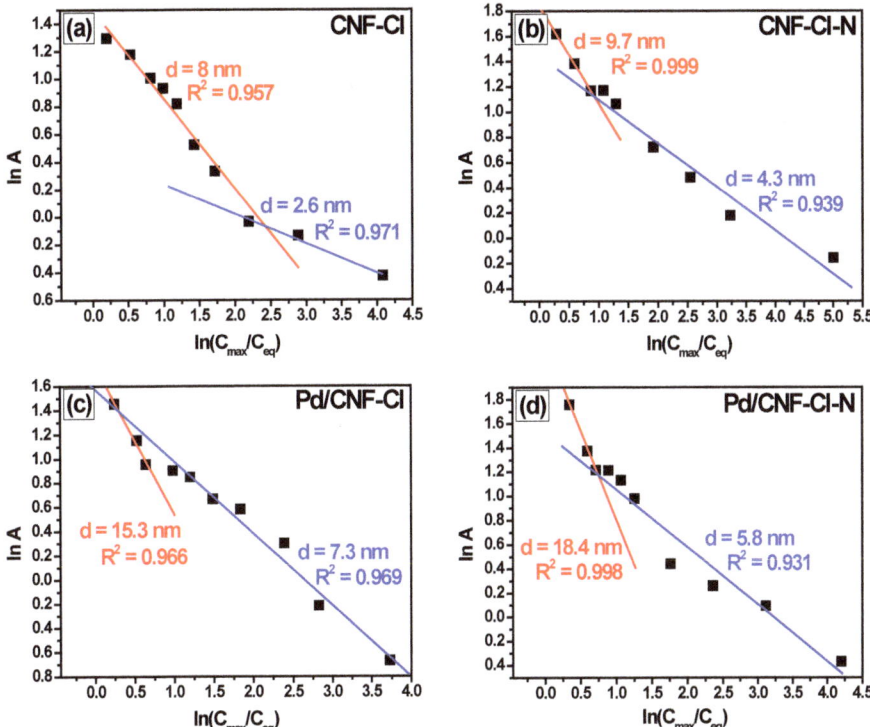

Figure 10. Dubinin–Astakhov model fitting results with $n = 1$ for 1,2-*DCB* adsorption on (**a**) CNF-Cl, (**b**) CNF-Cl-N, (**c**) Pd/CNF-Cl, and (**d**) Pd/CNF-Cl-N samples at 25 °C.

It is worth noting that the carbon nanofibers doped with nitrogen, CNF-Cl-N, have a greater adsorption capacity than the CNF-Cl sample. The achieved experimental maxima were 0.0051 mol·g^{-1} and 0.0037 mol·g^{-1} for CNF-Cl-N and CNF-Cl, respectively. It is assumed to be related to the larger pore volume and the more tailored pore structure in the case of the CNF-Cl-N sample (Table 3). At the same time, a high level of pore filling (F), 84% for CNF-Cl-N and 76% for CNF-Cl, distinguishes both the nanomaterials. Moreover, the synthesized CNFs have higher adsorption capacities in relation to 1,2-*DCB* if compared with other carbon materials described in the scientific literature. For example, a value of 0.00019 mol·g^{-1} was achieved for graphite nanosheets prepared by the wet ball milling of expanded graphite [38], a value of 0.00027 mol·g^{-1} for carbon nanotubes prepared by the catalytic pyrolysis of a propylene–hydrogen mixture [41], a value of 0.00179 mol·g^{-1} for the carbon material Sibunit [21], a value of 0.0019 mol·g^{-1} for industrial multiwalled carbon nanotubes NC7000 [61], and values of 0.00147 and 0.0028 mol·g^{-1} for commercially available activated carbons AG-2000 [21] and AG-5 [61], respectively.

In addition, the experimental adsorption isotherms of 1,2-*DCB* showed that the deposition of 1.5 wt% palladium has very slight effect on the adsorption capacity of CNF material (Figure 9). However, the average diameter of the pores filled by 1,2-*DCB* estimated using the Dubinin–Astakhov model is increased in this case (Figure 10). It appears that Pd nanoparticles (Figure 8) formed during the catalyst preparation partially block pores of a small diameter, and the adsorption of 1,2-*DCB* occurs in larger mesopores (Figure 10).

Thus, the adsorption of 1,2-*DCB* on the CNF-Cl-N sample occurs in a mode of the volume filling of pores, which are most likely located between the adjacent segments in the structure of carbon nanofibers.

3.3. Adsorptive-Catalytic Cycle of the Purification of 1,2-DCB Aqueous Emulsions

When a large amount of 1,2-*DCB* enters water, a stable emulsion is formed due to a low solubility of 1,2-*DCB* in water (0.986 mM at 25 °C). Thus, the development of the adsorption method for the purification of aqueous emulsions from 1,2-*DCB* with subsequent regeneration of adsorbent is of practical importance. In this stage of research, the efficiency of using the Pd-containing adsorbent-catalysts based on the synthesized CNF was studied for the purification of 1,2-*DCB* aqueous emulsions in the adsorptive-catalytic cycle. The process was carried out in two stages: (I) adsorption of 1,2-*DCB* on the adsorbent-catalyst at 25 °C; (II) regeneration of the adsorbent-catalyst by the liquid-phase hydrodechlorination of 1,2-*DCB* with molecular hydrogen [21] in the presence of KOH to remove formed HCl from the reaction medium (Figure 11).

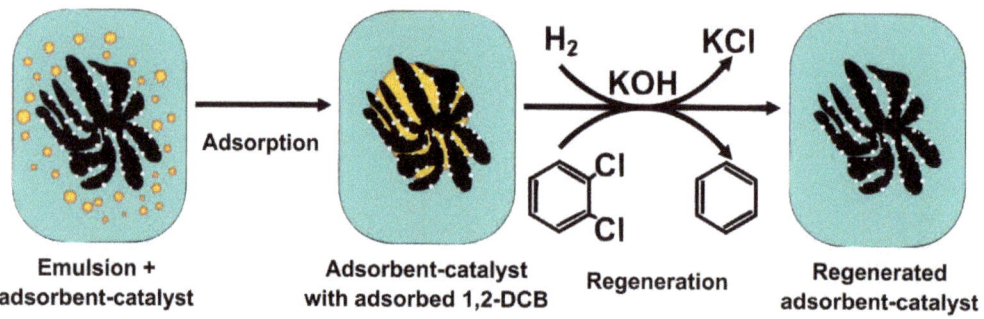

Figure 11. Scheme of the regeneration of adsorbent-catalysts by hydrodechlorination.

The adsorption of 1,2-*DCB* was performed at a ratio of the weight of carbon nanomaterial to the volume of emulsion equal to 5 $g_{Pd/CNF} \cdot L^{-1}$. As a result, the adsorbate concentration was reduced from 13 mM to 1.2 mM for Pd/CNF-Cl and to 1.1 mM for Pd/CNF-Cl-N. The achieved adsorption capacities were also quite similar, being 2.36 and 2.38 mmol·g^{-1} for Pd/CNF-Cl and Pd/CNF-Cl-N samples, respectively. Such a small difference is thought to be due to similar textural properties of the applied carbon materials.

After the adsorption of 1,2-*DCB* from an aqueous emulsion, the Pd/CNF adsorbent-catalysts were placed in 11 mL of 2-propanol (organic phase of the reaction medium) for the regeneration by liquid-phase hydrodechlorination with molecular hydrogen at 25 °C. It was shown that 1,2-*DCB* releases from the pore space of the adsorbent-catalyst by 2-propanol during the first minute (1,2-*DCB* concentration is 107 ± 1 mM) and then undergoes complete dechlorination. It should be noted that 2-propanol competes with 1,2-*DCB* for adsorption sites on the surface of carbon material [48].

According to the data shown in Figure 12, the regeneration of Pd/CNF-Cl occurs faster than that of the nitrogen-containing Pd/CNF-Cl-N sample. The presence of additional N-containing groups (Table 1) appears to have a negative impact on the regeneration of the adsorbent-catalyst. Meanwhile, it should be noted that the regeneration time was less than 1 h for both of the studied samples.

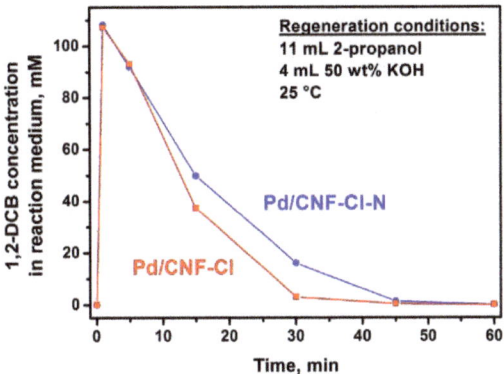

Figure 12. 1,2-*DCB* concentration changes in the reaction medium during the regeneration of Pd/CNF-Cl and Pd/CNF-Cl-N adsorbent-catalysts (25 °C, P(H$_2$) = 1 atm).

After regeneration, the adsorbent-catalysts were used for post-treatment of the solution with the remaining 1,2-*DCB* after the first adsorption procedure. According to UV-vis spectroscopy data (Figure 13), the repeated use of regenerated adsorbent-catalysts made it possible to achieve almost complete removal of 1,2-*DCB* from the aqueous medium. However, its amount was very small, and during the subsequent regeneration, it was possible to identify only trace amounts of benzene after 15 min.

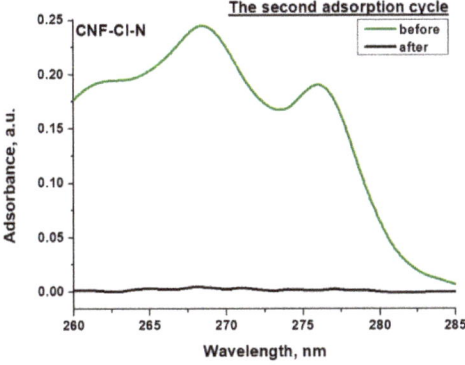

Figure 13. UV-vis spectrum of 1,2-*DCB* in aqueous solution before and after the second adsorption cycle for Pd/CNF-Cl-N adsorbent-catalyst at 25 °C.

Thereby, the carbon nanomaterials synthesized via combined catalytic pyrolysis of the Cl- and N-containing organic wastes could be also considered as promising adsorbents for water purification from chloroaromatic compounds. Their subsequent regeneration by means of the hydrodechlorination reaction enables one to obtain chemically valuable products that can be returned to the production cycle. Thus, this approach represents another step toward the rational utilization of hydrocarbon resources.

4. Conclusions

In this paper, carbon nanofibers were prepared by the catalytic decomposition of trichloroethylene (or a mixture of trichloroethylene with acetonitrile) over Ni-catalyst. In both cases, CNF showed a unique segmental structure of densely and loosely packed Cl-containing graphite layers, as well as high specific surface areas (295 ± 3 m$^2 \cdot$g^{-1}) and porosity. According to XPS data, nitrogen atoms were embedded on the CNF surface as N≡C and C-(NC)-C (graphitic nitrogen in aromatics) groups. Furthermore, there was an increase in chlorine content and a decrease in oxygen concentration. In addition, the N-containing CNF sample was characterized by a more tailored pore structure. It had a 20% higher total pore volume (0.69 $cm^3 \cdot g^{-1}$ vs. 0.55 $cm^3 \cdot g^{-1}$) due to the greater contribution of mesopores.

The synthesized Pd-containing adsorbent-catalysts based on the produced CNF samples were explored in the adsorption of 1,2-DCB from aqueous medium. It was found that carbon materials efficiently adsorb 1,2-DCB from aqueous solutions, and the adsorption process proceeds via the pore volume filling mechanism. The presence of palladium nanoparticles was found to have negligible effect on the adsorption capacity of CNF. The increased pore volume of N-containing CNF results in a higher adsorption degree of 1,2-DCB from its aqueous solution. However, when CNF samples were used for purification of emulsions with high concentrations of 1,2-DCB, the adsorption capacities achieved were quite similar. Further regeneration of the Pd/CNF adsorbent-catalysts by reductive liquid-phase hydrodechlorination was completed in 1 h with 100% yield. The use of regenerated adsorbent-catalysts in the second adsorption cycle made it possible to achieve almost complete removal of 1,2-DCB from the aqueous emulsion.

Thus, in this work, a complex approach to the disposal of chlorinated organic multi-component wastes was suggested. In the first stage, carbon materials were produced by utilizing a mixture of chlorine- and nitrogen-containing organic pollutants. The produced CNF materials were then effectively used for water purification from chloroaromatic compounds with further single-stage liquid-phase regeneration by hydrodechlorination at low temperature. The resulting products can be recycled as valuable chemicals.

Supplementary Materials: The following supporting information can be downloaded at: https://www.mdpi.com/article/10.3390/ma15238414/s1, Table S1: The quantification of XPS peaks for the sample CNF-Cl; Table S2: The quantification of XPS peaks for the sample CNF-Cl-N.

Author Contributions: Conceptualization, O.V.N. and I.V.M.; methodology, E.S.T., I.L.L. and Y.V.S.; software, A.V.N.; investigation, A.M.O., A.R.P., Y.I.B., A.V.N., E.S.T., I.L.L. and Y.V.S.; resources, Y.I.B.; writing—original draft preparation, A.M.O., A.R.P. and A.V.N.; writing—review and editing, O.V.N., I.V.M. and A.A.V.; visualization, A.R.P., O.V.N. and A.A.V.; supervision, A.A.V.; funding acquisition, I.V.M. All authors have read and agreed to the published version of the manuscript.

Funding: This work was financially supported by the Russian Science Foundation (project No. 22-13-00406, https://rscf.ru/en/project/22-13-00406/, accessed on 21 November 2022, BIC SB RAS).

Institutional Review Board Statement: Not applicable.

Informed Consent Statement: Not applicable.

Data Availability Statement: Data is contained within the article.

Acknowledgments: Characterization of the samples was performed using the equipment of the Center of Collective Use "National Center of Catalysts Research". TEM studies were performed in the Krasnoyarsk Regional Center of Research Equipment of the Federal Research Center "Krasnoyarsk Science Center SB RAS". The authors are grateful to M.N. Volochaev for help with the electron microscopy studies, and to A.B. Ayupov for the low-temperature nitrogen adsorption/desorption analysis.

Conflicts of Interest: The authors declare no conflict of interest. The funders had no role in the design of the study; in the collection, analyses, or interpretation of data; in the writing of the manuscript; or in the decision to publish the results.

References

1. Huang, B.; Lei, C.; Wei, C.; Zeng, G. Chlorinated volatile organic compounds (Cl-VOCs) in environment—Sources, potential human health impacts, and current remediation technologies. *Environ. Int.* **2014**, *71*, 118–138. [CrossRef] [PubMed]
2. Somma, S.; Reverchon, E.; Baldino, L. Water purification of classical and emerging organic pollutants: An extensive review. *ChemEngineering* **2021**, *5*, 47. [CrossRef]
3. Cazzolla, G.R. Why We Will Continue to Lose Our Battle with Cancers If We Do Not Stop Their Triggers from Environmental Pollution. *Int. J. Environ. Res. Public Health* **2021**, *18*, 6107. [CrossRef] [PubMed]

4. Shumbula, P.; Maswanganyi, C.; Shumbula, N. Methods and Treatment of Organic Pollutants in Wastewater. In *Persistent Organic Pollutants (POPs)*; Rashed, M.N., Ed.; IntechOpen: Rijeka, Croatia, 2021; Chapter 5.
5. Aranzabal, A.; Pereda-Ayo, B.; González-Marcos, M.; González-Marcos, J.; López-Fonseca, R.; González-Velasco, J. State of the art in catalytic oxidation of chlorinated volatile organic compounds. *Chem. Pap.* **2014**, *68*, 1169–1186. [CrossRef]
6. Dutta, N.; Usman, M.; Ashraf, M.A.; Luo, G.; Zhang, S. A Critical Review of Recent Advances in the Bio-remediation of Chlorinated Substances by Microbial Dechlorinators. *Chem. Eng. J. Adv.* **2022**, *12*, 100359. [CrossRef]
7. Nguyen, V.H.; Smith, S.M.; Wantala, K.; Kajitvichyanukul, P. Photocatalytic remediation of persistent organic pollutants (POPs): A review. *Arab. J. Chem.* **2020**, *13*, 8309–8337. [CrossRef]
8. Chen, Z.; Liu, Y.; Wei, W.; Ni, B.J. Recent advances in electrocatalysts for halogenated organic pollutant degradation. *Environ. Sci. Nano* **2019**, *6*, 2332–2366. [CrossRef]
9. Rodrigues, R.; Betelu, S.; Colombano, S.; Tzedakis, T.; Masselot, G.; Ignatiadis, I. In situ chemical reduction of chlorinated organic compounds. In *Environmental Soil Remediation and Rehabilitation*; van Hullebusch, E., Huguenot, D., Pechaud, Y., Simonnot, M.O., Colombano, S., Eds.; Springer: Lyon, France, 2020; pp. 283–398.
10. Des, L.E.; Dumée, L.F.; Kong, L. Nanofiber-based materials for persistent organic pollutants in water remediation by adsorption. *Appl. Sci.* **2018**, *8*, 166.
11. Nazir, M.A.; Najam, T.; Jabeen, S.; Wattoo, M.A.; Bashir, M.S.; Shah, S.S.A.; Rehman, U.A. Facile synthesis of Tri-metallic layered double hydroxides (NiZnAl-LDHs): Adsorption of Rhodamine-B and methyl orange from water. *Inorg. Chem. Commun.* **2022**, *145*, 110008. [CrossRef]
12. Nazir, M.A.; Najam, T.; Shahzad, K.; Wattoo, M.A.; Hussain, T.; Tufail, M.K.; Shah, S.S.A.; Rehman, U.A. Heterointerface engineering of water stable ZIF-8@ ZIF-67: Adsorption of rhodamine B from water. *Surf. Interfaces* **2022**, *34*, 102324. [CrossRef]
13. Su, Y.; Fu, K.; Pang, C.; Zheng, Y.; Song, C.; Ji, N.; Ma, D.; Lu, X.; Liu, C.; Han, R. Recent Advances of Chlorinated Volatile Organic Compounds' Oxidation Catalyzed by Multiple Catalysts: Reasonable Adjustment of Acidity and Redox Properties. *Environ. Sci. Technol.* **2022**, *56*, 9854–9871. [CrossRef] [PubMed]
14. Hiraoka, T.; Kawakubo, T.; Kimura, J.; Taniguchi, R.; Okamoto, A.; Okazaki, T.; Sugai, T.; Ozeki, Y.; Yoshikawa, M.; Shinohara, H. Selective synthesis of double-wall carbon nanotubes by CCVD of acetylene using zeolite supports. *Chem. Phys. Lett.* **2003**, *382*, 679–685. [CrossRef]
15. Kenzhin, R.M.; Bauman, Y.I.; Volodin, A.M.; Mishakov, I.V.; Vedyagin, A.A. Synthesis of carbon nanofibers by catalytic CVD of chlorobenzene over bulk nickel alloy. *Appl. Surf. Sci.* **2018**, *427*, 505–510. [CrossRef]
16. Mishakov, I.; Chesnokov, V.; Buyanov, R.; Pakhomov, N. Decomposition of chlorinated hydrocarbons on iron-group metals. *Kinet. Catal.* **2001**, *42*, 543–548. [CrossRef]
17. Mishakov, I.; Vedyagin, A.; Bauman, Y.; Shubin, Y.V.; Buyanov, R. Synthesis of carbon nanofibers via catalytic chemical vapor deposition of halogenated hydrocarbons. In *Carbon Nanofibers: Synthesis, Applications and Performance*; Nova Science Publishers, Inc.: Hauppauge, NY, USA, 2018; pp. 77–182.
18. Wang, J.; Cutright, T.J. Potential waste minimization of trichloroethylene and perchloroethylene via aerobic biodegradation. *J. Environ. Sci. Health A* **2005**, *40*, 1569–1584. [CrossRef]
19. Nasibulina, L.I.; Koltsova, T.S.; Joentakanen, T.; Nasibulin, A.G.; Tolochko, O.V.; Malm, J.E.M.; Karppinen, M.J.; Kauppinen, E.I. Direct synthesis of carbon nanofibers on the surface of copper powder. *Carbon* **2010**, *48*, 4559–4562. [CrossRef]
20. Shaikjee, A.; Coville, N.J. Catalyst restructuring studies: The facile synthesis of tripod-like carbon fibers by the decomposition of trichloroethylene. *Mater. Lett.* **2012**, *68*, 273–276. [CrossRef]
21. Netskina, O.V.; Tayban, E.S.; Moiseenko, A.P.; Komova, O.V.; Mukha, S.A.; Simagina, V.I. Removal of 1,2-dichlorobenzene from water emulsion using adsorbent catalysts and its regeneration. *J. Hazard. Mater.* **2015**, *285*, 84–93. [CrossRef]
22. Gusain, R.; Kumar, N.; Ray, S.S. Recent advances in carbon nanomaterial-based adsorbents for water purification. *Coord. Chem. Rev.* **2020**, *405*, 213111. [CrossRef]
23. Bauman, Y.I.; Lysakova, A.S.; Rudnev, A.V.; Mishakov, I.V.; Shubin, Y.V.; Vedyagin, A.A.; Buyanov, R.A. Synthesis of nanostructured carbon fibers from chlorohydrocarbons over Bulk Ni-Cr Alloys. *Nanotechnol. Russ.* **2014**, *9*, 380–385. [CrossRef]
24. Potylitsyna, A.R.; Mishakov, I.V.; Bauman, Y.I.; Kibis, L.S.; Shubin, Y.V.; Volochaev, M.N.; Melgunov, M.S.; Vedyagin, A.A. Metal dusting as a key route to produce functionalized carbon nanofibers. *React. Kinet. Mech. Catal.* **2022**, *135*, 1387–1404. [CrossRef]
25. Dhanya, V.; Arunraj, B.; Rajesh, N. Prospective application of phosphorylated carbon nanofibers with a high adsorption capacity for the sequestration of uranium from ground water. *RSC Adv.* **2022**, *12*, 13511–13522. [CrossRef] [PubMed]
26. Rodríguez, A.; Ovejero, G.; Sotelo, J.; Mestanza, M.; García, J. Adsorption of dyes on carbon nanomaterials from aqueous solutions. *J. Environ. Sci. Health* **2010**, *45*, 1642–1653. [CrossRef] [PubMed]
27. Machado, F.M.; Lima, É.C.; Jauris, I.M.; Adebayo, M.A. Carbon Nanomaterials for Environmental Applications. In *Carbon Nanomaterials as Adsorbents for Environmental and Biological Applications*; Bergmann, C.P., Machado, F.M., Eds.; Springer: Pelotas, RS, Brazil, 2015; pp. 85–105.
28. Tripathi, P.K.; Gan, L.; Liu, M.; Rao, N.N. Mesoporous carbon nanomaterials as environmental adsorbents. *J. Nanosci. Nanotechnol.* **2014**, *14*, 1823–1837. [CrossRef] [PubMed]
29. Mishakov, I.V.; Bauman, Y.I.; Brzhezinskaya, M.; Netskina, O.V.; Shubin, Y.V.; Kibis, L.S.; Stoyanovskii, V.O.; Larionov, K.B.; Serkova, A.N.; Vedyagin, A.A. Water Purification from Chlorobenzenes using Heteroatom-Functionalized Carbon Nanofibers Produced on Self-Organizing Ni-Pd Catalyst. *J. Environ. Chem. Eng.* **2022**, *10*, 107873. [CrossRef]

30. Bauman, Y.I.; Netskina, O.; Mukha, S.; Mishakov, I.; Shubin, Y.V.; Stoyanovskii, V.; Nalivaiko, A.Y.; Vedyagin, A.; Gromov, A. Adsorption of 1, 2-Dichlorobenzene on a Carbon Nanomaterial Prepared by Decomposition of 1, 2-Dichloroethane on Nickel Alloys. *Russ. J. Appl. Chem.* **2020**, *93*, 1873–1882. [CrossRef]
31. Aramendıa, M.; Borau, V.; Garcıa, I.; Jiménez, C.; Lafont, F.; Marinas, A.; Marinas, J.; Urbano, F. Liquid-phase hydrodechlorination of chlorobenzene over palladium-supported catalysts: Influence of HCl formation and NaOH addition. *J. Mol. Catal. A Chem.* **2002**, *184*, 237–245. [CrossRef]
32. Matatov-Meytal, Y.; Sheintuch, M. Catalytic regeneration of chloroorganics-saturated activated carbon using hydrodechlorination. *Ind. Eng. Chem. Res.* **2000**, *39*, 18–23. [CrossRef]
33. Ma, X.; Liu, S.; Liu, Y.; Gu, G.; Xia, C. Comparative study on catalytic hydrodehalogenation of halogenated aromatic com-pounds over Pd/C and Raney Ni catalysts. *Sci. Rep.* **2016**, *6*, 25068. [CrossRef]
34. Ruiz-García, C.; Heras, F.; Calvo, L.; Alonso-Morales, N.; Rodriguez, J.J.; Gilarranz, M.A. Platinum and N-doped carbon nanostructures as catalysts in hydrodechlorination reactions. *Appl. Catal. B* **2018**, *238*, 609–617. [CrossRef]
35. Janiak, T.; Okal, J. Effectiveness and stability of commercial Pd/C catalysts in the hydrodechlorination of meta-substituted chlorobenzenes. *Appl. Catal. B* **2009**, *92*, 384–392. [CrossRef]
36. Van, W.D.; Thompson, R.S.; Rooij, C.D.; Garny, V.; Lecloux, A.; Kanne, R. 1, 2-Dichlorobenzene marine risk assessment with special reference to the OSPARCOM region: North Sea. *Environ. Monit. Assess.* **2004**, *97*, 87–102.
37. Rudneva, Y.V.; Shubin, Y.V.; Plyusnin, P.E.; Bauman, Y.I.; Mishakov, I.V.; Korenev, S.V.; Vedyagin, A.A. Preparation of highly dispersed Ni1-xPdx alloys for the decomposition of chlorinated hydrocarbons. *J. Alloys Compd.* **2019**, *782*, 716–722. [CrossRef]
38. Li, X.; Chen, G. Surface modified graphite nanosheets used as adsorbent to remove 1, 2-dichlorobenzene from water. *Mater. Lett.* **2009**, *63*, 930–932. [CrossRef]
39. Negrea, P.; Sidea, F.; Negrea, A.; Lupa, L.; Ciopec, M.; Muntean, C. Studies regarding the benzene, toluene and o-xylene removal from waste water. *Chem. Bull. Politeh. Univ.* **2008**, *53*, 144–146.
40. Peng, X.; Li, Y.; Luan, Z.; Di, Z.; Wang, H.; Tian, B.; Jia, Z. Adsorption of 1, 2-dichlorobenzene from water to carbon nanotubes. *Chem. Phys. Lett.* **2003**, *376*, 154–158. [CrossRef]
41. Zagoruiko, N.I.; Rodzivilova, I.S.; Artemenko, S.E.; Glukhova, L.G. Sorption studies of the pore structure of carbon fibres. *Fibre Chem.* **2001**, *3*, 499–501. [CrossRef]
42. Langmuir, I. The Constitution and Fundamental Properties of Solids and Liquids. *J. Am. Chem. Soc.* **1917**, *183*, 102–105. [CrossRef]
43. Webb, P.A. Introduction to Chemical Adsorption Analytical Techniques and Their Applications to Catalysis. *MIC Tech. Publ.* **2003**, *13*, 1–4.
44. Podlesnyuk, V.; Levchenko, T.; Marutovskii, R.; Koganovskii, A. Adsorption of dissolved organic compounds on porous polymeric materials. *Theor. Exp. Chem.* **1985**, *21*, 363–365. [CrossRef]
45. Bering, B.; Dubinin, M.; Serpinsky, V. Theory of volume filling for vapor adsorption. *J. Colloid Interface Sci.* **1966**, *21*, 378–393. [CrossRef]
46. Dubinin, M. Adsorption properties and microporous structures of carbonaceous adsorbents. *Carbon* **1987**, *25*, 593–598. [CrossRef]
47. Dubinin, M.M.; Astakhov, V.A. Development of the Concepts of Volume Filling of Micropores in the Adsorption of Gases and Vapors by Microporous Adsorbents. *Bull. Acad. Sci. USSR Div. Chem. Sci.* **1971**, *20*, 8–12. [CrossRef]
48. Netskina, O.; Komova, O.; Tayban, E.; Oderova, G.; Mukha, S.; Kuvshinov, G.; Simagina, V. The influence of acid treatment of carbon nanofibers on the activity of palladium catalysts in the liquid-phase hydrodechlorination of dichlorobenzene. *Appl. Catal. A-Gen.* **2013**, *467*, 386–393. [CrossRef]
49. Simagina, V.I.; Tayban, E.S.; Grayfer, E.D.; Gentsler, A.G.; Komova, O.V.; Netskina, O.V. Liquid-phase hydrodechlorination of chlorobenzene by molecular hydrogen: The influence of reaction medium on process efficiency. *Pure Appl. Chem.* **2009**, *81*, 2107–2114. [CrossRef]
50. Gentsler, A.; Simagina, V.; Netskina, O.; Komova, O.; Tsybulya, S.; Abrosimov, O. Catalytic hydrodechlorination on palladium-containing catalysts. *Kinet. Catal.* **2007**, *48*, 60–66. [CrossRef]
51. Krumm, S. An interactive Windows program for profile fitting and size/strain analysis. *Mater. Sci. Forum.* **1996**, 183–190. [CrossRef]
52. Moulder, J.F.; Stickle, W.F.; Sobol, P.E.; Bomben, K.D. *Handbook of X-ray Photoelectron Spectroscopy*; Chastain, J., Ed.; Perkin-Elmer Corp.: Eden Prairie, MN, USA, 1992.
53. Chen, X.; Wang, X.; Fang, D. A review on C1s XPS-spectra for some kinds of carbon materials. *Fuller. Nanotub. Carbon Nanostructures* **2020**, *28*, 1048–1058. [CrossRef]
54. Eng, A.Y.S.; Sofer, Z.; Sedmidubsky, D.; Pumera, M. Synthesis of carboxylated-graphenes by the Kolbe–Schmitt process. *ACS Nano* **2017**, *11*, 1789–1797. [CrossRef]
55. Kuntumalla, M.K.; Attrash, M.; Akhvlediani, R.; Michaelson, S.; Hoffman, A. Nitrogen bonding, work function and thermal stability of nitrided graphite surface: An in situ XPS, UPS and HREELS study. *Appl. Surf. Sci.* **2020**, *525*, 146562. [CrossRef]
56. Dementjev, A.; Graaf, A.D.; Sanden, M.V.D.; Maslakov, K.; Naumkin, A.; Serov, A. X-ray photoelectron spectroscopy reference data for identification of the C_3N_4 phase in carbon–nitrogen films. *Diam. Relat. Mater.* **2000**, *9*, 1904–1907. [CrossRef]
57. Ayiania, M.; Smith, M.; Hensley, A.J.; Scudiero, L.; McEwen, J.S.; Garcia-Perez, M. Deconvoluting the XPS spectra for nitrogen-doped chars: An analysis from first principles. *Carbon* **2020**, *162*, 528–544. [CrossRef]

58. Chen, C.M.; Zhang, Q.; Yang, M.G.; Huang, C.H.; Yang, Y.G.; Wang, M.Z. Structural evolution during annealing of thermally reduced graphene nanosheets for application in supercapacitors. *Carbon* **2012**, *50*, 3572–3584. [CrossRef]
59. Oh, Y.J.; Yoo, J.J.; Kim, Y.I.; Yoon, J.K.; Yoon, H.N.; Kim, J.H.; Park, S.B. Oxygen functional groups and electrochemical capacitive behavior of incompletely reduced graphene oxides as a thin-film electrode of supercapacitor. *Electrochim. Acta* **2014**, *116*, 118–128. [CrossRef]
60. Simonov, P.A.; Likholobov, V.A. *Catalysis and Electrocatalysis at Nanoparticle Surfaces, In Physicochemical Aspects of Preparation of Carbon-Supported Noble Metal Catalysts*; Wieckowski, A., Savinova, E.R., Vayenas, C.G., Eds.; CRC Press: Boca Raton, FL, USA, 2003; pp. 422–423.
61. Jurkiewicz, M.; Pełech, R. Adsorption of 1, 2-Dichlorobenzene from the Aqueous Phase onto Activated Carbons and Modified Carbon Nanotubes. *Int. J. Mol. Sci.* **2021**, *22*, 13152. [CrossRef]
62. Dubinin, M.I. Physical adsorption of gases and vapors in micropores. In *Progress in Surface and Membrane Science*; Elsevier: Moscow, Russia, 1975; Volume 9, pp. 1–70.
63. Terzyk, A.P.; Gauden, P.A.; Kowalczyk, P. What kind of pore size distribution is assumed in the Dubinin–Astakhov adsorption isotherm equation? *Carbon* **2002**, *40*, 2879–2886. [CrossRef]

Article

Comparative Study on Carbon Erosion of Nickel Alloys in the Presence of Organic Compounds under Various Reaction Conditions

Alexander M. Volodin [1,*], Roman M. Kenzhin [1], Yury I. Bauman [1], Sofya D. Afonnikova [1], Arina R. Potylitsyna [1,2], Yury V. Shubin [3], Ilya V. Mishakov [1] and Aleksey A. Vedyagin [1]

1. Boreskov Institute of Catalysis SB RAS, 5 Lavrentyev ave., 630090 Novosibirsk, Russia
2. Department of Natural Sciences, Novosibirsk State University, 2 Pirogova str., 630090 Novosibirsk, Russia
3. Nikolaev Institute of Inorganic Chemistry SB RAS, 3 Lavrentyev ave., 630090 Novosibirsk, Russia
* Correspondence: volodin@catalysis.ru

Citation: Volodin, A.M.; Kenzhin, R.M.; Bauman, Y.I.; Afonnikova, S.D.; Potylitsyna, A.R.; Shubin, Y.V.; Mishakov, I.V.; Vedyagin, A.A. Comparative Study on Carbon Erosion of Nickel Alloys in the Presence of Organic Compounds under Various Reaction Conditions. *Materials* 2022, *15*, 9033. https://doi.org/10.3390/ma15249033

Academic Editor: Laura Calvillo

Received: 1 November 2022
Accepted: 15 December 2022
Published: 17 December 2022

Publisher's Note: MDPI stays neutral with regard to jurisdictional claims in published maps and institutional affiliations.

Copyright: © 2022 by the authors. Licensee MDPI, Basel, Switzerland. This article is an open access article distributed under the terms and conditions of the Creative Commons Attribution (CC BY) license (https://creativecommons.org/licenses/by/4.0/).

Abstract: The processes of carbon erosion of nickel alloys during the catalytic pyrolysis of organic compounds with the formation of carbon nanofibers in a flow-through reactor as well as under reaction conditions in a close volume (Reactions under Autogenic Pressure at Elevated Temperature, RAPET) were studied. The efficiency of the ferromagnetic resonance method to monitor the appearance of catalytically active nickel particles in these processes has been shown. As found, the interaction of bulk Ni-Cr alloy with the reaction medium containing halogenated hydrocarbons (1,2-dichloroethane, 1-iodobutane, 1-bromobutane) results in the appearance of ferromagnetic particles of similar dimensions (~200–300 nm). In the cases of hexachlorobenzene and hexafluorobenzene, the presence of a hydrogen source (hexamethylbenzene) in the reaction mixture was shown to be highly required. The microdispersed samples of Ni-Cu and Ni-Mo alloys were prepared by mechanochemical alloying of powders and by reductive thermolysis of salts-precursors, accordingly. Their interaction with polymers (polyethylene and polyvinyl chloride) under RAPET conditions and with ethylene and 1,2-dichloroethane in a flow-through reactor are comparatively studied as well. According to microscopic data, the morphology of the formed carbon nanofibers is affected by the alloy composition and by the nature of the used organic substrate.

Keywords: nickel alloys; carbon erosion; carbon nanofibers; ferromagnetic resonance; RAPET

1. Introduction

Carbon nanomaterials of various functional destinations are widely applied in modern science and technology [1,2]. Among the variety of these materials, two classes, carbon nanotubes (CNT) and carbon nanofibers (CNF) should be mentioned especially. Both CNT and CNF can be derived from hydrocarbon sources via catalytic methods [3–7]. As a rule, the catalytic growth of carbon nanostructures takes place over the dispersed particles of iron subgroup metals [8–10]. In particular, nickel and its alloys are of special interest [11–21].

It should be noted that the dispersed metal particles are easily oxidized being in contact with the air. Therefore, in order to obtain such dispersed particles, oxide precursors are often deposited on the supports with a high surface area (Al_2O_3, MgO, SiO_2, zeolites, etc.) and then reduced in a hydrogen flow immediately before the catalytic procedures [13–15,22,23]. One of the important issues is that only the dispersed particles of metals and alloys possess catalytic activity in the process of catalytic carbon growth. This process is also known as catalytic chemical vapor deposition (CCVD). The size and the chemical composition of the dispersed particles affect significantly the efficiency of the carbon nanostructures formation as well as their morphology. The commercially available bulk metals and alloys (foils, wires, and rods) produced on an industrial scale do not exhibit such a catalytic activity. One of the possible ways to form catalytically active particles from the bulk metal items is an initiation of the catalytic corrosion processes over their surface due to interaction with the reaction (aggressive) medium. Finally, the corrosion processes result in nano-/micro-structuring of the surface (the formation of the surface dispersed particles) or even complete disintegration of the initial items. The first scenario is well-known for the reductive-oxidative catalytic reactions [24–26], which are accompanied by the significant loosening of the metal surface and the formation of dispersed particles of a wide size distribution on it.

On the other hand, all the metals (Fe, Co, and Ni) traditionally used for the synthesis of nanostructured carbon possess ferromagnetism [27–30]. Therefore, the ferromagnetic resonance (FMR) method is applicable, in principle, to study the dispersed particles of these metals within the composition of catalysts. Despite many attempts by researchers to use this technique for catalyst characterization, it did not get a wide application in practice. Moreover, it was quite rarely applied to explore the catalysts for CNF synthesis. All this is connected with the difficulties in the interpretation of FMR spectra, which are caused by the presence of ferromagnets of high-spin states along with strong exchange interactions characterized by a high anisotropy. The most important factor complicating the application of FMR is the absence of any information regarding the spin of the studied samples that do not allow making even a qualitative estimation of the concentration of the ferromagnetic particles in the sample, contrary to the paramagnetic sites registered in different catalytic systems by the electron paramagnetic resonance (EPR) technique.

Nevertheless, as we have reported recently [31–34], taking into account the features of chemical and magnetic properties of the nickel-containing catalysts, the FMR method can be successfully applied to monitor the appearance of the dispersed metal particles during the catalytic reaction and to observe the noticeable changes in the stoichiometry of the alloys being used. The main feature of the catalytic systems described in these works deals with the phenomenon of self-organization of the catalyst when the relatively uniform in size metal particles (~200 nm) are being formed from different bulk metal precursors. Further, these particles act as sites for the catalytic growth of CNF. The appearance of such submicron ferromagnetic particles characterized by a narrow size distribution allows applying the FMR technique not only to detect them but to estimate the kinetics of their accumulation during the catalytic processes as well.

The FMR method is also useful to study the self-disintegration processes of nickel alloys in a closed volume in an in situ mode. Such a mode is known as Reactions under Autogenic Pressure at Elevated Temperature (RAPET). It was originally proposed by Prof. A. Gedanken et al. in their numerous works [35–40]. The RAPET approach allows studying the dynamics of the processes and revealing the experimental conditions required for the synthesis of a series of a new type of structured inorganic nanomaterials, which appeared as intermediates during the processes of catalytic transformation of various classes of organic molecules [31–33].

The present work is aimed to further develop this scientific direction and to widen the number of catalytic systems studied by means of FMR and used to synthesize CNF. The interaction of a wide range of organic substrates with nickel alloys was examined in a flow-through reactor and under RAPET conditions. In both cases, the FMR method was applied to monitor the appearance of dispersed nickel-containing particles. The resulting carbon nanostructures were investigated by electron microscopic techniques.

2. Materials and Methods

2.1. Chemicals and Materials

Nichrome wire (~75.5% Ni, ~23% Cr, ~1.5% Fe) of 0.1 mm in diameter was purchased from VZPS (Vladimir, Russia) and used without any further purification. Nickel powder (RusRedMet, Saint-Petersburg, Russia) and copper powder (NMK-Ural, Yekaterinburg, Russia) were used to synthesizing the Ni-Cu alloy.

$NiCl_2 \cdot 6H_2O$ (Reachem, Dzerzhinsk, Russia) was used to synthesize $[Ni(NH_3)_6]Cl_2$. A total of 5 g of $NiCl_2 \cdot 6H_2O$ was dissolved in a minimal amount of distilled water. About 30 mL of concentrated NH_3 along with 10 mL of saturated NH_4Cl solution was added to the solution of nickel chloride. The slow formation of a lilac precipitate was observed. The solution was maintained for 30 min. Then, the precipitate was filtered using a glass filter and washed with ethanol.

$(NH_4)_6Mo_7O_{24} \cdot 4H_2O$ was purchased from Reachem (Dzerzhinsk, Russia). All the chemicals were of chemical purity grade and were used without any preliminary purification.

2.2. Synthesis of Alloy Samples

The Ni-Cu alloy containing 12 wt.% of copper was prepared by the mechanochemical alloying (MCA) method as described elsewhere [41]. The preliminarily prepared mixture of nickel and copper powders (10 g) with a weight ratio Ni/Cu of 88/12 was loaded into stainless steel jars of 250 mL in volume along with stainless steel grinding balls (340 g). The diameter of the grinding balls was 5 mm. The MCA procedure was performed using a planetary mill Activator 2S (Activator LLC, Novosibirsk, Russia). The rotation frequency rate of the platform (956 rpm) and the jars (449 rpm) was controlled using an industrial frequency inverter VF-S15 (Toshiba Schneider Inverter Corp., Nagoya, Japan). The estimated acceleration of the grinding balls was 784 m/c^2 (~80 G). The jars were cooled with water in order to avoid overheating during the MCA procedure. Finally, the Ni-Cu alloy sample was unloaded in the air, separated from the grinding balls using a sieve, and weighed. The sample was labeled as Ni-Cu(12%).

The Ni-Mo alloy was prepared by the high-temperature reductive thermolysis of a multicomponent precursor. The latter was obtained as follows. The joint concentrated solution of [Ni(NH$_3$)$_6$]Cl$_2$ and (NH$_4$)$_6$Mo$_7$O$_{24}$, containing the calculated amounts of nickel and molybdenum to obtain 1 g of alloy, was added at intense stirring to a large volume of acetone cooled down to 0 °C. Due to the low solubility of these salts in acetone, the formation of an oversaturated solution leading to immediate precipitation of the microheterogeneous mixture of precursors occurs. The precipitate was filtered, dried in the air for ~20 h, and calcined in a hydrogen flow (130 mL/min) at 800 °C for 1 h. The temperature ramping rate was 20 °C/min. Then, the sample was cooled down in a hydrogen atmosphere and purged with helium. The Ni/Mo weight ratio was 92/8. The sample was labeled as Ni-Mo(8%).

2.3. Characterization of Alloys and Carbon Samples

Both the alloys and the carbon samples were studied by scanning electron microscopy (SEM). The images were collected using a JSM-6460 electron microscope (JEOL, Tokyo, Japan) at a magnification from 1000× to 100,000×.

The morphology of the carbon product was explored by transmission electron microscopy (TEM) using a Hitachi HT7700 TEM (Hitachi High-Technologies Corp., Tokyo, Japan) working at an acceleration voltage of 100 kV and equipped with a W source.

The low-temperature nitrogen adsorption method (Brunauer–Emmett–Teller, BET) was used to determine the specific surface area (SSA) of the samples. The samples were preliminary degassed under an oil-free vacuum at 300 °C for 5 h. The adsorption/desorption isotherms were recorded at 77 K on an ASAP-2400 automated instrument (Micromeritics, Norcross, GA, USA).

The ferromagnetic resonance (FMR) spectra were registered at room temperature using an ERS-221 spectrometer (Center of Scientific Instruments Engineering, Leipzig, Germany). The initial alloys as well as the samples after catalytic experiments in a flow-through reactor were studied. A specimen (0.3–0.5 mg for alloys, 5–10 mg for carbon samples) was loaded into a quartz ampule. Such a low sample loading is stipulated by an intensive absorption of microwave power by metals. The carbon samples were homogenized in an agate mortar before loading into the ampule.

2.4. Carbon Erosion of Alloys in a Flow-Through Reactor

The process of carbon erosion of the alloys followed by the growth of carbon nanofibers was studied in a flow-through reactor equipped with McBain balance, which allowed registering the carbon accumulation in real-time mode. A specimen of the alloy (~2 mg) was loaded into a quartz basket, which was then placed inside the reactor. Then, the reactor was purged with argon to eliminate any oxygen traces and heated up to the desired temperature in a range from 450 to 750 °C. The heated reactor was fed with hydrogen to reduce the catalyst (until the constant weight of the sample). After the reduction stage, the reactor was fed with the reaction mixture containing 1,2-dichloroethane or ethylene (6 vol%), hydrogen (56 vol%), and argon (38 vol%). The total flow rate of the reaction mixture was 267 mL/min. The duration of the experiments was 2 h. The weight of the sample was recorded every 2 min. Finally, the reactor was cooled down in an argon flow. The sample was unloaded, weighed, and characterized.

2.5. Carbon Erosion of Alloys under RAPET Conditions

In RAPET experiments, the following organic substrates were used: 1,2-dichloroethane (DCE); 1-iodobutane; 1-bromobutane; hexachlorobenzene (HCB); hexafluorobenzene (HFB); hexamethylbenzene (HMB); polyethylene, and polyvinyl chloride. The alloy sample (0.2–0.3 mg) along with organic substrate (2–3 mg) were loaded into the quartz ampule presented in Figure 1. The ampule was sealed. The FMR spectra were recorded before and after the ampule was heated to the desired temperature. After the experiment and the registration of the FMR spectrum, the ampule was opened, and the sample was characterized by microscopic techniques.

Figure 1. Quartz ampules used for the FMR studies: (**a**) initial ampule loaded with the sample (a piece of nichrome wire) and organic substrate (1,2-dichloroethane); (**b**) ampule after the RAPET experiment.

As already mentioned, the registration of the FMR spectra was carried out at room temperature. Under such conditions, only materials with a Curie temperature above room temperature can possess ferromagnetism. Table 1 shows the reference data on the Curie temperature for pure nickel and for the nickel alloys used in the present study.

Table 1. Reference data on a Curie temperature for nickel and its alloys [42].

#	Material	Curie temperature, °C
1	Ni	+360
2	Ni-Cr(23%)	−120
3	Ni-Cu(12%)	+250
4	Ni-Mo(8%)	−150

3. Results and Discussion

3.1. Interaction of Ni-Cr Alloy with Halogenated Hydrocarbons in a Flow-Through Reactor

It is evident that nickel alloys, not possessing ferromagnetism at the temperature of the spectra registration (in our case, room temperature), should not give any FMR signals. In this regard, the FMR method was recently shown to give a possibility for monitoring the evolution of the initial alloys during the catalytic synthesis of CNF [31]. Here, the bulk Ni-Cr alloy (a piece of wire) was examined before and after its interaction with the reaction medium containing DCE vapors. Figure 2a,b present typical SEM images of the initial alloy sample and the carbon product accumulated in a flow-through reactor at 650 °C for 2 h, accordingly. The corresponding FMR spectra are shown in Figure 2c. The submicron metal particles contained in the carbon product are characterized by an intensive FMR signal at room temperature. This testifies to the appearance of ferromagnetism in them due to a decrease in chromium concentration and a rise of the Curie temperature above room temperature (Table 1). The observed FMR spectrum can be appropriately fitted by a single Lorentzian-shaped line (see the simulated spectrum in Figure 2c). This indicates the narrow size distribution of the ferromagnetic particles in the studied sample and agrees well with the recently reported microscopic data regarding the character size of 200–300 nm for the similarly obtained particles [31].

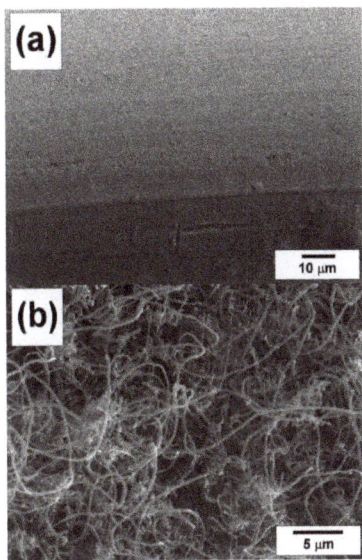
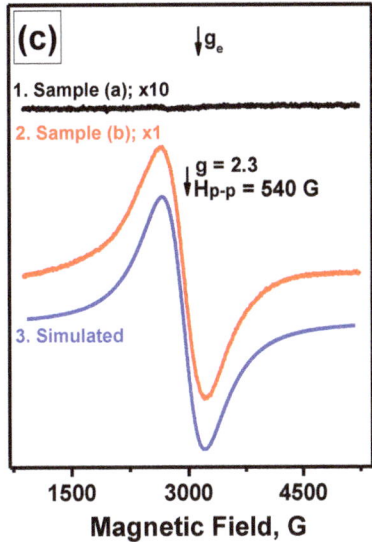

Figure 2. SEM images of the initial nichrome wire (**a**) and the carbon product obtained via the catalytic pyrolysis of DCE over this nichrome wire (**b**) and corresponding FMR spectra (**c**). Note that the spectra are normalized in intensity with regard to the nickel content in the samples. The spectrum of sample (**b**) is simulated with Lorentzian line shape and H_{p-p} = 540 G. The multipliers ×1 and ×10 indicate the relative amplification coefficients used at registering the spectra.

Similar situation was reported for the other bulk nickel alloys (chromel, alumel, etc.). Despite the difference in the shape of the FMR spectra of the initial alloys, they are uninformative ones, and, therefore, their interpretation is complicated. At the same time, after their contact with the aggressive reaction mixture and the formation of carbon nanostructures, the FMR spectra undergo noticeable changes. The wide singlet lines with g = 2.3, which is typical of the dispersed nickel particles, appeared [9,31]. It should be noted that in these works, the shapes of the FMR spectra cannot be fitted by a single Lorentzian-shaped line. This is presumably connected with a wider size distribution of the ferromagnetic particles in the cases of the described samples if compared with the sample obtained via the self-disintegration of nichrome (Figure 2).

The other two samples studied in the present work, Ni-Cu(12%) and Ni-Mo(8%), are dispersed powders of significantly different morphologies, which is stipulated by the differences in their preparation routes. As seen in Figure 3a, the Ni-Cu(12%) sample is represented by micro-dispersed particles of a wide size distribution (10–150 μm). Since these particles underwent a drastic mechanical action during the MCA procedure, they have a rough surface with a number of cracks and splits. This morphology is typical of the materials prepared by the MCA method [41]. The SSA value for this sample was found to be 0.6 m^2/g. The Ni-Mo(8%) sample was prepared by the reductive thermolysis technique [19,43,44]. Such a preparation method results in the formation of particles of a sponge-like morphology (Figure 3b) with a specific surface area of 0.2–0.5 m^2/g.

 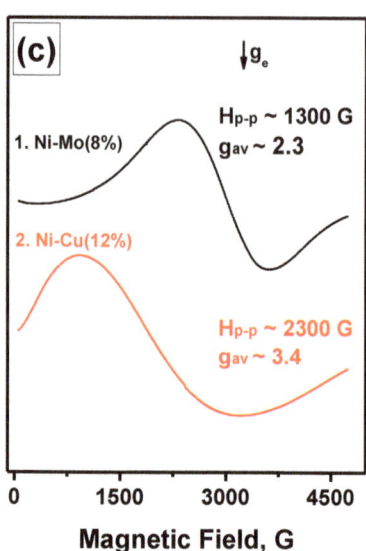

Figure 3. SEM images of the initial samples Ni-Cu (**a**) and Ni-Mo (**b**) and corresponding FMR spectra (**c**).

The FMR spectra of these samples are represented by relatively wide singlet lines (Figure 3c). Note that a Curie temperature for the Ni-Mo alloy with the Mo content of 8% lies below zero (−150 °C). This means that the Ni-Mo(8%) alloy should not give any FMR signal in the initial state. However, a sufficiently intensive signal is seen in the spectrum. The observation of the FMR signal can be explained by the presence of nickel-reached domains within the composition of the particles, which can give such a signal. A detailed analysis of the spectrum evolution during the catalytic process is presented in Section 3.3.

Thus, both the previously reported data and the presented in this section results convincingly show that the FMR method can be successfully applied to study the evolution of nickel-containing catalysts taking place during the catalytic synthesis of CNF from various organic precursors. The high sensitivity of this method allows for observing the self-dispersion of the bulk metals and alloys in the course of the induction period when no carbon accumulation is registered [31]. Another interesting application of this technique is the possibility for a quantitative diagnostics of the non-uniformity of nickel alloys, for which the Curie temperature lies below the spectrum registration temperature.

3.2. Interaction of Ni-Cr Alloy with Halogenated Hydrocarbons under RAPET Conditions

As reported by Gedanken et al. [35–40], the RAPET conditions give a unique possibility to obtain new classes of structurally organized materials. A similar approach was applied in our recent works to study catalytic pyrolysis in the presence of bulk nickel alloys [32–34,45]. The proposed characterization method is based on the registration of the FMR spectra in an in situ mode. A sealed quartz ampule loaded with an alloy sample and organic substrate (Figure 1) was used as a reactor in these experiments. The ampule was stepwise heated to the desired temperature (up to 850 °C) and the FMR spectra were recorded after each temperature step.

The catalytic corrosion processes under RAPET conditions when nickel and its alloys interact with halogen-containing organic molecules are quite general and in many cases are accompanied by the formation of new structured nanomaterials–nickel halogenides [32–34]. Such compounds can be considered as intermediates in the processes of the submicron metal particle formation from bulk precursors during the synthesis of CNF from halogen-containing organic substrates in a flow-through reactor as well. In such processes, the fragmentation (disintegration) of the bulk metal is accompanied by the formation of sufficiently homogeneous submicron (200–300 nm) metal particles, on which the growth of CNF occurs. Catalytic corrosion processes involve oxidation-reduction steps during the interaction of nickel with a reaction medium. For halogen-containing compounds, this interaction can be described by the following scheme [33]:

$$Ni + 2RX \rightarrow NiX_2 + 2R \qquad (1)$$

$$NiX_2 + H_2 \rightarrow Ni + 2HX \qquad (2)$$

where X is halogen contained in the initial organic substrate.

The results of the FMR comparative study on catalytic corrosion of nichrome interacting with various halogens (iodine, bromine, chlorine, and fluorine) under RAPET conditions are presented in Figure 4. The substrates were 1-iodobutane, 1-bromobutane, hexachlorobenzene (HCB), and hexafluorobenzene (HFB). In the last two cases, hexamethylbenzene (HMB) was loaded into the ampule as a hydrogen source for the reaction (2). As seen from the FMR spectra (Figure 4a), in the case of fluorine-containing medium, the formation of quite uniform ferromagnetic particles occurs already at 475 °C. The spectrum is characterized by the narrowest FMR signal (H_{p-p} = 550 G, g = 2.3), which indicates the formation of relatively small metal particles. An additional heating of the ampule at 640 °C results in changing the shape of the spectrum (Figure 4b). A significant enlargement of ferromagnetic particles can supposedly explain this. The appearance of an intensive narrow EPR signal in a $g \sim g_e$ region typical for paramagnetic sites of coke should also be noted.

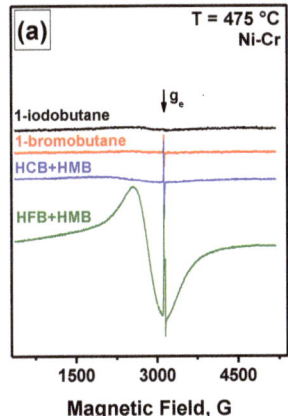 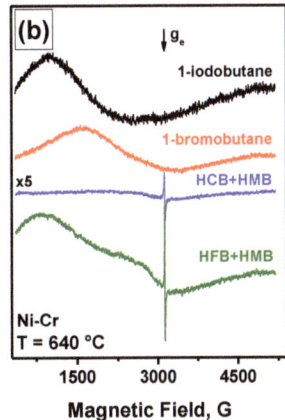

Figure 4. FMR spectra of nichrome interacting with various organic substrates under RAPET conditions at 475 °C (**a**) and 640 °C (**b**). The multiplier ×5 indicates the relative amplification coefficient used at registering the spectrum.

An unusual situation is observed for the chlorine-containing medium. A low-intensive FMR signal appears at the minimal temperature of 475 °C and then practically disappears at further heating. This can be connected with the formation of nickel chloride at elevated temperatures, which does not possess ferromagnetism.

These results clearly indicate the possibility to follow the evolution of the magnetic properties of nickel alloys by the FMR method during their interaction with various halogen-containing compounds. It should also be noted that the addition of HMB as a second component when using HFB and HCB as fluorine- and chlorine-containing compounds is due to the need to introduce hydrogen as a reducing agent. In the cases of individual substrates (HFB, HCB, and HMB) in similar experiments, no FMR signals have appeared throughout the studied temperature range.

It is obvious that other compounds that initiate catalytic corrosion of nichrome and contain, in particular, oxygen or nitrogen can act as an oxidizing agents in such reactions.

As such an N-containing reagent, melamine can be used. Interestingly, the reaction of bulk nichrome with this reagent under RAPET conditions not only made it possible to detect the appearance of dispersed particles of a ferromagnetic metal but also produced CNF with a high nitrogen content [46].

Note that the FMR technique is also applicable to explore the state of the Ni-Cr alloy subjected to an acid etching procedure, which is accompanied by significant enrichment of the surface metal layer with nickel due to more efficient etching of chromium from the alloy [31].

3.3. Interaction of Ni-Cu(12%) and Ni-Mo(8%) Alloys with Polymers under RAPET Conditions

In Section 3.1, data on the initial state of the Ni-Cu and Ni-Mo catalysts used in this study were presented. It should be noted again that the homogeneous Ni-Mo alloy containing 8% of molybdenum should have a Curie temperature of −150 °C and should not possess ferromagnetism at room temperature. The appearance of a sufficiently intense FMR signal in the spectrum (Figure 3c) is most likely due to the heterogeneity of the alloy and the presence of nickel-rich particles in it. This section presents the evolution of Ni-Cu and Ni-Mo catalysts during their interaction with chlorine-containing (polyvinyl chloride) and chlorine-free (polyethylene) polymers. The samples were studied under RAPET conditions in a temperature range of 450–850 °C with a step of 100 °C. At each temperature point, the ampule was maintained for 2 h.

The evolution of the FMR spectrum of the Ni-Mo(8%) sample during its interaction with polyethylene is shown in Figure 5a. As seen, the signal parameters (g and H_{p-p}) are not practically changed with a temperature increase. Only a rise in the amplitude of the FMR signal is observed, which may indicate an increase in the depth of carbon erosion of the Ni-Mo alloy. At the same time, when polyvinyl chloride is used as a reagent (Figure 5b), there is a noticeable shift of the g-factor towards large fields and a significant decrease in the width of the signal (H_{p-p}), similar to the evolution of the FMR spectra of nickel and its alloys described elsewhere [9,31].

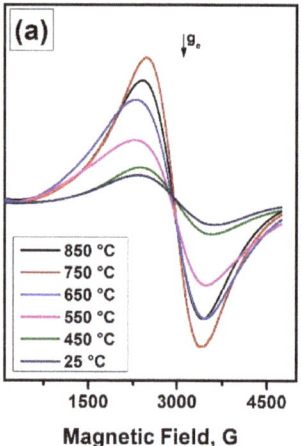
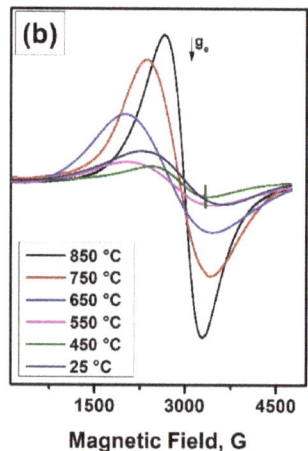
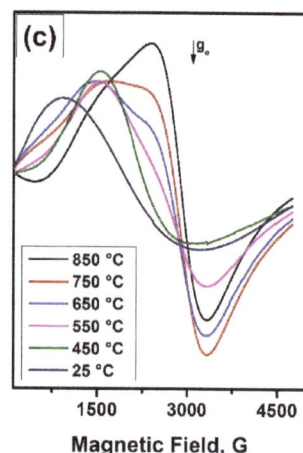

Figure 5. FMR spectra of Ni-Mo(8%) interacting with polyethylene (**a**) and polyvinyl chloride (**b**) under RAPET conditions at various temperatures. FMR spectrum of Ni-Cu(12%) interacting with polyvinyl chloride (**c**) under RAPET conditions at various temperatures.

Similar behavior was observed for the Ni-Cu(12%) alloy interacting with polyvinyl chloride. The Curie temperature for this alloy lies near +250 °C, therefore, a wide FMR signal is well seen at room temperature (Figure 3c). Figure 5c shows the evolution of the FMR spectra of this alloy during its interaction with polyvinyl chloride under RAPET conditions. With successive heating of the sample in a temperature range from 450 to 850 °C, the *g*-factor is shifted toward large fields. The results obtained for this sample also indicate a significant change in the magnetic properties of the initial alloy when interacting with a halogen-containing polymer.

Figure 6 shows characteristic SEM images of Ni-Mo(8%) and Ni-Cu(12%) samples after their interaction with the investigated polymers in RAPET mode at temperatures of 550 and 850 °C. In almost all cases, it is possible to observe the simultaneous presence of the nanostructured carbon product along with the dispersed metal particles formed as a result of the disintegration of the initial alloy. The presence of dispersed particles can be evidently seen in Figure 6b–d. In the case of polyvinyl chloride decomposition, for both alloys, there is a tendency to form carbon deposits in the shape of thin flakes (Figure 6d–f).

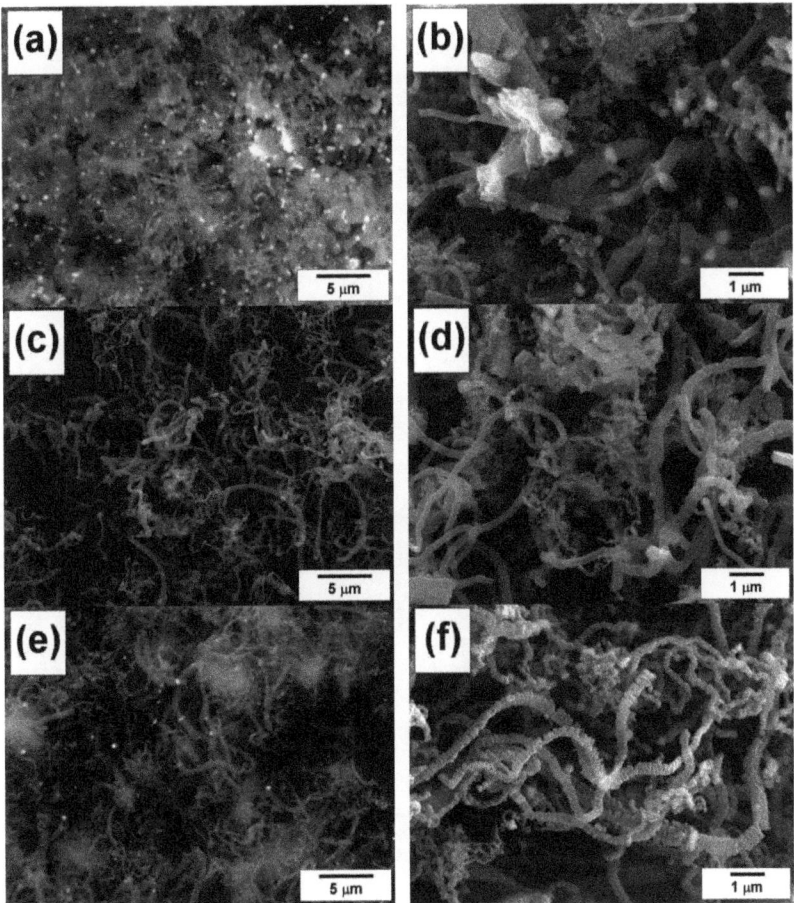

Figure 6. SEM images of the samples after the interaction of alloys with polymers under RAPET conditions: Ni-Mo(8%) with polyethylene at 550 °C (**a**) and 850 °C (**b**); Ni-Mo(8%) with polyvinyl chloride at 550 °C (**c**) and 850 °C (**d**); Ni-Cu(12%) with polyvinyl chloride at 550 °C (**e**) and 850 °C (**f**).

Thus, the FMR and SEM data indicate an effective process of carbon erosion of Ni-Cu(12%) and Ni-Mo(8%) alloys during their interaction in a closed volume with polymer substrates (polyethylene, polyvinyl chloride). Deep disintegration of the alloys results in the appearance of dispersed alloy particles involved in the further deposition of nanostructured carbon during catalytic pyrolysis of the substrates.

3.4. Synthesis of CNFs over Ni-Cu(12%) and Ni-Mo(8%) Alloys in a Flow-Through Reactor

It should be emphasized that the Ni-Cu and Ni-Mo alloys investigated in the present work are important and effective catalysts for CNF synthesis. Exactly the processes of catalytic corrosion under the action of the reaction medium lead to the formation of catalytically active particles from the starting bulk alloys.

Figure 7a,b show the result of the ethylene interaction with the Ni-Cu(12%) alloy at 550 °C for a short time (1 min). It can be seen that even after such a short time, bulk metal alloys (Figure 3a) undergo substantially complete disintegration to form dispersed particles (about 1 µm in size) catalyzing the further growth of CNF. Numerous active particles are evidently seen in SEM images in a backscattered-electron mode (Figure 7a,b). After 30 min of interaction with ethylene, long carbon filaments are formed, the diameter of which does not exceed 1 µm (Figure 7c,d).

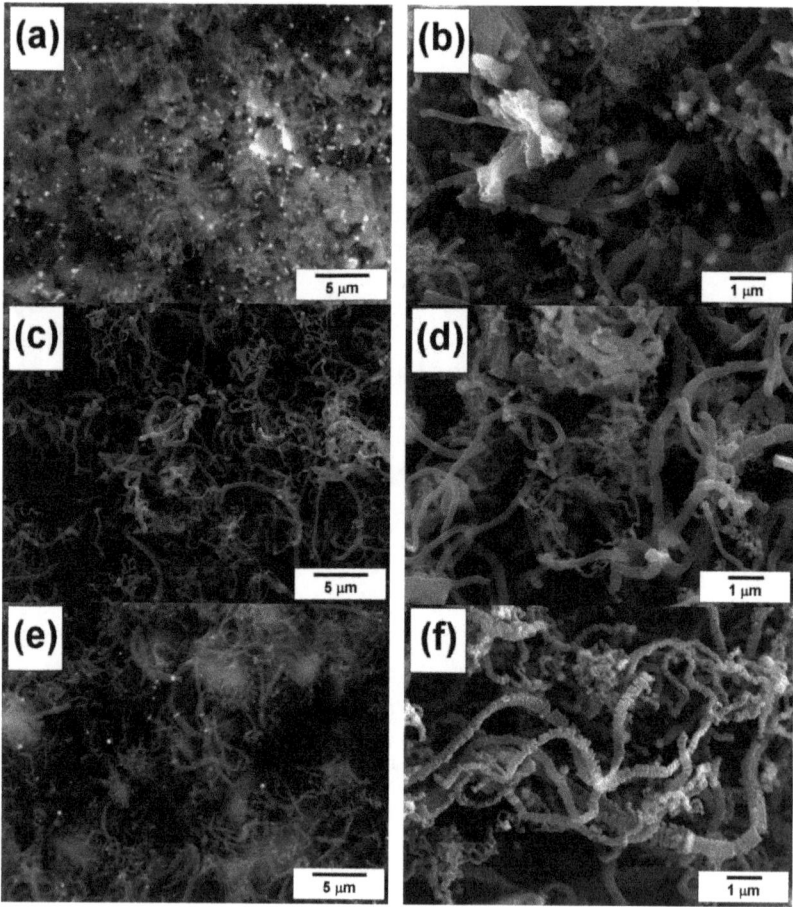

Figure 7. SEM images of the carbon product obtained via the interaction of alloys with the reaction mixture in a flow-through reactor: Ni-Cu(12%) with $C_2H_4/H_2/Ar$ at 550 °C for 1 min (**a,b**) and 30 min (**c,d**); Ni-Mo(8%) with $C_2H_4Cl_2/H_2/Ar$ at 550 °C for 120 min (**e,f**).

Similar self-disintegration effects are observed in the case of the Ni-Mo(8%) alloy [20]. Figure 7e,f show the result of its interaction with DCE vapors at 550 °C. It is clearly seen that carbon filaments of predominantly submicron diameter (200–350 µm) having a pronounced segmented structure are formed under these conditions.

Figure 8 demonstrates the dependence of the carbon yield on the temperature of the substrate decomposition over the self-organizing catalysts. In the cases of both reaction mixtures, $C_2H_4/H_2/Ar$ and $C_2H_4Cl_2/H_2/Ar$, there is an increase in the yield of CNF with a rise in the process temperature. In the case of ethylene decomposition at 650 °C, the productivity of the Ni-Cu(12%) catalyst after 30 min exceeds 160 g/g_{cat}. When processing DCE over the Ni-Mo(8%) alloy, the carbon yield is several times lower (55 g/g_{cat} after 120 min of the experiment). This difference is caused by the presence of chlorine in the reaction mixture, which partially deactivates the catalyst by chemisorption on its surface. The presence of chlorine also affects the morphology of the resulting carbon material (Figure 7e,f and Figure 9c,d).

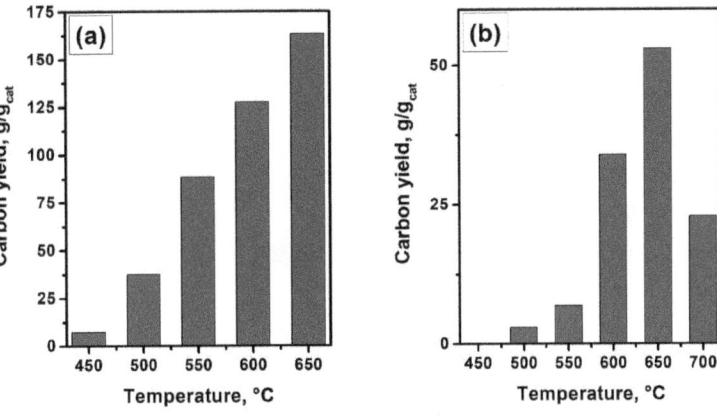

Figure 8. Temperature dependence of the carbon yield in a flow-through reactor: Ni-Cu(12%), $C_2H_4/H_2/Ar$, 30 min (**a**); Ni-Mo(8%), $C_2H_4Cl_2/H_2/Ar$, 120 min (**b**).

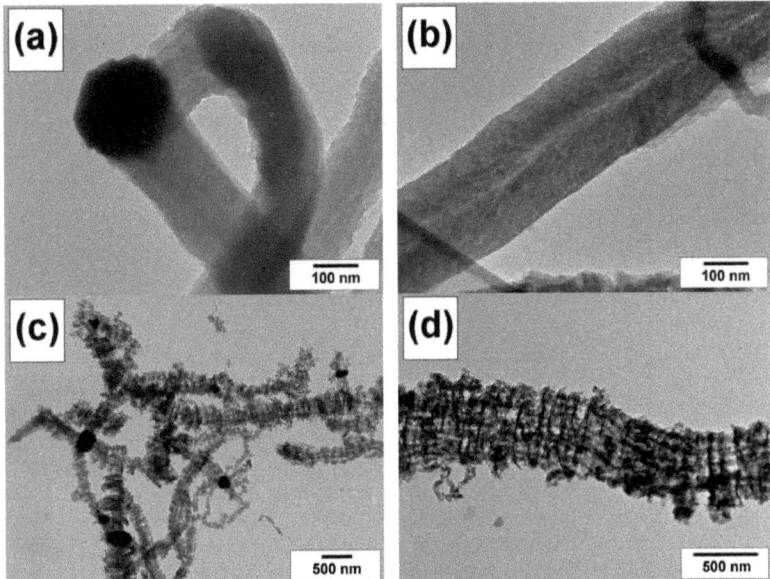

Figure 9. TEM images of the carbon product obtained via the interaction of alloys with the reaction mixture in a flow-through reactor: Ni-Cu(12%) with $C_2H_4/H_2/Ar$ at 600 °C for 30 min (**a**,**b**); Ni-Mo(8%) with $C_2H_4Cl_2/H_2/Ar$ at 600 °C for 120 min (**c**,**d**).

According to TEM data, the carbon nanofibers resulting from the ethylene decomposition have a tightly packed structure (Figure 9a,b). Both the types of packaging of graphene layers, the stacked (platelet) type (Figure 9a) and the coaxial-conic (fish-bone) type (Figure 9b), are equally found in the sample. In turn, the decomposition of the chlorine-substituted hydrocarbons leads to the formation of poorly ordered carbon filaments with a segmented structure (Figure 9c,d).

According to EDX data, the average concentration of Cl present on the surface of CNF is about 0.1 at.%, whereas the metallic particles exhibited a higher content of chlorine—up to 1 at.% (Figure 10, Table 2). The data obtained are in good agreement with the earlier works related to the synthesis of CNF from chlorinated hydrocarbons using different Ni-M alloys (M = Fe, Cu, Mo, etc.) [20,47,48]. It should be noted that the presence of halogen atoms over the surface of similarly prepared carbon nanomaterials was recently confirmed by means of X-ray photoelectron spectroscopy [19,49]. Thus, the chemisorbed chlorine species present on the surface of the active metallic particles exerts a determining impact on the formation of segmented carbon filaments due to a periodic blockage of their surface [18,20].

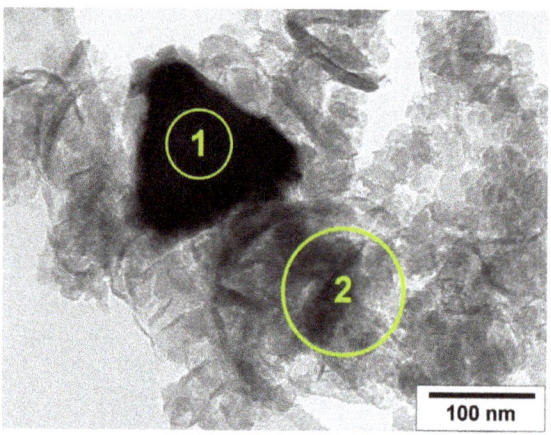

Figure 10. TEM image of the carbon product obtained via the interaction of Ni-Mo(8%) alloy with $C_2H_4Cl_2/H_2$/Ar at 600 °C in a flow-through reactor. The EDX analysis areas are highlighted by circles.

Table 2. EDX data for the Ni-Mo(8%) sample after its interaction with $C_2H_4Cl_2/H_2$/Ar at 600 °C in a flow-through reactor (see Figure 10).

Spectrum	Concentration, at.%			
	C	Cl	Ni	Mo
Area 1	-	0.8	94.5	4.7
Area 2	99.92	0.08	0	0

4. Conclusions

The present paper shows that the processes of self-organization of nickel-containing catalytic systems are accompanied by the formation of sufficiently homogeneous metal particles, acting as active sites in the CNF synthesis. It is important to mention that, in order to initiate the catalytic corrosion processes of nickel and its alloys, the presence of compounds capable of acting as both an oxidizing agent and a reducing agent in the reaction medium is a necessary condition. Thus, in the case of using halogen-containing molecules (HFB and HCB) as an oxidizer, hydrogen-containing molecules (in our case, HMB) must be added to the reaction medium to initiate such processes in RAPET mode.

It is very likely that the process of self-disintegration of nickel alloys studied within this work in a flow-through reactor during the pyrolysis of halogen-containing organic substances also proceeds through the formation of nickel halogenides as intermediates, which were directly detected under RAPET conditions among the reaction products.

The results presented in this work indicate a highly informative use of the FMR method for monitoring the evolution of nickel alloys in the reactions of catalytic pyrolysis of various organic molecules. It is believed that the proposed approaches can also be applied to investigate the processes of catalytic corrosion of Co- and Fe-containing catalysts widely used in the synthesis of structured carbon nanomaterials.

Author Contributions: Conceptualization, A.M.V., I.V.M. and A.A.V.; methodology, A.M.V., I.V.M., Y.I.B., Y.V.S. and A.A.V.; investigation, R.M.K., Y.I.B., Y.V.S., S.D.A. and A.R.P.; writing—original draft preparation, A.M.V., Y.I.B., S.D.A. and A.R.P.; writing—review and editing, I.V.M. and A.A.V.; funding acquisition, I.V.M. All authors have read and agreed to the published version of the manuscript.

Funding: This study was funded by the Russian Science Foundation (grant number 22-13-00406, https://rscf.ru/en/project/22-13-00406/, BIC SB RAS).

Institutional Review Board Statement: Not applicable.

Informed Consent Statement: Not applicable.

Data Availability Statement: Data are contained within the article.

Acknowledgments: Characterization of the samples was performed using the equipment of the Center of Collective Use "National Center of Catalysts Research".

Conflicts of Interest: The authors declare no conflict of interest. The funders had no role in the design of the study; in the collection, analyses, or interpretation of data; in the writing of the manuscript; or in the decision to publish the results.

References

1. Wang, X.; Yin, H.; Sheng, G.; Wang, W.; Zhang, X.; Lai, Z. Fabrication of self-entangled 3D carbon nanotube networks from metal–organic frameworks for Li-ion batteries. *ACS Appl. Nano Mater.* **2018**, *1*, 7075–7082. [CrossRef]
2. Yin, H.; Wang, X.; Sheng, G.; Chen, W.; Lai, Z. Facile single-step fabrication of robust superhydrophobic carbon nanotube films on different porous supports. *Ind. Eng. Chem. Res.* **2019**, *58*, 2976–2982. [CrossRef]
3. Baker, R.T.K. Carbon Nanofibers. In *Encyclopedia of Materials: Science and Technology*; Pergamon Press: Oxford, UK, 2001; pp. 932–941. [CrossRef]
4. Wang, H.; Baker, R.T.K. Decomposition of methane over a Ni–Cu–MgO catalyst to produce hydrogen and carbon nanofibers. *J. Phys. Chem. B* **2004**, *108*, 20273–20277. [CrossRef]
5. Yazdani, N.; Brown, E. Carbon nanofibers in cement composites. In *Innovative Developments of Advanced Multifunctional Nanocomposites in Civil and Structural Engineering*; Woodhead Publishing: Cambridge, UK, 2016; pp. 47–58. [CrossRef]
6. Yang, Z.; Wang, C.; Lu, X. Nanofibrous Materials. In *Electrospinning: Nanofabrication and Applications*; William Andrew: Norwich, NY, USA, 2019; pp. 53–92. [CrossRef]
7. Krasnikova, I.V.; Mishakov, I.V.; Vedyagin, A.A. Functionalization, modification, and characterization of carbon nanofibers. In *Carbon-Based Nanofillers and Their Rubber Nanocomposites*; Elsevier: Amsterdam, The Netherlands, 2019; pp. 75–137. [CrossRef]
8. Chesnokov, V.V.; Buyanov, R.A. The formation of carbon filaments upon decomposition of hydrocarbons catalysed by iron subgroup metals and their alloys. *Russ. Chem. Rev.* **2000**, *69*, 623–638. [CrossRef]
9. Bauman, Y.I.; Mishakov, I.V.; Buyanov, R.A.; Vedyagin, A.A.; Volodin, A.M. Catalytic properties of massive iron-subgroup metals in dichloroethane decomposition into carbon products. *Kinet. Catal.* **2011**, *52*, 547–554. [CrossRef]
10. Baker, R.T.K. Catalytic growth of carbon filaments. *Carbon* **1989**, *27*, 315–323. [CrossRef]
11. Ermakova, M.; Ermakov, D.Y. Ni/SiO$_2$ and Fe/SiO$_2$ catalysts for production of hydrogen and filamentous carbon via methane decomposition. *Catal. Today* **2002**, *77*, 225–235. [CrossRef]
12. Guevara, J.C.; Wang, J.A.; Chen, L.F.; Valenzuela, M.A.; Salas, P.; García-Ruiz, A.; Toledo, J.A.; Cortes-Jácome, M.A.; Angeles-Chavez, C.; Novaro, O. Ni/Ce-MCM-41 mesostructured catalysts for simultaneous production of hydrogen and nanocarbon via methane decomposition. *Int. J. Hydrogen Energy* **2010**, *35*, 3509–3521. [CrossRef]
13. Takenaka, S.; Kobayashi, S.; Ogihara, H.; Otsuka, K. Ni/SiO$_2$ catalyst effective for methane decomposition into hydrogen and carbon nanofiber. *J. Catal.* **2003**, *217*, 79–87. [CrossRef]
14. Streltsov, I.A.; Mishakov, I.V.; Vedyagin, A.A.; Melgunov, M.S. Synthesis of carbon nanomaterials from hydrocarbon raw material on Ni/SBA-15 catalyst. *Chem. Sustain. Develop.* **2014**, *22*, 185–192.
15. Keane, M.A.; Jacobs, G.; Patterson, P.M. Ni/SiO$_2$ promoted growth of carbon nanofibers from chlorobenzene: Characterization of the active metal sites. *J. Colloid Interf. Sci.* **2006**, *302*, 576–588. [CrossRef] [PubMed]

16. Bauman, Y.I.; Lysakova, A.S.; Rudnev, A.V.; Mishakov, I.V.; Shubin, Y.V.; Vedyagin, A.A.; Buyanov, R.A. Synthesis of nanostructured carbon fibers from chlorohydrocarbons over Bulk Ni-Cr Alloys. *Nanotechnol. Russ.* **2014**, *9*, 380–385. [CrossRef]
17. Bauman, Y.I.; Mishakov, I.V.; Vedyagin, A.A.; Ramakrishna, S. Synthesis of bimodal carbon structures via metal dusting of Ni-based alloys. *Mater. Lett.* **2017**, *201*, 70–73. [CrossRef]
18. Bauman, Y.I.; Shorstkaya, Y.V.; Mishakov, I.V.; Plyusnin, P.E.; Shubin, Y.V.; Korneev, D.V.; Stoyanovskii, V.O.; Vedyagin, A.A. Catalytic conversion of 1,2-dichloroethane over Ni-Pd system into filamentous carbon material. *Catal. Today* **2017**, *293–294*, 23–32. [CrossRef]
19. Bauman, Y.; Kibis, L.; Mishakov, I.; Rudneva, Y.; Stoyanovskii, V.O.; Vedyagin, A.A. Synthesis and functionalization of filamentous carbon material via decomposition of 1,2-dichlorethane over self-organizing Ni-Mo catalyst. *Mater. Sci. Forum* **2019**, *950*, 180–184. [CrossRef]
20. Bauman, Y.I.; Rudneva, Y.V.; Mishakov, I.V.; Plyusnin, P.E.; Shubin, Y.V.; Korneev, D.V.; Stoyanovskii, V.O.; Vedyagin, A.A.; Buyanov, R.A. Effect of Mo on the catalytic activity of Ni-based self-organizing catalysts for processing of dichloroethane into segmented carbon nanomaterials. *Heliyon* **2019**, *5*, e02428. [CrossRef]
21. Wang, C.; Bauman, Y.; Mishakov, I.; Vedyagin, A.A. Features of the carbon nanofibers growth over Ni-Pd catalyst depending on the reaction conditions. *Mater. Sci. Forum* **2019**, *950*, 144–148. [CrossRef]
22. Choi, Y.H.; Lee, W.Y. Effect of Ni loading and calcination temperature on catalyst performance and catalyst deactivation of Ni/SiO2 in the hydrodechlorination of 1,2-dichloropropane into propylene. *Catal. Lett.* **2000**, *67*, 155–161. [CrossRef]
23. Berndt, F.M.; Perez-Lopez, O.W. Catalytic decomposition of methane over Ni/SiO$_2$: Influence of Cu addition. *React. Kinet. Mech. Catal.* **2016**, *120*, 181–193. [CrossRef]
24. Grabke, H.J. Metal dusting. *Mater. Corros.* **2003**, *54*, 736–746. [CrossRef]
25. Schmid, B.; Aas, N.; Grong, Ø.; Ødegård, R. In situ environmental scanning electron microscope observations of catalytic processes encountered in metal dusting corrosion on iron and nickel. *Appl. Catal. A Gen.* **2001**, *215*, 257–270. [CrossRef]
26. Gladky, A.Y.; Kaichev, V.V.; Ermolaev, V.K.; Bukhtiyarov, V.I.; Parmon, V.N. Propane oxidation on nickel in a self-oscillation mode. *Kinet. Catal.* **2005**, *46*, 251–259. [CrossRef]
27. Slinkin, A.A. Application of the ferromagnetic resonance method in the study of heterogeneous catalysts. *Russ. Chem. Rev.* **1968**, *37*, 642–654. [CrossRef]
28. Sharma, V.K.; Baiker, A. Superparamagnetic effects in the ferromagnetic resonance of silica supported nickel particles. *J. Chem. Phys.* **1981**, *75*, 5596–5601. [CrossRef]
29. Sharma, V.K.; Baiker, A. Ferromagnetic resonance study of reduced NiCl$_2$ graphite intercalation compound. *Synth. Metals* **1985**, *11*, 1–8. [CrossRef]
30. Yulikov, M.M.; Abornev, I.S.; Mart'yanov, O.N.; Yudanov, V.F.; Isupov, V.P.; Chupakhina, L.E.; Tarasov, K.A.; Mitrofanova, R.P. Ferromagnetic resonance of nickel nanoparticles in an amorphous oxide matrix. *Kinet. Catal.* **2004**, *45*, 735–738. [CrossRef]
31. Bauman, Y.I.; Kenzhin, R.M.; Volodin, A.M.; Mishakov, I.V.; Vedyagin, A.A. Formation of growth centres of carbon nanofibres during self-dispersing Ni-containing alloys: Studies by means of ferromagnetic resonance. *Chem. Sustain. Dev.* **2012**, *20*, 119–127.
32. Kenzhin, R.M.; Bauman, Y.I.; Volodin, A.M.; Mishakov, I.V.; Vedyagin, A.A. Structural self-organization of solid-state products during interaction of halogenated compounds with bulk Ni-Cr alloy. *Mater. Lett.* **2016**, *179*, 30–33. [CrossRef]
33. Kenzhin, R.M.; Bauman, Y.I.; Volodin, A.M.; Mishakov, I.V.; Vedyagin, A.A. Interaction of bulk nickel and nichrome with halogenated butanes. *React. Kinet. Mech. Catal.* **2017**, *122*, 1203–1212. [CrossRef]
34. Kenzhin, R.M.; Bauman, Y.I.; Volodin, A.M.; Mishakov, I.V.; Vedyagin, A.A. Synthesis of carbon nanofibers by catalytic CVD of chlorobenzene over bulk nickel alloy. *Appl. Surf. Sci.* **2018**, *427*, 505–510. [CrossRef]
35. Pol, S.V.; Pol, V.G.; Gedanken, A. Reactions under Autogenic Pressure at Elevated Temperature (RAPET) of various alkoxides: Formation of metals/metal oxides-carbon core-shell structures. *Chem. Eur. J.* **2004**, *10*, 4467–4473. [CrossRef] [PubMed]
36. Pol, V.G.; Pol, S.V.; Gedanken, A. Novel synthesis of high surface area silicon carbide by RAPET (Reactions under Autogenic Pressure at Elevated Temperature) of organosilanes. *Chem. Mater.* **2005**, *17*, 1797–1802. [CrossRef]
37. Pol, V.G.; Pol, S.V.; Gedanken, A.; Lim, S.H.; Zhong, Z.; Lin, J. Thermal decomposition of commercial silicone oil to produce high yield high surface area SiC nanorods. *J. Phys. Chem. B* **2006**, *110*, 11237–11240. [CrossRef] [PubMed]
38. Gershi, H.; Gedanken, A.; Keppner, H.; Cohen, H. One-step synthesis of prolate spheroidal-shaped carbon produced by the thermolysis of octene under its autogenic pressure. *Carbon* **2011**, *49*, 1067–1074. [CrossRef]
39. Butovsky, E.; Perelshtein, I.; Nissan, I.; Gedanken, A. Fabrication, characterization, and printing of conductive ink based on multi core-shell nanoparticles synthesized by RAPET. *Adv. Funct. Mater.* **2013**, *23*, 5794–5799. [CrossRef]
40. Teller, H.; Krichevski, O.; Gur, M.; Gedanken, A.; Schechter, A. Ruthenium phosphide synthesis and electroactivity toward oxygen reduction in acid solutions. *ACS Catal.* **2015**, *5*, 4260–4267. [CrossRef]
41. Mishakov, I.V.; Afonnikova, S.D.; Bauman, Y.I.; Shubin, Y.V.; Trenikhin, M.V.; Serkova, A.N.; Vedyagin, A.A. Carbon erosion of a bulk nickel–copper alloy as an effective tool to synthesize carbon nanofibers from hydrocarbons. *Kinet. Catal.* **2022**, *63*, 97–107. [CrossRef]
42. Kikoin, I.K. *Tables of Physical Quantities*; Atomizdat: Moscow, Russia, 1976; p. 1006. (In Russian)
43. Rudneva, Y.V.; Shubin, Y.V.; Plyusnin, P.E.; Bauman, Y.I.; Mishakov, I.V.; Korenev, S.V.; Vedyagin, A.A. Preparation of highly dispersed Ni$_{1-x}$Pd$_x$ alloys for the decomposition of chlorinated hydrocarbons. *J. Alloys Compd.* **2019**, *782*, 716–722. [CrossRef]

44. Shubin, Y.V.; Bauman, Y.I.; Plyusnin, P.E.; Mishakov, I.V.; Tarasenko, M.S.; Mel'gunov, M.S.; Stoyanovskii, V.O.; Vedyagin, A.A. Facile synthesis of triple Ni-Mo-W alloys and their catalytic properties in chemical vapor deposition of chlorinated hydrocarbons. *J. Alloys Compd.* **2021**, *866*, 158778. [CrossRef]
45. Kenzhin, R.M.; Bauman, Y.I.; Volodin, A.M.; Mishakov, I.V.; Zaikovskii, V.I.; Vedyagin, A.A. Microscopic studies on the polymers decomposition in a closed volume at elevated temperatures in the presence of bulk NiCr alloy. *SN Appl. Sci.* **2018**, *1*, 139. [CrossRef]
46. Kenzhin, R.M.; Bauman, Y.I.; Volodin, A.M.; Mishakov, I.V.; Vedyagin, A.A. One-step synthesis of nitrogen-doped carbon nanofibers from melamine over nickel alloy in a closed system. *Chem. Phys. Lett.* **2017**, *685*, 259–262. [CrossRef]
47. Bauman, Y.I.; Kutaev, N.V.; Plyusnin, P.E.; Mishakov, I.V.; Shubin, Y.V.; Vedyagin, A.A.; Buyanov, R.A. Catalytic behavior of bimetallic Ni–Fe systems in the decomposition of 1,2-dichloroethane. Effect of iron doping and preparation route. *React. Kinet. Mech. Catal.* **2017**, *121*, 413–423. [CrossRef]
48. Bauman, Y.I.; Mishakov, I.V.; Korneev, D.V.; Shubin, Y.V.; Vedyagin, A.A.; Buyanov, R.A. Comparative study of 1,2-dichlorethane decomposition over Ni-based catalysts with formation of filamentous carbon. *Catal. Today* **2018**, *301*, 147–152. [CrossRef]
49. Mishakov, I.V.; Bauman, Y.I.; Shubin, Y.V.; Kibis, L.S.; Gerasimov, E.Y.; Mel'gunov, M.S.; Stoyanovskii, V.O.; Korenev, S.V.; Vedyagin, A.A. Synthesis of nitrogen doped segmented carbon nanofibers via metal dusting of Ni-Pd alloy. *Catal. Today* **2020**, *388–398*, 312–322. [CrossRef]

Article

Synthesis of Boron-Doped Carbon Nanomaterial

Vladimir V. Chesnokov *, Igor P. Prosvirin, Evgeny Yu. Gerasimov and Aleksandra S. Chichkan

Boreskov Institute of Catalysis SB RAS, Prospekt Akademika Lavrentieva 5, Novosibirsk 630090, Russia
* Correspondence: chesn@catalysis.ru

Abstract: A new method for the synthesis of boron-doped carbon nanomaterial (B-carbon nanomaterial) has been developed. First, graphene was synthesized using the template method. Magnesium oxide was used as the template that was dissolved with hydrochloric acid after the graphene deposition on its surface. The specific surface area of the synthesized graphene was equal to 1300 m^2/g. The suggested method includes the graphene synthesis via the template method, followed by the deposition of an additional graphene layer doped with boron in an autoclave at 650 °C, using a mixture of phenylboronic acid, acetone, and ethanol. After this carbonization procedure, the mass of the graphene sample increased by 70%. The properties of B-carbon nanomaterial were studied using X-ray photoelectron spectroscopy (XPS), high-resolution transmission electron microscopy (HRTEM), Raman spectroscopy, and adsorption-desorption techniques. The deposition of an additional graphene layer doped with boron led to an increase of the graphene layer thickness from 2–4 to 3–8 monolayers, and a decrease of the specific surface area from 1300 to 800 m^2/g. The boron concentration in B-carbon nanomaterial determined by different physical methods was about 4 wt.%.

Keywords: synthesis; graphene; doping; boron

1. Introduction

Graphene attracted the attention of researchers due to its exceptional structural, mechanical, and electronic properties [1–4]. The applications of graphene continue to grow due to its record thermal conductivity, excellent mechanical strength [5–7], remarkable ability to carry charges, outstanding thermal stability, and high surface area [8–10]. The applications of graphene-based nanoadsorbents, including graphenes, graphene oxides, reduced graphene oxides, and their nanocomposites in water purification were summarized in a recent review [11].

The properties of graphene-based materials can be improved by doping with other "I have checked".elements [12]. Boron, nitrogen, and phosphorous are the main elements used for doping graphene and graphene oxide [13,14]. Graphene doping with boron results in p-type conductivity, whereas phosphorus induces n-type conductivity. Doping with boron and phosphorous increases the surface area and concentration of defects in graphene materials. New applications for using graphene and graphene quantum dots doped with boron or phosphorous as sensors, sorbents, photocatalysts, and electrocatalysts for the detection and reduction of various pollutants were presented in a review [15].

The doping of carbon nanomaterials with boron atoms can alter their electronic properties. Recently, it was found that boron-doped carbon nanomaterials have remarkable properties in electrocatalytic oxygen reduction reactions (ORRs) [16–20]. Based on the obtained data, many researchers believe that B-doped carbon nanotubes and graphene can act as a substitute for expensive platinum catalysts in fuel cells.

In this article [21], ordered mesoporous carbon doped with boron was prepared by coimpregnation of SBA-15 silica with sucrose and 4-hydroxyphenylboric acid followed by carbonization. The authors found that mesoporous carbon doped with boron had a highly-ordered mesoporous structure, uniform pore size distribution and high surface area. Ordered mesoporous carbon doped with boron can be used as a potentially efficient

and cheap metal-free catalyst for ORRs with good stability in an alkaline solution. The boron concentration is the key factor in determining the catalytic activity in ORRs. The high surface area of ordered mesoporous carbon doped with boron makes active sites more accessible in ORRs.

There are several reports in the literature on the synthesis of boron-doped graphene. In these studies, the conventional CVD method was used for growing graphene on different polycrystalline foils. It was reported that boron-doped graphene could be synthesized using phenylboronic acid as a source of carbon and boron [22]. The effects of the phenylboronic acid mass and graphene growth time on the properties of graphene growing on polycrystalline copper foil in a three-zone CVD system were studied. The nanomaterial prepared by the substitution doping of graphene with boron has high stability, which is particularly important for applications in semiconductor technology, particularly in optoelectronics. However, this method has significant drawbacks. Boron atoms can penetrate both the graphene structure and the structure of polycrystalline foil. Therefore, subsequently, it is difficult to distinguish places where boron atoms are localized. In addition, it is difficult to scale up this method and synthesize uniform graphene films.

In the article [23], graphene oxide prepared by the Hammer's method was used as the precursor for the synthesis of B-graphene. Boron atoms were successfully introduced into the graphene oxide structure with the concentration of 1.64–1.89 at.%. Boron was introduced into the graphene oxide from boric acid by hydrothermal reduction at 250–300 °C. The presence of B–O, B–C and C–O bonds was confirmed by FTIR spectral analysis. The main drawback of this method for the synthesis of B-graphene is the high concentration of oxygen atoms in the graphene oxide precursor that affects the quality of the final product. The introduction of an additional graphene oxide reduction stage also does not solve this problem because it leads to the formation of many defects in the graphene structure.

The effect of oxygen and boron co-introduction in porous carbon on the activity of ORRs was studied in [24]. The synthesis of porous carbon doped with oxygen and boron was based on simple carbonization with CO_2 using $NaBH_4$ as a reducing agent, followed by thermal treatment. It was demonstrated that the presence of O–B–C fragments increased the activity of porous carbon in ORRs, and led to high stability in cycles.

In this study, graphene obtained via the template synthesis [25–27] was used as the precursor for the synthesis of B-carbon nanomaterial. The goals of this study were to deposit an additional boron-doped graphene layer on the surface of graphene prepared by the template method and to study the properties of the synthesized material.

2. Methods of Investigation

2.1. Graphene Synthesis

Magnesium oxide (Vekton, Russia, "pure for analysis" grade) was subjected to carbonization with 1,3-bitadiene at 600 °C in a quartz flow reactor. After this reaction the MgO particles were covered with a thin carbon film. Then, the carbon film was cleaned from the MgO template by treatment in a hydrochloric acid solution (Figure 1).

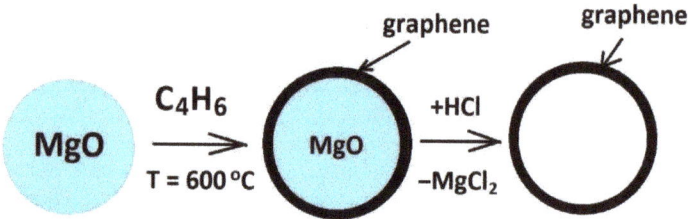

Figure 1. Scheme of graphene synthesis.

The synthesized graphene was used for doping with boron.

2.2. B-Carbon Nanomaterial Synthesis

Phenylboronic acid (C_6H_5-B-$(OH)_2$) (Aldrich, USA) was used as a boron-containing precursor. Its structure containing a phenyl group favored the introduction of boron into the synthesized graphene framework.

Graphene prepared using the template method was treated with phenylboronic acid dissolved in the mixture containing 80% acetone and 20% ethanol. Then, it was placed into an autoclave as described earlier [28]. The autoclave was heated to 650 °C and kept at this temperature for 2 h. A black carbon powder was obtained after the carbonization of the graphene precursor. The amount of carbon doped with boron deposited on graphene was controlled by the change of the sample weight. The synthesized B-carbon nanomaterial was studied using various physical methods.

2.3. Physical Methods for Investigation of B-Carbon Nanomaterial

High-resolution transmission electron microscopy (HRTEM) using a ThemisZ (Thermo Fisher Scientific, USA) electron microscope with an accelerating voltage of 200 kV and maximum lattice resolution of 0.07 nm was used for the investigation of the structure and microstructure of the catalysts. The TEM images were recorded with a Ceta 16 (Thermo Fisher Scientific, USA) CCD matrix. Elemental analysis was performed with Super-X EDS detector (Thermo Fisher Scientific, USA). High-angle annular dark-field images (HAADF STEM image) were recorded using a standard ThemisZ detector. The samples for the HRTEM study were deposited on a holey carbon film mounted on a copper grid by the ultrasonic dispersal of the B-graphene suspension in ethanol.

X-ray photoelectron spectra were recorded on a SPECS (Germany) photoelectron spectrometer using a hemispherical PHOIBOS-150-MCD-9 analyzer (Mg K_α radiation, hv = 1253.6 eV, 150 W). The binding energy (BE) scale was pre-calibrated using the positions of the peaks of $Au4f_{7/2}$ (BE = 84.0 eV) and $Cu2p_{3/2}$ (BE = 932.67 eV) core levels. The samples in the form of powder were loaded onto a conducting double-sided copper scotch. The survey and the narrow spectra were registered at the analyzer pass energy 20 eV. Atomic ratios of the elements were calculated from the integral photoelectron peak intensities, which were corrected by corresponding sensitivity factors based on Scofield photoionization cross-sections [29].

Raman spectra were measured using a Horiba Jobin Yvon Lab-Ram HR spectrometer. Excitation was supplied by an argon ion laser (λ = 488 nm) with 2 cm^{-1} spectral resolution.

The surface areas and porosity of the synthesized graphene and B-carbon nanomaterial were determined using an ASAP-2400 (Micromeritics, Norcross, GA, USA) instrument [30].

The boron concentration in B-carbon nanomaterial was determined by the atomic emission spectroscopy.

3. Results and Discussion

3.1. HRTEM Study of Graphene

Graphene synthesized by the template method was studied by electron microscopy. A HRTEM image of graphene is shown in Figure 2a.

The graphene aggregates have dimensions of about 1 μm (Figure 2a). The size of the carbon globules is determined by the geometrical dimensions of the MgO particles. The thickness of the graphene layer in the synthesized graphene sample was 2–4 layers. The graphene surface area was about 1300 m^2/g.

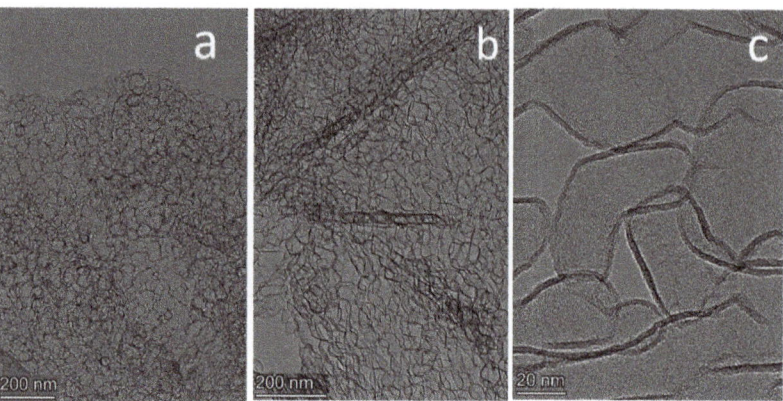

Figure 2. HRTEM images of graphene (**a**) and B-carbon nanomaterial at different magnifications (**b**,**c**).

3.2. B-Carbon Nanomaterial

Graphene was doped with boron atoms by heating in an autoclave in the presence of phenylboronic acid, acetone, and ethanol at 650 °C. After doping with boron, the mass of the graphene sample increased by 70%.

3.3. B-Carbon Nanomaterial Study by Electron Microscopy

HRTEM images of B-carbon nanomaterial are shown in Figure 2b,c. It also has globular structure (Figure 3b), and additional graphene carbonization led to an increasing of the graphene layer thickness to 3–8 layers.

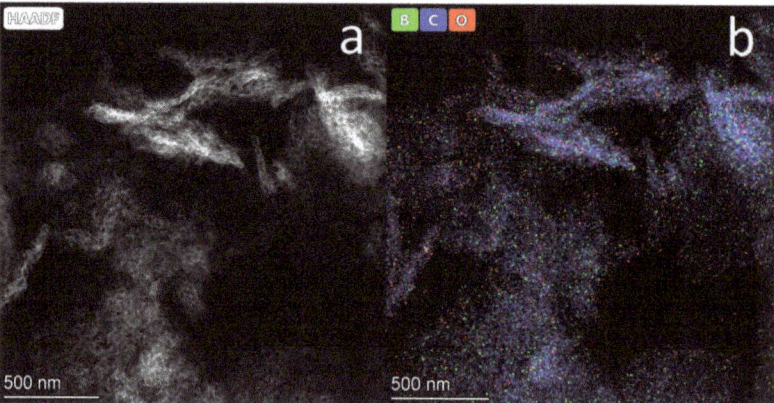

Figure 3. HAADF STEM image of B-carbon nanomaterial (**a**) and EDX mapping of boron, carbon and oxygen (**b**).

Figure 3 presents an HAADF STEM image of B-carbon nanomaterial (a) and EDX mapping of boron, carbon, and oxygen atoms (b).

The boron concentration in the B-carbon nanomaterial sample determined by EDX was equal to 4 wt.%. This boron concentration is in agreement with the results of the elemental analysis by atomic emission spectroscopy. According to the latter, the content of atomic boron in B-carbon nanomaterial was 4.3 wt.%.

3.4. Investigation of the B-Carbon Nanomaterial Pore Structure

The nitrogen adsorption-desorption isotherms measured on pristine graphene and B-carbon nanomaterial are presented in Figure 4.

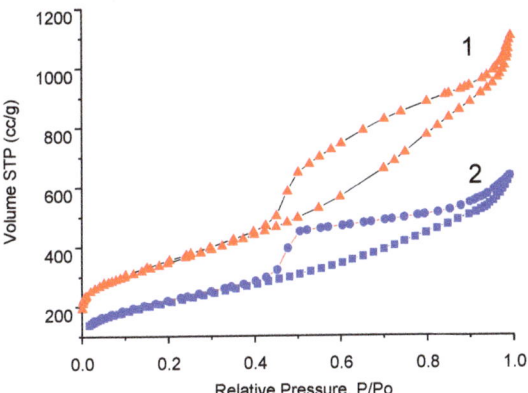

Figure 4. Nitrogen adsorption–desorption isotherms of pristine graphene (1) and B-carbon nanomaterial (2).

Specific surface area of the synthesized B-carbon nanomaterial was equal to 800 m^2/g. The deposition of the additional layer of graphene doped with boron led to the surface area decreasing from 1300 to 800 m^2/g.

The pore size distributions of pristine graphene and of the B-carbon nanomaterial sample are presented in Figure 5.

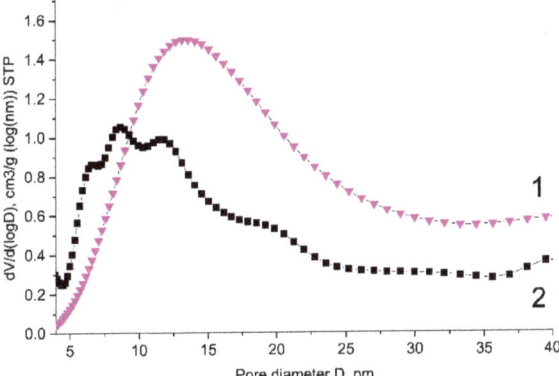

Figure 5. Pore size distributions of pristine graphene (1) and B-carbon nanomaterial (2) (cylindrical pores, QSDFT adsorption branch).

The data presented in Figures 4 and 5 demonstrate that the deposition of an additional boron-containing carbon layer on the surface of graphene leads to the decrease of both the volume and diameter of the pores.

B-carbon nanomaterial takes the shape of hollow spheres. According to the pore size distribution (Figure 5), its main pores are in the range of 5–25 nm. Note that the diameter of the graphene "spheres" approximately matches the size of the MgO template particles (5–20 nm), which were used for the synthesis of graphene.

3.5. B-Carbon Nanomaterial Study by Raman Spectroscopy

B-carbon nanomaterial samples were studied by Raman spectroscopy to determine specific features of their structure. The first-order Raman spectra of graphene and B-carbon nanomaterial are shown in Figure 6. The obtained results are also presented in Table 1. The data obtained for B-carbon nanomaterial were compared those determined for the graphene sample carbonized in the autoclave with the acetone-ethanol mixture at 650 °C in the absence of phenylboronic acid.

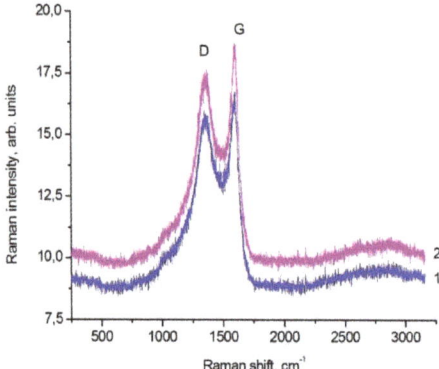

Figure 6. Raman spectra of graphene (1) and B-carbon nanomaterial (2) samples.

Table 1. Raman parameters of graphene and B-carbon nanomaterial samples.

Sample	Frequency, cm^{-1}		D/G
Graphene	1355	1597	1.74
B-carbon nanomaterial	1355	1597	1.81

Two typical modes were observed in the Raman spectra of the synthesized carbon nanomaterial samples: G mode (1597 cm^{-1}), related to vibrations of carbon atoms in the graphene layer, and D mode (1355 cm^{-1}), related to the presence of carbon atoms in sp^3 hybridization [31]. The appearance of the defect band in the spectrum of carbon materials is associated with the disordering of the graphite structure. The D/G ratio is commonly used to estimate the defectiveness of carbon nanomaterials. In the case of the studied graphene sample, the ratio is D/G=1.74 (Figure 6 and Table 1). Additional information on the graphene structure can be obtained from smaller peaks identified in the Raman spectra: G' ($\approx$2700 cm^{-1}), D + D' ($\approx$2950 cm^{-1}). The Raman spectrum confirmed that the synthesized carbon material had a graphene-like local structure. Therefore, according to the Raman data, B-carbon nanomaterial is a defective graphene-like carbon nanomaterial.

The data presented in Table 1 demonstrate that the introduction of boron into the graphene structure results in an increase of the D/G ratio from 1.74 to 1.81, i.e., the concentration of defects substantially increases.

Based on the results of B-carbon nanomaterial characterization by physical methods, the following general scheme of its synthesis can be suggested (Figure 7).

The suggested method includes the graphene synthesis using the template method, followed by deposition of an additional graphene layer doped with boron in an autoclave at 650 °C using a mixture of phenylboronic acid, acetone, and ethanol. After this carbonization procedure, the mass of the graphene sample increased by 70%. The boron concentration in B-carbon nanomaterial determined by different physical methods was about 4 wt.%. The deposition of an additional graphene layer doped with boron led to an increase of the graphene layer thickness from 2–4 to 3–8 monolayers, and a decrease of the specific surface area from 1300 to 800 m^2/g.

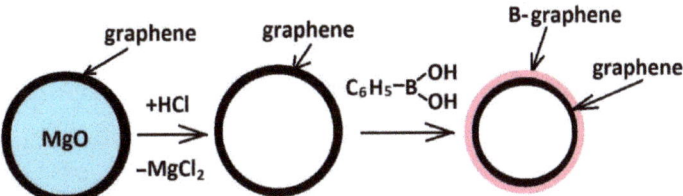

Figure 7. General scheme of B-carbon nanomaterial synthesis.

3.6. XPS Study of B-Carbon Nanomaterial

The state of carbon and boron in B-carbon nanomaterial samples was studied by X-ray photoelectron spectroscopy. Survey scans of pristine and B-carbon nanomaterial samples are presented in Figure 8.

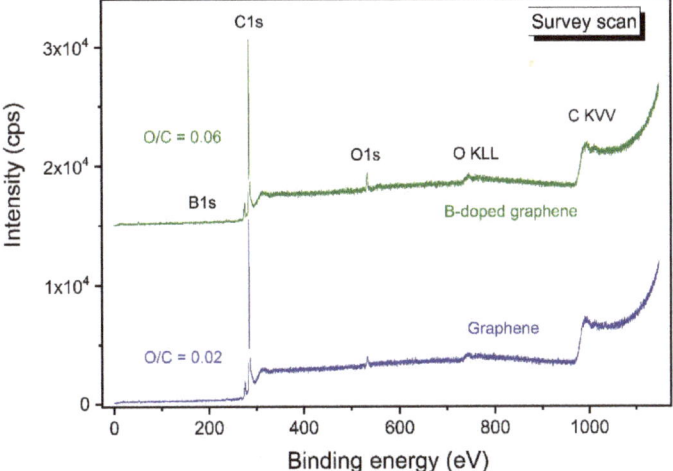

Figure 8. Survey scans of pristine and B-carbon nanomaterial samples.

Figure 8 demonstrates that the initial graphene contains 2% O and 98% C, making it is a high-quality nanomaterial. The oxygen concentration in B-carbon nanomaterial increases to 6% in addition to the appearance of 4% B. The observed changes are related to the incorporation of boron-oxygen fragments into the graphene structure.

The C1s spectrum of the B-carbon nanomaterial sample is presented in Figure 9. The high-resolution C1s spectrum can be divided into five individual peaks (Figure 9), respectively referring to C–B (283.6 eV) [32], C=C (284.5 eV), C–OH, C–O–C (285.8 eV), C=O (287.2 eV) and O–C=O (289.1 eV) functional groups [33–35]. The satellite peak at the binding energy of about 291 eV attributed to the π-π* transition.

The concentrations of different elements in B-carbon nanomaterial were determined by XPS. The results are presented in Table 2.

A more accurate conclusion on the state of boron can be made from the binding energy for the B1s peak of the B-carbon nanomaterial. The B1s peak of B-graphene sample is shown in Figure 10. According to the literature data, the binding energy of 193.1 eV is typical for the C-BO$_2$ bond [19,36,37].

Note that the binding energy of the B1s peak for the C–B bond is in the range of 188–189 eV. The shift of the B1s peak to higher binding energies indicates that the boron atom is bonded with an oxygen atom. So, the following model structure can be suggested for the C–BO$_2$ fragment in B-carbon nanomaterial (Figure 11).

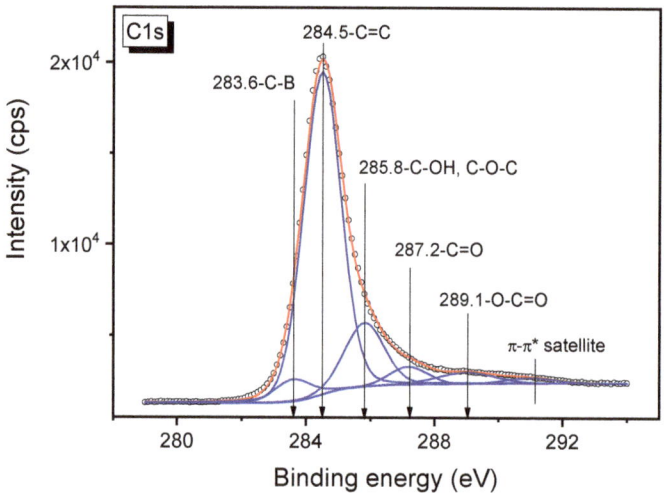

Figure 9. C1s XPS spectrum of B-carbon nanomaterial.

Table 2. Elemental analysis of the B-carbon nanomaterial sample by XPS.

Element	Concentration, wt.%
Boron	4
Carbon	90
Oxygen	6

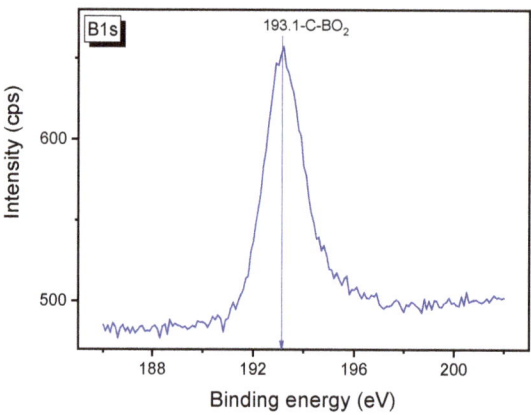

Figure 10. B1s XPS spectrum of B-carbon nanomaterial.

We believe that the incorporation of boron atoms into a formed carbon nanomaterial is difficult because boron atoms are larger that carbon atoms. It was shown [38] that the C–C distance in graphene is equal to 1.41 Å, whereas the C–B distance should be about 1.48 Å. Furthermore, boron atoms are firmly bound to oxygen.

The proposed structure of a fragment of the boron-doped carbon layer deposited on the surface of the initial graphene is shown in Figure 11. This layer is formed during the doping of the initial graphene. During this process the fragments are incorporated into the hexagonal lattice of the carbon monolayer. Therefore, there are two layers in Figure 7: a layer of the initial graphene and a layer of boron-doped graphene.

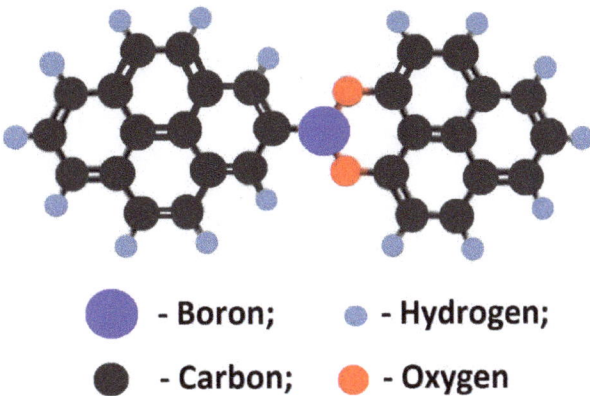

Figure 11. Model structure of the C-BO$_2$ fragment in B-carbon nanomaterial.

4. Conclusions

A new method for the synthesis of B-carbon nanomaterials has been developed. The method includes graphene synthesis using the template method, followed by the deposition of an additional layer of boron-doped graphene in an autoclave at 650 °C using a mixture of phenylboronic acid, acetone, and ethanol. Following the graphene carbonization in the autoclave, its weight increased by 70%. The presence of boron atoms in the graphene framework was confirmed by atomic emission spectroscopy, XPS, and EDX. The boron concentration in B- carbon nanomaterials was found to be about 4 wt.%. The deposition of an additional graphene layer doped with boron led to an increase of the graphene layer thickness from 2–4 to 3–8 monolayers, and a decrease of the surface area from 1300 to 800 m^2/g. The Raman spectrum confirmed that the synthesized carbon material had a graphene-like local structure. According to the XPS data, boron in B-carbon nanomaterials is present in the form of C–BO$_2$ fragments.

Author Contributions: V.V.C.: Conceptualization, Methodology, Data analysis and writing; I.P.P.: Investigation, photoelectron spectroscopy (XPS); E.Y.G.: Investigation, high-resolution transmission electron microscopy (HRTEM); A.S.C.: Investigation, Synthesis. All authors have read and agreed to the published version of the manuscript.

Funding: This research was funded by the Ministry of Science and Higher Education of the Russian Federation within the framework of a state contract of the Boreskov Institute of Catalysis, Siberian Branch, Russian Academy of Sciences (project no. AAAA-A21-121011390054-1).

Institutional Review Board Statement: Not applicable.

Informed Consent Statement: Not applicable.

Data Availability Statement: Not applicable.

Acknowledgments: The studies were performed using the equipment of the Center of Collective Use 'National Center of Catalyst Research'.

Conflicts of Interest: The authors declare that they have no known competing financial interests or personal relationships that could have appeared to influence the work reported in this paper.

References

1. Novoselov, K.S.; Geim, A.K.; Morozov, S.V.; Jiang, D.; Zhang, Y.; Dubonos, S.V.; Grigorieva, I.V.; Firsov, A.A. Electric Field Effect in Atomically Thin Carbon Films. *Science* **2004**, *306*, 666–669. [CrossRef] [PubMed]
2. Mi, B. Scaling up nanoporous graphene membranes. *Science* **2019**, *364*, 1033–1034. [CrossRef] [PubMed]
3. Pixley, J.H.; Andrei, E.Y. Ferromagnetism in magic-angle graphene. *Science* **2019**, *365*, 543–543. [CrossRef]
4. Sharpe, A.; Fox, E.; Barnard, A.; Finney, J.; Watanabe, K.; Taniguchi, T.; Kastner, M.; Goldhaber-Gordon, D. Emergent ferromagnetism near three-quarters filling in twisted bilayer graphene. *Science* **2019**, *365*, 605–608. [CrossRef]

5. Bunch, J.S.; Zande, A.M.; Verbridge, S.S.; Frank, I.W.; Tanenbaum, D.M.; Parpia, J.M.; Craighead, H.G.; McEuen, P.L. Electromechanical Resonators from Graphene Sheets. *Science* **2007**, *315*, 490–493. [CrossRef] [PubMed]
6. Xiang, J.; Drzal, L.T. Thermal Conductivity of a Monolayer of Exfoliated Graphite Nanoplatelets Prepared by Liquid-Liquid Interfacial Self-Assembly. *J. Nanomater.* **2010**, *2010*, 481753. [CrossRef]
7. Balandin, A.A.; Ghosh, S.; Bao, W.; Calizo, I.; Teweldebrhan, D.; Miao, F.; Lau, C.N. Superior thermal conductivity of single-layer graphene. *Nano Lett.* **2008**, *8*, 902–907. [CrossRef]
8. Nascimento, J.R.; D'Oliveira, M.R.; Veiga, A.G.; Chagas, C.A.; Schmal, M. Synthesis of reduced graphene oxide as a support for nano copper and palladium/copper catalysts for selective NO reduction by CO. *ACS Omega* **2020**, *5*, 25568–25581. [CrossRef]
9. Tiwari, S.K.; Sahoo, S.; Wang, N.; Huczko, A. Graphene research and their outputs: Status and prospect. *J. Sci. Adv. Mater. Devices* **2020**, *5*, 10–29. [CrossRef]
10. Chen, Z.; Lin, Y.-M.; Rooks, M.J.; Avouris, P. Graphene nano-ribbon electronics. *Phys. E Low-Dimens. Syst. Nanostructures* **2007**, *40*, 228–232. [CrossRef]
11. Kim, S.; Park, C.M.; Jang, M.; Son, A.; Her, N.; Yu, M.; Synder, S.; Kim, D.H.; Yoon, Y. Aqueous removal of inorganic and organic contaminants by graphene-based nanoadsorbents: A review. *Chemosphere* **2018**, *212*, 1104–1124. [CrossRef] [PubMed]
12. Lonkar, S.P.; Deshmukh, Y.S.; Abdala, A.A. Recent advances in chemical modifications of graphene. *Nano Res.* **2015**, *8*, 1039–1074. [CrossRef]
13. Liu, Z.; Wang, Q.; Zhang, B.; Wu, T.; Li, Y. Efficient Removal of Bisphenol A Using Nitrogen-Doped Graphene-Like Plates from Green Petroleum Coke. *Molecules* **2020**, *25*, 3543. [CrossRef] [PubMed]
14. Zhang, B.; Zhang, G.; Cheng, Z.; Ma, F.; Lu, Z. Atomic-scale friction adjustment enabled by doping-induced modification in graphene nanosheet. *Appl. Surf. Sci.* **2019**, *483*, 742–749. [CrossRef]
15. Kaur, M.; Ubhi, M.K.; Grewal, J.K.; Sharma, V.K. Boron- and phosphorous-doped graphene nanosheets and quantum dots as sensors and catalysts in environmental applications: A review. *Environ. Chem. Lett.* **2021**, *19*, 4375–4392. [CrossRef]
16. Wang, S.; Iyyamperumal, E.; Roy, A.; Xue, Y.; Yu, D.; Dai, L. Vertically Aligned BCN Nanotubes as Efficient Metal-Free Electrocatalysts for the Oxygen Reduction Reaction: A Synergetic Effect by Co-Doping with Boron and Nitrogen. *Angew. Chem.* **2011**, *50*, 11756–11760. [CrossRef] [PubMed]
17. Yang, L.; Jiang, S.; Zhao, Y.; Zhu, L.; Chen, S.; Wang, X.; Wu, Q.; Ma, J.; Ma, Y.; Hu, Z. Boron-Doped Carbon Nanotubes as Metal-Free Electrocatalysts for the Oxygen Reduction Reaction. *Angew. Chem.* **2011**, *50*, 7132–7135. [CrossRef]
18. Wang, S.; Zhang, L.; Xia, Z.; Roy, A.; Chang, D.W.; Baek, J.B.; Dai, L. Boron-Doped Carbon Nanotubes as Metal-Free Electrocatalysts for the Oxygen Reduction Reaction. *Angew. Chem.* **2012**, *51*, 4209–4212. [CrossRef] [PubMed]
19. Sheng, Z.-H.; Gao, H.-L.; Bao, W.-J.; Wang, F.-B.; Xia, X.-H. Synthesis of boron doped graphene for oxygen reduction reaction in fuel cells. *J. Mater. Chem.* **2012**, *22*, 390–395. [CrossRef]
20. Lilloja, J.; Kibena-Põldsepp, E.; Merisalu, M.; Rauwel, P.; Matisen, L.; Niilisk, A.; Cardoso, E.S.F.; Maia, G.; Sammelselg, V.; Tammeveski, K. An Oxygen Reduction Study of Graphene-Based Nanomaterials of Different Origin. *Catalysts* **2016**, *6*, 108. [CrossRef]
21. Bo, X.; Guo, L. Ordered mesoporous boron-doped carbons as metal-free electrocatalysts for the oxygen reduction reaction in alkaline solution. *Phys. Chem. Chem. Phys.* **2013**, *15*, 2459–2465. [CrossRef] [PubMed]
22. Altuntepe, A.; Zan, R. Permanent Boron Doped Graphene with high Homogeneity using Phenylboronic Acid. *J. Mol. Struct.* **2021**, *1230*, 129629. [CrossRef]
23. Mannan, M.A.; Hirano, Y.; Quitain, A.T.; Koinuma, M.; Kida, T. Boron Doped Graphene Oxide: Synthesis and Application to Glucose Responsive Reactivity. *J. Mater. Sci. Eng.* **2018**, *7*, 1000492. [CrossRef]
24. Byeon, A.; Park, J.; Baik, S.; Jung, Y.; Lee, J.W. Effects of boron oxidation state on electrocatalytic activity of carbons synthesized from CO_2. *J. Mater. Chem. A* **2015**, *3*, 5843–5849. [CrossRef]
25. Chesnokov, V.V.; Chichkan, A.S.; Bedilo, A.F.; Shuvarakova, E.I.; Parmon, V.N. Template Synthesis of Graphene. *Dokl. Phys. Chem.* **2019**, *488*, 154–157. [CrossRef]
26. Chesnokov, V.V.; Chichkan, A.S.; Bedilo, A.F.; Shuvarakova, E.I. Synthesis of Carbon-Mineral Composites and Graphene. *Fuller. Nanotub. Carbon Nanostructures* **2020**, *28*, 402–406. [CrossRef]
27. Buyanov, R.A. *Coking of Catalysts*; Nauka: Moskva, Russia, 1983; 207p. (In Russian)
28. Chesnokov, V.V.; Dik, P.P.; Chichkan, A.S. Formic Acid as a Hydrogen Donor for Catalytic Transformations of Tar. *Energies* **2020**, *13*, 4515. [CrossRef]
29. Chesnokov, V.V.; Kriventsov, V.V.; Prosvirin, I.P.; Gerasimov, E.Y. Effect of Platinum Precursor on the Properties of Pt/N-Graphene Catalysts in Formic Acid Decomposition. *Catalysts* **2022**, *12*, 1022. [CrossRef]
30. Gor, G.Y.; Thommes, M.; Cychosz, K.A.; Neimark, A.V. Quenched Solid Density Functional Theory Method for Characterization of Mesoporous Carbons by Nitrogen Adsorption. *Carbon* **2012**, *50*, 1583–1590. [CrossRef]
31. Sadezky, A.; Muckenhuber, H.; Grothe, H.; Niessner, R.; Poschl, U. Raman Microspectroscopy of Soot and Related Carbonaceous Materials: Spectral Analysis and Structural Information. *Carbon* **2005**, *43*, 1731–1742. [CrossRef]
32. Bleu, Y.; Bourquard, F.; Farre, C.; Chaix, C.; Galipaud, J.; Loir, A.-S.; Barnier, V.; Garrelie, F.; Donnet, C. Boron doped graphene synthesis using pulsed laser deposition and its electrochemical characterization. *Diam. Relat. Mater.* **2021**, *115*, 108382. [CrossRef]
33. Johra, F.T.; Jung, W.-G. Hydrothermally reduced graphene oxide as a supercapacitor. *Appl. Surf. Sci.* **2015**, *357*, 1911–1914. [CrossRef]

34. Yang, D.; Velamakanni, A.; Bozoklu, G.; Park, S.; Stoller, M.; Piner, R.D.; Stankovich, S.; Jung, I.; Field, D.A.; Ventrice, C.A., Jr.; et al. Chemical analysis of graphene oxide films after heat and chemical treatments by X-ray photoelectron and Micro-Raman spectroscopy. *Carbon* **2009**, *47*, 145–152. [CrossRef]
35. Ganguly, A.; Sharma, S.; Papakonstantinou, P.; Hamilton, J. Probing the Thermal Deoxygenation of Graphene Oxide Using High-Resolution In Situ X-ray-Based Spectroscopies. *J. Phys. Chem. C.* **2011**, *115*, 17009–17019. [CrossRef]
36. Junaid, M.; Md Khir, M.H.; Witjaksono, G.; Tansu, N.; Saheed, M.S.M.; Kumar, P.; Ullah, Z.; Yar, A.; Usman, F. Boron-Doped Reduced Graphene Oxide with Tunable Bandgap and Enhanced Surface Plasmon Resonance. *Molecules* **2020**, *25*, 3646. [CrossRef]
37. Bleu, Y.; Bourquard, F.; Barnier, V.; Lefkir, Y.; Reynaud, S.; Loir, A.-S.; Garrelie, F.; Donnet, C. Boron-doped graphene synthesis by pulsed laser co-deposition of carbon and boron. *Appl. Surf. Sci.* **2020**, *513*, 145843. [CrossRef]
38. Rani, P.; Jindal, V.K. Designing band gap of graphene by B and N dopant atoms. *RSC Adv.* **2013**, *3*, 802–812. [CrossRef]

Disclaimer/Publisher's Note: The statements, opinions and data contained in all publications are solely those of the individual author(s) and contributor(s) and not of MDPI and/or the editor(s). MDPI and/or the editor(s) disclaim responsibility for any injury to people or property resulting from any ideas, methods, instructions or products referred to in the content.

Article

Efficient Production of Segmented Carbon Nanofibers via Catalytic Decomposition of Trichloroethylene over Ni-W Catalyst

Arina R. Potylitsyna [1,2], Yuliya V. Rudneva [3], Yury I. Bauman [1], Pavel E. Plyusnin [3], Vladimir O. Stoyanovskii [1], Evgeny Y. Gerasimov [1], Aleksey A. Vedyagin [1], Yury V. Shubin [3] and Ilya V. Mishakov [1,*]

1 Boreskov Institute of Catalysis, Pr. Ac. Lavrentieva, 5, Novosibirsk 630090, Russia
2 Faculty of Natural Sciences, Novosibirsk State University, Str. Pirogova 2, Novosibirsk 630090, Russia
3 Nikolaev Institute of Inorganic Chemistry, Ac. Lavrentieva 3, Novosibirsk 630090, Russia
* Correspondence: mishakov@catalysis.ru

Abstract: The catalytic utilization of chlorine-organic wastes remains of extreme importance from an ecological point of view. Depending on the molecular structure of the chlorine-substituted hydrocarbon (presence of unsaturated bonds, intermolecular chlorine-to-hydrogen ratio), the features of its catalytic decomposition can be significantly different. Often, 1,2-dichloroethane is used as a model substrate. In the present work, the catalytic decomposition of trichloroethylene (C_2HCl_3) over microdispersed 100Ni and 96Ni-4W with the formation of carbon nanofibers (CNF) was studied. Catalysts were obtained by a co-precipitation of complex salts followed by reductive thermolysis. The disintegration of the initial bulk alloy driven by its interaction with the reaction mixture $C_2HCl_3/H_2/Ar$ entails the formation of submicron active particles. It has been established that the optimal activity of the pristine Ni catalyst and the 96Ni-4W alloy is provided in temperature ranges of 500–650 °C and 475–725 °C, respectively. The maximum yield of CNF for 2 h of reaction was 63 g/g_{cat} for 100Ni and 112 g/g_{cat} for 96Ni-4W catalyst. Longevity tests showed that nickel undergoes fast deactivation (after 3 h), whereas the 96Ni-4W catalyst remains active for 7 h of interaction. The effects of the catalyst's composition and the reaction temperature upon the structural and morphological characteristics of synthesized carbon nanofibers were investigated by X-ray diffraction analysis, Raman spectroscopy, and electron microscopies. The initial stages of the carbon erosion process were precisely examined by transmission electron microscopy coupled with elemental mapping. The segmented structure of CNF was found to be prevailing in a range of 500–650 °C. The textural parameters of carbon product (S_{BET} and V_{pore}) were shown to reach maximum values (374 m^2/g and 0.71 cm^3/g, respectively) at the reaction temperature of 550 °C.

Keywords: nickel; tungsten; carbon erosion; trichloroethylene; carbon nanofibers; carbon nanomaterials

Copyright: © 2023 by the authors. Licensee MDPI, Basel, Switzerland. This article is an open access article distributed under the terms and conditions of the Creative Commons Attribution (CC BY) license (https:// creativecommons.org/licenses/by/ 4.0/).

1. Introduction

The creation of carbon nanomaterials (CNM) with different structures, ranging from graphene and fullerene to nanofibers, seems to be a promising direction in the field of materials science. In terms of practice, the greatest interest is focused on an improvement of synthetic procedures to produce carbon nanotubes (CNT), carbon nanofibers (CNF), as well as diverse CNM-based composites with desired characteristics. Filamentous carbon materials (CNT and CNF) can be used as reinforcing additives in cement stone and concrete [1], antifriction agents for lubricants and oils [2], as fillers in polymer matrices to impart electrical conductivity and to increase the strength and resistance to abrasion [3], and also as carriers for catalysts used in the hydrodechlorination [4] and selective hydrogenation [5] processes.

The principle industrially relevant method for the synthesis of CNF is based on catalytic pyrolysis of hydrocarbons using metal catalysts (known as CCVD—Catalytic Chemical Vapor Deposition) [6,7]. As a rule, catalytic systems for the CCVD process, as well as for many other reactions (e.g., hydrogen evolution reaction), have to contain dispersed active particles, preparation of which usually requires passing through a number of stages [8]. There are a number of ways to disperse the initial catalytic systems in order to increase their specific surface area and improve the catalytic activity [9–11]. One possible option to simplify the preparative procedure is to apply the so-called "self-dispersion" method directly in the reactor, during the contact of metallic precursor with the reaction mixture. This approach is based on the process of carbon erosion (CE) or metal dusting (MD), which is well-known as a negative phenomenon leading to the destruction of iron- and nickel-based alloys operating in industrial reactors at temperatures of 400–800 °C in a carbon-containing atmosphere. Spontaneous disintegration of a bulk or coarse-dispersed metal (alloy) is accompanied by the formation of numerous submicron particles, which catalyze the growth of graphite-like carbon filaments [12–14]. For example, the complete wastage of 1 g of nichrome (Ni-Cr) in contact with 1,2-dichloroethane vapors produces about 10^{14} active particles with an average diameter of 250 nm, on which the CNF growth takes place [15]. Thereby, the target use of the carbon erosion phenomenon makes it possible to combine the stage of obtaining the catalyst in an active form and the synthesis of CNF, which greatly facilitates the overall procedure.

CO, CH_4, and C_{2+} hydrocarbons and their mixtures as well as the chlorinated hydrocarbons can serve as a carbon-containing source in the CCVD process [16–18]. Catalytic processing of organochlorine compounds is of particular research interest because it might solve a difficult environmental problem related to toxic waste disposal [19,20]. As known from the literature, the supported Ni-containing catalysts are the most commonly used for the pyrolysis of such chlorinated hydrocarbons as 1,2-dichloroethane (1,2-DCE) and trichloroethylene (TCE). At the same time, there are only a few works devoted to the use of TCE as a carbon source for CNF synthesis [21]. Our recent study related to the catalytic pyrolysis of TCE has demonstrated that the maximum yield of the resulting carbon product is 2.6 times greater than that obtained by the decomposition of 1,2-DCE under identical conditions [22]. Thus, the search for an effective catalyst and the study of features of the catalytic pyrolysis of TCE is appeared to be an important scientific task.

It should be emphasized that the catalytic processing of chlorinated hydrocarbons over the self-dispersing Ni-catalysts makes it possible to obtain carbon nanomaterial with unique structural peculiarities [23,24]. An opportunity to produce N-doped CNF materials via combined pyrolysis of the chlorinated hydrocarbons and N-containing precursors has been recently demonstrated as well [25,26]. The implementation of the periodic "chlorination-dechlorination" process on the active metallic surface was found to be responsible for the formation of a segmented secondary structure of growing carbon filaments [27]. The resulting carbon product is characterized by a high specific surface area (up to 400 m^2/g) and pore volume (up to 1 cm^3/g), which makes this material promising for various adsorption and catalytic applications [28–30].

Among the most frequently used metals to catalyze the synthesis of CNM (Fe, Co, and Ni) [31–33], nickel and its alloys demonstrate the greatest activity and resistance to deactivation in the case of catalytic pyrolysis of organochlorine substrates [34]. Meanwhile, the addition of the modifying metal M (up to 10 wt%) to the Ni-M catalyst can have a noticeable promoting impact on the catalytic performance of nickel. For example, Ni-Mo (up to 8 wt% Mo) and Ni-W (4 wt% W) alloys exhibited the greatest activity in the decomposition of 1,2-DCE: the yield of CNF increased by 1.5–2.5 times if compared to pure nickel [35,36]. The observed positive effect is associated primarily with an increase in the rate of decomposition of the chlorinated hydrocarbons. On the other hand, the replacement of nickel by larger Mo and W atoms results in the formation of the solid solutions Ni-M with an increased crystal lattice, which ultimately leads to an enhancement of the carbon capacity of the Ni-M alloy [37–39]. It should be noted that the opportunity of using the

Ni-W alloy system as a catalyst for hydrocarbon pyrolysis remains practically unrevealed. At the same time, very recent works report a promising effect of tungsten on the activity of nickel catalysts in the synthesis of carbon nanofibers [40] and nanotubes [41]. Therefore, of particular interest is the exploration of the impact of the small addition of W (4 wt%) upon the catalytic performance of a self-dispersing Ni-catalyst. The choice of the promoter metal concentration was based on the results of previous research, in which the optimal catalyst composition for the pyrolysis of 1,2-DCE to produce CNF has been defined [36]. Moreover, the possible high activity of Ni-W catalysts in CCVD of chlorinated hydrocarbons makes this approach very attractive from the ecologic point of view (processing of the organochlorine wastes).

The aim of the present work was to estimate the efficacy of catalytic decomposition of TCE (C_2HCl_3), containing the unsaturated C=C bond and characterized by the intermolecular chlorine-to-hydrogen ratio of 3:1, over metallic Ni and Ni-W catalysts. TCE was selected as an insufficiently studied, accessible substrate known as a principal constituent in a number of organochlorine wastes. The synthesized carbon nanomaterials were thoroughly characterized by a set of physicochemical methods, including X-ray diffraction analysis, Raman spectroscopy, and transmission electron microscopy coupled with elemental mapping. Special attention was paid to the early stages of the carbon erosion process when catalytically active particles are being formed. The unique segmental structure of the carbon product was characterized by scanning transmission electron microscopies. The textural characteristics were measured by the low-temperature nitrogen adsorption technique. In addition, the results of comparative longevity tests for 96Ni-4W and 100Ni catalysts are also presented.

2. Materials and Methods

2.1. Materials and Reagents

The following chemicals used for the synthesis of catalysts were purchased from Vekton (Saint-Petersburg, Russia): H_2WO_4 (pure), ammonia solution (25%, high purity grade), and acetone (chemically pure). The precursor salt [$Ni(NH_3)_6$]Cl_2 was synthesized as described elsewhere [42]. Chemically pure TCE (Komponent-reactive, Moscow, Russia), high-purity argon, and hydrogen were used in the catalytic experiments. All the gases were of chemical purity grade and were used without any preliminary purification.

2.2. Synthesis of 100Ni and 96Ni-4W Catalysts

A calculated amount of H_2WO_4 (0.054 g) was added to 10 mL of 25% ammonia solution, heated (60–70 °C) with stirring until almost complete evaporation of the solution, and left for a day. Next, 20 mL of 25% ammonia solution was added to H_2WO_4 and heated with stirring until the powder dissolved. The solution was pale yellow and cloudy. 10 mL of H_2O and $Ni(NH_3)_6Cl_2$ taken in a certain ratio (3.789 g) were added to the resulting solution, stirred until complete dissolution, and cooled to room temperature. The solution was poured into 300 mL of acetone and cooled down to T~0 °C with stirring. The resulting sediment of light violet color was filtered, washed abundantly with acetone, and dried at room temperature for 10 h. The dried sample was then reduced in a hydrogen flow of 130 mL/min at 800 °C for 1 h. The reduced 96Ni-4W alloy sample was cooled down to room temperature in a helium flow. A pure nickel catalyst (100Ni, reference sample) was prepared by a similar procedure, excluding the addition of H_2WO_4.

2.3. Studies on the Metal Dusting Process and Carbon Deposition

The catalytic studies were performed in a flow-through quartz reactor equipped with McBain balances, thus allowing one to follow the accumulation of the carbon product over the catalyst in a real-time mode [43]. The weight of the initial bulk alloy (2.0 ± 0.05 mg) was placed in a quartz basket, which was hooked to a calibrated quartz spring. Before the experiment, the reactor was purged with argon (150 mL/min) and heated to the reaction temperature (450–725 °C) in an argon flow. After that, the sample was reduced in a hydrogen flow (100 mL/min) until the catalyst weight was stabilized. Next, the reaction mixture TCE/Ar/H$_2$ (TCE—6 vol%, Ar—56 vol%, H$_2$—38 vol%) was fed to the reactor. The total flow rate of this mixture was 267 mL/min. Each experiment lasted for 2 h. In the case of the longevity tests, the experiments continued for 7 h. During the experiment, the cathetometer was used to follow the extension of the quartz spring caused by the process of carbon deposition and then to calculate the weight gain with a time on stream. At the end of the experiment, the reactor was cooled in an argon flow to room temperature. The carbon product was unloaded and weighed in order to measure the carbon yield (Y_C, g/g$_{cat}$).

2.4. Characterization of Catalysts and Carbon Nanomaterials

The powder X-ray diffraction (XRD) analysis of 96Ni-4W and pure Ni samples has been performed at room temperature on a Shimadzu XRD-7000 diffractometer (Shimadzu, Tokyo, Japan) using CuKα radiation, and graphite monochromator. The patterns were recorded in the step mode within the angular range 2θ = 15–80°, step 0.1° (survey diffraction pattern) and 2θ = 140–148°, step 0.05° (for precise determination of a lattice parameter). Data from the PDF database were used as references [44]. The lattice parameters were determined by the position of 331 diffraction reflection (at 2θ ≈ 144°) using the PowderCell 2.4 software (BAM, Berlin, Germany) [45]. The volume-averaged crystallite sizes were calculated from the broadening of the (111), (200), and (220) peaks using the Scherrer equation [46], after the separation of the contribution from the instrumental broadening. The deconvolution and fitting of the X-ray diffraction lines based on the Pearson (PVII) function were performed using the WinFit 1.2.1 software (Institute of Geology and Mineralogy, Erlangen, Germany) [47].

The chemical composition of synthesized 96Ni–4W alloy was determined by inductively coupled plasma atomic emission spectroscopy (ICP-AES) on a Thermo Scientific iCAP-6500 spectrometer (Thermo Fisher Scientific Inc., Waltham, MA, USA). Prior to the measurement procedure, the sample was dissolved in a mixture of nitric and hydrofluoric acid.

Raman spectra of obtained carbon nanomaterials were recorded on a Horiba Jobin Yvon LabRAM HR Ultraviolet-Visible-Near Infrared (UV-VIS-NIR) Evolution Raman spectrometer (Horiba, Kyoto, Japan) equipped with Olympus BX41 microscope (Olympus Corp., Tokyo, Japan) and 514.5-nm line of Ar ion laser. In order to avoid the thermal decomposition of the sample, the power of light focused on a spot with a diameter of ~2 μm was less than 0.8 mW.

The secondary structure and morphology of pristine catalysts and synthesized carbon nanomaterials were examined by scanning electron microscopy (SEM) on a JSM-6460 instrument (JEOL Ltd., Tokyo, Japan) at magnifications of 1000× to 100,000×.

Additionally, the morphology of the carbon nanomaterials was examined on a two-beam scanning electron microscope TESCAN SOLARIS FE-SEM (TESCAN, Brno, Czech Republic) working with an acceleration voltage of 20 kV in a secondary electron mode.

The transmission electron microscopy (TEM) studies coupled with elemental mapping were performed using a Hitachi HT7700 TEM (acceleration voltage 100 kV, W source, Hitachi Ltd., Tokyo, Japan) equipped with a STEM system and a Bruker Nano XFlash 6T/60 energy dispersive X-ray (EDX) spectrometer (Bruker Nano GmbH, Berlin, Germany).

The textural characteristics of the obtained carbon nanomaterials were determined by low-temperature nitrogen adsorption, Brunauer–Emmett–Teller (BET) method. The adsorption/desorption isotherms were measured at 77 K on an automated ASAP-2400 (Micromeritics, Norcross, GA, USA) device. The temperature of preliminary degassing of carbon nanomaterial samples was 300 °C.

3. Results and Discussion

3.1. Study of Microdispersed 100Ni and 96Ni-4W Catalysts by XRD and SEM

The samples of the synthesized catalysts were explored by SEM and XRD methods. As can be seen from SEM images (Figure 1), both samples have a porous structure, which is built from the fused particles of ~1–2 μm in diameter connected by bridges. The specific surface area of the pristine samples is rather developed and achieves the value of 10 m^2/g. It was found that the addition of W has no effect on the morphology and structure of the nickel catalyst. According to ICP-AES analysis data, the W content in the composition of 96Ni-4W alloy was about 4.4 wt%.

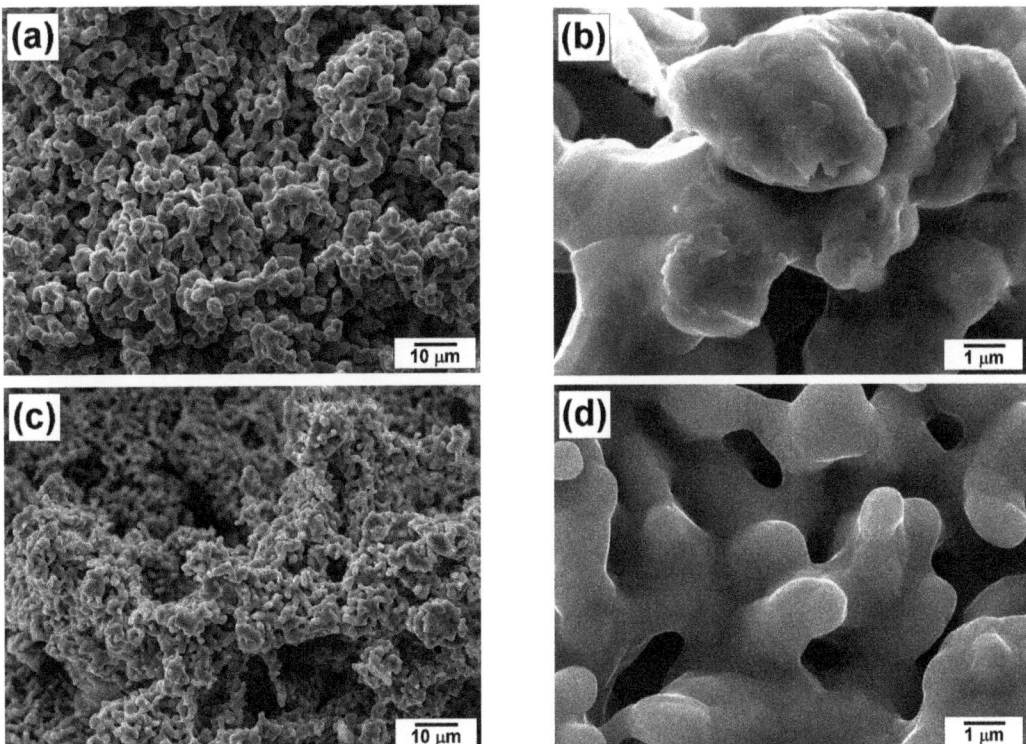

Figure 1. SEM images of the pristine samples prepared by co-precipitation and subsequent reductive thermolysis: (**a**,**b**) 100Ni; (**c**,**d**) 96Ni-4W.

The phase composition of the synthesized 96Ni-4W and pure Ni samples was examined by a powder XRD analysis. The XRD patterns recorded in various 2θ ranges are shown in Figure 2. In the diffraction patterns within 2θ = 15–100°, a set of reflections typical of a face-centered cubic (fcc) lattice can be observed (Figure 2a). Note that no impurity peaks were identified. The reflections for the 96Ni-4W sample are seen to be shifted to the low-angle range with respect to pure nickel (100Ni). The observed shift of the peaks is more pronounced in the far-angle range (2θ = 140–150°). Thus, the developed synthetic procedure allows obtaining the single-phase alloy having the lattice parameter (a) of 3.529 Å (for comparison, a = 3.524 Å for 100Ni). The observed data, along with the absence of any extra peaks in the XRD patterns, permits one to claim that the prepared bimetallic 96Ni-4W sample is represented by a single-phase solid solution. It is worth noting that the addition of W leads to a certain broadening of the (331) reflection (Figure 2b), which is associated with a decrease in the crystallite size. The crystallite sizes for the 96Ni-4W and 100Ni samples calculated from the XRD data were found to be 35 and 70 nm, respectively.

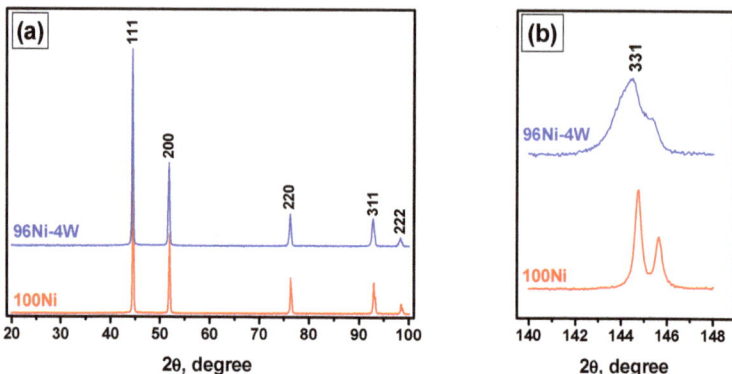

Figure 2. XRD profiles for 96Ni-4W alloy and 100Ni: survey patterns (**a**) and 331 reflections in the far angle range (**b**).

3.2. Catalytic Decomposition of TCE over 100Ni and 96Ni-4W Catalysts

The synthesized 100Ni and 96Ni-4W microdispersed samples were studied in the catalytic pyrolysis of TCE vapors in an excess of hydrogen to produce CNM. The reaction temperature was varied within a range from 475 to 725 °C. The results were plotted as kinetic curves representing the dependence of the carbon nanomaterial weight gain on the time of exposure to the reaction mixture (Figure 3). At the end of the experiment, the yield of CNM (or catalyst productivity) was calculated as a ratio of the weight of the synthesized carbon product to the weight of the catalyst sample used (g/g$_{cat}$).

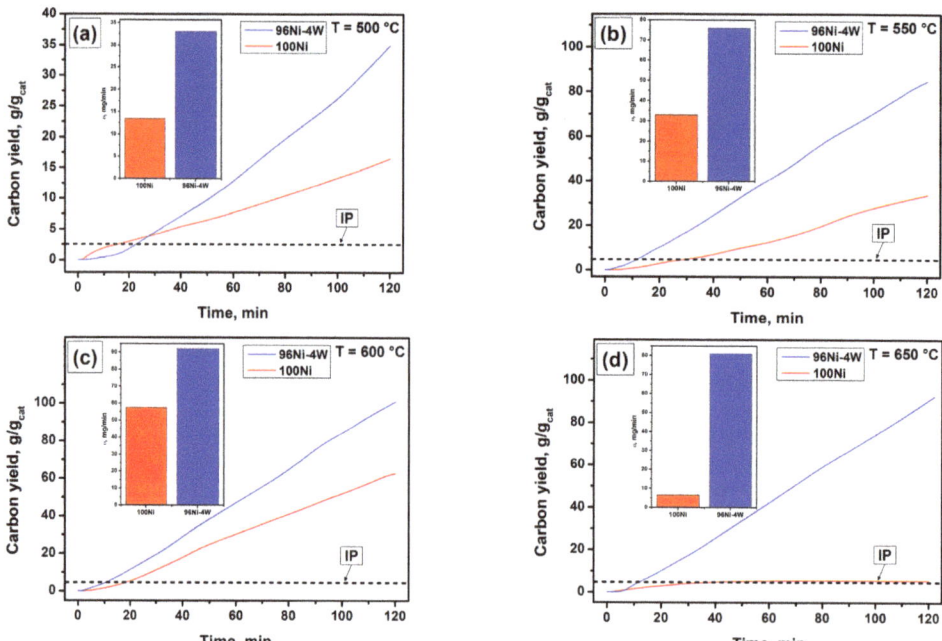

Figure 3. Carbon yield and corresponding carbon deposition rate (v, mg/min) during the 2-h decomposition of TCE vapors in excess of H$_2$ over 100Ni and 96Ni-4W catalysts at temperatures: (**a**) 500 °C; (**b**) 550 °C; (**c**) 600 °C; (**d**) 650 °C.

The choice of the temperature interval used for the catalytic pyrolysis of TCE is explained by the following reasons. Both the upper and lower temperature limits are determined by a rapid deactivation of the catalyst. At temperatures below 500 °C, the decomposition of TCE is suppressed due to irreversible chlorination of the Ni surface (to form $NiCl_2$), while at T > 700 °C the catalyst deactivation is caused by an encapsulation of the alloy surface with amorphous carbon deposits. In both cases, the yield of carbon product does not exceed the level of 10 g/g_{cat}. The observed regularities are in good agreement with the reported results of previous works devoted to the study of temperature regimes of the carbon erosion of bulk metals and alloys [35,48].

It can be clearly seen from Figure 3 that the rate of accumulation of the carbon product is insignificant at the initial stage (first 10–20 min). A typical "delay", known in the literature as the "induction period" (IP), is explained by a slow process of carbon erosion of the alloy [49]. In this regard, all the kinetic curves of the carbon product accumulation can be nominally divided into 2 stages: (i) induction period, and (ii) sustainable growth of CNM. The value of the carbon product yield of 5 g/g_{cat} (which corresponds to a 500% increase in the weight of the catalyst loading) was conventionally chosen as the boundary separating these two stages. As will be shown further, the time to reach a 500% weight gain (i.e., the duration of IP) depends on the catalyst composition and the reaction temperature.

During the IP of the reaction, the initial alloy undergoes disintegration under the action of carbon erosion, which is accompanied by the emergence of numerous active dispersed particles. Simultaneously, the nucleation of the graphite-like phase and the subsequent growth of carbon filaments on top of the active particles are observed in course of the first phase. The intense growth of the carbon nanomaterial (the beginning of a significant increase in the sample weight) is considered the transfer to the second stage. In order to characterize the second stage of the interaction, such parameters as the carbon product yield (P, g/g_{cat}) and the carbon deposition rate (v, mg/min) can be used. The v parameter was calculated as the slope of the kinetic curve in a region of 20–120 min. The values of these parameters calculated for 100Ni and 96Ni-4W samples at different temperatures are summarized in Table 1. It is worth noting that the experiments for the reference sample (100Ni) were carried out at four temperature points (500, 550, 600, and 650 °C), which made it possible to compare 100Ni with 96Ni-4W alloy and establish the effect of the tungsten addition on the catalytic performance of nickel. As for the 96Ni-4W alloy catalyst, its behavior in the TCE decomposition was thoroughly studied at some extra temperature points.

Table 1. Catalytic decomposition of TCE over 100Ni and 96Ni-4W samples within the temperature range of 450–650 °C.

N	Temperature, °C	100Ni			96Ni-4W		
		τ(IP), min	Yield of CNF, g/g_{cat}	v, mg/min	τ(IP), min	Yield of CNF, g/g_{cat}	v, mg/min
1	475	-	-	-	93	8	9
2	500	37	16	13	32	35	33
3	525	-	-	-	25	43	40
4	550	31	34	33	12	85	76
5	575	-	-	-	15	100	92
6	600	20	63	57	12	101	92
7	625	-	-	-	10	112	99
8	650	40 *	5	-	12	92	81
9	675	-	-	-	12	58	47
10	700	-	-	-	10	48	35
11	725	-	-	-	91	7	-

* deactivation.

Figure 4 shows the dependence of IP duration on temperature for both catalyst samples. As can be seen from the plot, for the 96Ni-4W alloy, the duration of IP varies insignificantly (10–15 min) in a temperature interval of 550–700 °C. This indicates that the carbon erosion process proceeds at approximately the same rates within a wide temperature range. In turn, for the monometallic reference sample (100Ni), the minimum duration of IP at 600 °C was equal to 20 min, which is 2 times higher than that for the 96Ni-4W alloy. At higher temperatures (650 °C), it takes a much longer time (40 min) for a pure nickel sample to disintegrate, which testifies to a much lower rate of the CE process. The observed effect can be explained by the deposition of amorphous carbon due to the decomposition of TCE on the outer surface of metal agglomerates, which leads to a surface blockage and a decrease in the rate of carbon transfer into the bulk of nickel. In the case of the 100Ni monometallic sample, the high-temperature deactivation occurs at T~650 °C. Meanwhile, the introduction of W into the Ni-alloy results in a noticeable rise in the deactivation temperature (up to 700 °C). It should be noted that the IP duration for 96Ni-4W alloy has very close values in the temperature range of 550–700 °C. Most probably, the addition of W makes the alloy more resistant to chlorination at lower temperatures, and this fact determines an increase in the CE rate.

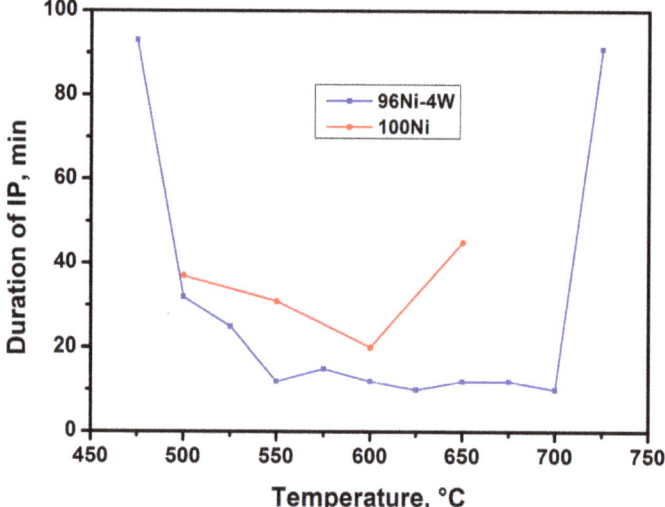

Figure 4. Dependence of IP duration on temperature for 100Ni and 96Ni-4W catalysts tested in the reaction of catalytic decomposition of TCE in excess of H_2.

The process of carbon erosion of the 96Ni-4W sample was precisely studied by TEM coupled with elemental mapping (Figures 5 and 6). Since the CE process proceeds very rapidly in the TCE atmosphere, the temperature of the reactor was decreased to acquire the possibility to register all the changes taking place at the initial stages. This allows deceleration of the decomposition of the substrate and the disintegration of the catalyst. A temperature of 500 °C was chosen for this purpose. At this temperature, the IP duration exceeds 30 min (Figure 4), and no catalyst deactivation by amorphous carbon is observed. Therefore, in these experiments, the 96Ni-4W was exposed to the reaction mixture for 2, 5, 10, and 15 min. The 2-h experiment represents the steady-state operation of the catalyst.

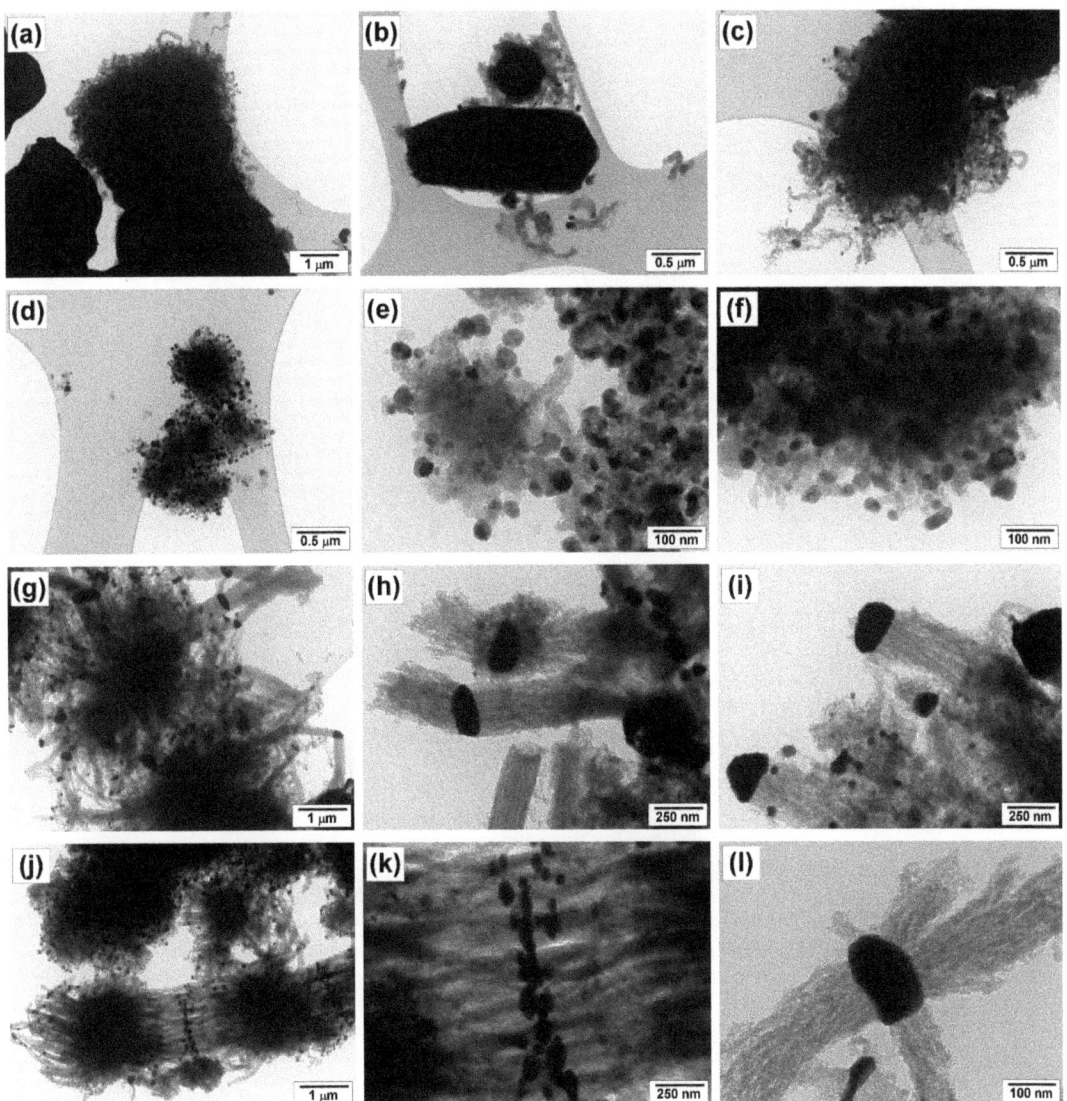

Figure 5. TEM images of the carbon nanomaterials and the catalytic particles formed as a result of carbon erosion of 96Ni-4W alloy during its interaction with TCE at 500 °C for 2 min (**a–c**), 5 min (**d–f**), 10 min (**g–i**), and 15 min (**j–l**).

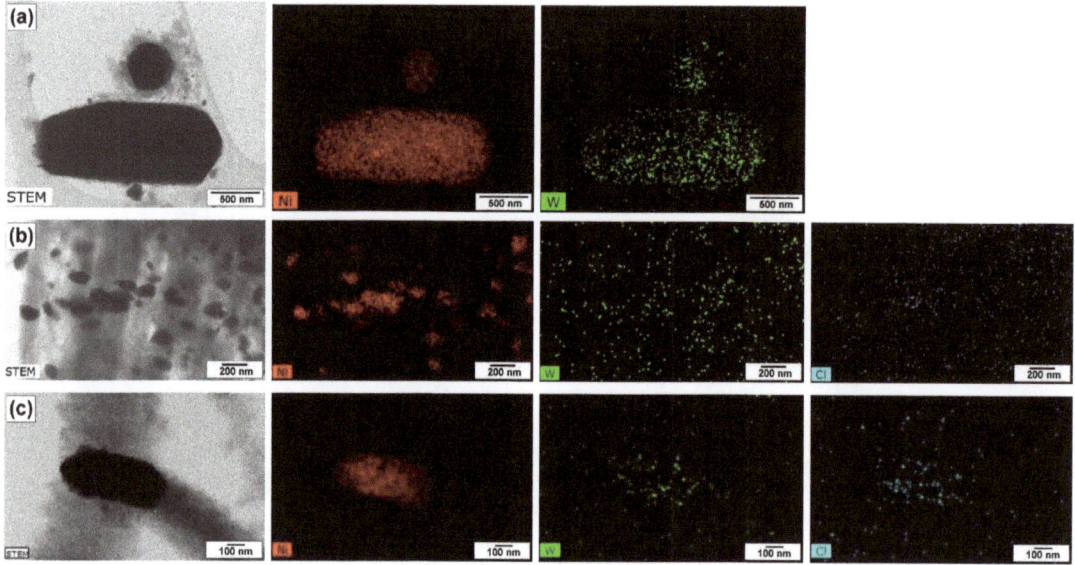

Figure 6. Elemental mapping of the catalytic particles formed as a result of carbon erosion of 96Ni-4W alloy during its interaction with TCE at 500 °C for 2 min (**a**), 15 min (**b**), and 2 h (**c**).

As is seen from Figure 5a, after 2 min of exposure, the surface of the microdispersed alloy particles, which serve as a precursor of the active catalyst, undergoes loosening and disintegration under the action of the carbon nanostructures being formed. Some small particles of up to 100 nm in diameter are detached from the bulk of the sample. At this stage, the carbon product has a fibrous structure, and the diameter of the fibers varies in a range from 10 to 200 nm. Figure 5d–f shows the TEM images for the sample after 5 min of exposure. The catalyst particles of spherical or oval shape not exceeding 10 nm in size are evidently seen. No agglomerated areas of the initial bulk alloy are seen. Thereby, a complete disintegration of the precursor with the formation of a variety of active particles can be stated. After 10 min of the CE process, the larger particles (100–250 nm in size) catalyzing the growth of the nanostructured carbon material are observable. Supposedly, they were formed via the sintering of smaller particles that appeared after the first 5 min of CE. As a rule, one of the faces of such large particles is responsible for the process of TCE decomposition. Next, carbon diffuses to the other face where the release and graphitization of carbon take place. The relatively small particles of the active component (up to 50 nm in size) are also seen. Some of them are separated from the large ones and incorporated into the structure of the growing fibers (Figure 5i). Therefore, it can be concluded that during the first 10 min of exposure, active catalytic particles of optimal size are being formed. Such particles provide an efficient growth of CNF. After 15 min, the majority of thus formed particles operate in a steady-state regime (Figure 5j–l).

According to the elemental mapping of the samples, nickel and tungsten are evenly distributed over the samples. The chlorine species appeared on the surface of the samples exposed to the reaction mixture for 15 min and more. Based on EDX analysis data it is possible to conclude that the concentration of chlorine in the obtained CNF samples does not exceed the value of 1.2 at%. The observed fact is in good agreement with earlier reported data obtained by EDX and XPS methods for CNF materials produced via catalytic pyrolysis of DCE over Ni-M alloys [26,35,48]. It can also be seen from Figure 6c that chlorine species are mainly adsorbed on the surface of active metal particles, where they appear to bind preferentially to tungsten atoms. The observed fact might have a significant impact on the enhanced catalytic performance of the Ni-W system with respect to decomposition of chlorinated hydrocarbons.

In the next step, a comparative analysis of the second stage of the TCE decomposition over the studied catalysts was performed. Figure 7 shows a diagram demonstrating the dependence of the 2-h carbon product yield on the temperature. As can be seen from the experimental results, the maximum productivity of pure 100Ni was 63 g/g_{cat} at a temperature of 600 °C. It should be reminded that the minimum duration of IP (20 min) for pure nickel is also observed at this temperature point. Meanwhile, the maximum yield of CNF for the 96Ni-4W catalyst was as high as 112 g/g_{cat}, which is almost two times higher than that for the unmodified nickel. As can be seen from the data presented in Figure 7, the alloyed Ni-W catalyst outperforms the reference sample (pure nickel) in productivity within the entire 475–725 °C temperature range. Based on the literature data concerning the decomposition of 1,2-dichloroethane ($C_2H_4Cl_2$) at the same conditions, one can find that the maximum yield, in this case, did not exceed 45 g/g_{cat} [35]. It is also worth emphasizing that the productivity of the 96Ni-4W alloy at temperatures from 575 to 650 °C remains very high and changes very slightly (102 ± 10 g/g_{cat}). The observed fact might be taken into account when scaling the process up since the lowest possible temperature is one of the important criteria to ensure the efficient realization of the TCE processing.

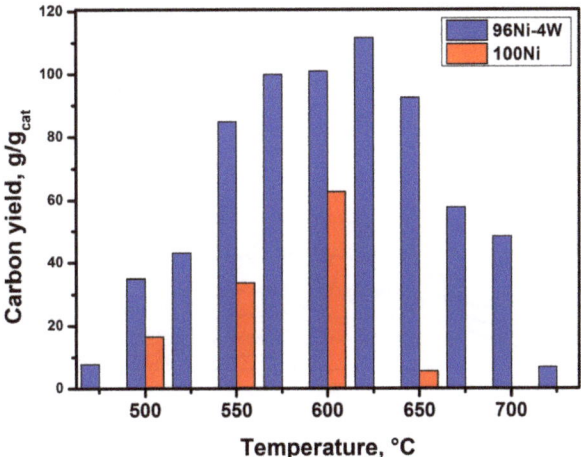

Figure 7. Dependence of the CNF yield on the temperature for 100Ni and 96Ni-4W catalysts tested in the reaction of catalytic decomposition of TCE in an excess of H_2.

Thereby, a comparison of the catalytic activity of 100Ni and alloyed 96Ni-4W catalyst revealed that even a small amount of tungsten (4 wt%) has a significant positive impact upon the catalyst performance of nickel, which is reflected by a remarkable shortening of the IP duration as well as by an increase in the carbon product growth rate (Table 1).

As it was mentioned above, there is a lack of information in the scientific literature concerning the impact of tungsten on the catalytic performance of Ni-based catalysts used for the decomposition of hydrocarbons and their derivatives [40,41]. In the recently reported paper [41] it was claimed that W is capable of increasing the catalytic activity and stability of nickel particles during the hydrocarbon decomposition due to the partial transfer of electrons from Ni to W, along with the formation of the W_2C phase. The latter serves as a regulator of carbon atoms from the surrounding atmosphere to Ni. These assumptions can also be applied to the systems studied in this article. Meanwhile, it seems reasonable to consider W as an analog to Mo. In turn, Mo is very well known in the literature as one of the most effective promotors of Ni-catalysts used for the production of carbon nanotubes and nanofibers [22,50]. The emergence of a strong synergistic effect can be explained by the ability of Mo to enhance greatly the carbon capacity of nickel and to accelerate the diffusion of carbon during the CNM growth. The revealed impact of W addition is very significant in the case of TCE decomposition, thus showing a good potential for the Ni-W system to be further applied for the processing of polychlorinated aliphatic hydrocarbons.

3.3. Study of Morphology and Structure of Carbon Product

3.3.1. XRD Data

As mentioned above, the solid-phase product of the catalytic pyrolysis of C_2HCl_3 vapors is represented by a nanostructured carbon material grown on dispersed active particles originating from the disintegration of the pristine samples. The results of the XRD analysis of the carbon materials are presented in Figure 8. In the diffraction patterns of studied samples, the graphite-like phase can be identified as the predominant phase (2Θ~26°). The second phase is represented by the dispersed metal particles (Ni or Ni-W alloy) present in the sample after the reaction. In addition, there are traces of the Ni_3C phase registered for the CNF samples, obtained at 500–550 °C (Figure 8). The presence of nickel carbide is consistent with the supposed mechanism of the catalytic growth of carbon nanofibers known as the "carbide cycle mechanism" [51].

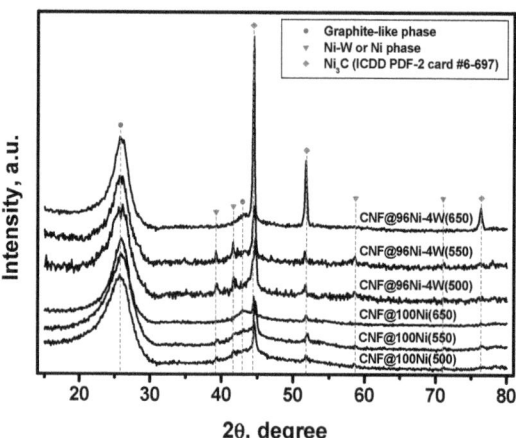

Figure 8. X-ray diffraction profiles for 96Ni–4W alloys and pure 100Ni samples after 2h reaction with TCE (500, 550, and 650 °C).

The content of metal particles of the catalyst within the composition of the samples was estimated as a ratio of the peak (111) area for the *fcc* phase at 2θ of 44.9° to the peak (002) area for graphite-like phase at 2θ of 25.9° (S_{111}/S_{002}). The biggest amount of the residue catalyst was found in the sample obtained via the decomposition of TCE over the 96Ni-4W catalyst at 650 °C (Table 2). It is worth noting that the lattice parameter for the samples obtained over pure nickel (100Ni) does not expectedly change at varying temperature conditions of the reaction. For the samples synthesized using 96Ni-4W alloy, the lattice parameters are different due to a big measurement error. In the cases of CNF@96Ni-4W(500) and CNF@96Ni-4W(550) samples, no reflection (331) in a 2θ range of 140–148°, which is used for the most precise estimation of the lattice parameter, was identified. Therefore, another reflection (220) in a 2θ range of 75–78° was used for the calculations. Moreover, the CNF@96Ni-4W(500) sample contains a very low amount of the metallic phase (S_{111}/S_{002} = 0.03). This complicates the calculation of the lattice parameter using the reflection (220) as well and contributes to the overall uncertainty of measurement.

Table 2. XRD data for the carbon nanomaterials produced over 100Ni and 96Ni-4W samples in a temperature range of 500–650 °C.

Sample	Carbon				Metal Particles (Ni or Ni-W)			S_{111}/S_{002}
	S_{002} (*)	d_{002}, Å	D, nm	n	a, Å	D, nm	S_{111}	
CNF@100Ni(500)	11400	3.46	1.7	5	3.526(2) (140–148°) 3.522(3) (75–78°)	25	600	0.05
CNF@100Ni(550)	9500	3.41	1.8	5	3.525(2) (140–148°) 3.523(3) (75–78°)	25	500	0.05
CNF@100Ni(650)	6400	3.44	2.3	6	3.526(2) (140–148°) 3.527(3) (75–78°)	>100	100	0.02
CNF@96Ni-4W(500)	2400	3.42	1.8	5	3.541(4) (75–78°)	40	70	0.03
CNF@96Ni-4W(550)	1300	3.45	2.7	7	3.533(3) (75–78°)	30	120	0.09
CNF@96Ni-4W(650)	5400	3.43	2.5	7	3.529(2) (140–148°) 3.531(3) (75–78°)	35	1000	0.19

* S is the peak area.

3.3.2. Raman Data

Raman spectra of the carbon product formed over the 96Ni-4W(T) samples, where T is the synthesis temperature, for a region of the bands of first and second orders are presented in Figure 9. The first-order spectra are characterized by the G bands at ~1590–1600 cm^{-1}, corresponding to allowed vibrations E_{2g} of the hexagonal lattice of graphite [52], and by the disorder-induced D line of an activated A_{1g} mode due to finite crystal size [53,54] at ~1340 cm^{-1}.

The D_2 bands at ~1618 cm^{-1}, corresponding to the disordered graphitic lattice (surface graphene layers, E_{2g} symmetry) [55], appeared for the samples obtained at temperatures of 600 °C and above. The bands D_3 at ~1500 cm^{-1} and D_4 at ~1200 cm^{-1}, assigned to the amorphous carbon and the disordered graphitic lattice (A_{1g} symmetry) or polyenes [56] and typical for soot and related carbon materials, are present in spectra of all the samples.

Among the group of second-order bands, the most intensive ones are the bands 2D at ~2677 cm^{-1} and D + D_2 at ~2927 cm^{-1}. The other bands, $2D_2$ and G*~D_4 + D, are of noticeably lower intensity. In order to approximate the second-order lines, excluding the 96Ni-4W(475) and 96Ni-4W(500) samples, a set of bands 2D and D + D_2 of lower intensity and sufficiently higher half-width should be used. It is worth noting that for comparative analysis of the data obtained by Raman spectroscopy and other methods, it should be taken into account that the information provided by Raman spectroscopy for carbon materials corresponds to a laser penetration depth of ~0.1–0.2 μm [57].

The dependences of the main parameters (I_D/I_G, I_{D3}/I_G, and half-width HWHM G) on the temperature of the synthesis are demonstrated in Figure 10a. As is seen, an increase in the synthesis temperature results in a rise in the I_D/I_G ratio. Considering the equation $I_D/I_G = C'(\lambda) \cdot L_a^2$ proposed by Ferrari and Robertson [54], where C' is about 0.0055 for the wavelength of 514.5 nm, this corresponds to an increase in the in-plane crystallite sizes (L_a) from 13.4 to 17.44 Å. The decrease in HWHM G, which usually occurs simultaneously with the increase in the in-plane crystallite sizes (L_a), shows an ill-defined minimum at 600–650 °C followed by a rise at 700 °C. The dependence of the amorphous carbon portion (I_{D3}/I_G ratio) on the synthesis temperature has an explicit minimum at 550–600 °C.

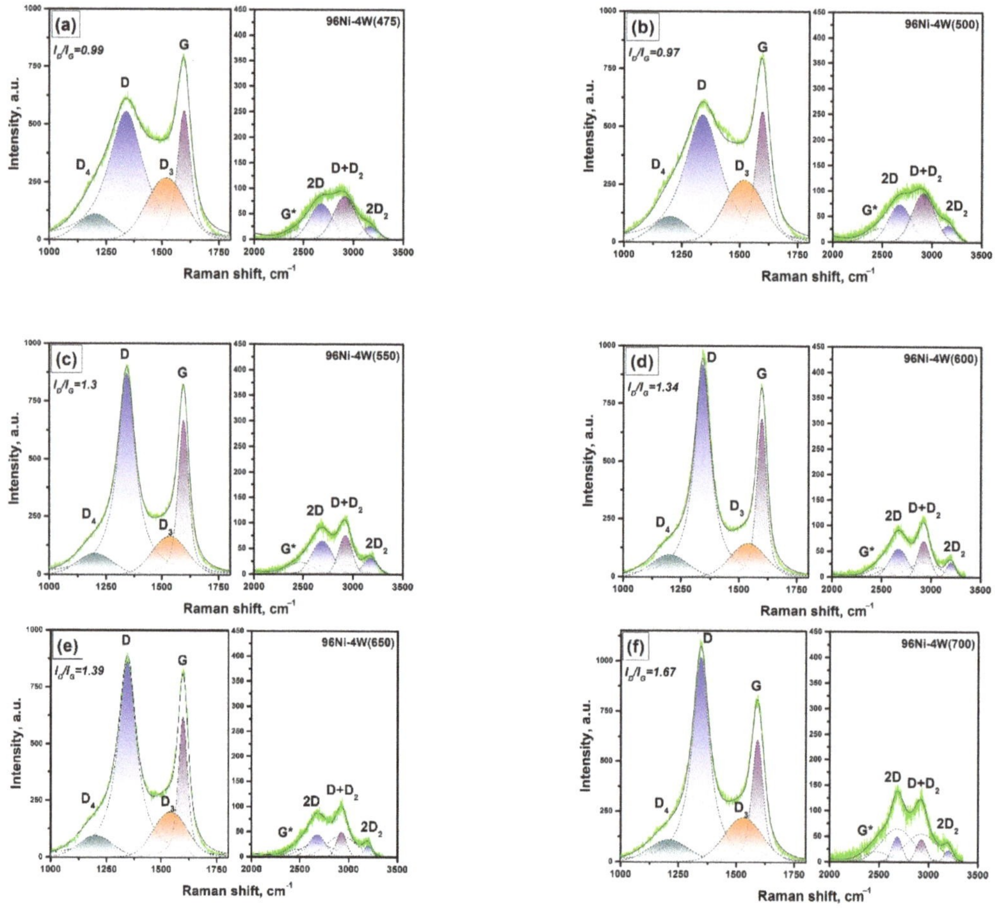

Figure 9. Raman spectra of the carbon product formed over the 96Ni-4W samples for a region of the bands of first and second orders: (**a**) 96Ni-4W(475); (**b**) 96Ni-4W(500); (**c**) 96Ni-4W(550); (**d**) 96Ni-4W(600); (**e**) 96Ni-4W(650); (**f**) 96Ni-4W(700).

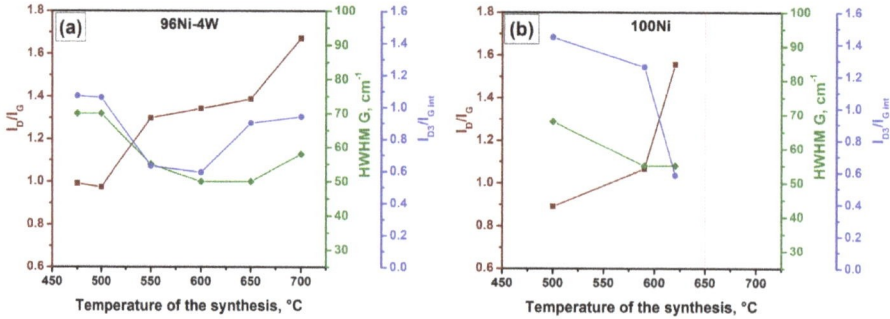

Figure 10. Dependences of the parameters (ID/IG, ID3/IGint, and HWHM G) on the synthesis temperature for the samples 96Ni-4W (**a**) and 100Ni (**b**).

Since, in our case, the samples were obtained at close temperatures, the described diversity is not quantitative. This is mostly a qualitative difference defined by the catalytic process of CNF growth. Therefore, there are three temperature points, which should be marked out in terms of the catalytic growth of CNF over 96Ni-4W catalysts: 475–500 °C, near 600 °C, and near 700 °C.

Raman spectra of the carbon product formed over the 100Ni(T) samples for T = 500, 590, and 620 °C are shown in Figure 11. The corresponding dependences of the main parameters on the synthesis temperature are summarized in Figure 10b. As can be seen, these samples are characterized by a higher portion of the amorphous carbon (I_{D3}/I_G ratio) at 500 °C if compared with the case of 96Ni-4W. An increase in temperature leads to the rapid growth of the in-plane crystallite sizes (L_a) along with a decrease in the amorphous carbon portion (I_{D3}/I_G ratio). However, this process is also accompanied by a deceleration of the catalytic reaction leading to its complete cessation at 650 °C. This allows concluding that the observable phenomenon of carbon ordering is connected with the decelerated carbon growth near the edge of the temperature window.

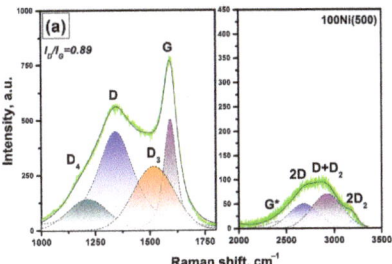
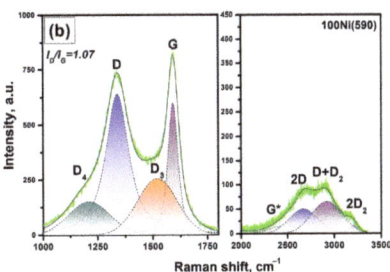
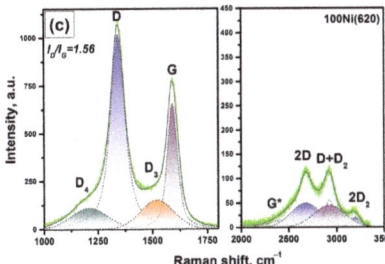

Figure 11. Raman spectra of the carbon product formed over the 100Ni samples for a region of the bands of first and second orders: (**a**) 100Ni(500); (**b**) 100Ni(590); (**c**) 100Ni(620).

3.3.3. SEM and TEM Data

The morphology and structure of the carbon product obtained at different reaction temperatures were examined by scanning and transmission electron microscopies (Figures 12–14). The material resulting from the catalytic decomposition of TCE over 100Ni and 96Ni-4W is predominantly represented by long carbon nanofibers. Figure 12 compares SEM and TEM images of the carbon fibers obtained at the same temperatures over different catalytic systems. It can be seen that the morphology of CNM is almost independent of the catalyst composition. The secondary structure of carbon filaments is characterized by a segmental arrangement regardless of the catalyst's composition.

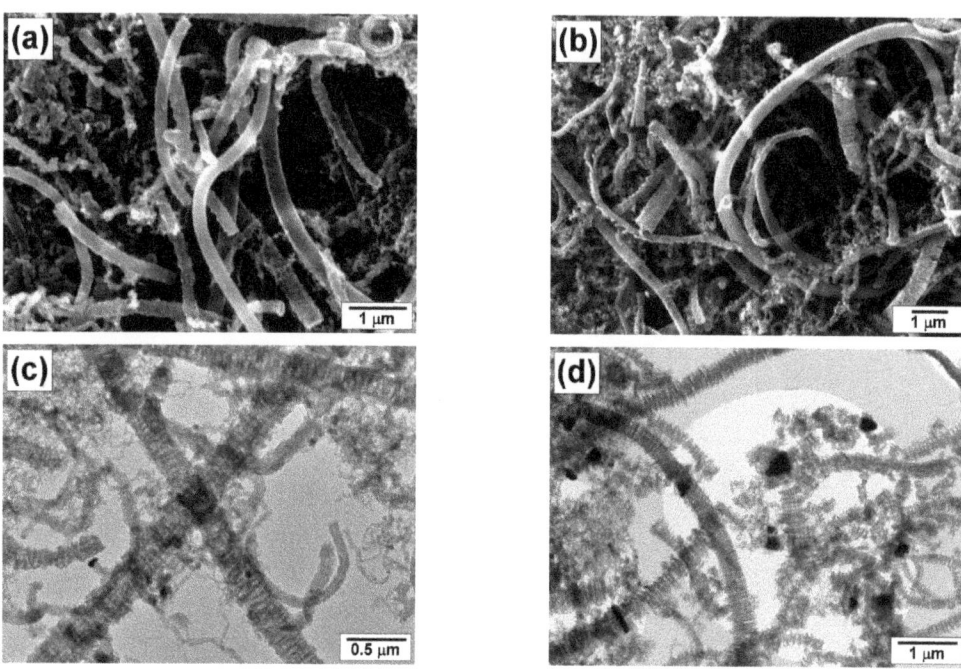

Figure 12. SEM (**a**,**b**) and TEM (**c**,**d**) images of the carbon product obtained by the decomposition of TCE at 600 °C over catalysts: (**a**,**c**) 100Ni; (**b**,**d**) 96Ni-4W.

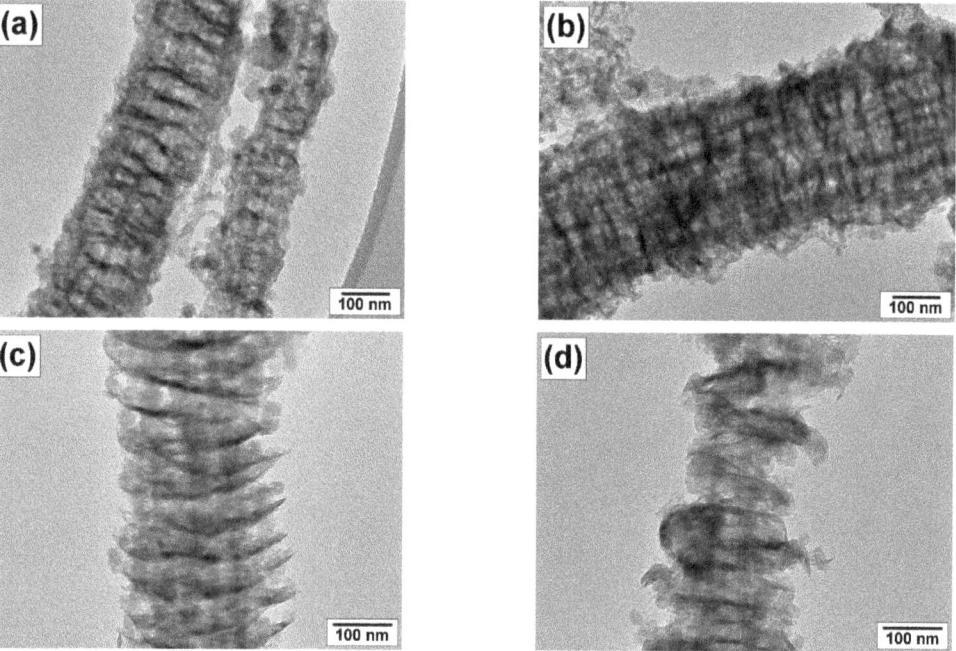

Figure 13. HR TEM images of the carbon product obtained by the decomposition of TCE at 600 °C over catalysts: (**a**,**b**) 100Ni; (**c**,**d**) 96Ni-4W.

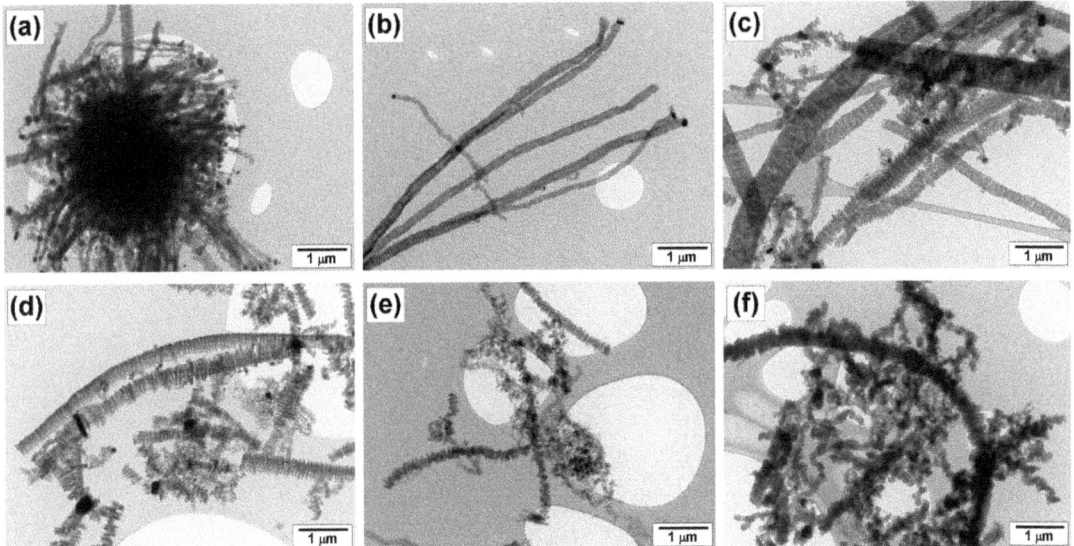

Figure 14. TEM images of the carbon product obtained via decomposition of TCE over 96Ni-4W alloy at temperatures: (**a**) 475 °C; (**b**) 500 °C; (**c**) 550 °C; (**d**) 600 °C; (**e**) 650 °C; (**f**) 700 °C.

High-resolution TEM images show in more detail the morphology of the carbon nanomaterials (Figure 13). It can be seen that both samples are represented by alternating graphene packets of different densities. The diameter of the fibers ranges from 100 to 250 nm.

According to TEM data presented in Figure 14, the catalyst is represented by rounded particles of submicron size (200–400 nm) connected to the grown carbon filaments. Depending on the geometric shape of the particle, the growth of CNF occurs simultaneously in 2–4 directions. It should be noted that the catalyst composition has a very slight effect on the morphology of the resulting carbon product.

At the same time, the main factor influencing the morphological and structural features of carbon fibers is the temperature of the TCE decomposition. For example, one can see that the process of carbon erosion was not accomplished at T = 475 °C: only the surface layer of the metal was subjected to fragmentation to form the active particles (Figure 14a). As mentioned above, this is due to the almost complete chlorination of metallic surface happening in course of the TCE decomposition at low temperatures. As the reaction temperature increases, the growth of carbon filaments accelerates but their structure becomes more defective. Thus, most of the fibers produced at T = 500–550 °C have rather regular segmented structures (Figure 14b,c), while the rise of temperature by 50 °C results in a growth of filaments with noticeably "damaged" segmentation (Figure 14d). This is accompanied by a growth of short fragments and the appearance of smaller catalyst particles (50 nm), which are most likely the product of the secondary fragmentation of large submicron particles. The metallic particles derived from the secondary disintegration also play the role of growth centers for the thinner carbon filaments.

A further increase in the reaction temperature (650 °C) leads to a higher extent of disordering and defectiveness of the resulting carbon product. In this case, the short lateral branches appear within the structure of carbon filaments, the contribution of non-segmental fibers increases, and the direction of the fiber growth becomes tortuous and winding, making it difficult to trace the beginning and the end of a single filament (Figure 14e,f). Finally, the view of CNM obtained at the maximum reaction temperature (700 °C) testifies to the presence of the partially amorphous product (Figure 14f).

According to the data of low-temperature N_2 adsorption (BET method), the specific surface area and pore volume of the carbon material produced at 600 °C were 360 m^2/g and 0.68 cm^3/g (for the 100Ni sample) and 354 m^2/g and 0.68 cm^3/g (for the 96Ni-4W alloy). The closeness of values of the textural parameters allows one to infer that the composition of the catalyst used for TCE pyrolysis has no significant effect on the properties of the resulting carbon nanomaterial.

3.4. Results of Longevity Tests of 100Ni and 96Ni-4W Catalysts

In this study, the longevity (resource) tests of the catalysts were carried out at 600 °C for 7 h. The experiments were performed in a gravimetric flow-through setup equipped with McBain balances. The results of the test are shown in Figure 15 and Table 3. It can be seen that the carbon deposition rate for the 100Ni catalyst gradually decreases. It is only 11 mg/min by the end of the 7th hour, which is almost 4 times less than the rate at the initial stage. The alloyed 96Ni-4W catalyst also exhibited a certain decrease in the carbon deposition rate, but at the end of the test, it was equal to 39 mg/min, which is only 2.2 times less than the initial value (Table 3). In summary, the results of the longevity tests showed that, after the 7-h experiment, the rate of carbon accumulation over the 96Ni-4W alloyed catalyst is 3 times higher compared to the reference sample. The productivity of the 96Ni-4W alloy catalyst at the end of the test was as high as 256 g/g_{cat}, which is 2 times superior with respect to the 100Ni reference sample (114 g/g_{cat}).

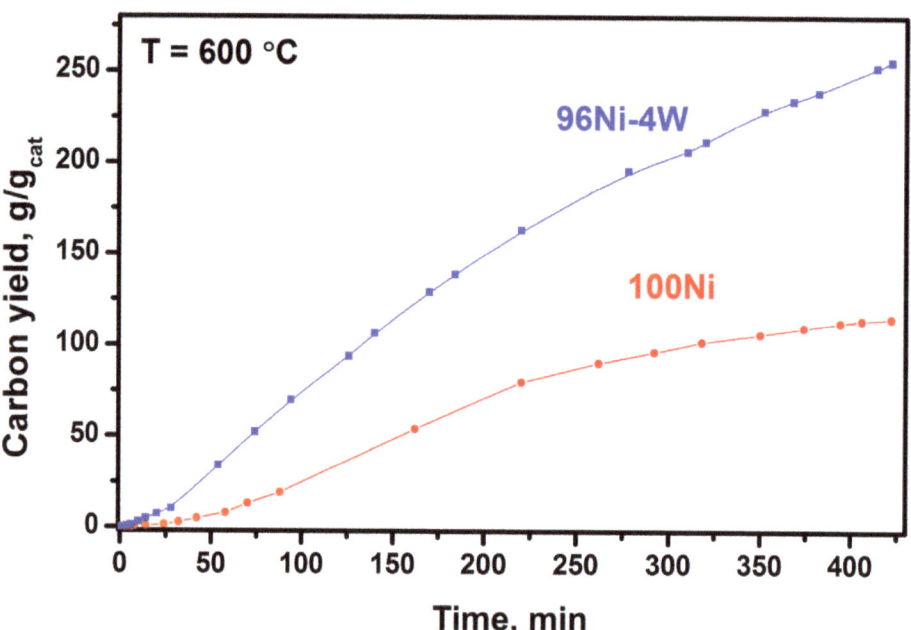

Figure 15. Accumulation of the carbon product during the decomposition of TCE over 100Ni and 96Ni-4W catalysts for 7 h at 600 °C.

Table 3. Parameters of the TCE decomposition reaction carried out over 100Ni and 96Ni-4W catalysts for 7 h at 600 °C.

Reaction Time, min	100Ni		96Ni-4W	
	Yield of CNF, g/g$_{cat}$	v, mg/min	Yield of CNF, g/g$_{cat}$	v, mg/min
120	35	35	91	88
180	61	44	136	77
240	85	44	174	64
300	99	23	204	51
360	108	15	232	49
420	114	11	256	39

4. Conclusions

In the present work, the catalytic decomposition of TCE over microdispersed 100Ni and 96Ni-4W catalysts was studied in a wide temperature range for the first time. Such a process results in the formation of carbon nanomaterial represented by long nanofibers attached to catalytic particles. The obtained carbon nanofibers possess a segmented secondary structure, regardless of the catalyst composition. Such a structure provides both a high specific surface area (360 m^2/g) and a pore volume (0.68 cm^3/g). It was revealed by Raman spectroscopy that the ordering degree of the carbon product strongly depends on the temperature of the process and the composition of the catalyst.

A precise study of the initial stages of the carbon erosion process (disintegration of the microdispersed particles under the action of the aggressive carburizing atmosphere) has shown that the formation of catalytically active particles of optimal size occurs during the first ten minutes. Next, the formed particles operate in a steady-state regime. Elemental mapping of the samples exposed to contact with the reaction medium for different times confirmed the uniform distribution of nickel and tungsten and detected the appearance of chlorine species on the surface of the samples after 15 min of interaction.

Additionally, a significant promoting effect of the tungsten addition was found, which is manifested in: (i) a 2-fold shortening of the IP duration; (ii) a 3-fold increase in the carbon deposition rate; and iii) an increase in the carbon productivity from 63 g/g$_{cat}$ to 101 g/g$_{cat}$ (600 °C, 2 h). The longevity tests revealed that the Ni-W alloy showed much better tolerance to deactivation, demonstrating the extremely high yield of 256 g/g$_{cat}$ for 7 h, which is two times higher than that for the unmodified nickel.

Author Contributions: Conceptualization, A.R.P., Y.V.S. and I.V.M.; methodology, Y.I.B., Y.V.S., Y.V.R., V.O.S. and P.E.P.; investigation, A.R.P., Y.I.B., Y.V.R., V.O.S., E.Y.G. and P.E.P.; visualization, A.R.P., Y.V.R., E.Y.G. and A.A.V.; writing—original draft preparation, A.R.P., I.V.M. and Y.V.S.; writing—review and editing, A.A.V. and I.V.M.; supervision, Y.V.S. and I.V.M. All authors have read and agreed to the published version of the manuscript.

Funding: The research was supported by the Ministry of Science and Higher Education of the Russian Federation [projects No. 121031700315-2 and AAAA-A21-121011390054-1]. The XRD studies were supported by the Russian Science Foundation (project No. 21-13-00414), https://rscf.ru/en/project/21-13-00414/, NIIC SB RAS.

Institutional Review Board Statement: Not applicable.

Informed Consent Statement: Not applicable.

Data Availability Statement: Data are contained within the article.

Acknowledgments: Analysis of the physicochemical properties of the samples was performed using the equipment of the "National Center for Catalyst Research". The TEM studies were carried out using the equipment of the Krasnoyarsk Regional Center for Collective Use of the Federal Research Center Krasnoyarsk Scientific Center SB RAS. The authors are grateful to A.N. Serkova and M.N. Volochaev for their help in the electron microscopy studies, and to A.B. Ayupov for the low-temperature nitrogen adsorption/desorption analysis.

Conflicts of Interest: The authors declare that they have no conflict of interest. The funders had no role in the design of the study; in the collection, analyses, or interpretation of data; in the writing of the manuscript; or in the decision to publish the results.

References

1. Meng, W.; Khayat, K.H. Mechanical properties of ultra-high-performance concrete enhanced with graphite nanoplatelets and carbon nanofibers. *Compos. B. Eng.* **2016**, *107*, 113–122. [CrossRef]
2. Eswaraiah, V.; Sankaranarayanan, V.; Ramaprabhu, S. Graphene-based engine oil nanofluids for tribological applications. *ACS Appl. Mater. Interfaces* **2011**, *3*, 4221–4227. [CrossRef] [PubMed]
3. Santiago-Calvo, M.; Tirado-Mediavilla, J.; Rauhe, J.C.; Jensen, L.R.; Ruiz-Herrero, J.L.; Villafañe, F.; Rodríguez-Pérez, M.Á. Evaluation of the thermal conductivity and mechanical properties of water blown polyurethane rigid foams reinforced with carbon nanofibers. *Eur. Polym. J.* **2018**, *108*, 98–106. [CrossRef]
4. Netskina, O.V.; Komova, O.V.; Tayban, E.S.; Oderova, G.V.; Mukha, S.A.; Kuvshinov, G.G.; Simagina, V.I. The influence of acid treatment of carbon nanofibers on the activity of palladium catalysts in the liquid-phase hydrodechlorination of dichlorobenzene. *Appl. Catal. A-Gen.* **2013**, *467*, 386–393. [CrossRef]
5. Qu, Z.; Mao, C.; Zhu, X.; Zhang, J.; Jiang, H.; Chen, R. Pd-Decorated Hierarchically Porous Carbon Nanofibers for Enhanced Selective Hydrogenation of Phenol. *Ind. Eng. Chem. Res.* **2022**, *61*, 13416–13430. [CrossRef]
6. Mishakov, I.V.; Vedyagin, A.A.; Bauman, Y.I.; Shubin, Y.V.; Buyanov, R.A. Synthesis of carbon nanofibers via catalytic chemical vapor deposition of halogenated hydrocarbons. In *Carbon Nanofibers: Synthesis, Applications and Performance*; Nova Science Publishers, Inc.: New York, NY, USA, 2018; pp. 77–182.
7. Nasibulina, L.I.; Koltsova, T.S.; Joentakanen, T.; Nasibulin, A.G.; Tolochko, O.V.; Malm, J.E.M.; Karppinen, M.J.; Kauppinen, E.I. Direct synthesis of carbon nanofibers on the surface of copper powder. *Carbon* **2010**, *48*, 4559–4562. [CrossRef]
8. Ermakova, M.A.; Ermakov, D.Y.; Kuvshinov, G.G. Effective catalysts for direct cracking of methane to produce hydrogen and filamentous carbon: Part I. Nickel catalysts. *Appl. Catal. A-Gen.* **2000**, *201*, 61–70. [CrossRef]
9. Lin, Z.; Xiao, B.; Huang, M.; Yan, L.; Wang, Z.; Huang, Y.; Shen, S.; Zhang, Q.; Gu, L.; Zhong, W. Realizing Negatively Charged Metal Atoms through Controllable d-Electron Transfer in Ternary $Ir_{1-x}Rh_xSb$ Intermetallic Alloy for Hydrogen Evolution Reaction. *Adv. Energy Mater.* **2022**, *12*, 2200855. [CrossRef]
10. Ma, S.; Deng, J.; Xu, Y.; Tao, W.; Wang, X.; Lin, Z.; Zhang, Q.; Gu, L.; Zhong, W. Pollen-like self-supported FeIr alloy for improved hydrogen evolution reaction in acid electrolyte. *J. Energy Chem.* **2022**, *66*, 560–565. [CrossRef]
11. Shen, S.; Hu, Z.; Zhang, H.; Song, K.; Wang, Z.; Lin, Z.; Zhang, Q.; Gu, L.; Zhong, W. Highly active Si sites enabled by negative valent Ru for electrocatalytic hydrogen evolution in LaRuSi. *Angew. Chem. Int. Ed.* **2022**, *134*, e202206460. [CrossRef]
12. Chang, J.K.; Tsai, H.Y.; Tsai, W.-T. A metal dusting process for preparing nano-sized carbon materials and the effects of acid post-treatment on their hydrogen storage performance. *Int. J. Hydrog. Energy* **2008**, *33*, 6734–6742. [CrossRef]
13. Mishakov, I.V.; Afonnikova, S.D.; Bauman, Y.I.; Shubin, Y.V.; Trenikhin, M.V.; Serkova, A.N.; Vedyagin, A.A. Carbon Erosion of a Bulk Nickel–Copper Alloy as an Effective Tool to Synthesize Carbon Nanofibers from Hydrocarbons. *Kinet. Catal.* **2022**, *63*, 97–107. [CrossRef]
14. Grabke, H.J.; Schütze, M. *Corrosion by Carbon and Nitrogen: Metal Dusting, Carburisation and Nitridation*; Elsevier: Amsterdam, The Netherlands, 2007; pp. 1–308.
15. Mishakov, I.V.; Korneev, D.V.; Bauman, Y.I.; Vedyagin, A.A.; Nalivaiko, A.Y.; Shubin, Y.V.; Gromov, A.A. Interaction of chlorinated hydrocarbons with nichrome alloy: From surface transformations to complete dusting. *Surf. Interfaces* **2022**, *30*, 101914. [CrossRef]
16. Maboya, W.K.; Coville, N.J.; Mhlanga, S.D. The synthesis of carbon nanomaterials using chlorinated hydrocarbons over a Fe-Co/$CaCO_3$ catalyst. *S. Afr. J. Chem.* **2016**, *69*, 15–26. [CrossRef]
17. Setayesh, S.R.; Waugh, K.C. Kinetic studies of carbon nanofibre and hydrogen evolution via ethane decomposition over fresh and steam regenerated Ni/La2O3 catalyst. *Appl. Catal. A-Gen.* **2012**, *417*, 174–182. [CrossRef]
18. Toebes, M.L.; Bitter, J.H.; Van Dillen, A.J.; de Jong, K.P. Impact of the structure and reactivity of nickel particles on the catalytic growth of carbon nanofibers. *Catal. Today* **2002**, *76*, 33–42. [CrossRef]
19. Malik, J.K.; Aggarwal, M.; Kalpana, S.; Gupta, R.C. Chlorinated hydrocarbons and pyrethrins/pyrethroids. In *Reproductive and Developmental Toxicology*; Elsevier: Amsterdam, The Netherlands, 2022; pp. 641–664.
20. Rosner, D.; Markowitz, G. Persistent pollutants: A brief history of the discovery of the widespread toxicity of chlorinated hydrocarbons. *Environ. Res.* **2013**, *120*, 126–133. [CrossRef] [PubMed]
21. Shaikjee, A.; Coville, N.J. Catalyst restructuring studies: The facile synthesis of tripod-like carbon fibers by the decomposition of trichloroethylene. *Mater. Lett.* **2012**, *68*, 273–276. [CrossRef]
22. Potylitsyna, A.R.; Bauman, Y.I.; Mishakov, I.V.; Plyusnin, P.E.; Vedyagin, A.A.; Shubin, Y.V. The Features of the CCVD of Trichloroethylene Over Microdispersed Ni and Ni–Mo Catalysts. *Top. Catal.* **2022**. [CrossRef]
23. Bauman, Y.I.; Mishakov, I.V.; Rudneva, Y.V.; Popov, A.A.; Rieder, D.; Korneev, D.V.; Serkova, A.N.; Shubin, Y.V.; Vedyagin, A.A. Catalytic synthesis of segmented carbon filaments via decomposition of chlorinated hydrocarbons on Ni-Pt alloys. *Catal. Today* **2020**, *348*, 102–110. [CrossRef]
24. Nieto-Márquez, A.; Valverde, J.L.; Keane, M.A. Catalytic growth of structured carbon from chloro-hydrocarbons. *Appl. Catal. A-Gen.* **2007**, *332*, 237–246. [CrossRef]

25. Potylitsyna, A.R.; Mishakov, I.V.; Bauman, Y.I.; Kibis, L.S.; Shubin, Y.V.; Volochaev, M.N.; Melgunov, M.S.; Vedyagin, A.A. Metal dusting as a key route to produce functionalized carbon nanofibers. *React. Kinet. Mech. Catal.* **2022**, *135*, 1387–1404. [CrossRef]
26. Brzhezinskaya, M.; Mishakov, I.V.; Bauman, Y.I.; Shubin, Y.V.; Maksimova, T.A.; Stoyanovskii, V.O.; Gerasimov, E.Y.; Vedyagin, A.A. One-pot functionalization of catalytically derived carbon nanostructures with heteroatoms for toxic-free environment. *Appl. Surf. Sci.* **2022**, *590*, 153055. [CrossRef]
27. Bauman, Y.I.; Lysakova, A.S.; Rudnev, A.V.; Mishakov, I.V.; Shubin, Y.V.; Vedyagin, A.A.; Buyanov, R.A. Synthesis of nanostructured carbon fibers from chlorohydrocarbons over Bulk Ni-Cr Alloys. *Nanotechnol. Russ.* **2014**, *9*, 380–385. [CrossRef]
28. Mishakov, I.V.; Bauman, Y.I.; Brzhezinskaya, M.; Netskina, O.V.; Shubin, Y.V.; Kibis, L.S.; Stoyanovskii, V.O.; Larionov, K.B.; Serkova, A.N.; Vedyagin, A.A. Water Purification from Chlorobenzenes using Heteroatom-Functionalized Carbon Nanofibers Produced on Self-Organizing Ni-Pd Catalyst. *J. Environ. Chem. Eng.* **2022**, *10*, 107873. [CrossRef]
29. Modi, A.; Bhaduri, B.; Verma, N. Facile one-step synthesis of nitrogen-doped carbon nanofibers for the removal of potentially toxic metals from water. *Ind. Eng. Chem. Res.* **2015**, *54*, 5172–5178. [CrossRef]
30. Zhang, C.; Zhang, J.; Shao, Y.; Jiang, H.; Chen, R.; Xing, W. Controllable synthesis of 1D Pd@ N-CNFs with high catalytic performance for phenol hydrogenation. *Catal. Lett.* **2021**, *151*, 1013–1024. [CrossRef]
31. Iwasaki, T.; Makino, Y.; Fukukawa, M.; Nakamura, H.; Watano, S. Low-temperature growth of nitrogen-doped carbon nanofibers by acetonitrile catalytic CVD using Ni-based catalysts. *Appl. Nanosci.* **2016**, *6*, 1211–1218. [CrossRef]
32. Shalagina, A.E.; Ismagilov, Z.R.; Podyacheva, O.Y.; Kvon, R.I.; Ushakov, V.A. Synthesis of nitrogen-containing carbon nanofibers by catalytic decomposition of ethylene/ammonia mixture. *Carbon* **2007**, *45*, 1808–1820. [CrossRef]
33. Takenaka, S.; Ishida, M.; Serizawa, M.; Tanabe, E.; Otsuka, K. Formation of carbon nanofibers and carbon nanotubes through methane decomposition over supported cobalt catalysts. *J. Phys. Chem. B* **2004**, *108*, 11464–11472. [CrossRef]
34. Mishakov, I.V.; Chesnokov, V.V.; Buyanov, R.A.; Pakhomov, N.A. Decomposition of chlorinated hydrocarbons on iron-group metals. *Kinet. Catal.* **2001**, *42*, 543–548. [CrossRef]
35. Bauman, Y.I.; Rudneva, Y.V.; Mishakov, I.V.; Plyusnin, P.E.; Shubin, Y.V.; Korneev, D.V.; Stoyanovskii, V.O.; Vedyagin, A.A.; Buyanov, R.A. Effect of Mo on the catalytic activity of Ni-based self-organizing catalysts for processing of dichloroethane into segmented carbon nanomaterials. *Heliyon* **2019**, *5*, e02428. [CrossRef] [PubMed]
36. Mishakov, I.V.; Bauman, Y.I.; Potylitsyna, A.R.; Shubin, Y.V.; Plyusnin, P.E.; Stoyanovskii, V.O.; Vedyagin, A.A. Catalytic properties of bulk (1−x) Ni−xW alloys in the decomposition of 1, 2-Dichloroethane with the production of carbon nanomaterials. *Kinet. Catal.* **2022**, *63*, 75–86. [CrossRef]
37. Allahyarzadeh, M.H.; Aliofkhazraei, M.; Rezvanian, A.R.; Torabinejad, V.; Rouhaghdam, A.R.S. Ni-W electrodeposited coatings: Characterization, properties and applications. *Surf. Coat. Technol.* **2016**, *307*, 978–1010. [CrossRef]
38. Yang, R.; Du, X.; Zhang, X.; Xin, H.; Zhou, K.; Li, D.; Hu, C. Transformation of jatropha oil into high-quality biofuel over Ni-W bimetallic catalysts. *ACS Omega* **2019**, *4*, 10580–10592. [CrossRef]
39. Zhang, S.; Shi, C.; Chen, B.; Zhang, Y.; Qiu, J. An active and coke-resistant dry reforming catalyst comprising nickel–tungsten alloy nanoparticles. *Catal. Commun.* **2015**, *69*, 123–128. [CrossRef]
40. Shekunova, V.M.; Sinyapkin, Y.T.; Didenkulova, I.I.; Tsyganova, E.I.; Aleksandrov, Y.A.; Sinyapkin, D.Y. Catalytic pyrolysis of light hydrocarbons in the presence of ultrafine particles formed by electrically induced explosive dispersion of metal wires. *Pet. Chem.* **2013**, *53*, 92–96. [CrossRef]
41. Jia, J.; Veksha, A.; Lim, T.-T.; Lisak, G.; Zhang, R.; Wei, Y. Modulating local environment of Ni with W for synthesis of carbon nanotubes and hydrogen from plastics. *J. Clean. Prod.* **2022**, *352*, 131620. [CrossRef]
42. Von Brauer, G. *Handbuch Der Präparativen Anorganischen Chemie in Drei Bänden*; Enke: Omaha, NE, USA, 1978.
43. Wang, C.; Bauman, Y.I.; Mishakov, I.V.; Stoyanovskii, V.O.; Shelepova, E.V.; Vedyagin, A.A. Scaling up the Process of Catalytic Decomposition of Chlorinated Hydrocarbons with the Formation of Carbon Nanostructures. *Processes* **2022**, *10*, 506. [CrossRef]
44. The Powder Diffraction File 2 (PDF-2): International Centre for Diffraction Data; Newtown Square, PA, USA, 2009.
45. Kraus, W.; Nolze, G. PowderCell 2.0 for Windows. *Powder Diffr.* **1998**, *13*, 256.
46. Cullity, B.D. *Elements of X-ray Diffraction, Addison*; Wesley Mass: Reading, MA, USA, 1978; pp. 127–131.
47. Krumm, S. An interactive Windows program for profile fitting and size/strain analysis. *Mater. Sci. Forum* **1996**, *228–231*, 183–190. [CrossRef]
48. Bauman, Y.I.; Shorstkaya, Y.V.; Mishakov, I.V.; Plyusnin, P.E.; Shubin, Y.V.; Korneev, D.V.; Stoyanovskii, V.O.; Vedyagin, A.A. Catalytic conversion of 1, 2-dichloroethane over Ni-Pd system into filamentous carbon material. *Catal. Today* **2017**, *293*, 23–32. [CrossRef]
49. Bauman, Y.I.; Mishakov, I.V.; Buyanov, R.A.; Vedyagin, A.A.; Volodin, A.M. Catalytic properties of massive iron-subgroup metals in dichloroethane decomposition into carbon products. *Kinet. Catal.* **2011**, *52*, 547–554. [CrossRef]
50. Modekwe, H.U.; Mamo, M.A.; Moothi, K.; Daramola, M.O. Effect of different catalyst supports on the quality, yield and morphology of carbon nanotubes produced from waste polypropylene plastics. *Catalysts* **2021**, *11*, 692. [CrossRef]
51. Buyanov, R.; Chesnokov, V.; Afanas' ev, A.; Babenko, V. Carbide mechanism of formation of carbonaceous deposits and their properties on iron-chromium dehydrogenation catalysts. *Kin. Cat.* **1977**, *18*, 839–845.
52. Nemanich, R.J.; Solin, S.A. First- and second-order Raman scattering from finite-size crystals of graphite. *Phys. Rev. B* **1979**, *20*, 392–401. [CrossRef]
53. Tuinstra, F.; Koenig, J.L. Raman spectrum of graphite. *J. Chem. Phys.* **1970**, *53*, 1126–1130. [CrossRef]

54. Ferrari, A.C.; Robertson, J. Interpretation of Raman spectra of disordered and amorphous carbon. *Phys. Rev. B* **2000**, *61*, 14095–14107. [CrossRef]
55. Wang, Y.; Alsmeyer, D.C.; McCreery, R.L. Raman spectroscopy of carbon materials: Structural basis of observed spectra. *Chem. Mater.* **1990**, *2*, 557–563. [CrossRef]
56. Sadezky, A.; Muckenhuber, H.; Grothe, H.; Niessner, R.; Pöschl, U. Raman microspectroscopy of soot and related carbonaceous materials: Spectral analysis and structural information. *Carbon* **2005**, *43*, 1731–1742. [CrossRef]
57. Smausz, T.; Kondász, B.; Gera, T.; Ajtai, T.; Utry, N.; Pintér, M.; Kiss-Albert, G.; Budai, J.; Bozóki, Z.; Szabó, G.; et al. Determination of UV–visible–NIR absorption coefficient of graphite bulk using direct and indirect methods. *Appl. Phys. A* **2017**, *123*, 633. [CrossRef]

Disclaimer/Publisher's Note: The statements, opinions and data contained in all publications are solely those of the individual author(s) and contributor(s) and not of MDPI and/or the editor(s). MDPI and/or the editor(s) disclaim responsibility for any injury to people or property resulting from any ideas, methods, instructions or products referred to in the content.

Article

Porous Co-Pt Nanoalloys for Production of Carbon Nanofibers and Composites

Sofya D. Afonnikova [1], Anton A. Popov [2], Yury I. Bauman [1], Pavel E. Plyusnin [2], Ilya V. Mishakov [1], Mikhail V. Trenikhin [3], Yury V. Shubin [2], Aleksey A. Vedyagin [1] and Sergey V. Korenev [2,*]

[1] Boreskov Institute of Catalysis SB RAS, 5 Lavrentyev Ave., 630090 Novosibirsk, Russia
[2] Nikolaev Institute of Inorganic Chemistry of SB RAS, 3 Lavrentyev Ave., 630090 Novosibirsk, Russia
[3] Center of New Chemical Technologies BIC SB RAS, 54 Neftezavodskaya St., 644060 Omsk, Russia
* Correspondence: korenev@niic.nsc.ru

Abstract: The controllable synthesis of carbon nanofibers (CNF) and composites based on CNF (Metals/CNF) is of particular interest. In the present work, the samples of CNF were produced via ethylene decomposition over Co-Pt (0–100 at.% Pt) microdispersed alloys prepared by a reductive thermolysis of multicomponent precursors. XRD analysis showed that the crystal structure of alloys in the composition range of 5–35 at.% Pt corresponds to a *fcc* lattice based on cobalt (*Fm-3m*), while the CoPt (50 at.% Pt) and CoPt$_3$ (75 at.% Pt) samples are intermetallics with the structure *P4/mmm* and *Pm-3m*, respectively. The microstructure of the alloys is represented by agglomerates of polycrystalline particles (50–150 nm) interconnected by the filaments. The impact of Pt content in the Co$_{1-x}$Pt$_x$ samples on their activity in CNF production was revealed. The interaction of alloys with ethylene is accompanied by the generation of active particles on which the growth of nanofibers occurs. Plane Co showed low productivity (~5.5 g/g$_{cat}$), while Pt itself exhibited no activity at all. The addition of 15–25 at.% Pt to cobalt catalyst leads to an increase in activity by 3–5 times. The maximum yield of CNF reached 40 g/g$_{cat}$ for Co$_{0.75}$Pt$_{0.25}$ sample. The local composition of the active alloyed particles and the structural features of CNF were explored.

Keywords: cobalt; platinum; porous nanoalloys; X-ray diffraction; alloy catalysts; carbon nanofibers

1. Introduction

The production of composites based on carbon nanomaterials (CNMs) attracts great interest due to the continuous expansion of their practical application. In particular, much attention of researchers is focused on the development of metal-carbon M/CNM composites, which find application in various electrochemical reactions [1–4], catalytic processes [5–7] and adsorption [8,9]. Noble metals, such as Pt, Pd, Rh, and Ir, are often used as an active component within the composition of catalysts. Currently, the search for the ways of complete or partial replacement of such expensive metals with much cheaper analogues (Co, Ni, etc.) seems to be reasonable. At the same time, such aspects as dispersion of active particles, uniformity of their distribution within the carbon matrix, as well as structural and textural properties of CNM itself (specific surface area, bulk density, etc.) are important for the target synthesis of M/CNM composites.

The impregnation of the carbon support is the most often used method to prepare the M/CNM composites [4,10–12]. To obtain alloy particles consisting of several metals (M$_1$, M$_2$, etc.), the sequential impregnation of the support with solutions of the precursor salts can be used. However, this approach might lead to an irregular distribution of metals in the alloy particle's composition. A wide range of carbon nanomaterials, such as graphene [13], fullerenes [14], carbon nanofibers (CNFs) [1–5,7,10–12], and carbon nanotubes (CNTs) [15] could be used as a support. Nevertheless, a common disadvantage of such supported composites and catalysts is their susceptibility to rapid deactivation due to mechanical

Citation: Afonnikova, S.D.; Popov, A.A.; Bauman, Y.I.; Plyusnin, P.E.; Mishakov, I.V.; Trenikhin, M.V.; Shubin, Y.V.; Vedyagin, A.A.; Korenev, S.V. Porous Co-Pt Nanoalloys for Production of Carbon Nanofibers and Composites. *Materials* 2022, 15, 7456. https://doi.org/10.3390/ma15217456

Academic Editors: Klára Hernádi and Carlos Javier Duran-Valle

Received: 3 October 2022
Accepted: 21 October 2022
Published: 24 October 2022

Publisher's Note: MDPI stays neutral with regard to jurisdictional claims in published maps and institutional affiliations.

Copyright: © 2022 by the authors. Licensee MDPI, Basel, Switzerland. This article is an open access article distributed under the terms and conditions of the Creative Commons Attribution (CC BY) license (https://creativecommons.org/licenses/by/4.0/).

entrainment, washout, or migration of active component particles over the support's surface with a subsequent loss of dispersion (agglomeration).

The catalytic pyrolysis of hydrocarbons (or catalytic chemical vapor deposition, CCVD) can serve as an appropriate platform for the synthesis of M/CNM composites. Iron subgroup metals (Fe, Co, Ni) are the most commonly used catalysts for CNM synthesis, as well as their numerous alloys with other metals [16–19]. Among the metals of the iron triad, cobalt is probably the least studied. It is well known that the Co-based composite materials are of a great demand in different areas of catalysis and materials sciences. In particular, composite catalysts containing cobalt and its alloys are applied in such processes as Fischer–Tropsch synthesis [20,21], low-temperature CO oxidation [22], hydrogenation and dehydrogenation of organic substrates [23,24]. At the same time, it should be noted that, despite a fairly uniform distribution of particles within the structure of carbon support, cobalt-carbon composite catalysts can be deactivated due to sintering, surface coking, or encapsulation of the active particles in CNT channels [25].

The relatively new route based on the phenomenon of carbon erosion (CE) of bulk metals and alloys can be proposed as an alternative approach to the preparation of cobalt-carbon (Co/CNM) composites. In the course of the CE process, the spontaneous disintegration of a bulk alloy exposed to a carbon-containing atmosphere occurs. The fragmentation of bulk metal is accompanied by the emergence of active metal particles on which the graphite-like filaments grow [26–28]. As a result of the alloy disintegration, a multitude of similar sized active particles directly embedded into the structure of growing carbon nanofibers is formed. The disperse metallic crystals anchored within the bodies of CNF turned to be mechanically separated, which makes them absolutely resistant to migration and inevitable sintering. A significant advantage of CNF-based composites is in stabilization of Co-M alloy particles in the metallic state due to the tight contact with the carbon graphitic matrix, which prevents such catalyst from deactivation caused by chemical modification of active component (e.g., oxidation). The described effect was previously demonstrated for the case of carbon–carbon hierarchical Co-Cu/CNF/(Carbon Cloth) composites in which the fixed Co-Cu particles showed a higher productivity and stability in the ethanol dehydrogenation reaction [29].

It is known that the addition of such metals as Ag, Cu, Sn, and Pt might exert a considerable promoting and stabilizing effect on the performance of cobalt catalyst in the process of hydrocarbons decomposition to produce CNM [30–33]. In the case of modification with platinum, for instance, the resulting Co-Pt/CNM composite can be used as a catalyst for electrocatalytic reactions (similar to Pt/CNM catalysts), which would lead to a significant savings due to lower cost. In turn, the bulk Co-Pt alloy can be obtained using the previously developed method of a reductive thermolysis of multicomponent precursor salts [33]. This approach makes it possible to obtain single-phase porous nanoalloys with a developed specific surface area, which ensures their ability to interact rapidly with a carbon-containing gas and undergo disintegration with the subsequent growth of CNF. Similar nickel-based alloys have already exhibited their efficiency in the decomposition of various chlorinated hydrocarbons to produce CNFs [26,28].

Thus, the purpose of the present research was to investigate in detail the opportunity to produce Co-Pt/CNF composite materials based on spontaneous fragmentation of bulk Co-Pt nanoalloys used as catalysts. The phase composition and the structure of the obtained Co-Pt alloys were studied by X-ray diffraction analysis and electron microscopy techniques. The synthesized $Co_{1-x}Pt_x$ alloys of wide composition (x = 0.0–1.0) were subjected to carbon erosion in the course of catalytic pyrolysis of ethylene resulting in the growth of CNF serving as a carbon support. The influence of the Pt concentration in the alloy composition on the catalytic activity of Co in the CNF synthesis was explored. The initial stages of the formation of active particles were studied in detail by SEM, TEM, and XRD methods, and the local elemental composition of the obtained metallic particles (EDX) embedded in the structure of carbon nanofibers was determined.

2. Materials and Methods

2.1. Synthesis of Co-Pt Alloys

Co-Pt alloys were synthesized, as described elsewhere [33]. The complex salts [Co(NH$_3$)$_6$]Cl$_3$ (Reachem, Moscow, Russia), [Pt(NH$_3$)$_4$]Cl$_2$·H$_2$O (Aurat, Moscow, Russia), and acetone (puriss.) were used without further purification. In the typical synthesis, 6 mL of aqueous solution containing [Co(NH$_3$)$_6$]Cl$_3$ (0.223 g, 0.8 mmol) and [Pt(NH$_3$)$_4$]Cl$_2$·H$_2$O (0.293 g, 0.8 mmol) were added to 120 mL of cold acetone (~5 °C) with vigorous stirring. The precipitate was collected on a glass filter and dried in the air. The product weight was 0.490 g (95%). Then, the reductive thermolysis of the sample was carried out at 600 °C for 1 h in a hydrogen atmosphere. The thermolysis temperature was chosen based on the in situ XRD study of Co-Pt alloy formation performed in previous research [33]. Afterwards, the sample was cooled in a helium flow to room temperature.

2.2. Synthesis of Co-Pt/CNF Composites

Catalytic testing of the porous nanoalloy samples was carried out in a flow-through quartz horizontal reactor. Four specimens of the catalysts, 30.00 ± 0.02 mg each, were located one by one over a quartz plate. The plate was then placed inside the reactor installed in the high-temperature furnace (Zhengzhou Brother Furnace Co., Ltd., Zhengzhou, Henan, China), which is characterized by a stable temperature profile along the reactor length (±5 °C). Then, the samples were heated in an inert atmosphere (Ar) up to reaction temperature (600 °C). The heating rate was 10 °C/min. Next, H$_2$ was fed for 15 min in order to remove the oxide film from the surface of the alloy samples. Then, a gas mixture containing high-purity 40 vol.% ethylene (Nizhnekamskneftekhim, Nizhnekamsk, Russia), 20 vol.% hydrogen, and 40 vol.% argon was purged through the reactor at the same temperature. The flow rate of the reaction mixture was 54 L/h. At the end of the reaction, the obtained carbon product was cooled in an argon flow and unloaded from the reactor. After weighing the product, the carbon yield (g/g$_{cat}$) and the bulk density (g/L) of the resulting composite samples were calculated. The experiments were repeated several times to refine the data on the catalytic activity of the samples. The measurement error of the carbon yield was 10%.

2.3. Characterization of Co-Pt Alloys and Co-Pt/CNF Composites

The elemental analysis of the samples was performed by atomic emission spectrometry (AES) on a Thermo Scientific iCAP-6500 spectrometer (Thermo Scientific, Waltham, MA, USA). The sample was dissolved in aqua regia at heating and then evaporated with hydrochloric acid up to the complete removal of nitric acid. The relative standard deviation of the Co and Pt determination was 0.03.

X-ray powder diffraction (XRD) analysis of the alloy samples and the obtained carbon product was carried out on a Shimadzu XRD-7000 diffractometer (CuKα- radiation, graphite monochromator, 2Θ angles range of 20–100°, step 0.05°) (Shimadzu, Tokyo, Japan). The phases were identified by comparing the positions and intensities of the diffraction peaks with the PDF-2 data base [34]. Determination of the unit cell parameters was performed by the full-profile method using the PowderCell 2.4 program [35].

The microstructure of carbon product was studied by transmission electron microscopy (TEM) using a JEOL JEM 2100 microscope (JEOL Ltd., Tokyo, Japan) equipped with an energy dispersive X-ray spectrometer INCA-250 "Oxford Instruments" (Oxford Instruments, Abingdon, Oxfordshire, UK). The accelerating voltage was 200 kV, and the resolution was 0.14 nm. The samples were deposited on perforated carbon films fixed on a copper grid. An image of the crystal lattice of gold monocrystals with the Miller index (111)–0.235 nm was used as a standard for the linear size calibration on electron microscopic images. Computer processing of the obtained electron microscopic images was carried out using the Digital Micrograph "Gatan" program, as well as the FFT technique [36,37].

The alloys and the obtained carbon materials were studied by scanning electron microscopy (SEM) using a JSM-5100LV microscope (JEOL Ltd., Tokyo, Japan) equipped with an EX-23000BU EDX spectrometer (SPECTRO Analytical Instruments, Kleve, Germany). The operating voltage of the microscope was 15 kV, the magnification was in a range of 1000–100,000×.

The textural characteristics of the alloys and the carbon nanomaterials were determined using a Sorbtometer analyzer (PromEnergoLab, Moscow, Russia) by the Brunauer, Emmet, and Teller (BET) method. Nitrogen was used as an adsorbate. Before analysis, the samples were degassed in a helium flow at 150 °C for 10 min.

3. Results and Discussion

3.1. Characterization of Co-Pt Alloys

A series of Co-Pt alloys with a wide range of Pt concentration (0–100 at.%) was synthesized at the first stage of this work. Typical diffraction patterns of the obtained Co-Pt samples are shown in Figure 1. The diffraction patterns of the samples containing 15 and 35 at.% Pt exhibit a set of reflexes corresponding to the *fcc* phase based on the cubic modification of cobalt (*Fm3-m*, ICDD PDF-2 #15-0806), as seen in Figure 1. The positions of the characteristic peaks are shifted with respect to the reflexes for pure Co and Pt, thus indicating the formation of $Co_{1-x}Pt_x$ solid solutions. It should be noted that, besides the main set of reflexes attributed to the *fcc*-phase, there are also the low-intensity peaks at $2\Theta \sim 41$ and $46°$ observed on the diffraction patterns of the samples with the Pt content up to 15 at.% (Figure 1). These peaks are most likely related to the solid solution based on the hexagonal (*hcp*) modification of cobalt ($P6_3/mmc$, ICDD PDF-2 #05-0727).

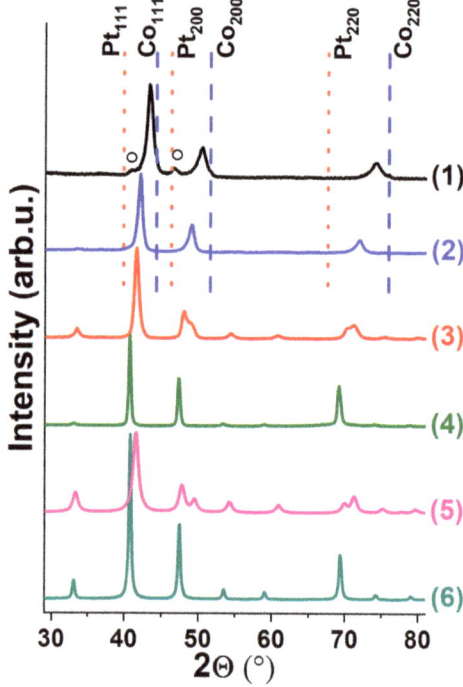

Figure 1. XRD patterns of Co-Pt samples: 1—15 at.% Pt; 2—35 at.% Pt; 3—50 at.% Pt; 4—75 at.% Pt; and the references for ordered phases: 5—CoPt (ICDD PDF-2 #65-8969); 6—$CoPt_3$ (ICDD PDF-2 #29-0499). °—reflexes of solid solution based on hexagonal modification of cobalt. The position of reflexes for 100 Co and 100 Pt are shown by dash lines for comparison.

Further increase in Pt concentration (>15 at.%) results in principle changes of the diffraction profiles. As follows from the XRD data, the Co-Pt sample (50 at.% Pt) is composed of a single tetragonal phase with a CoPt intermetallic structure (*P4/mmm*, ICDD PDF-2 #65-8969). The sample with a Pt content of 75 at.% is represented by the single-phase $CoPt_3$ intermetallic structure with a primitive cubic crystal lattice (*Pm3m*, ICDD PDF-2 #29-0499). The analysis of XRD data makes it possible to conclude that the phase composition of the synthesized samples corresponds well to the structure of the phase diagram known as the {Co-Pt} binary system [38].

The results of the SEM technique combined with the EDX mapping demonstrate a rather uniform distribution of cobalt and platinum atoms within the obtained alloys (Figure 2). The chemical composition of the synthesized samples determined by the AES method was found to be in good agreement with a target ratio Co/Pt specified during the synthesis (Table 1).

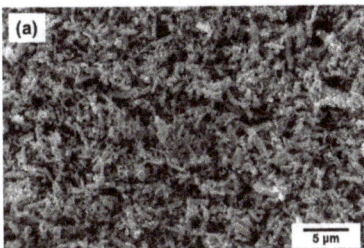 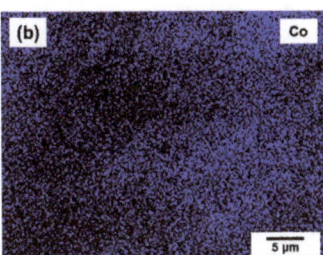 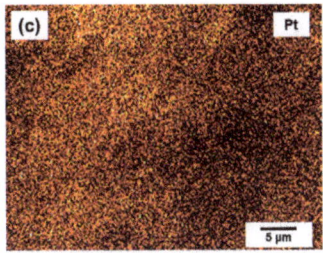

Figure 2. SEM micrograph and EDX element mapping for CoPt (50 at. % Pt) alloy. (**a**)—SEM data; (**b**)—EDX mapping of Co; (**c**)—EDX mapping of Pt.

Table 1. Chemical compositions of the synthesized Co-Pt alloys.

Sample	$Co_{0.95}Pt_{0.05}$	$Co_{0.90}Pt_{0.10}$	$Co_{0.85}Pt_{0.15}$	$Co_{0.75}Pt_{0.25}$	$Co_{0.65}Pt_{0.35}$	CoPt	$CoPt_3$
Pt (preset), at.%	5	10	15	25	35	50	75
Pt (AES), at.%	4.4 (3)	9 (1)	14 (1)	23 (2)	34 (3)	47 (3)	73 (5)

The as-prepared Co-Pt alloys are microdispersed powders of dark grey color. Figure 3 demonstrates the SEM data for the Co-Pt sample with different content of Pt (25, 50, and 75 at.%). It is clear from the SEM image taken at a low magnification (Figure 3a) that the Co-Pt alloys are represented by agglomerates of different size. There are both the small (<10 μm) and large (>100 μm) particles. Further investigation of the secondary structure of the samples by SEM showed that these agglomerates have pores (and possess a sponge-like structure consisting of polycrystalline fragments (50–150 nm in size) interconnected by crosspieces. It is possible to see from the images presented in Figure 3c–h that there are no principle differences in the secondary structure for the Co-Pt alloys with different concentration of Pt dopant. Nevertheless, the sample with high concentration of Pt (75 at.) is seen to have the thinnest structure and smallest size of structural elements (Figure 3g,h). It is worth noting that such a morphology is rather typical for the alloys obtained by the reductive thermolysis of multicomponent precursors [26,28].

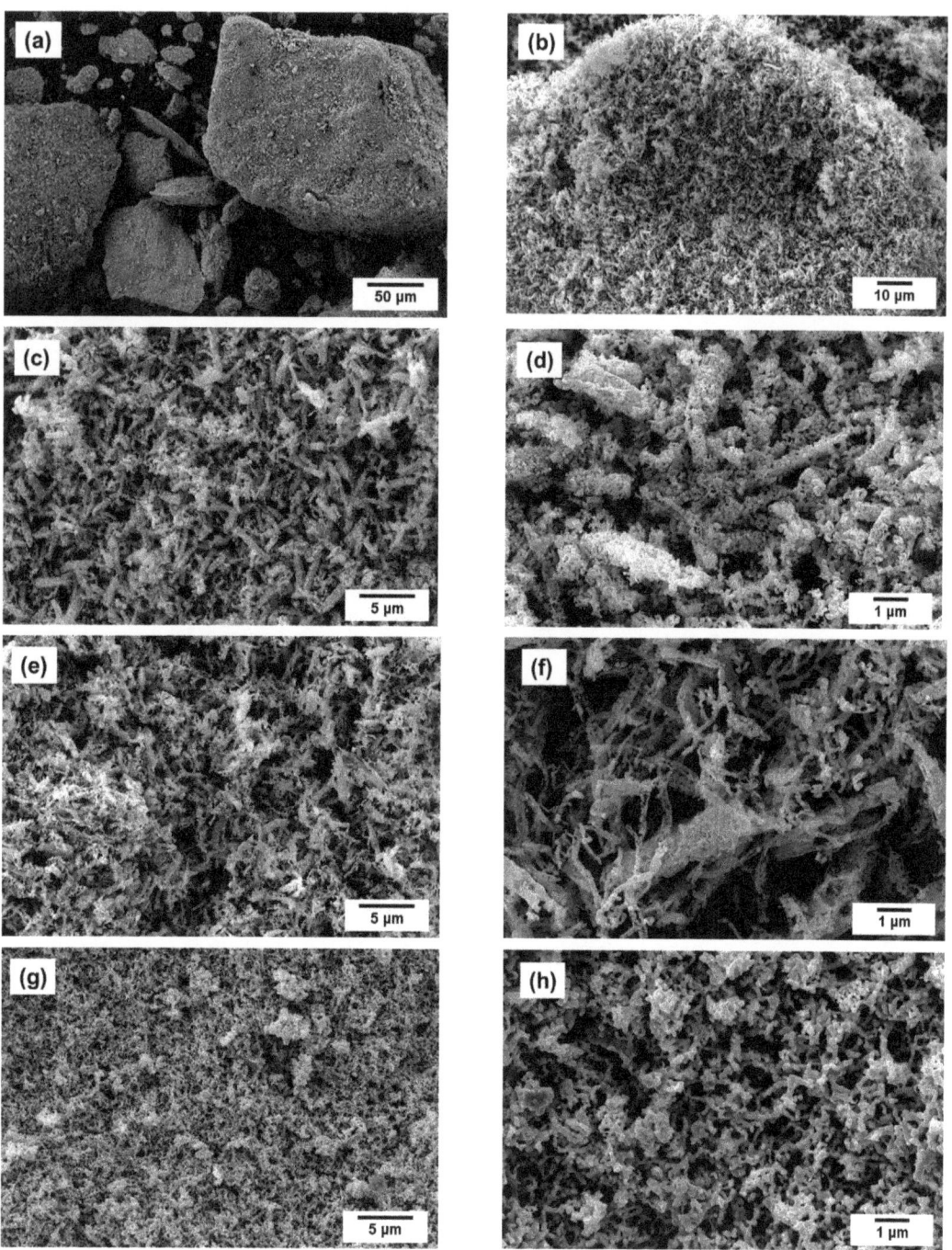

Figure 3. SEM micrographs of the CoPt alloy: (**a–d**)—50 at.% Pt; (**e,f**)—25 at.% Pt; (**g,h**)—75 at.% Pt.

Selected TEM images of the bimetallic samples are presented separately for a better demonstration of their primary structure (Figure 4a–c). As such, individual particles are poorly pronounced; only the individual irregularly shaped fragments can be seen. The size of the fragments does not exceed 150 nm due to the specific synthesis procedure. As it was shown earlier, the size of the crosspieces is determined by the temperature and the duration of the reducing stage of thermolysis [39]. It is also possible to observe the presence of a thin layer on the surface of the alloys, which presumably belongs to the oxide shell (Figure 4b–e), resulting from the interaction of the metallic surface with an atmospheric air. It should be noted that the thickness of this layer becomes smaller with an increase in Pt content, which is a well-known effect for alloys in general. Thus, the addition of an inert component to the composition of the alloy increases its resistance to oxidation [40]. The higher the content of the noble metal (Pt) in the alloy, the more resistant the Co-Pt alloy to oxidation. One can see from Figure 4d that the oxide layer on the alloy surface is practically absent in case of the $CoPt_3$ sample (75 at.% Pt). The specific surface area of the alloy samples determined by the BET method is varied within a range of 5–10 m^2/g. The porous structure along with the rather developed surface area makes such alloys more reactive and "qualified" for the process of carbon erosion during the contact with hydrocarbons (ethylene).

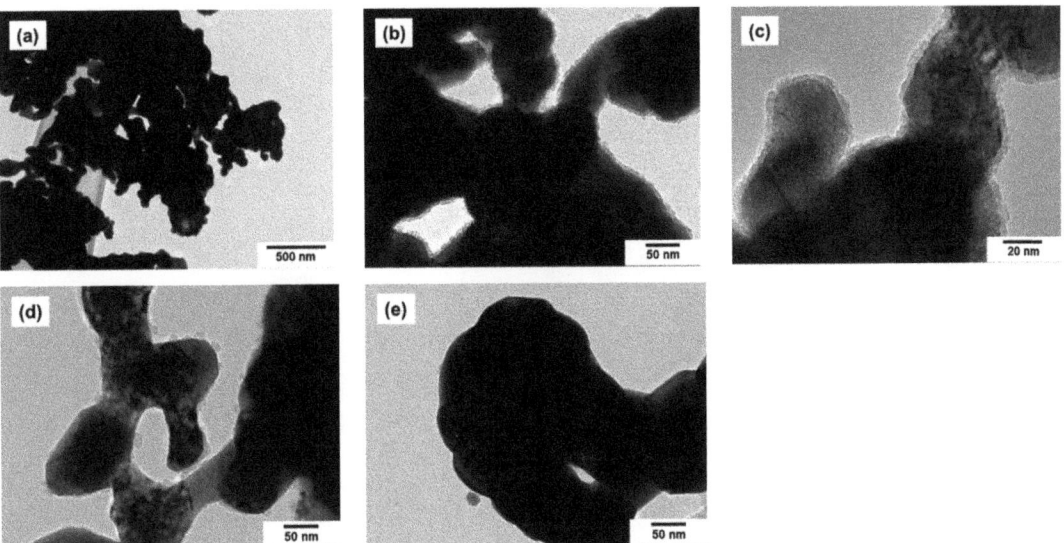

Figure 4. TEM images of Co-Pt alloy samples: 25 at.% Pt (**a–c**), 50 at.% Pt (**d**), 75 at.% Pt (**e**).

3.2. Catalytic Decomposition of Ethylene on Co-Pt Alloys

The obtained samples of Co-Pt alloys are not catalysts themselves but play the role of their precursors. The interaction of the alloys with the carbon-containing atmosphere at high temperature (400–800 °C) results in their disintegration and consequently leads to an appearance of the active particles catalyzing the growth of CNF. The process of CE of the alloys occurs very rapidly due to the developed surface of the initial Co-Pt alloys referred to as catalyst precursors.

Figure 5 shows the methodology of catalytic tests carried out in a laboratory reactor. Decomposition of ethylene on a metallic catalyst with formation of CNF proceeds according to the reaction:

$$C_2H_4 \rightarrow 2C\ (CNF) + 2H_2$$

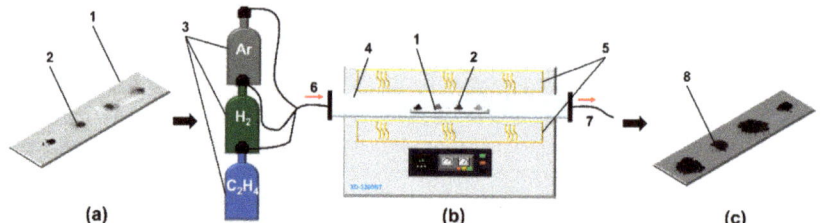

Figure 5. Schematic of catalytic testing of the Co-Pt samples. (**a**)—Co-Pt alloy samples on quartz plate before testing; (**b**)—horizontal flow quartz reactor; (**c**)—CNF samples produced via ethylene decomposition. 1—quartz plate for samples; 2—samples of alloys; 3—gases; 4—quartz reactor; 5—heating element; 6—input of reaction mixture ($C_2H_4/H_2/Ar$); 7—outlet of gaseous products; 8—obtained samples of composites.

During the experiment, an additional hydrogen is fed along with C_2H_4 in order to accelerate the CE process [41].

The presented testing procedure allows one to acquire the catalytic data for several samples at the same time during the single experiment, which makes it possible to rank the samples according to their catalyst performance. The results of the catalytic tests are summarized in the diagram in terms of their productivity (Figure 6). It shows the amount of carbon product (g) grown per 1 g of the alloy during the experiment of 30 min.

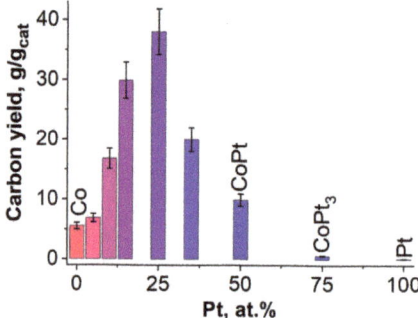

Figure 6. Dependence of the carbon yield on the platinum content in the ethylene decomposition reaction. The data for Co (100%) and Pt (100%) are given for comparison. The reaction conditions are as follows: $C_2H_4/H_2/Ar$, T = 600 °C, 30 min.

It can be seen that the reference samples (100% Co and 100% Pt) are characterized by a very low productivity. For the pure Co-catalyst, the yield of CNF does not exceed the value of 5 g/g_{cat}, which is most likely associated with a rapid deactivation of cobalt particles due to their encapsulation with amorphous carbon [42]. It is also clear that the Pt itself (which is supposed to play the role of an activating additive [43]) does not show any activity at all (yield of carbon ~0 g/g_{cat}). This result is quite expected since pure Pt does not belong to the family of metals capable of catalyzing the growth of carbon nanostructures [27].

The analysis of the catalytic testing results revealed that the Pt concentration range of 15–50 at.% in Co-Pt alloy seems to be the most interesting from a practical point of view. It can be seen that inside this interval, there is a significant increase in the catalytic performance of the system. The addition of Pt in the amount of 15 at.% leads to a significant increase in carbon yield if compared with the reference Co (Figure 6). A further increase in Pt concentration gives even greater gain in productivity. The sample of $Co_{0.75}Pt_{0.25}$ alloy was found to exhibit the highest productivity among the tested samples (40 g/g_{cat}). Thus, as a result, it is possible to obtain a composite with the content of Co-Pt active particles ranging from 2.5 to 10 wt%.

At the same time, a continuous increase in Pt concentration (over 25 at.%) results in a drastic worsening of the catalytic performance. For example, the sample with an equimolar ratio of Co and Pt (50/50) shows the same level of productivity as the reference Co sample (< 10 g/g$_{cat}$). Further increase in Pt content up to 75 at.% turns the Co-Pt catalyst to almost inactive form.

When comparing the obtained results with the literature data, it can be concluded that the catalytic behavior of the Co-Pt samples resembles that of the Ni-Pt system previously studied in decomposition of chlorinated hydrocarbons [44]. In particular, it was found that the performance of Ni-Pt catalysts depending on the Pt content passes through a maximum of 27 g/g$_{cat}$ at 4.3 wt % Pt [44]. This can be explained by a dual action of Pt. On the one hand, Pt enhances the catalytic ability of Co to decompose hydrocarbons. On the other hand, the presence of platinum in high concentration inhibits further diffusion of carbon atoms through the bulk of the catalytic Co-Pt particles.

3.3. Study of the Carbon Erosion Process of Co-Pt Alloys

The samples of the obtained composite were characterized by XRD, SEM and TEM methods. Figure 7 shows the change in the phase composition of the Co$_{0.85}$Pt$_{0.15}$ sample in the course of reaction (before the interaction with the reaction medium, as well as after 6 and 30 min of reaction). In comparison with the pristine sample (diffraction pattern 1), the reflections at 2Θ = 26.2°, 44.7°, 54.0°, 77.0° appeared in pattern 2, corresponding to the family of graphite crystal reflecting planes (002), (101), (004), (110) correspondingly. The marked reflections become more intensive (pattern 3) as the carbon content in the sample increases. At the same time, a decrease in the intensities of the reflexes related to the initial alloy can be observed.

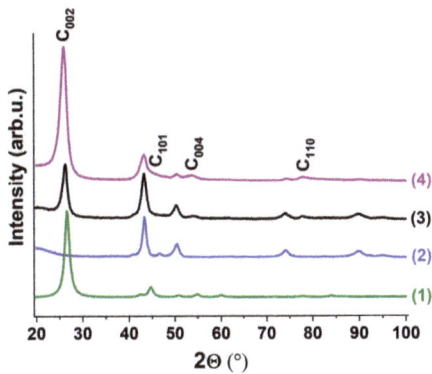

Figure 7. XRD patterns of graphite (reference, ICDD PDF-2 #41-1487) (1), Co$_{0.85}$Pt$_{0.15}$ alloy before (2) and after interaction with reaction mixture: (3)—6 min; (4)—30 min. C$_2$H$_4$/H$_2$/Ar, T = 600 °C.

It is obvious from the SEM data that the interaction of Co-Pt alloys with the reaction mixture is accompanied by their fragmentation (Figures 3 and 4), which is accomplished by the emergence of numerous dispersed particles (Figure 8a,b) playing the role of catalytically active centers for the CNF synthesis. No significant effect of the Pt content on the thickness of the obtained carbon filaments (200–500 nm) was revealed. However, in the case of pure Co, all the grown carbon filaments appeared to be very short, and their average length did not exceed 2 µm. It can be seen that all the observed catalytic particles are covered with a layer of carbonaceous deposits completely preventing them from further interaction with ethylene. Thus, low productivity of pure cobalt should be explained by a rapid deactivation of the as-formed active particles (Figure 6). In Figure 9a,b one can see the TEM images showing the characteristic appearance of the particles responsible for the growth of carbon nanofibers. The observed data also confirm that the active particles are rather quickly "shrouded" by the carbon deposits, which eventually leads to their deactivation.

Figure 8. SEM micrographs for Co (100%) (**a,b**) and Pt (100%) (**c,d**) reference samples after being exposed to reaction mixture ($C_2H_4/H_2/Ar$) for 30 min.

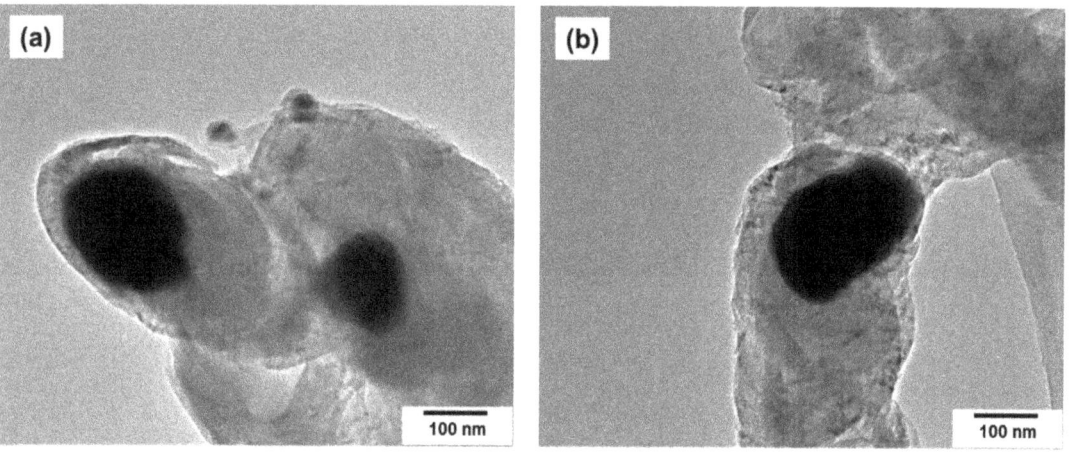

Figure 9. TEM data for the comparison sample Co (100%). The reaction conditions are as follows: $C_2H_4/H_2/Ar$, T = 600 °C, 30 min, (**a,b**)—active particles.

The result of the interaction of pure 100% Pt sample with ethylene is presented in Figure 8c,d. As follows from the SEM data, there is no evidence of the sample's destruction forced by the CE process. After being exposed to 30 min of contact with the reaction mixture, the sample is seen to be covered by a thin layer of non-catalytic carbon, fully retaining its original morphology.

The general view and the structure of CNF product obtained over the most active Co-Pt alloy samples are presented in Figure 10. It can be seen that a classical carbon nanomaterial of a filamentous morphology is formed in this case. It should be stressed that the filamentous morphology of carbon product is predominant. In contrast to the reference Co sample, the catalyst promoted by the addition of Pt produces significantly longer carbon filaments (Figure 10a,b). With increasing Pt concentration, the resulting product is represented by very long and less tangled carbon nanofibers (Figure 10c,d). According to the microscopic data taken in the back-scattered electron mode (Figure 10e,f), it is clear that the metallic particles are embedded to the structure of carbon filaments. The majority of particles generate the CNF growth in two opposite directions.

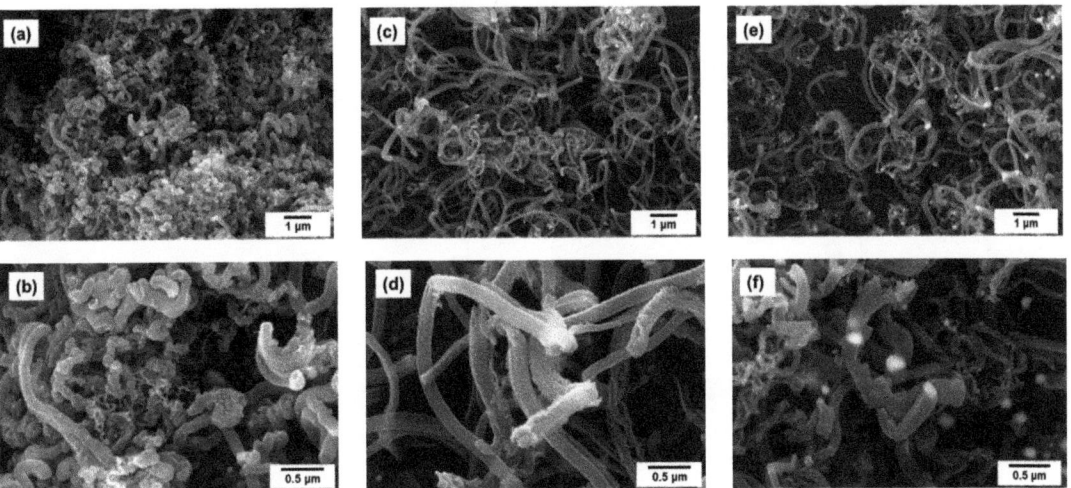

Figure 10. SEM micrographs of composites obtained on Co-Pt alloys: 25 at.% Pt (**a,b**), 50 at.% Pt (**c,d**), 75 at.% Pt (**e,f**) back-scattered electron beam mode. The reaction conditions are as follows: $C_2H_4/H_2/Ar$, T = 600 °C, 30 min.

Examination of the obtained composites under transmission microscope reveals that the surface of the active particles is open and free from amorphous carbon deposits (Figure 11a–d). Thus, the proposed method makes it possible to synthesize in one step the composite systems containing Co-Pt alloy particles distributed within the structure of the carbon support. It is important to note that the peculiarity of this approach is the use of a bulk alloy that is capable of growing the carbon product, which further bears the function of a support for the active particles. As a result, one can get a nano-dispersed fixed catalyst where Co-Pt alloy particles of desired composition are separated from each other and physically incapable of agglomeration. Such properties make this material attractive for a wide range of applications.

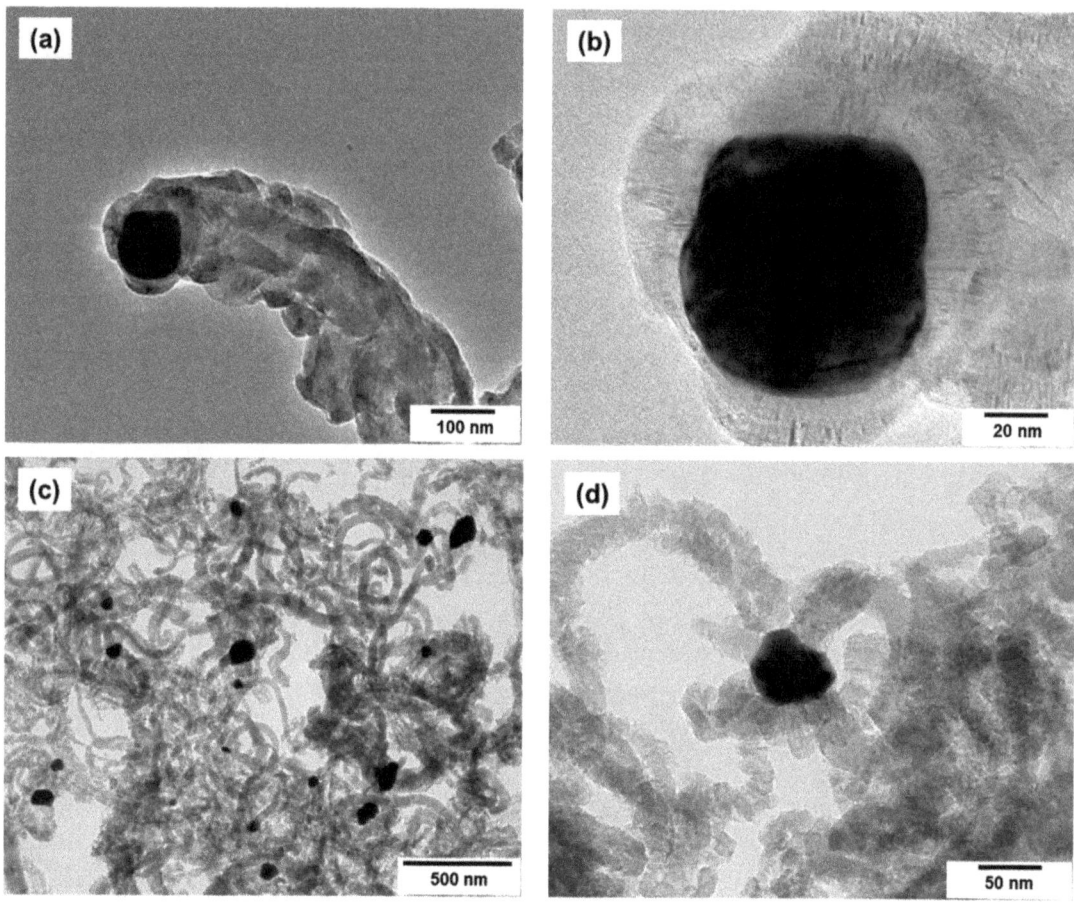

Figure 11. TEM images of Co-Pt/CNF composite samples: (**a,b**)—25 at.% Pt; (**c,d**)—35 at.% Pt. The reaction conditions are as follows: $C_2H_4/H_2/Ar$, T = 600 °C, 30 min.

Then, the character of the Pt distribution throughout the obtained composite (as well as in the initial alloy) was explored and compared. As seen from the EDX data obtained for a number of samples, both metals are evenly distributed in the structure of alloyed particles. Hence, there is no redistribution of metal atoms during the process of carbon erosion of $Co_{0.75}Pt_{0.25}$ alloy, which agrees well with the XRD data. The formed active particles are characterized by the chemical composition similar to that of the starting alloy. This allows one to assert that Pt is not inclined to redistribution during the CE process, in contrast to, for example, Cr and Mo present in Ni-Cr and Ni-Mo alloys [45,46].

Figure 12 demonstrates the TEM-EDX data, according to which platinum and cobalt atoms are both localized within the active particles. This fact is also confirmed by the EDX results for the active particles obtained in course of the interaction of CoPt (50 at.% Pt) with ethylene for 30 min (Table 2). Taking into account the experimental error of the EDX method, one can conclude that the ratio of Co/Pt remains the same as for the initial alloy (close to 50/50).

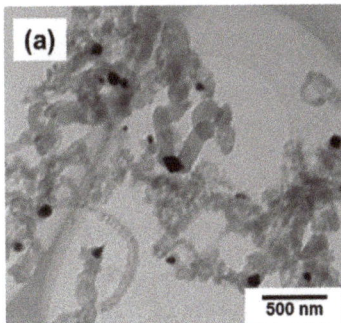 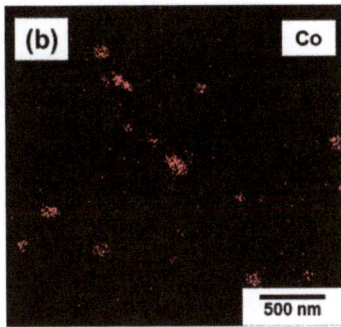 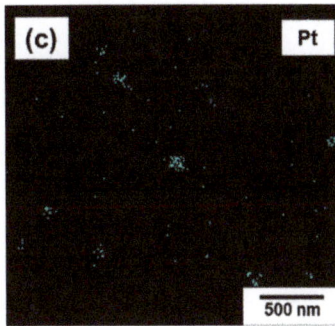

Figure 12. EDX data for $Co_{0.75}Pt_{0.25}$ sample after interaction with its reaction mixture ($C_2H_4/H_2/Ar$, T = 600 °C) for 30 min. (**a**)—image of CNF, (**b**)—distribution of Co atoms (**c**)—distribution of Pt atoms.

Table 2. EDX data for CoPt (50 at.% Pt) at different areas after being exposed to $C_2H_4/H_2/Ar$ reaction mixture at T = 600 °C for 30 min.

Spectrum	Co, at.%	Pt, at.%
1	46.3	53.7
2	56.7	43.3
3	45.4	54.6
4	52.3	47.7
5	45.4	54.6
Average value	49.2	50.8

4. Conclusions

The method proposed in the present work permits one to synthesize a number of single-phase Co-Pt alloys serving as catalyst precursors. The influence of Pt concentration on the ability of Co-Pt alloys to disintegrate under the action of CE under the reaction conditions of ethylene decomposition was established. The addition of Pt in an amount of 15–50 at.% was found to boost the catalytic performance of Co with respect to CNF synthesis (increase of ~8 times). As a result, the Co-Pt/CNF composites containing metallic particles in a concentration of 2.5–10 wt% can be produced. The ratio of Co and Pt in the obtained fixed particles correspond to the value that was present during the synthesis of the initial alloys. The synthesized Co-Pt/CNF composite materials attract great interest for application in the area of heterogeneous catalysis, including electrocatalytic applications. In particular, this approach makes it possible to reduce the Pt content by replacing it with Co and thus reducing the cost of the Pt-based materials.

Author Contributions: Conceptualization, I.V.M. and S.D.A.; methodology, A.A.P., Y.I.B. and M.V.T.; investigation, Y.I.B., A.A.P., M.V.T. and P.E.P.; writing—original draft preparation, S.D.A. and A.A.P.; writing—review and editing, I.V.M., Y.V.S., S.V.K. and A.A.V.; visualization, S.D.A. and A.A.P.; supervision, Y.V.S., A.A.V. and S.V.K.; funding acquisition, Y.V.S. All authors have read and agreed to the published version of the manuscript.

Funding: This work was financially supported by the Ministry of Science and Higher Education of the Russian Federation (project No. AAAA-A21-121011390054-1). The synthesis and X-ray diffraction analysis of the starting materials were supported by the Russian Science Foundation (project No. 21-13-00414), https://rscf.ru/en/project/21-13-00414/, NIIC SB RAS.

Institutional Review Board Statement: Not applicable.

Informed Consent Statement: Not applicable.

Data Availability Statement: Data are contained within the article.

Acknowledgments: Analysis of the physicochemical properties of the samples was performed using the equipment of the "National Center for Catalyst Research" and the Omsk Regional Center for Collective Use of the Siberian Branch of the Russian Academy of Sciences.

Conflicts of Interest: The authors declare that they have no conflict of interest.

References

1. Bui, H.T.; Kim, D.Y.; Kim, D.W.; Suk, J.; Kang, Y. Carbon nanofiber@platinum by a coaxial electrospinning and their improved electrochemical performance as a Li−O_2 battery cathode. *Carbon* **2018**, *130*, 94–104. [CrossRef]
2. Huang, J.; Wang, D.; Hou, H.; You, T. Electrospun Palladium Nanoparticle-Loaded Carbon Nanofibers and Their Electrocatalytic Activities towards Hydrogen Peroxide and NADH. *Adv. Funct. Mater.* **2008**, *18*, 441–448. [CrossRef]
3. Karuppannan, M.; Kim, Y.; Gok, S.; Lee, E.; Hwang, J.Y.; Jang, J.-H.; Cho, Y.-H.; Lim, T.; Sung, Y.-E.; Kwon, O.J. A highly durable carbon-nanofiber-supported Pt–C core–shell cathode catalyst for ultra-low Pt loading proton exchange membrane fuel cells: Facile carbon encapsulation. *Energy Environ. Sci.* **2019**, *12*, 2820–2829. [CrossRef]
4. Lee, C.-H.; Park, H.-N.; Lee, Y.-K.; Chung, Y.S.; Lee, S.; Joh, H.-I. Palladium on yttrium-embedded carbon nanofibers as electrocatalyst for oxygen reduction reaction in acidic media. *Electrochem. Commun.* **2019**, *106*, 106516. [CrossRef]
5. Chen, W.; Chen, S.; Qian, G.; Song, L.; Chen, D.; Zhou, X.; Duan, X. On the nature of Pt-carbon interactions for enhanced hydrogen generation. *J. Catal.* **2020**, *389*, 492–501. [CrossRef]
6. Ledoux, M.-J.; Pham-Huu, C. Carbon nanostructures with macroscopic shaping for catalytic applications. *Catal. Today* **2005**, *102–103*, 2–14. [CrossRef]
7. Maiyalagan, T. Pt–Ru nanoparticles supported PAMAM dendrimer functionalized carbon nanofiber composite catalysts and their application to methanol oxidation. *J. Solid State Electrochem.* **2008**, *13*, 1561–1566. [CrossRef]
8. Pelech, R.; Milchert, E.; Wrobel, R. Adsorption dynamics of chlorinated hydrocarbons from multi-component aqueous solution onto activated carbon. *J. Hazard Mater.* **2006**, *137*, 1479–1487. [CrossRef]
9. Taboada, C.D.; Batista, J.; Pintar, A.; Levec, J. Preparation, characterization and catalytic properties of carbon nanofiber-supported Pt, Pd, Ru monometallic particles in aqueous-phase reactions. *Appl. Catal. B Environ.* **2009**, *89*, 375–382. [CrossRef]
10. Jiménez, V.; Jiménez-Borja, C.; Sánchez, P.; Romero, A.; Papaioannou, E.I.; Theleritis, D.; Souentie, S.; Brosda, S.; Valverde, J.L. Electrochemical promotion of the CO_2 hydrogenation reaction on composite Ni or Ru impregnated carbon nanofiber catalyst-electrodes deposited on YSZ. *Appl. Catal. B Environ.* **2011**, *107*, 210–220. [CrossRef]
11. Liang, C.; Xia, W.; van den Berg, M.; Wang, Y.; Soltani-Ahmadi, H.; Schlüter, O.; Fischer, R.A.; Muhler, M. Synthesis and Catalytic Performance of Pd Nanoparticle/Functionalized CNF Composites by a Two-Step Chemical Vapor Deposition of Pd(allyl)(Cp) Precursor. *Chem. Mater.* **2009**, *21*, 2360–2366. [CrossRef]
12. Zhou, J.H.; Zhang, M.G.; Zhao, L.; Li, P.; Zhou, X.G.; Yuan, W.K. Carbon nanofiber/graphite-felt composite supported Ru catalysts for hydrogenolysis of sorbitol. *Catal. Today* **2009**, *147*, S225–S229. [CrossRef]
13. Kumar, R.; Malik, S.; Mehta, B.R. Interface induced hydrogen sensing in Pd nanoparticle/graphene composite layers. *Sens. Actuators B Chem.* **2015**, *209*, 919–926. [CrossRef]
14. Zhang, X.; Ma, J.; Yan, R.; Cheng, W.; Zheng, J.; Jin, B. Pt-Ru/polyaniline/carbon nanotube composites with three-layer tubular structure for efficient methanol oxidation. *J. Alloys Compd.* **2021**, *867*, 159017. [CrossRef]
15. Zhu, Y.; Huai, S.; Jiao, J.; Xu, Q.; Wu, H.; Zhang, H. Fullerene and platinum composite-based electrochemical sensor for the selective determination of catechol and hydroquinone. *J. Electroanal. Chem.* **2020**, *878*, 114726. [CrossRef]
16. Ahmad, S.; Liao, Y.; Hussain, A.; Zhang, Q.; Ding, E.-X.; Jiang, H.; Kauppinen, E.I. Systematic investigation of the catalyst composition effects on single-walled carbon nanotubes synthesis in floating-catalyst CVD. *Carbon* **2019**, *149*, 318–327. [CrossRef]
17. Ghaemi, F.; Ali, M.; Yunus, R.; Othman, R.N. Chapter 1—Synthesis of Carbon Nanomaterials Using Catalytic Chemical Vapor Deposition Technique. In *Synthesis, Technology and Applications of Carbon Nanomaterials*; Rashid, S.A., Raja Othman, R.N.I., Hussein, M.Z., Eds.; Elsevier: Amsterdam, The Netherlands, 2019; pp. 1–27.
18. Narkiewicz, U.; Podsiadły, M.; Jędrzejewski, R.; Pełech, I. Catalytic decomposition of hydrocarbons on cobalt, nickel and iron catalysts to obtain carbon nanomaterials. *Appl. Catal. A Gen.* **2010**, *384*, 27–35. [CrossRef]
19. Zhou, X.; Liu, B.; Chen, Y.; Guo, L.; Wei, G. Carbon nanofiber-based three-dimensional nanomaterials for energy and environmental applications. *Mater. Adv.* **2020**, *1*, 2163–2181. [CrossRef]
20. Bezemer, G.L.; Radstake, P.B.; Koot, V.; van Dillen, A.J.; Geus, J.W.; de Jong, K.P. Preparation of Fischer–Tropsch cobalt catalysts supported on carbon nanofibers and silica using homogeneous deposition-precipitation. *J. Catal.* **2006**, *237*, 291–302. [CrossRef]
21. Fu, T.; Jiang, Y.; Lv, J.; Li, Z. Effect of carbon support on Fischer–Tropsch synthesis activity and product distribution over Co-based catalysts. *Fuel Process. Technol.* **2013**, *110*, 141–149. [CrossRef]
22. Ferencz, Z.; Erdőhelyi, A.; Baán, K.; Oszkó, A.; Óvári, L.; Kónya, Z.; Papp, C.; Steinrück, H.P.; Kiss, J. Effects of Support and Rh Additive on Co-Based Catalysts in the Ethanol Steam Reforming Reaction. *ACS Catal.* **2014**, *4*, 1205–1218. [CrossRef]
23. Firdous, N.; Janjua, N.K.; Wattoo, M.H.S. Promoting effect of ruthenium, platinum and palladium on alumina supported cobalt catalysts for ultimate generation of hydrogen from hydrazine. *Int. J. Hydrogen Energy* **2020**, *45*, 21573–21587. [CrossRef]
24. Gallezot, P.; Richard, D. Characterization and selectivity in cinnamaldehyde hydrogenation of graphitesupported platinum catalysts prepared from a zero-valent platinum complex. *Catal. Rev. Sci. Eng.* **1998**, *40*, 81–126. [CrossRef]

25. Chen, W.; Pan, X.; Bao, X. Tuning of Redox Properties of Iron and Iron Oxides via Encapsulation within Carbon Nanotubes. *J. Am. Chem. Soc.* **2007**, *129*, 7421–7426. [CrossRef]
26. Bauman, Y.I.; Mishakov, I.V.; Vedyagin, A.A.; Ramakrishna, S. Synthesis of bimodal carbon structures via metal dusting of Ni-based alloys. *Mater. Lett.* **2017**, *201*, 70–73. [CrossRef]
27. Grabke, H.J. Metal dusting. *Mater. Corros.* **2003**, *54*, 736–746. [CrossRef]
28. Mishakov, I.V.; Bauman, Y.I.; Korneev, D.V.; Vedyagin, A.A. Metal dusting as a route to produce active catalyst for processing chlorinated hydrocarbons into carbon nanomaterials. *Top. Catal.* **2013**, *56*, 1026–1032. [CrossRef]
29. Ponomareva, E.A.; Krasnikova, I.V.; Egorova, E.V.; Mishakov, I.V.; Vedyagin, A.A. Dehydrogenation of ethanol over carbon-supported Cu–Co catalysts modified by catalytic chemical vapor deposition. *React. Kinet. Mech. Catal.* **2017**, *122*, 399–408. [CrossRef]
30. Chambers, A.; Rodriguez, N.M.; Baker, R.T.K. Influence of silver addition on the catalytic behavior of cobalt. *J. Phys. Chem.* **1996**, *100*, 4229–4236. [CrossRef]
31. Chambers, A.; Rodriguez, N.M.; Baker, R.T.K. Influence of copper on the structural characteristics of carbon nanofibers produced from the cobalt-catalyzed decomposition of ethylene. *J. Mater. Res.* **1996**, *11*, 430–438. [CrossRef]
32. Nemes, T.; Chambers, A.; Baker, R.T.K. Characteristics of carbon filament formation from the interaction of Cobalt– Tin particles with ethylene. *J. Phys. Chem. B* **1998**, *102*, 6323–6330. [CrossRef]
33. Popov, A.A.; Shubin, Y.V.; Bauman, Y.I.; Plyusnin, P.E.; Mishakov, I.V.; Sharafutdinov, M.R.; Maksimovskiy, E.A.; Korenev, S.V.; Vedyagin, A.A. Preparation of porous Co-Pt alloys for catalytic synthesis of carbon nanofibers. *Nanotechnology* **2020**, *31*, 495604. [CrossRef] [PubMed]
34. International Centre for Diffraction Data; Joint Committee on Powder Diffraction Standards. *Powder Diffraction File*; International Centre for Diffraction Data: Newtown Square, PA, USA, 2009.
35. Kraus, W.; Nolze, G. PowderCell—A program to visualize crystal structures, calculate the corresponding powder patterns and refine experimental curves. *J. Appl. Cryst.* **1996**, *29*, 301–303. [CrossRef]
36. Shim, H.-S.; Hurt, R.H.; Yang, N.Y.C. A methodology for analysis of 002 lattice fringe images and its application to combustion-derived carbons. *Carbon* **2000**, *38*, 29–45. [CrossRef]
37. Zhu, W.; Miser, D.E.; Chan, W.G.; Hajaligol, M.R. HRTEM investigation of some commercially available furnace carbon blacks. *Carbon* **2004**, *42*, 1841–1845. [CrossRef]
38. Massalski, T.B.; Okamoto, H.; Subramanian, P.R.; Kacprzak, L. *Binary Alloy Phase Diagrams*; ASM International: Almere, The Netherlands, 1990.
39. Shubin, Y.; Plyusnin, P.; Sharafutdinov, M. In situ synchrotron study of Au–Pd nanoporous alloy formation by single-source precursor thermolysis. *Nanotechnology* **2012**, *23*, 405302. [CrossRef]
40. Corti, C.W.; Coupland, D.R.; Selman, G.L. Platinum-Enriched Superalloys. *Platin. Met. Rev.* **1980**, *24*, 2–11.
41. Vlassiouk, I.; Regmi, M.; Fulvio, P.; Dai, S.; Datskos, P.; Eres, G.; Smirnov, S. Role of hydrogen in chemical vapor deposition growth of large single-crystal graphene. *ACS Nano* **2011**, *5*, 6069–6076. [CrossRef]
42. Henao, W.; Cazaña, F.; Tarifa, P.; Romeo, E.; Latorre, N.; Sebastian, V.; Delgado, J.J.; Monzón, A. Selective synthesis of carbon nanotubes by catalytic decomposition of methane using Co-Cu/cellulose derived carbon catalysts: A comprehensive kinetic study. *Chem. Eng. J.* **2021**, *404*, 126103. [CrossRef]
43. Choi, D.S.; Robertson, A.W.; Warner, J.H.; Kim, S.O.; Kim, H. Low-temperature chemical vapor deposition synthesis of Pt–Co alloyed nanoparticles with enhanced oxygen reduction reaction catalysis. *Adv. Mater.* **2016**, *28*, 7115–7122. [CrossRef]
44. Bauman, Y.I.; Mishakov, I.V.; Rudneva, Y.V.; Popov, A.A.; Rieder, D.; Korneev, D.V.; Serkova, A.N.; Shubin, Y.V.; Vedyagin, A.A. Catalytic synthesis of segmented carbon filaments via decomposition of chlorinated hydrocarbons on Ni-Pt alloys. *Catal. Today* **2020**, *348*, 102–110. [CrossRef]
45. Bauman, Y.I.; Mishakov, I.V.; Vedyagin, A.A.; Rudnev, A.V.; Plyusnin, P.E.; Shubin, Y.V.; Buyanov, R.A. Promoting effect of Co, Cu, Cr and Fe on activity of Ni-based alloys in catalytic processing of chlorinated hydrocarbons. *Top. Catal.* **2017**, *60*, 171–177. [CrossRef]
46. Bauman, Y.I.; Rudneva, Y.V.; Mishakov, I.V.; Plyusnin, P.E.; Shubin, Y.V.; Korneev, D.V.; Stoyanovskii, V.O.; Vedyagin, A.A.; Buyanov, R.A. Effect of Mo on the catalytic activity of Ni-based self-organizing catalysts for processing of dichloroethane into segmented carbon nanomaterials. *Heliyon* **2019**, *5*, e02428. [CrossRef] [PubMed]

Article

Effect of Pretreatment with Acids on the N-Functionalization of Carbon Nanofibers Using Melamine

Tatyana A. Maksimova [1], Ilya V. Mishakov [1], Yury I. Bauman [1], Artem B. Ayupov [1,2], Maksim S. Mel'gunov [1,2], Aleksey M. Dmitrachkov [1,2], Anna V. Nartova [1,2], Vladimir O. Stoyanovskii [1] and Aleksey A. Vedyagin [1,*]

[1] Boreskov Institute of Catalysis, 630090 Novosibirsk, Russia
[2] Department of Natural Sciences, Novosibirsk State University, Pirogova Str. 2, 630090 Novosibirsk, Russia
* Correspondence: vedyagin@catalysis.ru

Abstract: Nowadays, N-functionalized carbon nanomaterials attract a growing interest. The use of melamine as a functionalizing agent looks prospective from environmental and cost points of view. Moreover, the melamine molecule contains a high amount of nitrogen with an atomic ratio C/N of 1/2. In present work, the initial carbon nanofibers (CNFs) were synthesized via catalytic pyrolysis of ethylene over microdispersed Ni–Cu alloy. The CNF materials were pretreated with 12% hydrochloric acid or with a mixture of concentrated nitric and sulfuric acids, which allowed etching of the metals from the fibers and oxidizing of the fibers' surface. Finally, the CNFs were N-functionalized via their impregnation with a melamine solution and thermolysis in an inert atmosphere. According to the microscopic data, the initial structure of the CNFs remained the same after the pretreatment and post-functionalization procedures. At the same time, the surface of the N-functionalized CNFs became more defective. The textural properties of the materials were also affected. In the case of the oxidative treatment with a mixture of acids, the highest content of the surface oxygen of 11.8% was registered by X-ray photoelectron spectroscopy. The amount of nitrogen introduced during the post-functionalization of CNFs with melamine increased from 1.4 to 4.3%. Along with this, the surface oxygen concentration diminished to 6.4%.

Keywords: CCVD; Ni–Cu catalyst; carbon nanofibers; oxidative treatment; N-functionalization; melamine

1. Introduction

In recent decades, researchers have paid special attention to the improvement of methods for the introduction of various heteroatoms into the structure of carbon materials (CMs). The growing interest in this research field was due to the fact that heteroatoms introduced into CM structures noticeably change the properties of the CMs and widen the areas of their application [1–3]. For instance, nitrogen-containing CMs (N-CMs) are widely used in production of supercapacitors [4–13] and in catalysis (as catalysts or catalysts' supports) [14–20] as well as in biomedicine [21] and even in criminalistics [22].

All the approaches to prepare N-doped CMs can be classified onto two groups: (i) one-pot functionalization, where nitrogen is introduced at the stage of the CM's synthesis and (ii) post-functionalization, where preliminarily prepared CMs are chemically modified with nitrogen. The one-pot functionalization considers the preparation of N-CMs via a simultaneous pyrolysis or thermal catalytic decomposition of precursors serving as carbon and nitrogen sources [7,8,10,21,23–28]. Often, one substrate containing both carbon and nitrogen atoms can be chosen as such a precursor. In the case of post-functionalization, previously prepared (or purchased from a supplier) CMs are treated with some N-containing reagents. This results in bonding of the N-containing functional groups to the CM's surface or in embedding of nitrogen atoms into the near-surface carbon structure [4–6,9,12,14–17,20,29–37]. The following N-containing compounds are usually used for the functionalization of CMs: ammonia [9,12,29,30,34], aniline in a mixture with ammonium persulphate [32], 3-hydroxyaniline [23], acetonitrile [24], glucosamine hydrochloride [31], dicyandiamide [18], melamine [4,6–12,14–20,22,23,37], urea [4,9,16,18,21], nitrogen (II) oxide [12], pyridine [23,36], pyrrole [35,38]; tiourea [20], 1,10-phenantroline [36], and ethylene diamine [13,33]. Among them, melamine, due to its high nitrogen content (atomic ratio C/N = 1/2), relative non-toxicity, and cheapness, is more preferable.

On the other hand, carbon nanofibers (CNFs) represent an important class of CMs with unique properties [39]. Therefore, the search for an appropriate method for their controllable synthesis and N-functionalization is an actual direction in modern materials sciences. In the present work, the synthesis of N-functionalized carbon nanofibers (N-CNFs) was performed via the post-functionalization route using melamine as a precursor. The initial CNF samples were prepared by the process of catalytic chemical vapor deposition (CCVD) of ethylene at 550 °C over the microdispersed Ni–Cu alloy (12 wt% Cu) obtained by the method of mechanochemical alloying of metals [40]. The CCVD process starts with a spontaneous disintegration of the bulk Ni–Cu alloy under the action of carbon erosion. This leads to the formation of dispersed active particles catalyzing the growth of the morphologically uniform carbon filaments of submicron diameter [27,40–42]. The subsequent acidic treatment of the thus-obtained CNFs results in a complete elimination of mineral impurities such as metallic particles of the catalyst, since no oxide supports (silica, alumina, etc.) were used in the composition of initial catalyst.

Among the diverse and ever-growing areas of practical application, the N-functionalized carbon nanomaterials attract attention in the field of heterogeneous catalysis. N-doped CNTs and CNFs are now considered as prospective substrates for supported metal catalysts. The metallic clusters (or single atoms) can be stabilized in a highly dispersed state by their anchoring at the surface N-sites [14,15,43]. For instance, the catalysts based on N-doped carbon nanomaterials can be effectively used in such processes as hydrodechlorination [36,44,45], selective hydrogenation reactions [46–48], formic-acid decomposition to produce hydrogen [49,50], and reversible (de)hydrogenation of liquid organic hydrogen carriers (LOHCs) [51–53]. In the latter methods, catalysts assist both to absorb hydrogen during the hydrogenation of unsaturated bonds of a substrate and to release stored hydrogen via catalytic dehydrogenation (reverse reaction). Thus, the development of a simple, versatile, and cost-effective technique for the N-functionalization of CNMs remains of particular interest.

The aim of the present study was to investigate the effect of the preliminary acidic treatment of CNFs on the characteristics of the N-CNF samples obtained by the post-functionalization of the former CNFs with melamine. The CNFs were treated in two regimes: mild washing out from the catalyst using hydrochloric acid and oxidative treatment in a mixture of concentrated sulfuric and nitric acids. The second regime allows not only washing out the catalyst's particles but oxidizing of the CNF surface as well [4,29,30,35,54–56]. The morphology, structure, and chemical composition of the N-functionalized CNF samples were examined by a number of characterization techniques, including electron microscopy, Raman spectroscopy, and X-ray photoelectron spectroscopy.

2. Materials and Methods

2.1. Chemicals and Materials

Nickel powder (PNK-UT3) and copper powder (PMS-1) used to synthesize the Ni–Cu catalyst were purchased from RusRedMet (Saint-Petersburg, Russia) and NMK-Ural (Yekaterinburg, Russia), respectively. High-purity ethylene (Nizhnekamskneftekhim, Nizhnekamsk, Russia) and high-purity hydrogen (GasProduct, Yekaterinburg, Russia) were used to produce the initial carbon nanofibers. Hydrochloric acid (ultrapure), nitric acid (ultrapure), and sulfuric acid (ultrapure) used for the pretreatment procedures were purchased from SigmaTek LLC (Khimki, Russia). Melamine (Shanxi Fenghe Melamine Co., Yuncheng, China) was used as a nitrogen source for the post-functionalization.

2.2. Synthesis of the Catalyst

The Ni–Cu catalyst was prepared by a mechanochemical alloying of metal powders using an Activator-2S planetary mill (Activator LLC, Novosibirsk, Russia), as described elsewhere [40,57]. Initially, nickel and copper powders were premixed in a weight ratio of Ni/Cu = 88/12. Then, 1 wt% of recently prepared CNF was added to the mixture. This was required to accelerate the carbon erosion stage at the beginning of the CCVD process. A specimen of this mixture (10 g) along with grinding balls (340 g; stainless steel; 5 mm in diameter) was loaded into stainless-steel bowls (250 mL in volume). The rotation frequency of the bowls and the platform was controlled using a VF-S15 industrial frequency inverter (Toshiba Schneider Inverter Corp., Nagoya, Japan). To avoid overheating, the bowls were water-cooled. The acceleration of the grinding balls was 780 m/c^2 (~80 G). The mechanochemical alloying was performed for 5 min. Finally, the bowls were unloaded in air, and the product was separated from the grinding balls and weighed.

2.3. Synthesis, Pretreatment and N-Functionalization of CNFs

Initial carbon nanofibers (labeled as cat/CNF, see Table 1) were synthesized via CCVD, as described elsewhere [40]. A specimen of the Ni–Cu catalyst (125 mg) was place into a horizontal quartz tubular reactor installed inside an XD-1200NT high-temperature furnace (Zhengzhou Brother Furnace Co., Zhengzhou, China). The reactor was fed with an argon flow and heated up to 550 °C with a heating rate of 10 °C/min. Then, a reaction mixture composed of ethylene (25 L/h), hydrogen (10 L/h), and argon (20 L/h) was passed through the reactor. The CCVD process was performed at 550 °C for 1 h. The carbon yield was ~110 g$_{CNF}$/g$_{cat}$.

Table 1. Designation of the prepared and studied samples.

#	Sample Designation	Description
1	cat/CNF	As-prepared CNFs obtained via CCVD of ethylene at 550 °C over Ni–Cu alloy catalyst
2	ha-CNF	cat/CNF sample pretreated in hydrochloric acid (elimination of metal particles)
3	ox-CNF	cat/CNF sample pretreated in a mixture of sulfur and nitric acids (elimination of metal particles, oxidation of the CNF's surface)
4	N/ha-CNF	ha-CNF sample post-functionalized with melamine
5	N/ox-CNF	ox-CNF sample post-functionalized with melamine

Pretreatment of the cat/CNF sample in hydrochloric acid (labeled as ha-CNF, see Table 1) was carried out by soaking it in an aqueous solution of HCl (200 mL of concentrated HCl mixed with 400 mL of distilled water) for 1 day at room temperature. Then, the sample was washed with distilled water on a glass filter until the pH value of the rinse water was neutral.

Oxidative pretreatment of the cat/CNF samples in a mixture of sulfuric and nitric acids (labeled as ox-CNF, see Table 1) was performed on a water bath at 60 °C for 2 h. The cat/CNF sample was placed into a flask along with 150 mL of distilled water, 100 mL of concentrated HNO_3, and 100 mL of concentrated H_2SO_4. Finally, the sample was washed with distilled water on a glass filter until the pH value of the rinse water was neutral.

Post-functionalization of the ha-CNF and ox-CNF samples with melamine (labeled as N/ha-CNF and N/ox-CNF, respectively, see Table 1) was performed as follows. A specimen (1 g) of the ha-CNF or ox-CNF sample was placed into a cylindrical vessel with 0.476 g of melamine dissolved in 20 mL of distilled water in a water bath at 60 °C. The mixture was maintained for 1.5 h and then dried in a drying oven at 150 °C. The dried sample was placed into a muffle, slowly heated up to 400 °C during 1 h, and calcined at this temperature for another 1 h.

2.4. Characterization of the Samples

The N_2 adsorption/desorption isotherms were recorded at 77 K using an Autosorb-6B-Kr automated adsorption analyzer (Quantachrome Instruments, Boynton Beach, FL, USA). In order to remove the adsorbed moisture and other impurities, before the measurements, a specimen (~150 mg) of each sample was treated in a vacuum (1 Pa) at 250 °C for 20 h. The characteristics of the porous structure were calculated using the ASWin 2.02 software package (Quantachrome Instruments, Boynton Beach, FL, USA). The specific surface area (SSA) values were obtained in accordance with a MA-BET approach described elsewhere [58]. To calculate the outer SSA of the CNF samples, a method of α_S-curves [59] was applied. An adsorption isotherm for Cabot-BP 280 was used as a reference equation [60]. The pore size distribution was evaluated by a quenched solid density functional theory (QSDFT) method using an integrated model with the following parameters: the adsorbate was nitrogen at 77 K, the pores were slotted, the material was carbon, and the calculation was applied to the adsorption branch of the isotherm.

Scanning electron microscopy (SEM, JEOL, Tokyo, Japan) studies of CNFs were performed using a JSM-6460 electron microscope (JEOL, Tokyo, Japan) at a magnification of 1000–100,000×. The morphology of CNFs was examined by transmission electron microscopy (TEM, Hitachi High-Technologies Corp., Tokyo, Japan) using an Hitachi HT7700 TEM (Hitachi High-Technologies Corp., Tokyo, Japan) working at an acceleration voltage of 100 kV and equipped with a W source.

Raman spectra of CNFs were registered on a Horiba Jobin Yvon LabRAM HR UV-VIS-NIR Evolution Raman spectrometer (Horiba, Kyoto, Japan) equipped with an Olympus BX41 microscope (Olympus Corp., Tokyo, Japan) and a 514.5 nm line Ar-ion laser. To avoid thermal decomposition of the CNF samples, the power of light focused in a spot with a diameter of ~2 μm did not exceed 0.8 mW.

The X-ray photoelectron spectroscopy (XPS) experiments were performed using a SPECS spectrometer (SPECS Surface Nano Analysis GmbH, Berlin, Germany) equipped with a hemispherical PHOIBOS-150-MCD-9 analyzer (SPECS Surface Nano Analysis GmbH, Berlin, Germany). The non-monochromatic MgK_α radiation (hν = 1253.6 eV) at 200 W and monochromatic AlK_α (hν = 1486.7 eV) at 150 W were used for the primary excitation. The spectrometer was calibrated using the peaks $Au4f_{7/2}$ (84.0 eV) and $Cu2p_{3/2}$ (932.7 eV) attributed to metallic gold and copper foils [61]. The samples were fixed on a holder with 3 M double-sided adhesive copper conducting tape. Peak 4.1 XPS software was applied for spectral analysis and data processing. In order to determine the binding energy values and the areas of XPS peaks, a Shirley background was subtracted. The XPS spectra were fitted with Gaussian–Lorentzian functions for each XPS region. The atomic ratios of elements were calculated from the integral photoelectron peak intensities corrected by the corresponding relative atomic sensitivity factors [61] and the transmission function of the analyzer.

3. Results and Discussion

3.1. Textural Properties of CNFs

It is well known that any modification or functionalization of the surface of carbon materials should affect their textural properties. To follow such changes, the low-temperature nitrogen adsorption/desorption technique was applied. The obtained adsorption/desorption isotherms for the CNF samples are shown in Figure 1. The shape of the isotherms indicates the presence of large micropores and small mesopores in all studied CNFs. According to IUPAC classification [62], the isotherms are close to Type I(b). They exhibit a narrow hysteresis loop with a shape close to Types H3 or H4. The adsorption and desorption branches do not connect with each other until the relative pressure value reaches 0.025, thus indicating the non-equilibrium character of adsorption on these samples, which can be explained by the swelling of carbon materials. Apparently, the desorption branch is not equilibrium. Therefore, the adsorption branch was used for the further analysis.

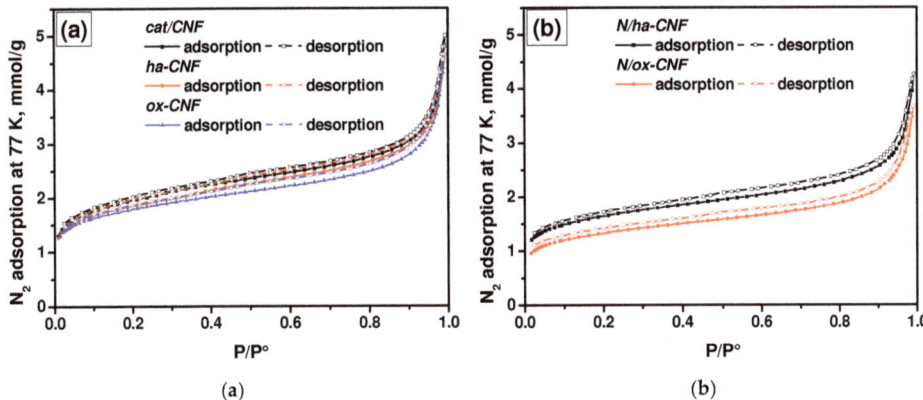

Figure 1. Isotherms of nitrogen adsorption/desorption at 77 K for the studied CNF samples: (**a**) as-prepared and pretreated CNFs and (**b**) N-functionalized CNFs. The samples were treated in a vacuum (1 Pa) at 250 °C for 20 h.

The main characteristics of the porous structure of CNFs are summarized in Table 2. As seen from the presented data, the pretreatment and post-functionalization procedures resulted in a decrease in both the SSA and the total pore volume. The average pore diameter, calculated as $4V/A$, where V is the total pore volume and A is the specific surface area, had changed insignificantly. It should be also mentioned that the effect of the post-functionalization with melamine on the textural characteristics of CNFs was more noticeable than the effect of their pretreatment with acids.

Table 2. The main textural characteristics of CNFs.

Sample	SSA, m^2/g	Pore Volume, cm^3/g	Average Pore Diameter $(4V/A)$, nm
cat/CNF	161.9	0.173	4.3
ha-CNF	154.3	0.161	4.2
ox-CNF	147.4	0.161	4.4
N/ha-CNF	137.1	0.147	4.3
N/ox-CNF	110.2	0.127	4.6

The calculation results for the relative pressure region of 0.5–0.9 of the adsorption branch in comparison with the corresponding data of the QSDFT method are presented in Table 3. As recently reported [63], the comparative method can be used to calculate the SSA of the rough micropores. Comparing the presented characteristics, it can be supposed that the method of α_S-curves allows the volume of inner pores located inside CNFs to be obtained. Indeed, the values of pore volumes estimated by the α_S-curves method were close to those calculated by the QSDFT method for the pores less than 2 nm. Therefore, the value of 2 nm seems to be a border value of width for pores located inside and outside the fibers. Almost all the studied samples, except N/ox-CNF, had similar values for the volume of micropores with a diameter < 2 nm to those calculated by the QSDFT method.

The pore size distributions calculated by the QSDFT method are shown in Figure 2. As seen, all the samples were characterized by the presence of micropores of ~0.8 nm in width. The samples cat/CNF and ha-CNF also contained small mesopores of 2–3 nm in width. It is evident that the post-functionalization with melamine changed the pore size distribution. Thus, the number of pores of 0.8 nm in size decreased noticeably. It is interesting to note that the pretreatment with hydrochloric acid led to a slight shift of the distribution peak at ~2 nm towards higher sizes without a significant change in its value.

Table 3. The characteristics of micropores in CNFs.

Sample	SSA of the 1st Adsorption layer *, m^2/g	Volume of Micropores in CNFs *, cm^3/g	Outer SSA of CNFs *, m^2/g	Volume of Micropores < 2 nm [†], cm^3/g	Outer SSA of Pores > 2 nm [†], m^2/g
cat/CNF	155.6	0.059	48.5	0.057	20.4
ha-CNF	150.3	0.055	38.5	0.053	20.7
ox-CNF	137.7	0.051	36.4	0.053	17.2
N/ha-CNF	134.5	0.046	34.2	0.048	16.4
N/ox-CNF	109.5	0.035	31.7	0.038	14.1

* calculated by a method of α_S-curves; [†] calculated by a QSDFT method.

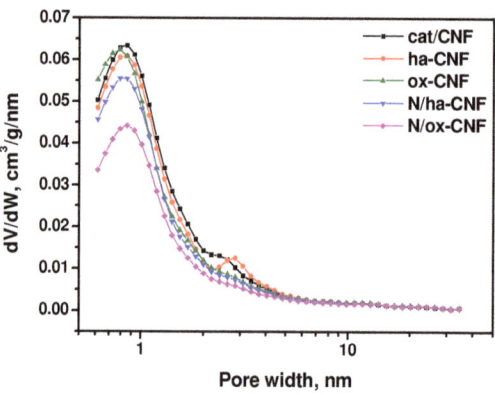

Figure 2. Pore size distribution calculated by the QSDFT method using ASWin 2.02 software.

In general, each of modification procedures significantly affected the textural characteristics of CNFs. The main tendencies are a decrease in the contribution of small mesopores and an appearance of macropores with a distribution maximum at ~100 nm.

3.2. SEM Characterization of CNFs

The modified CNF samples were explored by the SEM technique. Figures 3 and 4 demonstrate the microscopic images for CNFs treated in hydrochloric acid (ha-CNF) and post-functionalized with melamine (N/ha-CNF), respectively. As seen, the ha-CNF sample before the post-functionalization (Figure 3) possessed a fibrous architecture represented by the assemblage of carbon nanofibers. The diameter of the carbon filaments was varied in a range of 100–500 nm. It should be noted that this parameter is defined by the size of the active Ni–Cu particles, catalyzing the growth of CNFs. The images show that all the fibers were of submicron diameter. This is related to the catalyst's formation process. During the first minutes of interaction with ethylene, the microdispersed Ni–Cu alloy used as a catalyst's precursor undergoes a rapid disintegration under the action of carbon erosion [40–42]. Such a disintegration of the alloy results in the formation of active particles responsible for the growth of CNFs. It is worth noting that the carbon erosion process predetermines, at earlier stages, the formation of a very fluffy carbon product. The carbon filaments within this product do not tangle or form dense agglomerates.

The individual fibers and their fragments are shown in Figure 3d,e. As seen, the fibers are characterized by a relatively dense packing with a few slit-like pores. The mutual orientation of graphene layers defines the following structural types of CNFs: the stacked "pile of plates" structure and the coaxial cones "fishbone" structure [64,65]. The surface of CNF looks rough.

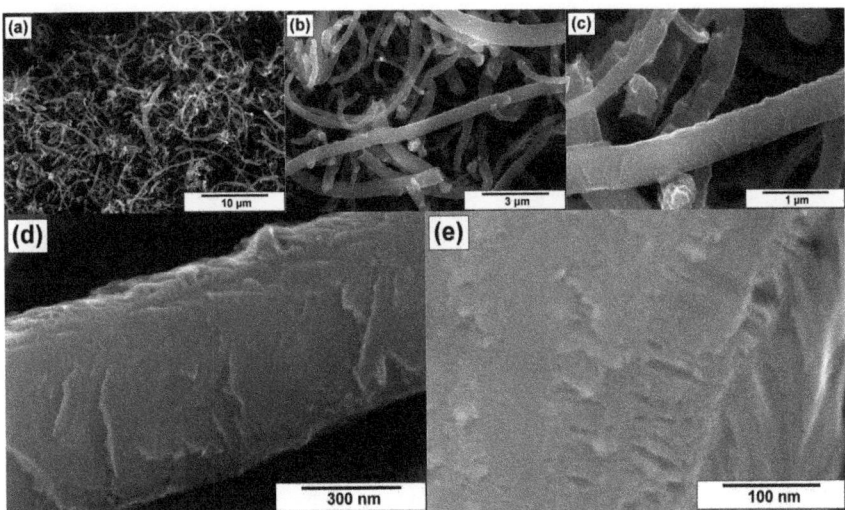

Figure 3. SEM images of the ha-CNF sample at various magnifications: (**a**) 3000×, (**b**) 10,000×, (**c**) 30,000×, (**d**) 60,000×, and (**e**) 100,000×.

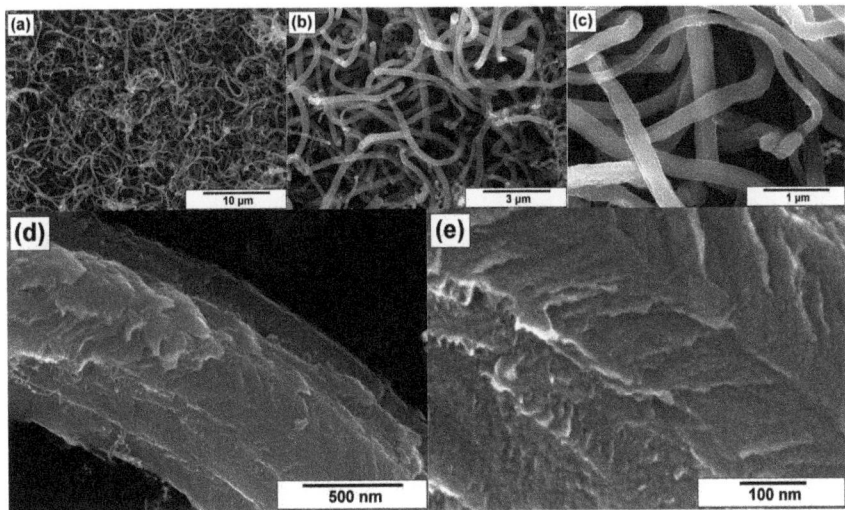

Figure 4. SEM images of the N/ha-CNF sample at various magnifications: (**a**) 3000×, (**b**) 10,000×, (**c**) 30,000×, (**d**) 60,000×, and (**e**) 100,000×.

Comparing the ha-CNF samples before (Figure 3) and after the post-functionalization (Figure 4), it can be seen that this procedure (the impregnation with a solution of melamine followed by the calcination at 400 °C) did not practically affect the morphology and the structure of CNFs and did not lead to their destruction.

On the contrary, pretreatment by heating in a mixture of sulfuric and nitric acids resulted in loosening of the outer surface of CNFs, thus making them more rough (Figure 5c,d). Figure 5e illustrates the packing character of the flakes within the fiber's body with eliminated metal particle of the catalyst. Nevertheless, the structure integrity of initial CNFs was not broken during such a drastic conditions of the oxidative treatment.

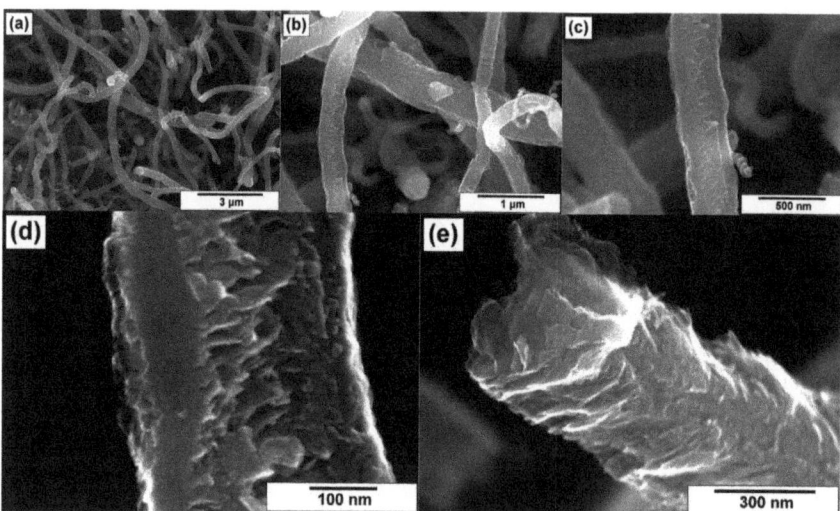

Figure 5. SEM images of the ox-CNF sample at various magnifications: (**a**) 3000×, (**b**) 10,000×, (**c**) 30,000×, (**d**) 60,000×, and (**e**) 100,000×.

The influence of the subsequent N-functionalization of the ox-CNF sample on its secondary structure is shown in Figure 6. The fibrous structure of the carbon material remained mainly the same. At the same time, the surface of fibers was maximally loosened (Figure 6c,d). Despite no sufficient changes in the secondary structure being observed, many splits, cleaved facets, and cracks are clearly seen.

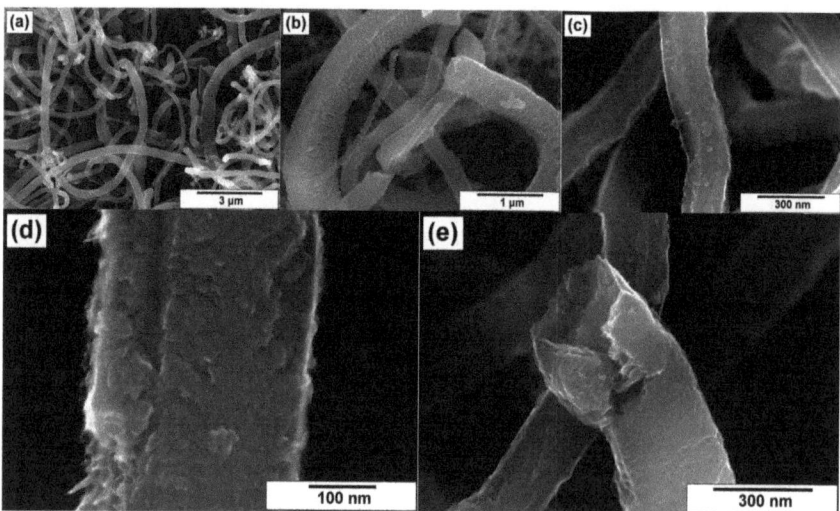

Figure 6. SEM images of the N/ox-CNF sample at various magnifications: (**a**) 3000×, (**b**) 10,000×, (**c**) 30,000×, (**d**) 60,000×, and (**e**) 100,000×.

3.3. TEM Characterization of CNFs

Transmission electron microscopy gives more detailed information regarding the morphological and structural features of CNFs. Therefore, the samples were examined by this method as well. The resulting TEM images are collected in Figures 7–9.

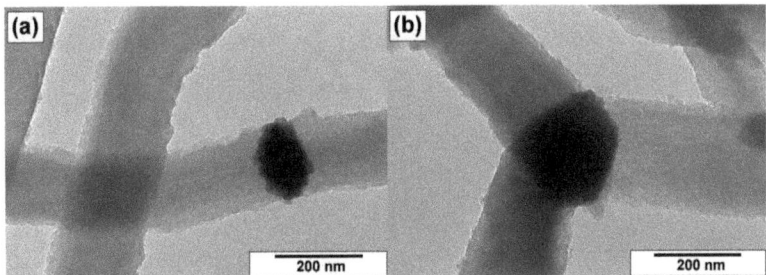

Figure 7. TEM images of the cat/CNF sample prepared via CCVD of ethylene at 550 °C over Ni–Cu alloy: (**a**) two-directional growth of carbon filaments and (**b**) three-directional growth of carbon filaments.

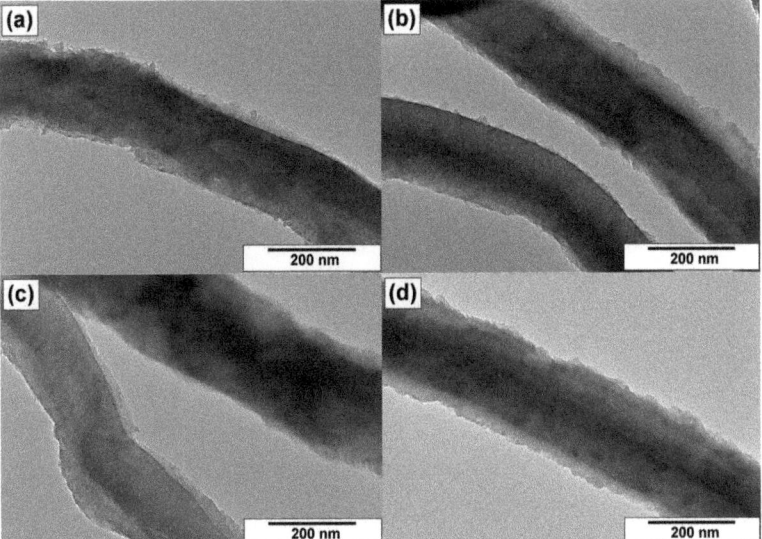

Figure 8. TEM images of the studied samples: (**a**) cat/CNF, (**b**) ox-CNF, (**c**) N/ha-CNF, and (**d**) N/ox-CNF.

Figure 7 shows the TEM images for the as-prepared cat/CNF sample obtained via CCVD of ethylene over Ni–Cu alloy. As already mentioned, the growth of carbon filaments occurs as a result of the catalytic action of the dispersed active particles that appear due to the rapid disintegration of the initial Ni–Cu alloy. Such submicron particles of 100–200 nm in size can be clearly seen in Figure 7. One type of particle has a relatively symmetric shape and carries on the growth of carbon filaments in two opposite directions (Figure 7a). At the same time, these particles possess a relief surface without expressed faceting. The second type of particle is alloyed crystallites connected by three or more carbon filaments (Figure 7b). Such particles are clearly faceted. The crystallite plates, where the CNF growth takes place, are perpendicular to the axes of carbon filaments, thus indicating the formation of the stacked structure of CNFs.

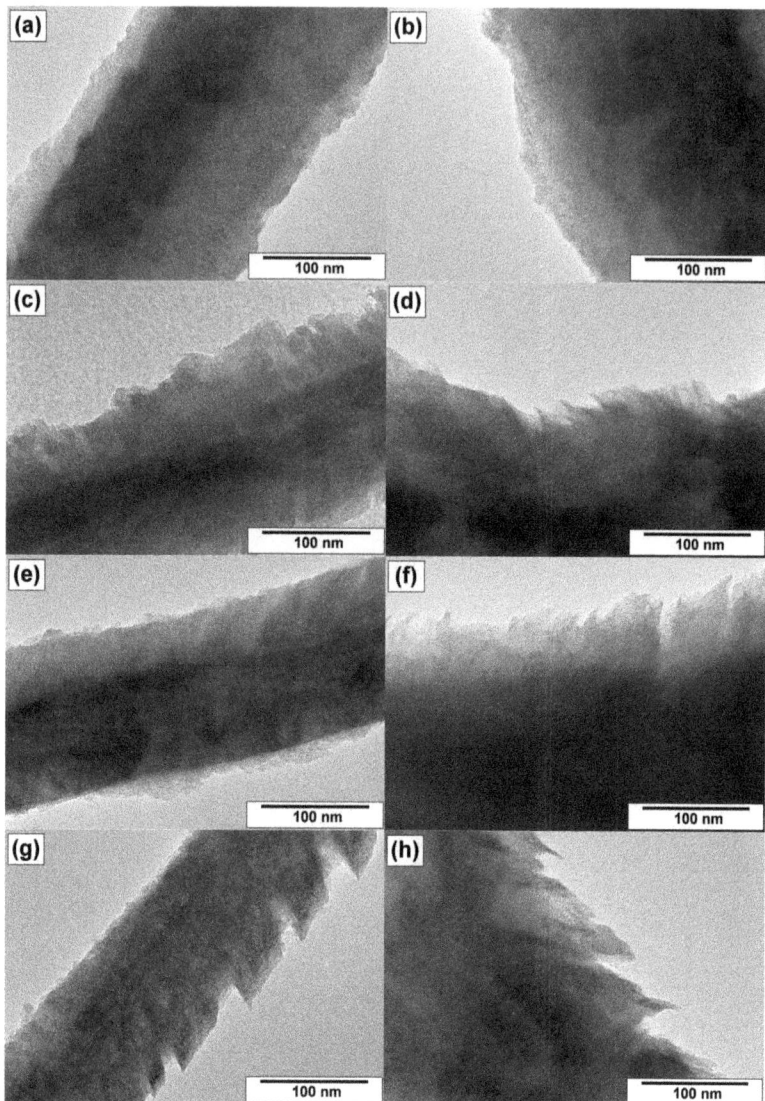

Figure 9. Magnified TEM images of the filament's surface for the studied samples: (**a**,**b**) cat/CNF, (**c**,**d**) ox-CNF (**e**,**f**) N/ha-CNF, and (**g**,**h**) N/ox-CNF.

It is important to note that in all case, the particles are not entirely covered with carbon. There are places on the surface that are free of carbon, which are responsible for the catalytic decomposition of ethylene molecules and the supply of carbon atoms [66]. These particles are easily accessible for any reagents and, therefore, they can be completely removed from the composition of CNFs even under mild treatment conditions. On the other hand, the acidic pretreatment breaks the integrity of CNFs. As expected, these procedures lead to the breakage of the filaments at the places of removed metal particles and their crushing into two and more pieces. At the same time, the filaments themselves are not fragile. They can stand even boiling in acids without further lengthwise destruction.

Figures 8 and 9 present the TEM images of the same magnification for the samples cat/CNF, ox-CNF, N/ha-CNF, and N/ox-CNF, which allows their structural features to be compared. It is evident that the structure of filaments was not significantly changed during the acidic pretreatment and the subsequent N-functionalization. All the fibers are characterized by a dense packing in the bulk and a loose rough surface.

In some cases, the surface of filaments is covered with a thin layer of carbon with a different character of the graphite packing (Figures 8a,c and 9a,e). The boundary interface is obviously seen in these images. The loose surface of CNFs seems to be more liable for the structural changes under the action of oxidative treatment or N-functionalization. Thus, the carbon fibers with a fringe on the surface were found in the case of CNFs post-functionalized with melamine (Figure 9f,h). At the same time, despite the surface loosening, the structure of the core part of carbon filaments within the composition of N/ha-CNF and N/ox-CNF did not suffer any visual changes.

Based on numerous SEM and TEM data acquired for all studied samples, it was possible to statistically measure the diameter distributions for the carbon nanofibers. The obtained results are presented as diagrams in Figure 10. It is obvious from the comparison of given data that the acidic treatment (both methods) and subsequent N-functionalization had an insignificant effect on the CNF diameter distribution. The measured average diameter of CNFs for all the samples was around 200 nm, regardless to the treatment conditions. The obtained result confirms that the proposed method for N-functionalization of carbon nanofibers appears to be non-destructive.

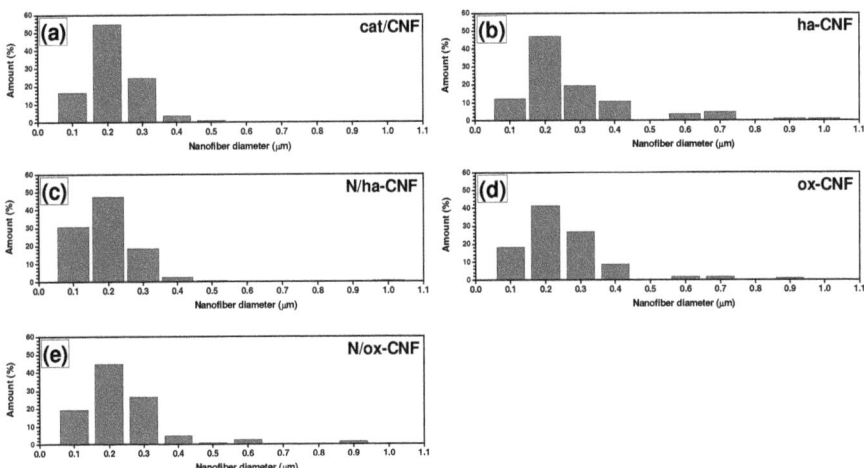

Figure 10. CNF diameter distributions depending on the treatment conditions (based on SEM and TEM data): (**a**) cat/CNF, (**b**) ha-CNF, (**c**) N/ha-CNF, (**d**) ox-CNF, and (**e**) N/ox-CNF.

Therefore, it can be concluded that the proposed approach for the pretreatment of CNFs before their N-functionalization allows the metal particles of the catalysts to be effectively eliminated from the fiber's structure. The structure and the integrity of the initial filaments as well as their diameter distribution remained almost the same. The post-modification of these CNFs with melamine also did not affect these characteristics noticeably; however, the surface structure was being changed and the surface layer of the filaments became significantly looser.

3.4. Raman Spectroscopy Data

Another informative method to study the structure and orderliness of CNFs is Raman spectroscopy. The typical Raman spectra of the samples in a region of the first order bands are shown in Figure 11. The spectra are characterized by the G bands at ~1595 cm^{-1}

corresponding to the allowed vibrations E_{2g} of the hexagonal graphite lattice [67,68] and by the disorder-induced D band (activated A_{1g} mode due to the finite crystal size) at 1345 cm^{-1}, which is a characteristic of disordered carbon materials [68,69]. Using a phenomenological three-stage model for the ordering trajectory from tetrahedral amorphous carbon to graphite or the amorphization trajectory considering the change in the G band position for the wavelength of 514.5 nm from 1581 to 1600, 1510, and 1560 cm^{-1}, respectively, the studied samples can be assigned closer to nanocrystalline graphite [68].

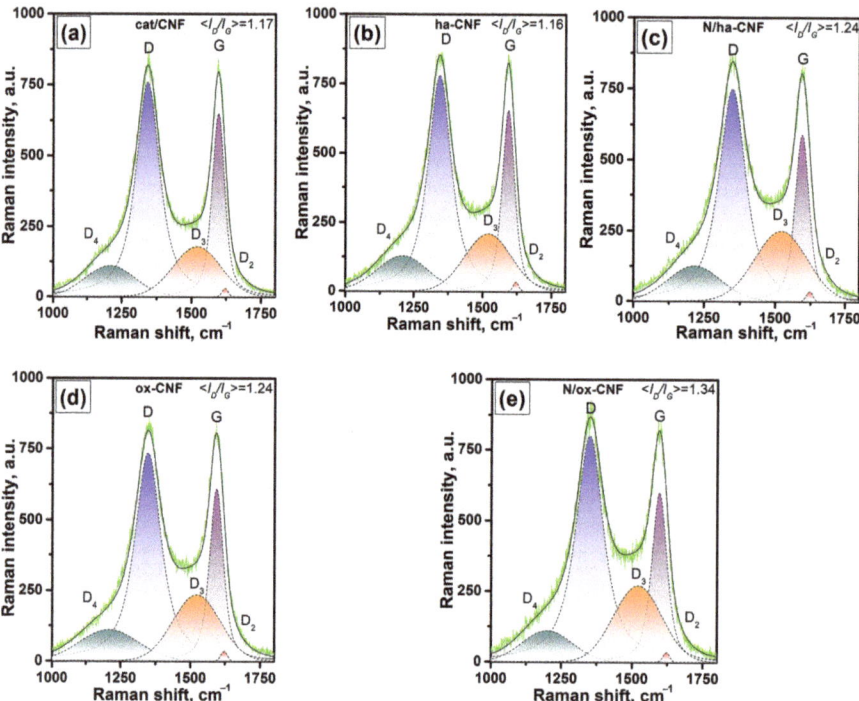

Figure 11. Raman spectra in a region of the first order bands: (**a**) cat/CNF (**b**) ha-CNF, (**c**) N/ha-CNF, (**d**) ox-CNF, and (**e**) N/ox-CNF.

For all the samples, D_2 bands can be observed at ~1618 cm^{-1}, D_3 at ~1520 cm^{-1}, and D_4 at ~1208 cm^{-1}. The D_2 bands correspond to a disordered graphitic lattice (surface graphene layers, E_{2g}-symmetry) [70], while the two other bands can be assigned to amorphous carbon and disordered graphitic lattice (A_{1g}-symmetry) or polyenes [71], which is typical for soot and related carbonaceous materials.

The second order bands are ill-defined and very close to each other in intensity for all the studied samples. The characteristic intensities of the 2D bands at ~2700 cm^{-1} and D + D_2 at 2934 cm^{-1} were I_{2D}/I_G ~ 0.11 and I_{D+D2}/I_G ~ 0.14, respectively. Taking into account the half-width values (HWHM) of the G band > 45 cm^{-1} and according to Ferrari and Robertson [68], the cluster diameter (or in-plane correlation length) L_a can be approximated from the I_D/I_G ratio using the equation $I_D/I_G = C'(\lambda)\cdot L_a^2$, where C' is of ~0.0055 for the wavelength of 514.5 nm. The obtained L_a values lie in a range of 13.3–15.7 Å.

The changes of the main parameters (I_D/I_G, HWHM G, and I_{D3}/I_G) resulting from the pretreatment and post-functionalization of CNFs are shown in Figure 12. Note that each value is an average for four points. As seen, the pretreatment of the initial cat/CNF sample in hydrochloric acid led to an insignificant increase in the portion of amorphous carbon (I_{D3}/I_G). The oxidative treatment in a mixture of sulfuric and nitric acids (ox-CNF sample), in addition to an increase in the I_{D3}/I_G ratio, also increased the HWHM G value and the I_D/I_G ratio. This indicates a higher number of defects along with an increased portion of crystalline carbon. In other words, this acidic treatment procedure decreased the portion of disordered carbon on the surface of CNFs. In should be noted that a similar effect of etching-out the disordered carbon from the surface of carbon paper using a solution of sulfuric acid being accompanied by an increase in the I_D/I_G ratio has been reported in the literature [72]. However, the authors interpreted this change in the I_D/I_G ratio oppositely—as an increase in the disorder degree of carbon, which is the only explanation if no analysis of the HWHM G parameter and no approximation of the L_a value are considered.

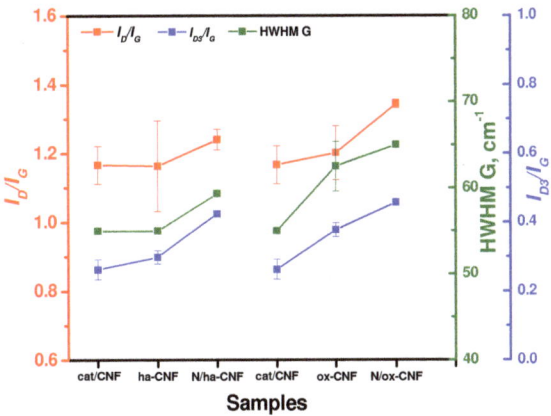

Figure 12. Changes of the I_D/I_G, HWHM G, and I_{D3}/I_G parameters for the studied samples.

The post-functionalization of CNFs with melamine resulted in a synchronous increase in the I_D/I_G ratio, the HWHM G parameter, and the portion of amorphous carbon I_{D3}/I_G (Figure 12). This observation can be interpreted as a rise of the disordered carbon portion on the surface of CNFs. Therefore, it can be concluded that all the modification procedures dealt only with the amorphous carbon and the surface layer of CNFs.

3.5. XPS Study of CNFs

The qualitative and quantitative estimations of the number of introduced nitrogen and oxygen atoms can be made by means of XPS analysis. According to the survey spectra presented in Figure 13a, the studied CNF samples contain atoms of carbon, oxygen, and nitrogen only. The atomic ratios O/C and N/C are summarized in Table 4.

Figure 13b,c shows the spectra of the N 1s and O 1s regions recorded using the monochromatic radiation AlK$_\alpha$. The binding energy (BE) values, the corresponding species, and their ratios to carbon for all the registered components in these spectra are compared in Tables 5 and 6. In cases of the ha-CNF and ox-CNF samples, the spectra contain low-intensive peaks at ~ 400 eV, whose appearance could have been caused by the presence of a trace amount of nitrogen. Such a low intensity of these peaks does not allow identifying the state of these species. In addition, in the spectrum of ox-CNF, a state with BE of ~405.8 eV is predominant. This state can be definitely identified as C-NO$_2$ groups [73,74]. For both the samples treated with melamine, three states of nitrogen are observed with BE at 398.4, 399.0, and 399.8–400.0 eV. The first (398.4 eV) and second (399.0 eV) states correspond to C-NH$_2$ groups [75,76] and C=N-C fragments [75,77], respectively. Note that the state with BE of 398.4 eV can be also attributed to C≡N species [61,75]. The state with BE of ~400 eV is usually assigned to pyrrolic nitrogen [76–78]. However, the peak of pyrrolic N is overlapped in XPS spectra by nitrogen of diazine and triazine rings [76].

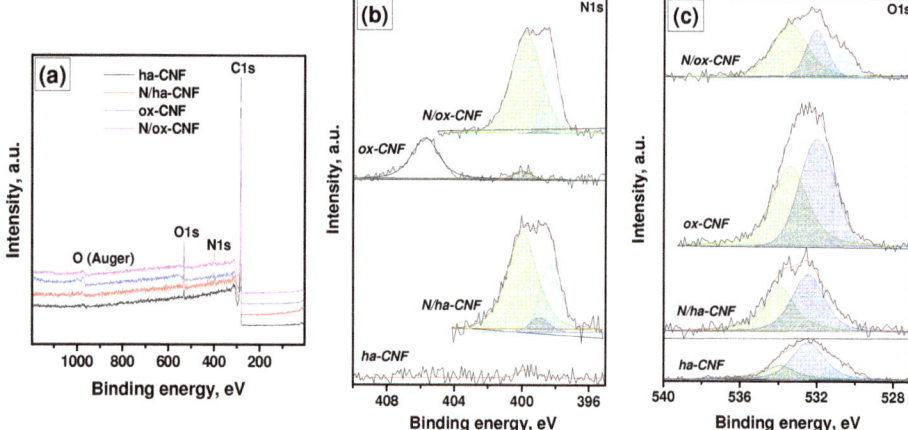

Figure 13. XPS spectra of the studied samples: (**a**) survey spectra, (**b**) N 1s region, and (**c**) O 1s region.

Table 4. Atomic ratios for the studied CNF samples.

Sample	N/C	O/C
ha-CNF	0.0007	0.025
N/ha-CNF	0.014	0.046
ox-CNF	0.014	0.118
N/ox-CNF	0.043	0.064

Table 5. Components of the N 1s spectra and their atomic ratios N/C.

| BE, eV | Sample | | | | Species | Ref. |
	ha-CNF	N/ha-CNF	ox-CNF	N/ox-CNF		
398.4	-	0.0036	-	0.011	$C-NH_2$	[75,76]
399.0	-	0.00078	-	0.00048	$C \equiv N$	[61,75]
399.8	-	-	-	0.032	$C=N-C$	[75,77]
400.0	-	0.0098	-	-	Pyrrolic N	[76–78]
~400.0	0.0007	-	0.0013	-	Not identified	
405.8	-	-	0.013	-	$C-NO_2$	[74]

Table 6. Components of the O 1s spectra and their atomic ratios O/C.

| BE, eV | Sample | | | | Species | Ref. |
	ha-CNF	N/ha-CNF	ox-CNF	N/ox-CNF		
530.7	-	-	-	0.0095		
530.8	0.0019	-	-	-	C(O)O	
530.9	-	0.0017	-	-		
532.0	-	-	0.061	0.018		
532.4	0.019	0.024	-	-	C=O	[79–81]
533.4	-	-	0.057	0.037		
533.9	0.0044	-	-	-	$C-OH-NO_2$	
534.0	-	0.02	-	-		

The presence of an intensive peak at BE near 399 eV in the spectrum of the N/ox-CNF sample allows for the assumption that there was a large amount of residual melamine (or its fragments) in this sample. Note that in the melamine spectrum, nitrogen is in two states with BE of 398.6 eV ($C-NH_2$) and 399.2 eV (C=N-C) [75].

The O 1s spectra for the samples ha-CNF, N/ha-CNF, and N/ox-CNF contain components that can be attributed to the following groups (Figure 13c, Table 6): C(O)O with BE of 530.7–530.8 eV; C=O with BE of 532.0–532.4 eV; and C-OH with BE of 533.4–534.0 eV [79,80]. In the case of the ox-CNF sample, the components in the XPS spectrum correspond to C=O and C-OH groups, and the latter ones are overlapped with –NO$_2$ groups [79–81].

Therefore, the performed XPS study confirmed the introduction of heteroatoms (O and N) into the structure of carbon nanofibers during the pretreatment and post-functionalization procedures. Both of the pretreatment regimes caused the complete elimination of metal particles of the catalyst from the composition of CNFs. No components assigned to nickel or copper were observed in XPS spectra. The drastic oxidative treatment (ox-CNF sample) resulted in the modification of the surface of the CNFs with ~11.8 at% of oxygen. Under the mild conditions (ha-CNF sample), the oxidation of the CNF's surface also took place, but to a lower degree (~2.5 at% of oxygen). The N-functionalization of these samples with melamine allowed for the introduction of ~1.4 and ~4.3 at% of nitrogen for the N/ha-CNF and N/ox-CNF samples, respectively. It is important to note that the oxygen content decreased to 6.5 at% for N/ox-CNF and increased to 4.6 at% for N/ha-CNF. Such an opposite behavior is supposedly connected with the formation of C-NO$_2$ groups in the latter case.

By comparing the results of the present research with the literature data related to the synthesis and characterization of the N-doped CNFs, it can be seen that the produced N-CNFs contained a nitrogen amount of several percent which is rather typical for N-CNM materials [4,7,9,12,16,20,29,36]. This is probably related to the comparatively low porosity and flexible structure of CNFs as well as to the application of a post-functionalization method, which mainly affects the surface of carbon fibers. The higher amount of nitrogen can be introduced into the structure of more porous and/or ordered carbons [6,11,14], microspheres [10], and tubes [15], including the use of severe conditions for the functionalization [25].

4. Conclusions

In present work, the carbon nanofibers obtained via CCVD of ethylene at 550 °C over an Ni–Cu alloy were modified via the introduction of O and N heteroatoms. The application of a low-temperature nitrogen-adsorption method, SEM and TEM, Raman spectroscopy, and XPS technique gave an opportunity to perform an in-depth study of the modification process. The first stage of this process was the pretreatment with hydrochloric acid (mild conditions) or with a mixture of sulfuric and nitric acids under heating (drastic conditions). Both these procedures allowed the complete elimination of metal particles of the catalysts from the structure of carbon filaments. The elimination of metal particles, in its turn, resulted in a breakage of filaments in places where the particles were located. Thus, the filaments became shorter, and the slight compaction of the carbon material occurred. Therefore, the specific surface area of the samples after the acidic pretreatment diminished insignificantly. According to TEM and Raman spectroscopy data, the modification process dealt mostly with the surface changes. The appearance of oxygen species on the surface of CNFs was registered in both the pretreatment cases. The drastic oxidative conditions gave an oxygen content as high as 11.8 at%. The subsequent post-functionalization with melamine did not practically affect the morphology of the CNF samples. At the same time, it allowed for the introduction of 1.4 and 4.3 at% of nitrogen into the structure of ha-CNF and ox-CNF samples, respectively. Therefore, the proposed approach proved to be applicable for the modification of carbon nanofibers with heteroatoms (O and N), which are primarily demanded as a support for heterogeneous catalysts used in hydrodechlorination, selective hydrogenation, and other processes.

Author Contributions: Conceptualization, I.V.M. and A.A.V.; methodology, I.V.M., M.S.M., A.V.N. and V.O.S.; investigation, T.A.M., Y.I.B., A.B.A. and A.M.D.; writing—original draft preparation, T.A.M., I.V.M., A.B.A., A.V.N. and V.O.S.; writing—review and editing, I.V.M., M.S.M. and A.A.V.; funding acquisition, I.V.M. All authors have read and agreed to the published version of the manuscript.

Funding: This work was supported by the Ministry of Science and Higher Education of the Russian Federation (project No. AAAA-A21-121011390054-1).

Institutional Review Board Statement: Not applicable.

Informed Consent Statement: Not applicable.

Data Availability Statement: Data are contained within the article.

Acknowledgments: Characterization of the samples was performed using the equipment of the Center of Collective Use "National Center of Catalysts Research". TEM studies were performed in the Krasnoyarsk Regional Center of Research Equipment of the Federal Research Center "Krasnoyarsk Science Center SB RAS". The authors are grateful to A. N. Serkova for her help in SEM studies, to M. N. Volochaev for his help in TEM studies, and to A. S. Kadtsyna for preparing the CNF samples.

Conflicts of Interest: The authors declare no conflict of interest. The funders had no role in the design of the study; in the collection, analyses, or interpretation of data; in the writing of the manuscript; or in the decision to publish the results.

References

1. Lee, W.J.; Maiti, U.N.; Lee, J.M.; Lim, J.; Han, T.H.; Kim, S.O. Nitrogen-doped carbon nanotubes and graphene composite structures for energy and catalytic applications. *Chem. Commun.* **2014**, *50*, 6818–6830. [CrossRef] [PubMed]
2. Podyacheva, O.Y.; Ismagilov, Z.R. Nitrogen-doped carbon nanomaterials: To the mechanism of growth, electrical conductivity and application in catalysis. *Catal. Today* **2015**, *249*, 12–22. [CrossRef]
3. Inagaki, M.; Toyoda, M.; Soneda, Y.; Morishita, T. Nitrogen-doped carbon materials. *Carbon* **2018**, *132*, 104–140. [CrossRef]
4. Hulicova-Jurcakova, D.; Seredych, M.; Lu, G.Q.; Bandosz, T.J. Combined effect of nitrogen- and oxygen-containing functional groups of microporous activated carbon on its electrochemical performance in supercapacitors. *Adv. Func. Mater.* **2009**, *19*, 438–447. [CrossRef]
5. Nasini, U.B.; Bairi, V.G.; Ramasahayam, S.K.; Bourdo, S.E.; Viswanathan, T.; Shaikh, A.U. Phosphorous and nitrogen dual heteroatom doped mesoporous carbon synthesized via microwave method for supercapacitor application. *J. Power Sources* **2014**, *250*, 257–265. [CrossRef]
6. Gao, F.; Shao, G.; Qu, J.; Lv, S.; Li, Y.; Wu, M. Tailoring of porous and nitrogen-rich carbons derived from hydrochar for high-performance supercapacitor electrodes. *Electrochim. Acta* **2015**, *155*, 201–208. [CrossRef]
7. Liu, C.; Wang, J.; Li, J.; Zeng, M.; Luo, R.; Shen, J.; Sun, X.; Han, W.; Wang, L. Synthesis of N-doped hollow-structured mesoporous carbon nanospheres for high-performance supercapacitors. *ACS Appl. Mater. Interf.* **2016**, *8*, 7194–7204. [CrossRef]
8. Yang, Y.; Yang, F.; Lee, S.; Li, X.; Zhao, H.; Wang, Y.; Hao, S.; Zhang, X. Facile fabrication of MnO_x and N co-doped hierarchically porous carbon microspheres for high-performance supercapacitors. *Electrochim. Acta* **2016**, *191*, 1018–1025. [CrossRef]
9. Kim, J.; Lim, H.; Jyoung, J.-Y.; Lee, E.-S.; Yi, J.S.; Lee, D. Effects of doping methods and kinetic relevance of N and O atomic co-functionalization on carbon electrode for V(IV)/V(V) redox reactions in vanadium redox flow battery. *Electrochim. Acta* **2017**, *245*, 724–733. [CrossRef]
10. Ma, C.; Chen, X.; Long, D.; Wang, J.; Qiao, W.; Ling, L. High-surface-area and high-nitrogen-content carbon microspheres prepared by a pre-oxidation and mild KOH activation for superior supercapacitor. *Carbon* **2017**, *118*, 699–708. [CrossRef]
11. Chen, Y.; Xiao, Z.; Liu, Y.; Fan, L.-Z. A simple strategy toward hierarchically porous graphene/nitrogen-rich carbon foams for high-performance supercapacitors. *J. Mater. Chem. A* **2017**, *5*, 24178–24184. [CrossRef]
12. Sutarsis; Patra, J.; Su, C.-Y.; Li, J.; Bresser, D.; Passerini, S.; Chang, J.-K. Manipulation of nitrogen-heteroatom configuration for enhanced charge-storage performance and reliability of nanoporous carbon electrodes. *ACS Appl. Mater. Interf.* **2020**, *12*, 32797–32805. [CrossRef] [PubMed]
13. Luo, Y.; Lu, Y.; Wang, Q.; Xu, F.; Sun, L.; Wang, Y.; Lao, J.; Liao, L.; Zhang, K.; Zhang, H.; et al. Porous carbon with facial tuning of the heteroatom N by polyvinylidene chloride dehalogenation toward enhanced supercapacitor performance. *J. Electrochem. Energy Conv. Stor.* **2023**, *20*, 011004. [CrossRef]
14. Li, Z.; Liu, J.; Xia, C.; Li, F. Nitrogen-functionalized ordered mesoporous carbons as multifunctional supports of ultrasmall Pd nanoparticles for hydrogenation of phenol. *ACS Catal.* **2013**, *3*, 2440–2448. [CrossRef]
15. Chen, S.; Qi, P.; Chen, J.; Yuan, Y. Platinum nanoparticles supported on N-doped carbon nanotubes for the selective oxidation of glycerol to glyceric acid in a base-free aqueous solution. *RSC Adv.* **2015**, *5*, 31566–31574. [CrossRef]
16. Soares, O.S.G.P.; Rocha, R.P.; Gonçalves, A.G.; Figueiredo, J.L.; Órfão, J.J.M.; Pereira, M.F.R. Highly active N-doped carbon nanotubes prepared by an easy ball milling method for advanced oxidation processes. *Appl. Catal. B Environ.* **2016**, *192*, 296–303. [CrossRef]

17. Ferrero, G.A.; Fuertes, A.B.; Sevilla, M.; Titirici, M.-M. Efficient metal-free N-doped mesoporous carbon catalysts for ORR by a template-free approach. *Carbon* **2016**, *106*, 179–187. [CrossRef]
18. Liang, P.; Zhang, C.; Duan, X.; Sun, H.; Liu, S.; Tade, M.O.; Wang, S. N-doped graphene from metal–organic frameworks for catalytic oxidation of p-hydroxylbenzoic acid: N-functionality and mechanism. *ACS Sustain. Chem. Eng.* **2017**, *5*, 2693–2701. [CrossRef]
19. Mian, M.M.; Liu, G.; Zhou, H. Preparation of N-doped biochar from sewage sludge and melamine for peroxymonosulfate activation: N-functionality and catalytic mechanisms. *Sci. Total Environ.* **2020**, *744*, 140862. [CrossRef]
20. Anfar, Z.; El Fakir, A.A.; Zbair, M.; Hafidi, Z.; Amedlous, A.; Majdoub, M.; Farsad, S.; Amjlef, A.; Jada, A.; El Alem, N. New functionalization approach synthesis of Sulfur doped, Nitrogen doped and Co-doped porous carbon: Superior metal-free Carbocatalyst for the catalytic oxidation of aqueous organics pollutants. *Chem. Eng. J.* **2021**, *405*, 126660. [CrossRef]
21. Lee, B.; Tian, S.; Xiong, G.; Yang, Y.; Zhu, X. Solvothermal synthesis of transition metal (iron/copper) and nitrogen co−doped carbon nanomaterials: Comparing their peroxidase—Like properties. *J. Nanopart. Res.* **2022**, *24*, 85. [CrossRef]
22. Prabakaran, E.; Pillay, K. Synthesis and characterization of fluorescent N-CDs/ZnONPs nanocomposite for latent fingerprint detection by using powder brushing method. *Arabian J. Chem.* **2020**, *13*, 3817–3835. [CrossRef]
23. Pérez-Cadenas, M.; Moreno-Castilla, C.; Carrasco-Marín, F.; Pérez-Cadenas, A.F. Surface chemistry, porous texture, and morphology of N-doped carbon xerogels. *Langmuir* **2009**, *25*, 466–470. [CrossRef] [PubMed]
24. Maboya, W.K.; Coville, N.J.; Mhlanga, S.D. Fabrication of chlorine nitrogen co-doped carbon nanomaterials by an injection catalytic vapor deposition method. *Mater. Res. Express* **2021**, *8*, 015007. [CrossRef]
25. Kenzhin, R.M.; Bauman, Y.I.; Volodin, A.M.; Mishakov, I.V.; Vedyagin, A.A. One-step synthesis of nitrogen-doped carbon nanofibers from melamine over nickel alloy in a closed system. *Chem. Phys. Lett.* **2017**, *685*, 259–262. [CrossRef]
26. Kenzhin, R.M.; Bauman, Y.I.; Volodin, A.M.; Mishakov, I.V.; Vedyagin, A.A. Interaction of heteroatom-containing organic compounds with bulk nickel alloy in a closed reactor system. *Juniper Online J. Mater. Sci.* **2018**, *4*, 555633. [CrossRef]
27. Potylitsyna, A.R.; Mishakov, I.V.; Bauman, Y.I.; Kibis, L.S.; Shubin, Y.V.; Volochaev, M.N.; Melgunov, M.S.; Vedyagin, A.A. Metal dusting as a key route to produce functionalized carbon nanofibers. *Reac. Kinet. Mech. Catal.* **2022**, *135*, 1387–1404. [CrossRef]
28. Brzhezinskaya, M.; Mishakov, I.V.; Bauman, Y.I.; Shubin, Y.V.; Maksimova, T.A.; Stoyanovskii, V.O.; Gerasimov, E.Y.; Vedyagin, A.A. One-pot functionalization of catalytically derived carbon nanostructures with heteroatoms for toxic-free environment. *Appl. Surf. Sci.* **2022**, *590*, 153055. [CrossRef]
29. Jiang, Y.; Zhang, J.; Qin, Y.-H.; Niu, D.-F.; Zhang, X.-S.; Niu, L.; Zhou, X.-G.; Lu, T.-H.; Yuan, W.-K. Ultrasonic synthesis of nitrogen-doped carbon nanofibers as platinum catalyst support for oxygen reduction. *J. Power Sources* **2011**, *196*, 9356–9360. [CrossRef]
30. Yin, J.; Qiu, Y.; Yu, J.; Zhou, X.; Wu, W. Enhancement of electrocatalytic activity for oxygen reduction reaction in alkaline and acid media from electrospun nitrogen-doped carbon nanofibers by surface modification. *RSC Adv.* **2013**, *3*, 15655–15663. [CrossRef]
31. Sevilla, M.; Yu, L.; Zhao, L.; Ania, C.O.; Titiricic, M.-M. Surface modification of CNTs with N-doped carbon: An effective way of enhancing their performance in supercapacitors. *ACS Sustain. Chem. Eng.* **2014**, *2*, 1049–1055. [CrossRef]
32. Zhang, J.; Zhang, X.; Zhou, Y.; Guo, S.; Wang, K.; Liang, Z.; Xu, Q. Nitrogen-doped hierarchical porous carbon nanowhisker ensembles on carbon nanofiber for high-performance supercapacitors. *ACS Sustain. Chem. Eng.* **2014**, *2*, 1525–1533. [CrossRef]
33. Kerdi, F.; Ait Rass, H.; Pinel, C.; Besson, M.; Peru, G.; Leger, B.; Rio, S.; Monflier, E.; Ponchel, A. Evaluation of surface properties and pore structure of carbon on the activity of supported Ru catalysts in the aqueous-phase aerobic oxidation of HMF to FDCA. *Appl. Catal. A Gen.* **2015**, *506*, 206–219. [CrossRef]
34. Wang, T.; Chen, Z.-X.; Chen, Y.-G.; Yang, L.-J.; Yang, X.-D.; Ye, J.-Y.; Xia, H.-P.; Zhou, Z.-Y.; Sun, S.-G. Identifying the active site of N-doped graphene for oxygen reduction by selective chemical modification. *ACS Energy Lett.* **2018**, *3*, 986–991. [CrossRef]
35. Butsyk, O.; Olejnik, P.; Romero, E.; Plonska-Brzezinska, M.E. Postsynthetic treatment of carbon nano-onions: Surface modification by heteroatoms to enhance their capacitive and electrocatalytic properties. *Carbon* **2019**, *147*, 90–104. [CrossRef]
36. Ruiz-Garcia, C.; Heras, F.; Calvo, L.; Alonso-Morales, N.; Rodriguez, J.J.; Gilarranz, M.A. Improving the activity in hydrodechlorination of Pd/C catalysts by nitrogen doping of activated carbon supports. *J. Environ. Chem. Eng.* **2020**, *8*, 103689. [CrossRef]
37. Golub, F.S.; Beloshapkin, S.; Gusel'nikov, A.V.; Bolotov, V.A.; Parmon, V.N.; Bulushev, D.A. Boosting hydrogen production from formic acid over Pd catalysts by deposition of N-containing precursors on the carbon support. *Energies* **2019**, *12*, 3885. [CrossRef]
38. Li, S.; Bian, F.; Wu, X.; Sun, L.; Yang, H.; Meng, X.; Qin, G. Microstructure evolution and its correlation with performance in nitrogen-containing porous carbon prepared by polypyrrole carbonization: Insights from hybrid calculations. *Materials* **2022**, *15*, 3705. [CrossRef]
39. Mishakov, I.V.; Vedyagin, A.A.; Bauman, Y.I.; Shubin, Y.V.; Buyanov, R.A. Synthesis of carbon nanofibers via catalytic chemical vapor deposition of halogenated hydrocarbons. In *Carbon Nanofibers: Synthesis, Applications and Performance*; Chang-Seop, L., Ed.; Nova Science Publishers: Hauppauge, NY, USA, 2018; pp. 77–181.
40. Mishakov, I.V.; Afonnikova, S.D.; Bauman, Y.I.; Shubin, Y.V.; Trenikhin, M.V.; Serkova, A.N.; Vedyagin, A.A. Carbon erosion of a bulk nickel–copper alloy as an effective tool to synthesize carbon nanofibers from hydrocarbons. *Kinet. Catal.* **2022**, *63*, 97–107. [CrossRef]
41. Atwater, M.A.; Guevara, L.N.; Knauss, S.J. Multifunctional porous catalyst produced by mechanical alloying. *Mater. Res. Lett.* **2019**, *7*, 131–136. [CrossRef]

42. Mishakov, I.V.; Bauman, Y.I.; Korneev, D.V.; Vedyagin, A.A. Metal dusting as a route to produce active catalyst for processing chlorinated hydrocarbons into carbon nanomaterials. *Top. Catal.* **2013**, *56*, 1026–1032. [CrossRef]
43. Zhu, Q.-L.; Xu, Q. Immobilization of ultrafine metal nanoparticles to high-surface-area materials and their catalytic applications. *Chem* **2016**, *1*, 220–245. [CrossRef]
44. Saadun, A.J.; Ruiz–Ferrando, A.; Büchele, S.; Faust Akl, D.; López, N.; Pérez–Ramírez, J. Structure sensitivity of nitrogen–doped carbon–supported metal catalysts in dihalomethane hydrodehalogenation. *J. Catal.* **2021**, *404*, 291–305. [CrossRef]
45. Liu, S.; Otero, J.A.; Martin-Martinez, M.; Rodriguez-Franco, D.; Rodriguez, J.J.; Gómez-Sainero, L.M. Understanding hydrodechlorination of chloromethanes. Past and future of the technology. *Catalysts* **2020**, *10*, 1462. [CrossRef]
46. Chesnokov, V.V.; Kriventsov, V.V.; Malykhin, S.E.; Svintsitskiy, D.A.; Podyacheva, O.Y.; Lisitsyn, A.S.; Richards, R.M. Nature of active palladium sites on nitrogen doped carbon nanofibers in selective hydrogenation of acetylene. *Diam. Relat. Mater.* **2018**, *89*, 67–73. [CrossRef]
47. Hu, F.; Leng, L.; Zhang, M.; Chen, W.; Yu, Y.; Wang, J.; Horton, J.H.; Li, Z. Direct synthesis of atomically dispersed palladium atoms supported on graphitic carbon nitride for efficient selective hydrogenation reactions. *ACS Appl. Mater. Interf.* **2020**, *12*, 54146–54154. [CrossRef]
48. Glyzdova, D.V.; Vedyagin, A.A.; Tsapina, A.M.; Kaichev, V.V.; Trigub, A.L.; Trenikhin, M.V.; Shlyapin, D.A.; Tsyrulnikov, P.G.; Lavrenov, A.V. A study on structural features of bimetallic Pd-M/C (M: Zn, Ga, Ag) catalysts for liquid-phase selective hydrogenation of acetylene. *Appl. Catal. A-Gen.* **2018**, *563*, 18–27. [CrossRef]
49. Bulushev, D.A.; Bulusheva, L.G. Catalysts with single metal atoms for the hydrogen production from formic acid. *Catal. Rev.* **2022**, *64*, 835–874. [CrossRef]
50. Suboch, A.N.; Podyacheva, O.Y. Pd Catalysts supported on bamboo-like nitrogen-doped carbon nanotubes for hydrogen production. *Energies* **2021**, *14*, 1501. [CrossRef]
51. Kwak, Y.; Kirk, J.; Moon, S.; Ohm, T.; Lee, Y.-J.; Jang, M.; Park, L.-H.; Ahn, C.-i.; Jeong, H.; Sohn, H.; et al. Hydrogen production from homocyclic liquid organic hydrogen carriers (LOHCs): Benchmarking studies and energy-economic analyses. *Energ. Conv. Manag.* **2021**, *239*, 114124. [CrossRef]
52. Abdin, Z.; Tang, C.; Liu, Y.; Catchpole, K. Large-scale stationary hydrogen storage via liquid organic hydrogen carriers. *iScience* **2021**, *24*, 102966. [CrossRef] [PubMed]
53. Niermann, M.; Drünert, S.; Kaltschmitt, M.; Bonhoff, K. Liquid organic hydrogen carriers (LOHCs)—Techno-economic analysis of LOHCs in a defined process chain. *Energy Environ. Sci.* **2019**, *12*, 290–307. [CrossRef]
54. Gómez, S.; Rendtorff, N.M.; Aglietti, E.F.; Sakka, Y.; Suárez, G. Surface modification of multiwall carbon nanotubes by sulfonitric treatment. *Appl. Surf. Sci.* **2016**, *379*, 264–269. [CrossRef]
55. Turan, K.; Kaur, P.; Manhas, D.; Sharma, J.; Verma, G. Novel insights into the dispersed and acid-mediated surface modification of the carbon nanofibers. *Mater. Chem. Phys.* **2020**, *239*, 121978. [CrossRef]
56. Ud Din, I.; Shaharun, M.S.; Subbarao, D.; Naeem, A. Surface modification of carbon nanofibers by HNO_3 treatment. *Ceram. Int.* **2016**, *42*, 966–970. [CrossRef]
57. Afonnikova, S.D.; Mishakov, I.V.; Bauman, Y.I.; Trenikhin, M.V.; Shubin, Y.V.; Serkova, A.N.; Vedyagin, A.A. Preparation of Ni-Cu Catalyst for carbon nanofiber production by the mechanochemical route. *Top. Catal.* **2022**. [CrossRef]
58. Mel'gunov, M.S.; Ayupov, A.B. Direct method for evaluation of BET adsorbed monolayer capacity. *Micropor. Mesopor. Mat.* **2017**, *243*, 147–153. [CrossRef]
59. Sing, K. The use of nitrogen adsorption for the characterisation of porous materials. *Colloids Surf. A* **2001**, *187–188*, 3–9. [CrossRef]
60. Gor, G.Y.; Thommes, M.; Cychosz, K.A.; Neimark, A.V. Quenched solid density functional theory method for characterization of mesoporous carbons by nitrogen adsorption. *Carbon* **2012**, *50*, 1583–1590. [CrossRef]
61. Moulder, J.F.; Stickle, W.F.; Sobol, W.M.; Bomben, K.D. (Eds.) *Handbook of X-Ray Photoelectron Spectroscopy*; Perkin-Elmer Corporation Physical Electronics Division: Eden Prairie, MN, USA, 1992; p. 261.
62. Thommes, M.; Kaneko, K.; Neimark, A.V.; Olivier, J.P.; Rodriguez-Reinoso, F.; Rouquerol, J.; Sing, K.S.W. Physisorption of gases, with special reference to the evaluation of surface area and pore size distribution (IUPAC Technical Report). *Pure Appl. Chem.* **2015**, *87*, 1051–1069. [CrossRef]
63. Shubin, Y.V.; Bauman, Y.I.; Plyusnin, P.E.; Mishakov, I.V.; Tarasenko, M.S.; Mel'gunov, M.S.; Stoyanovskii, V.O.; Vedyagin, A.A. Facile synthesis of triple Ni-Mo-W alloys and their catalytic properties in chemical vapor deposition of chlorinated hydrocarbons. *J. Alloys Compnd.* **2021**, *866*, 158778. [CrossRef]
64. Mishakov, I.V.; Buyanov, R.A.; Zaikovskii, V.I.; Strel'tsov, I.A.; Vedyagin, A.A. Catalytic synthesis of nanosized feathery carbon structures via the carbide cycle mechanism. *Kinet. Catal.* **2008**, *49*, 868–872. [CrossRef]
65. Monthioux, M.; Noé, L.; Dussault, L.; Dupin, J.C.; Latorre, N.; Ubieto, T.; Romeo, E.; Royo, C.; Monzón, A.; Guimon, C. Texturising and structurising mechanisms of carbon nanofilaments during growth. *J. Mater. Chem.* **2007**, *17*, 4611–4618. [CrossRef]
66. Chesnokov, V.V.; Buyanov, R.A. The formation of carbon filaments upon decomposition of hydrocarbons catalysed by iron subgroup metals and their alloys. *Russ. Chem. Rev.* **2000**, *69*, 623–638. [CrossRef]
67. Nemanich, R.J.; Solin, S.A. First- and second-order Raman scattering from finite-size crystals of graphite. *Phys. Rev. B* **1979**, *20*, 392–401. [CrossRef]
68. Ferrari, A.C.; Robertson, J. Interpretation of Raman spectra of disordered and amorphous carbon. *Phys. Rev. B* **2000**, *61*, 14095–14107. [CrossRef]

69. Tuinstra, F.; Koenig, J.L. Raman Spectrum of Graphite. *J. Chem. Phys.* **1970**, *53*, 1126–1130. [CrossRef]
70. Wang, Y.; Alsmeyer, D.C.; McCreery, R.L. Raman spectroscopy of carbon materials: Structural basis of observed spectra. *Chem. Mater.* **1990**, *2*, 557–563. [CrossRef]
71. Sadezky, A.; Muckenhuber, H.; Grothe, H.; Niessner, R.; Pöschl, U. Raman microspectroscopy of soot and related carbonaceous materials: Spectral analysis and structural information. *Carbon* **2005**, *43*, 1731–1742. [CrossRef]
72. Liu, C.; Sun, C.; Gao, Y.; Lan, W.; Chen, S. Improving the electrochemical properties of carbon paper as cathodes for microfluidic fuel cells by the electrochemical activation in different solutions. *ACS Omega* **2021**, *6*, 19153–19161. [CrossRef]
73. Fu, W.Y.; Wang, K.F.; Lv, X.S.; Fu, H.L.; Dong, X.G.; Chen, L.; Zhang, X.M.; Jiang, G.M. Palladium nanoparticles assembled on titanium nitride for enhanced electrochemical hydrodechlorination of 2,4-dichlorophenol in water. *Chin. J. Catal.* **2018**, *39*, 693–700. [CrossRef]
74. Lindberg, B.J.; Hedman, J. Molecular spectroscopy by means of ESCA. *Chem. Scripta* **1975**, *7*, 155–166.
75. Dementjev, A.P.; de Graaf, A.; van de Sanden, M.C.M.; Maslakov, K.I.; Naumkin, A.V.; Serov, A.A. X-Ray photoelectron spectroscopy reference data for identification of the C3N4 phase in carbon–nitrogen films. *Diam. Relat. Mater.* **2000**, *9*, 1904–1907. [CrossRef]
76. Kuntumalla, M.K.; Attrash, M.; Akhvlediani, R.; Michaelson, S.; Hoffman, A. Nitrogen bonding, work function and thermal stability of nitrided graphite surface: An in situ XPS, UPS and HREELS study. *Appl. Surf. Sci.* **2020**, *525*, 146562. [CrossRef]
77. Ayiania, M.; Smith, M.; Hensley, A.J.R.; Scudiero, L.; McEwen, J.-S.; Garcia-Perez, M. Deconvoluting the XPS spectra for nitrogen-doped chars: An analysis from first principles. *Carbon* **2020**, *162*, 528–544. [CrossRef]
78. Xu, Y.; Mo, Y.; Tian, J.; Wang, P.; Yu, H.; Yu, J. The synergistic effect of graphitic N and pyrrolic N for the enhanced photocatalytic performance of nitrogen-doped graphene/TiO$_2$ nanocomposites. *Appl. Catal. B Environ.* **2016**, *181*, 810–817. [CrossRef]
79. Chen, C.-M.; Zhang, Q.; Yang, M.-G.; Huang, C.-H.; Yang, Y.-G.; Wang, M.-Z. Structural evolution during annealing of thermally reduced graphene nanosheets for application in supercapacitors. *Carbon* **2012**, *50*, 3572–3584. [CrossRef]
80. Oh, Y.J.; Yoo, J.J.; Kim, Y.I.; Yoon, J.K.; Yoon, H.N.; Kim, J.-H.; Park, S.B. Oxygen functional groups and electrochemical capacitive behavior of incompletely reduced graphene oxides as a thin-film electrode of supercapacitor. *Electrochim. Acta* **2014**, *116*, 118–128. [CrossRef]
81. Folkesson, B.; Sundberg, P. A Reinvestigation of the binding energy versus atomic charge relation for oxygen from X-ray photoelectron spectroscopy. *Spectroscopy Lett.* **1987**, *20*, 193–200. [CrossRef]

Article

The Effect of Sibunit Carbon Surface Modification with Diazonium Tosylate Salts of Pd and Pd-Au Catalysts on Furfural Hydrogenation

Dmitrii German [1], Ekaterina Kolobova [1], Ekaterina Pakrieva [1], Sónia A. C. Carabineiro [2,3], Elizaveta Sviridova [1], Sergey Perevezentsev [4], Shahram Alijani [5], Alberto Villa [5], Laura Prati [5], Pavel Postnikov [1], Nina Bogdanchikova [6] and Alexey Pestryakov [1,7,*]

[1] Research School of Chemistry and Applied Biomedical Sciences, National Research Tomsk Polytechnic University, Lenin Av. 30, 634050 Tomsk, Russia; germandmitry93@gmail.com (D.G.); ekaterina_kolobova@mail.ru (E.K.); epakrieva@mail.ru (E.P.); evs31@tpu.ru (E.S.); postnikov@tpu.ru (P.P.)
[2] Centro de Química Estrutural, Institute of Molecular Sciences, Departamento de Engenharia Química, Instituto Superior Técnico, Universidade de Lisboa, Av. Rovisco Pais, 1049-001 Lisboa, Portugal; sonia.carabineiro@fct.unl.pt
[3] LAQV-REQUIMTE, Department of Chemistry, NOVA School of Science and Technology, Universidade NOVA de Lisboa, 2829-516 Caparica, Portugal
[4] Institute of Petroleum Chemistry, Russian Academy of Science, Akademichesky Av. 4, 634021 Tomsk, Russia; sap311@yandex.ru
[5] Dipartimento di Chimica, Università degli Studi di Milano, via Camillo Golgi 19, 20133 Milano, Italy; sli@dinex.fi (S.A.); alberto.villa@unimi.it (A.V.); laura.prati@unimi.it (L.P.)
[6] Centro de Nanociencias y Nanotecnología, Universidad Nacional Autónoma de México, Ensenada 22800, Mexico; nina@cnyn.unam.mx
[7] Laboratory of Catalytic and Biomedical Technologies, Sevastopol State University, 299053 Sevastopol, Russia
* Correspondence: pestryakov2005@yandex.ru

Citation: German, D.; Kolobova, E.; Pakrieva, E.; Carabineiro, S.A.C.; Sviridova, E.; Perevezentsev, S.; Alijani, S.; Villa, A.; Prati, L.; Postnikov, P.; et al. The Effect of Sibunit Carbon Surface Modification with Diazonium Tosylate Salts of Pd and Pd-Au Catalysts on Furfural Hydrogenation. *Materials* 2022, 15, 4695. https://doi.org/10.3390/ma15134695

Academic Editor: Ilya V. Mishakov

Received: 21 May 2022
Accepted: 28 June 2022
Published: 4 July 2022

Publisher's Note: MDPI stays neutral with regard to jurisdictional claims in published maps and institutional affiliations.

Copyright: © 2022 by the authors. Licensee MDPI, Basel, Switzerland. This article is an open access article distributed under the terms and conditions of the Creative Commons Attribution (CC BY) license (https://creativecommons.org/licenses/by/4.0/).

Abstract: Herein, we investigated the effect of the support modification (Sibunit carbon) with diazonium salts of Pd and Pd-Au catalysts on furfural hydrogenation under 5 bars of H_2 and 50 °C. To this end, the surface of Sibunit (Cp) was modified with butyl (Cp-Butyl), carboxyl (Cp-COOH) and amino groups (Cp-NH_2) using corresponding diazonium salts. The catalysts were synthesized by the sol immobilization method. The catalysts as well as the corresponding supports were characterized by Fourier transform infrared spectroscopy, N_2 adsorption-desorption, inductively coupled plasma atomic emission spectroscopy, high resolution transmission electron microscopy, energy dispersive spectroscopy, X-ray diffraction, Hammet indicator method and X-ray photoelectron spectroscopy. The analysis of the results allowed us to determine the crucial influence of surface chemistry on the catalytic behavior of the studied catalysts, especially regarding selectivity. At the same time, the structural, textural, electronic and acid–base properties of the catalysts were practically unaffected. Thus, it can be assumed that the modification of Sibunit with various functional groups leads to changes in the hydrophobic/hydrophilic and/or electrostatic properties of the surface, which influenced the selectivity of the process.

Keywords: gold; palladium; Sibunit carbon; bimetallic catalysts; surface modification; tosylate salts; hydrogenation; furfural; furfuryl alcohol

1. Introduction

Carbon-based materials have captured significant attention of the scientific community for decades due to their interesting properties, such as superb chemical and mechanical dependability, large surface areas and pore volumes, lightweight nature, variable structural and morphological combinations, mass-scale availability, excellent recyclability, and low production cost. This is reflected in the targeted synthesis of allotropic forms of carbon (carbines, fullerenes, nanotubes, circulenes, etc.), as well as in the creation of a wide range of porous materials in a series of mixed (transitional) forms of carbon [1–3].

Carbon-based materials can be successfully used as structural modifiers of construction materials, electronic elements, hydrogen accumulators, and additives to lubricants, varnishes and paints, gas distribution layers of fuel cells, and high-performance adsorbents. The use of carbon nanostructures in fine chemical synthesis, biology, and medicine is also widely discussed [4–6].

Notably, carbon materials are suitable catalyst supports for metal nanoparticles, (NPs) owing to advantages such as developed porous space to transfer reactants and reaction products, chemical inertness (especially in the presence of strong acids and bases), mechanical stability, structural diversity, and controlled chemical surface properties [7–9]. Among the variety of carbon supports for heterogeneous catalysts, the Sibunit material attracts particular attention because it has a unique combination of properties of graphite (chemical stability) and activated carbon (high specific surface and sorption capacity). Palladium heterogeneous catalysts deposited on Sibunit were applied in industrial processes of hydropurification of terephthalic acid, hydrogenation of m-nitrobenzene trifluoride and o-nitrophenol, and in rosin disproportionation [10].

At the same time, carbon materials free from surface functional groups are known to be hydrophobic or non-polar, which can lead to weak stabilization of metal NPs on the carbon support. Therefore, these factors determine the sorption interaction of the active component with the support. Besides metal dispersion enhancement, the functionalities on the carbon surface can act as anchoring sites in the synthesis of carbon-based composite materials; and carbon surface modification can change the electronic surface state and the contribution of the active state of the deposited metal, modify the acid–base properties, etc., [11,12].

Thus, chemical functionalization can be used to tailor the surface physicochemical properties by applying appropriate thermal or chemical treatments or by anchoring the desired functional groups.

Among the functionalization approaches, diazonium chemistry is becoming increasingly attractive as this promising method, which combines the ease of preparation of diazonium salts, rapid reduction, and strong aryl–substrate surface atoms covalent bonding, allows grafting different organic moieties onto various solid supports, such as carbon material [13–15]. Among all types of diazonium salts, arenediazonium tosylates can be considered as the most convenient reagents for the surface modification, due to their stability, inexplosive properties and good solubility in water [16]. Moreover, the arenediazonium tosylates have been widely applied for the surface modification of metal and carbon surfaces [17,18].

Surface functionalization by incorporation of other elements (heteroatoms) or functional groups (which are formed from these heteroatoms) is believed to be essential for the enhancement of the catalytic activity of carbon-containing materials. For instance, an increase in the basic properties via introducing nitrogen-containing groups of the support might be beneficial in liquid phase oxidation or hydrogenation. The previous paper demonstrated that the presence of N on the carbon surface can influence the oxidation state of active sites (Pd^{2+} and Au^+) and, consequently, led to higher desired acid production, derived from 5-hydroxymetylfurfurol oxidation [19]. Amadou et al. investigated Pd on nitrogen-doped carbon nanotubes for the selective hydrogenation of cinnamaldehyde into hydrocinnamaldehyde [20]. The high TOF and high selectivity towards the C=C bond hydrogenation were attributed to possible electronic or morphological modifications that occurred after nitrogen atom incorporation. Acidic groups, such as carboxyl, quinone and lactone, provide interaction between the carbon surface and positively charged cations of the precursor metal. In addition, they reduce the hydrophobicity of carbon, thus making the surface more accessible for aqueous solutions of precursors. Liang et al. found that the acidic functional groups (carboxyl groups) of activated carbon, used as the support for Pd catalysts, enhanced the content of Pd^{2+}, benefited the dispersion of Pd, and eventually improved the H_2O_2 selectivity from H_2 and O_2 [21]. Phenol, carbonyl and ether groups are slightly acidic or neutral. For example, Bianchi et al. found that the activity in the liquid-phase oxidation of ethylene glycol on Au/C with similarly sized Au particles increases by increasing the amount phenol-type groups, indicating that a specific metal–support interaction does exist [22].

In this study, the hydrogenation of furfural was used as a model reaction due to the wide range of possible products, which allows us to clearly demonstrate the possibility of varying the selectivity by changing the surface chemistry. Moreover, by hydrogenation of furfural, organic compounds, such as furfuryl alcohol, tetrahydrofurfuryl alcohol, 2-methylfuran and 2-methyltetrahydrofuran, can be obtained, which are alternative sources of organic substances [23]. Furfural hydrogenation products are in demand as environmentally friendly solvents, in the polymer and coating industry, fuel additives, pharmaceuticals, etc [24–26]. Table 1 shows results for hydrogenation of furfural over various heteroheneous catalysts.

Table 1. Catalytic hydrogenation of furfural on different palladium catalysts.

Catalyst	Reaction Conditions	Conv., %	Selectivity, %				Ref.
			FA	THFA	2-MF	2-MTHF	
1% Pd/TiO$_2$	RT, 3 bar H$_2$, 0.1 g of catalyst, 1 g of furfural, octane—solvent, 2 h	20.7	73	14	7	-	[27]
2.5 % Pd— 2.5 % Ru/TiO$_2$	RT, 3 bar H$_2$, 0.1 g of catalyst, 1 g of furfural, octane—solvent, 2 h	33.8	58	0.4	14	-	[27]
5% Pd/C	150 °C, 2.0 MPa H$_2$, 800 rpm, 0.4 g of catalysts, 9.6 g of furfural, 6 g of acetic acid, toluene—solvent, 4 h	41.2	35	-	21.6	-	[28]
5 % Pd/Al$_2$(SiO$_3$)$_3$	150 °C, 2.0 MPa H$_2$, 800 rpm, 0.4 g of catalysts, 9.6 g of furfural, 6 g of acetic acid, toluene—solvent, 4 h	56.9	52.7	-	-	-	[28]
1% Pd/CNT	180 °C, 2.0 MPa H$_2$, 600 rpm, 0.01 g of catalysts, 1 g of furfural, 2-propanol—solvent, 5 h	29.1	14.9	2.5	1.8	-	[29]
1% Pd/Vulcan	180 °C, 2.0 MPa H$_2$, 600 rpm, 0.01 g of catalysts, 1 g of furfural, 2-propanol—solvent, 5 h	46.7	24.3	7.2	1.8	-	[29]
1% Pd/CMK-5	180 °C, 2.0 MPa H$_2$, 600 rpm, 0.01 g of catalysts, 1 g of furfural, 2-propanol—solvent, 5 h	100	20.3	31.5	13.4	-	[29]
Pd/C + CuO	170 °C, 3.0 MPa H$_2$, 500 rpm, 0.3 g of furfural, 1,4-dioxane—solvent, 3 h	24.1	61.4	5.8	-	-	[30]
0.9 % Pd/ZrO$_2$	30 °C, 0.5 MPa H$_2$, 0.05 g of catalysts, 0.1 mL of furfural, water—solvent, 3 h	34	21	16	-	-	[31]
0.8 % Pd/SiO$_2$	30 °C, 0.5 MPa H$_2$, 0.05 g of catalysts, 0.1 mL of furfural, water—solvent, 3 h	63	39	21	-	-	[31]
1 % Pd/MgAlO$_x$	150 °C, 3.0 MPa H$_2$, 0.5 g of catalysts, 5.0 cm^3 of furfural, water—solvent, 1 h	62.9	48.6	32.4	-	-	[32]

Thus, the preparation of highly dispersed, efficient and selective metal catalysts relies heavily on the proper control of the surface chemistry of the carbon support, which is a great scientific and practical interest in the last few years for the development of advanced catalytic systems for different conversions. This study aims to show how the catalytic behavior of mono- and bimetallic catalysts for furfural hydrogenation can be influenced by modifying the surface of the carbon material with diazonium tosylate salts.

2. Materials and Methods

The reagents (furfural, sodium tetrachloropalladate (II) (Na$_2$PdCl$_4$), sodium tetrachloroaurate (III) dehydrate (NaAuCl$_4$·H$_2$O), polyvinyl alcohol (PVA), sulfuric acid (H$_2$SO$_4$), sodium borohydride (NaBH$_4$)) from Merck (Darmstadt, Germany) were used. The carbon material Sibunit was purchased from the Center of New Chemical Technologies of Boreskov Institute of Catalysis (Omsk, Russia).

The diazonium tosylate (4-carboxybenzenediazonium tosylate, 4-butylbenzenediazonium tosylate or 4-aminobenzenediazonium tosylate) was prepared according to the published procedure [33,34]. The covalent modification was carried out by the immersion of Sibunit in the solution of diazonium salts as follows: the carbon material Sibunit (denoted hereinafter as Cp) (1 g) was dispersed in 5 mL of water/methanol (4/1) solution and 1 mmol of the corresponding diazonium tosylate (4-carboxybenzenediazonium tosylate, 4-butylbenzenediazonium tosylate or 4-aminobenzenediazonium tosylate) was added to solution. The mixture was sonicated for 20 min and then stirred for 60 min at 60 °C. Afterwards, the modified powders were sequentially rinsed under sonication with deionized water, ethanol, and methanol for 10 min and dried in a desiccator for 3 h (Figure 1).

The monometallic catalysts (Pd/Cp, Pd/Cp-COOH, Pd/Cp-butyl and Pd/Cp-NH_2) were synthesized by the sol immobilization method [35]. A total of 1 mL Na_2PdCl_4 solution (10 mg Pd/mL H_2O) and 0.5 mL PVA solution (1 wt. %) were added to 100 mL miliQ H_2O under vigorous stirring. After a few minutes, a solution of $NaBH_4$ (Pd/$NaBH_4$ = 1/8 mol/mol) was added to form a brown palladium sol. Then, the support was added into the sol solution (assuming that the catalyst would contain 1 wt. % palladium) and a few drops of H_2SO_4 (to immobilize palladium nanoparticles on the support). The synthesis was carried out for 1 h. The resulting catalyst washed and dried at 80 °C for 2 h under air.

The bimetallic catalysts (Pd-Au/Cp, Pd-Au/Cp-COOH, Pd-Au/Cp-butyl and Pd-Au/Cp-NH_2) were prepared by the same method as the monometallics. A total of 0.684 mL Na_2PdCl_4 (10 mg Pd/mL H_2O), 0.316 mL $NaAuCl_4 \cdot H_2O$ (10 mg Au/mL H_2O) and 0.5 mL PVA solution (1 wt. %) were added to 100 mL miliQ H_2O under vigorous stirring. After a few minutes, a solution of $NaBH_4$ (metal/$NaBH_4$ = 1/8 mol/mol) was added and a brownish-red palladium-gold sol was formed. Then, the support was added in the sol solution (assuming that the catalyst would contain 1 wt.% of metals with a ratio of Pd:Au = 4:1 mol/mol) and several drops of H_2SO_4 (for immobilization of palladium and gold nanoparticles on the support). The synthesis was carried out for 1 h. The resulting catalyst was washed and dried at 80 °C for 2 h under air.

Fourier transform infrared (FTIR) spectra were recorded using a ThermoScientific FTIR instrument (Nicolet 8700) to confirm the functionalization of the Sibunite surface. Palladium and gold content were measured by inductively coupled plasma atomic emission spectroscopy (ICP-AES) using iCAP 6300 Duo (Thermo Fisher Scientific, Waltham, MA, USA). XRD patterns were registered by Bruker D8 X-ray diffractometer (Bruker Corporation, Billerica, MA, USA) for the identification of the phase composition of catalysts. The textural properties were measured using an ASAP 2060 (Micromeritics Instrument Corporation, Norcross, GA, USA) apparatus. The acid–base properties of the supports and the corresponding catalysts were studied by the Hammett indicator method. The sizes and distribution of Pd or Pd-Au particles were estimated by high resolution transmission electron microscopy (HRTEM) using a JEOL JEM-2100F (JEOL Ltd., Tokyo, Japan). Surface composition and the chemical state of each element were determined by X-ray photoelectron spectroscopy (XPS), performed on a VG Scientific ESCALAB 200A (Thermo Fisher Scientific, Waltham, MA, USA). Detailed information is available in the Supplementary Materials.

The reactions of hydrogenation were carried out under 0.5 MPa pressure of H_2, at 50 °C and stirring at 1200 rpm (Figure 2). The molar ratio of metal: furfural was 1:500 for monometallic and 1:580 for bimetallic systems. The catalytic reactions were performed in a 50 mL stainless steel batch reactor equipped with a heater and a stirrer. Firstly, 10 mL of a 0.3 M solution of furfural in isopropanol and catalyst were mixed into the reactor, and then the system was purged several times with nitrogen and hydrogen. To monitor the progress of the reaction, aliquots were taken at certain time intervals and analyzed using an Agilent 6890 gas chromatograph (Agilent, Santa Clara, CA, USA) equipped with a Zebron ZB-5 column 60 m × 0.32 mm × 1 μm.

3. Results

3.1. FTIR Results

A scheme dealing with the functionalization is shown in Figure 1. The surface modification by butylphenyl, aminophenyl and carboxyphenyl groups was proved by Fourier transform infrared (FTIR) spectroscopy. FTIR spectrum of Cp exhibits vibration bands associated with the graphitic carbon structure (C=C bond at 1600 cm^{-1}, CH$_3$ group at 1440–40 cm^{-1}). After modification, novel vibration bands associated with the covalent attachment of amino, carboxyl or butyl groups appeared, related to the stretching vibrations of C=O (1700 cm^{-1}), butyl aliphatic stretch (1100–1250 cm^{-1}) and C–N stretch (1300–1100 cm^{-1}) (Figure 3).

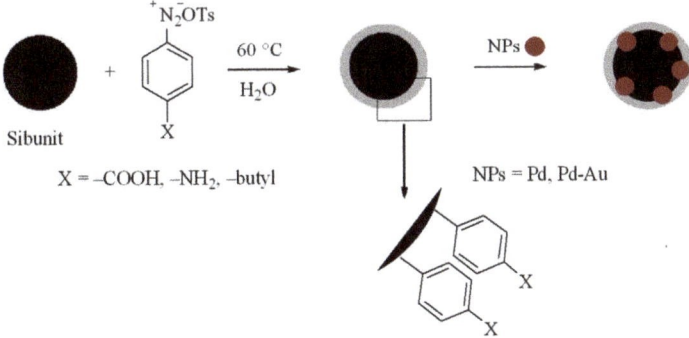

Figure 1. Scheme of support modification by diazonium salts and immobilization of Pd and Au nanoparticles.

Figure 2. Schematic representation of the catalytic process.

3.2. ICP-AES and XEDS Results

Palladium and gold contents (Table 2) were determined by inductively coupled plasma atomic emission spectroscopy (ICP-AES) and energy dispersive spectroscopy (XEDS). The presented data show that the Pd and Au contents are close to the nominal values for the entire series of catalysts.

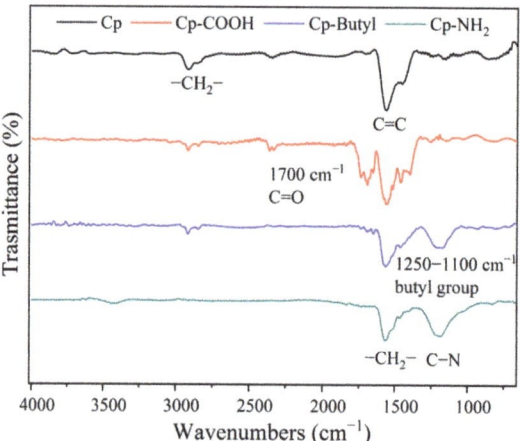

Figure 3. FTIR spectra of Cp, Cp−COOH, Cp−butyl and Cp−NH$_2$.

Table 2. Pd and Au contents in the catalysts.

Catalyst	Content of Element, %	
	Pd	Au [a]
Nominal	0.7–1.0 [b]	0–0.3 [c]
Pd/Cp	0.72	0
Pd-Au/Cp	0.73	0.32
Pd/Cp-COOH	0.93	0
Pd-Au/Cp-COOH	0.63	0.36
Pd/Cp-butyl	0.71	0
Pd-Au/Cp-butyl	0.58	0.42
Pd/Cp-NH$_2$	0.92	0
Pd-Au/Cp-NH$_2$	0.78	0.28

[a] determined by XEDS; [b] content of Pd is 1.0 and 0.7% for mono- and bimetallic catalysts, respectively; [c] content of Au is 0 and 0.3% for mono- and bimetallic catalysts, respectively.

3.3. XRD Results

The phase composition of the catalysts and corresponding supports was studied using X-ray diffraction (Figure 4). The analysis of the spectra showed the absence of palladium and gold reflections for all samples, except Pd/Cp. For the latter, maxima were found in the XRD pattern at 2θ = 40.1°, 46.6°, and 68.1°, which are typical for palladium with a face-centered cubic lattice [36]. The appearance of the reflex for Pd/Cp is probably related to the non-uniform distribution of Pd NPs on the support surface resulting in their local accumulation, as evidenced by EDX maps (Figure 5). The absence of reflections in the X-ray diffraction patterns of the remaining samples is probably due to the small size of palladium and gold particles (below sensitivity of the XRD method) or their X-ray amorphous structure. The reflections of the support (2θ = 25.7°, 44.3°) correspond to multi-walled carbon nanotubes, which is consistent with the literature data [37,38].

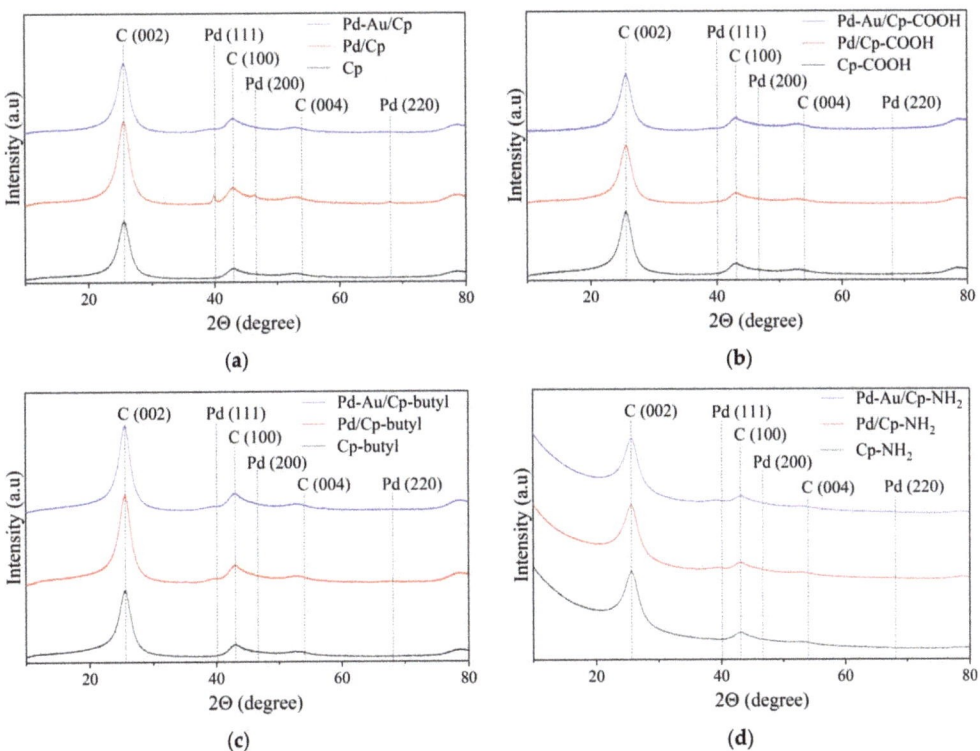

Figure 4. XRD patterns of catalysts and corresponding supports: (**a**) Cp, Pd/Cp and Pd-Au/Cp; (**b**) Cp-COOH, Pd/Cp-COOH and Pd-Au/Cp-COOH; (**c**) Cp-butyl, Pd/Cp-butyl and Pd-Au/Cp-butyl; (**d**) Cp-NH$_2$, Pd/Cp-NH$_2$ and Pd-Au/Cp-NH$_2$.

3.4. Textural Properties

According to the data presented in Table 3, the specific surface area (S_{BET}) is reduced by 26, 12 and 16%, respectively (Table 3, entries 1, 4, 7 and 10) as a result of the Sibunit modification by carboxyl, butyl and amino groups. Taking into account that the average cross section of a nitrogen molecule is 3.2 Å, the presence of functional groups at the entrance to the pores will prevent their filling with gas, thereby leading to a decrease in the surface available for adsorption [39–41]. This effect is more pronounced for pores lower than 2 nm. An important role in the manifestation of this effect is played by the nature of the functional groups. In the present study, it is clearly demonstrated by the example of a decrease in the specific surface of micro- and mesopores, as well as an increase in the average size of mesopores, indicating the blockage of small mesopores (Table 3). Separately, it should be noted that in the case of modification of Sibunit by carboxyl groups, all micropores become inaccessible to nitrogen molecules (Table 3, entry 4).

The deposition of palladium on the surface of unmodified and carboxyl-modified Sibunit resulted in an increase in S_{BET} by 10% and 8%, respectively, compared to the corresponding supports (Table 3, entries 1, 2, 4 and 5). An increase in the specific surface area after the deposition of palladium may indicate a small size of Pd nanoparticles (Pd NPs), a defective structure of NPs and their uniform distribution over the support surface, thereby providing new adsorption sites. In the case of a support modified by butyl groups, the application of palladium does not significantly change the value of the specific surface area compared to the corresponding support (Table 3, entries 7 and 8). When palladium is supported on Sibunit modified by amino groups, the specific surface area does not change (Table 3, entries 10 and 11).

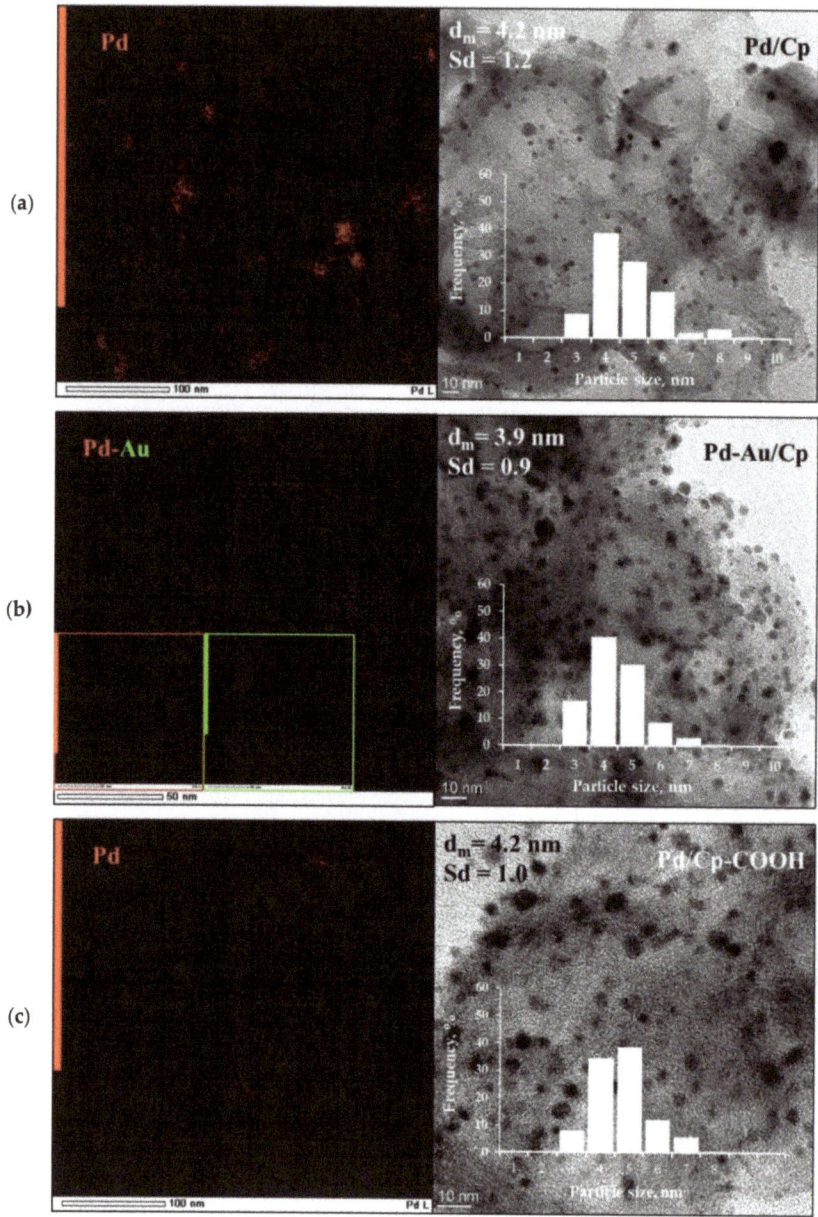

Figure 5. *Cont.*

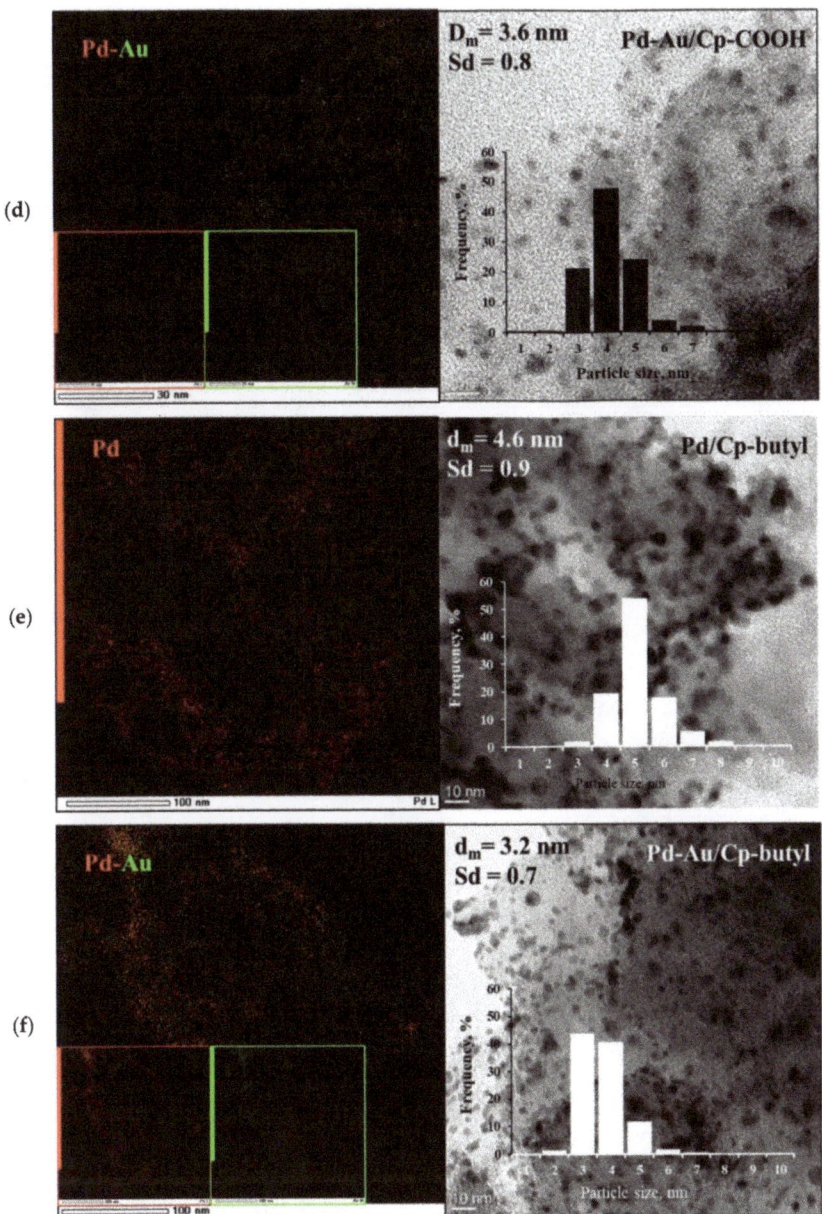

Figure 5. Cont.

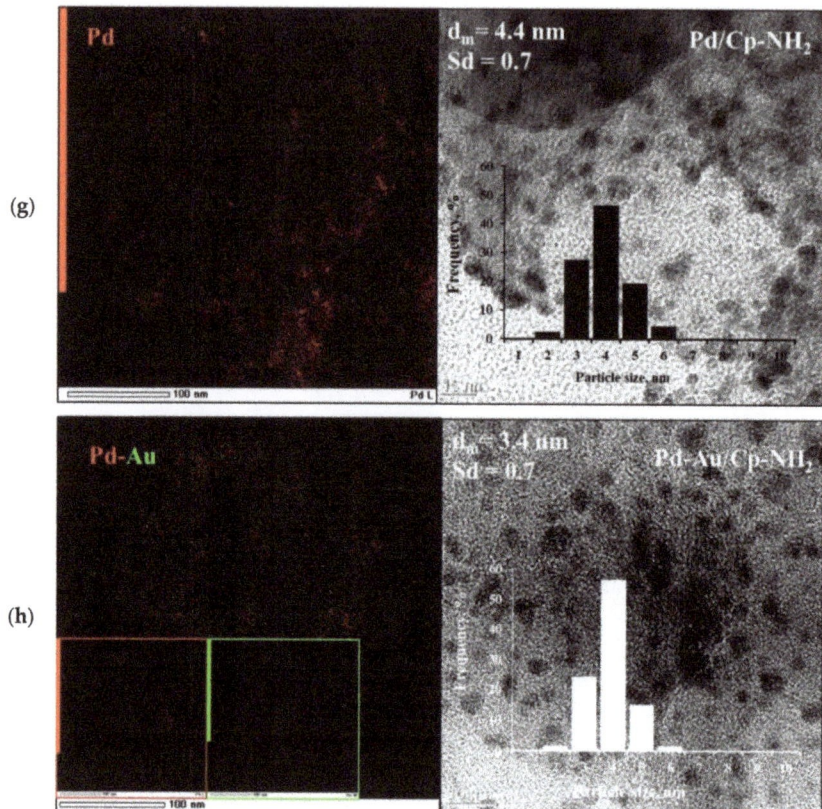

Figure 5. TEM images and EDX maps of the studied catalysts, as well as Pd and Pd-Au NPs distribution: (**a**) Pd/Cp; (**b**) Pd-Au/Cp; (**c**) Pd/Cp-COOH; (**d**) Pd-Au/Cp-COOH; (**e**) Pd/Cp-butyl; (**f**) Pd-Au/Cp-butyl; (**g**) Pd/Cp-NH$_2$; (**h**) Pd-Au/Cp-NH$_2$.

Table 3. Textural properties of supports and corresponding catalysts.

Entry	Sample	BET Surface Area, m^2 g^{-1}	Surface Area of Mesopores, m^2 g^{-1}	Surface Area of Micropores, m^2 g^{-1}	Mesopore Volume, cm^3 g^{-1}	Micropore Volume, cm^3 g^{-1}	Mesopore Size, nm	Micropore Size, nm
1	Cp	281	310	7	0.48	0.003	6.2	1.9
2	Pd/Cp	314	357	5	0.59	0.002	6.6	2.0
3	Pd-Au/Cp	327	368	5	0.60	0.003	6.5	2.0
4	Cp-COOH	209	264	-	0.45	-	6.8	-
5	Pd/Cp-COOH	228	280	2	0.45	0.001	6.4	1.9
6	Pd-Au/Cp-COOH	252	302	4	0.52	0.002	6.9	1.9
7	Cp-butyl	246	307	4	0.53	0.002	6.9	1.8
8	Pd/Cp-butyl	237	299	3	0.49	0.002	6.5	1.8
9	Pd-Au/Cp-butyl	258	321	3	0.52	0.001	6.5	1.8
10	Cp-NH$_2$	235	294	4	0.48	0.002	6.5	1.8
11	Pd/Cp-NH$_2$	235	291	3	0.48	0.002	6.0	1.6
12	Pd-Au/Cp-NH$_2$	274	326	6	0.48	0.003	5.9	1.8

Bimetallic catalysts are characterized by higher S$_{BET}$ compared to the corresponding monometallic catalysts and supports (Table 3, entries 3, 6, 9 and 12). This may be a consequence of the formation of additional adsorption sites, due to the uniform distribution and small size of Pd-Au NPs. It is also necessary to consider possible changes in the surface of the supports under the action of the reagents used in the process of catalyst synthesis. For all investigated catalysts (Table 3, entries 2, 3, 5, 6, 8, 9, 11 and 12), the average sizes and volumes of micro- and mesopores vary from 1.8 to 2.0 nm and from 5.9 to 6.9 nm, from 0.001 to 0.003 cm^3/g and from 0.45 to 0.60 cm^3/g, respectively. In this case, the observed changes in the size and volume of the pores may be due to the impact of the reagents used for the synthesis of catalysts, as well as pore blocking by metal nanoparticles, which can serve as new adsorption sites.

3.5. Acid-Base Properties

The distribution and concentration of acid–base sites on the surface of supports and corresponding catalysts are presented in Table 4.

Table 4. The concentration of acid–base centers on the surface of the studied catalysts and corresponding supports (q, μmol/g): 1—Cp; 2—Pd/Cp; 3—Pd-Au/Cp; 4—Cp-COOH; 5—Pd/Cp-COOH; 6—Pd-Au/Cp-COOH; 7—Cp-butyl; 8—Pd/Cp-butyl; 9—Pd-Au/Cp-butyl; 10—Cp-NH$_2$; 11—Pd/Cp-NH$_2$; 12—Pd-Au/Cp-NH$_2$.

pK$_a$	SAMPLE											
	1	2	3	4	5	6	7	8	9	10	11	12
LEWIS BASE												
−0.29	13.7	14.5	13.4	5.9	13.8	13.3	10.9	11,5	10.3	14.3	14.5	14
BRØNSTED ACID												
0.71	7.0	6.8	6.9	5.1	5.1	5.7	4.9	6.1	7.0	7.2	7.2	7.1
1.30	3.5	2.6	3.5	2.0	3.4	3.4	2.0	1.6	2.0	3.1	2.7	3.6
2.50	6.7	7.1	7.0	0.0	2.5	4.5	5.6	1.7	2.8	0.1	5.9	5
3.46	6.6	6.7	7.1	7.1	5.8	7.3	6.9	6.1	6.4	7.2	7	7
5.00	6.8	6.7	6.2	7.3	6.5	6.7	6.5	5.7	5.9	2.4	5.5	5.4
6.40	2.2	1.9	0.3	0.1	0.1	0.1	0.1	0.1	0.4	0.5	2	2.1
Σ	32.8	31.8	31.0	21.6	23.4	27.7	26.0	21.3	24.5	20.5	30.3	30.2
BRØNSTED BASE												
9.45	14.9	14.4	14.2	4.2	6.8	9.9	6.1	9.2	10.3	13.2	0.8	0.5
10.50	3.2	3.6	3.6	3.1	3.4	3.1	3.3	3.0	2.7	3	2.9	3
12.00	8.3	7.2	8.9	6.3	7.8	7.7	5.9	5.6	8.2	9.4	8.4	8.7
Σ	26.4	25.2	26.7	13.6	18.0	20.7	15.3	17.8	21.2	25.6	12.1	12.2
LEWIS ACID												
17.20	9.0	7.8	8.3	0.4	4.3	0.9	6.5	8.4	9.6	8.2	7.7	8.1
TOTAL												
Σ	81.9	79.3	79.4	41.5	59.5	62.6	58.7	59	65.6	68.6	64.6	64.5

For Lewis basic sites (LBS), the highest content of LBS was detected on the surface of samples modified by amino groups, including the support and the corresponding mono- and bimetallic catalysts, followed by unmodified samples, materials modified by carboxyl groups, except the support, and samples containing butyl groups. Within the same group of samples. The concentration of LBS is practically unchanged.

For Brønsted acid sites (BAS), the highest content of BAS was detected on the surface of unmodified and amino-modified samples, except for Cp-NH$_2$, followed by samples modified by carboxyl and butyl groups. The concentration of BAS for the unmodified samples varies slightly between the support and the corresponding catalysts. For samples modified by carboxyl and amino groups, the BAS concentration increases after palladium or bimetallic Pd-Au system deposition and decreases for samples containing a butyl group.

For Brønsted basic sites (BBS), the highest content of BBS was detected on the surface of unmodified samples. At the same time, the concentration of BBS for support and corresponding mono- and bimetallic catalysts is practically the same. The samples modified by carboxyl and butyl groups follow in terms of BBS content. The concentration of BBS on the surface of these samples increases in the following order: supports < monometallic catalysts < bimetallic catalysts. In the case of samples modified by amino groups, the highest BBS content is observed on the surface of the support; after palladium or bimetallic deposition, it decreases by more than 2 fold.

For Lewis acid sites (LAS), the highest LAS content was detected on the surface of unmodified and amino-modified samples. Moreover, the LAS concentration changes slightly after Sibunit modification by amino groups or deposition of palladium and bimetallic Pd-Au system. For samples containing a butyl functional group, the LAS concentration increases in the following order: support < monometallic catalyst < bimetallic catalyst. Modification of Sibunit by carboxyl groups leads to almost complete abolishment of LAS. However, after palladium deposition, their concentration increases by 11 fold. At the same time, application of a bimetallic system leads to an increase in LAS concentration by only 2.3 fold in comparison with the corresponding support.

The analysis of the obtained results indicates the predominance of Brønsted acid and basic sites (BAS and BBS) on the surface of all the studied samples. In the case of unmodified samples, the concentration of all types of sites slightly varies between the support and the corresponding mono- and bimetallic catalysts. However, for the modified supports, the redistribution of acid and basic sites is observed after palladium or bimetallic Pd-Au system deposition. At the same time, the character of the change in acid–base properties depends on the nature of the functional groups (COOH, butyl or NH_2).

3.6. HRTEM Results

Figure 5 shows HRTEM images and EDX maps of the investigated catalysts, as well as histograms of palladium nanoparticles (Pd NPs) and bimetallic Pd-Au nanoparticle (Pd-Au NP) distribution on the surface of these catalysts.

The distribution of Pd NPs on the unmodified Sibunit surface, as well as modified by butyl, carboxyl and amino groups, is approximately the same and ranges from 2 to 9 nm, with the average size of Pd NPs being 4.2 nm for Pd/Cp and Pd/Cp-COOH catalysts, 4.6 nm for Pd/Cp-butyl and 4.4 nm for Pd/Cp-NH_2. The bimetallic catalysts are characterized by a smaller average nanoparticle size and narrower distribution compared to the corresponding monometallic systems. The smallest average size of NPs was found on the surface of the Pd-Au/Cp-butyl sample (3.2 nm). The average size of Pd-Au NPs in the case of Pd-Au/Cp, Pd-Au/Cp-COOH and Pd-Au/Cp-NH_2 catalysts was 3.9, 3.6 and 3.4 nm, respectively. A possible explanation for the formation of smaller NPs in the case of bimetallic catalysts may be the shorter metal–metal bond length for Pd-Au (2.50 Å) compared with the bond length for Pd-Pd (2.74 Å), which in turn leads to the formation of NPs with higher tightly packed crystal lattice [42,43]. The EDX maps show clear evidence of the formation of bimetallic Pd-Au NPs. According to the data presented, palladium and gold are localized on the support surface in close proximity to each other.

3.7. XPS Results

The electronic states of palladium, gold (in the case of bimetallic catalysts), oxygen and carbon on the surface of the investigated materials were assessed by XPS.

The Pd3d XPS spectra are shown in Figure 6. Analysis of the spectra demonstrates that palladium is present in the following three states on the surface of all catalysts: Pd^0, Pd^{2+} and Pd^{4+} with binding energies ($Pd3d_{5/2}$) 335.9—336.1, 337.7—337.8 and 338.7 eV, respectively [44–47]. It is worth noting that the binding energy values of 335.9 and 336.0 eV, which refers to the Pd^0 state in the current study, are 0.5 and 0.6 eV higher than the standard BE value characterizing the Pd^0 state (335.4 eV), which indicates the presence of highly dispersed metal particles on the surface of the studied samples, for which a shift toward higher binding energies up to 1 eV is possible [48–51].

The contribution of different electronic states of palladium on the surface of the investigated samples, determined by deconvolution of the Pd3d spectrum, is presented in Table 5. The data show that, for most catalysts, the ratio between different states of palladium is approximately the same, except for the Pd/Cp sample, for which 41% of palladium is in the oxidized state (Pd^{2+} and Pd^{4+}), while for other catalysts, this value does not exceed 22%, and increases in the following order: $Pd/Cp-NH_2$ and $Pd-Au/Cp-NH_2 < Pd-Au/Cp < Pd/Cp$-butyl and $Pd-Au/Cp$-butyl < Pd/Cp-COOH and $Pd-Au/Cp$-COOH. It should be separately noted that surface palladium concentration for unmodified catalysts (Table 6) is only 0.1–0.2 at.%, whereas, for other samples, this value varies from 0.8 to 2.7 at.% and increases in the following order: COOH < butyl < NH_2. The low surface concentration of palladium on the surface of Pd/Cp and, accordingly, the low intensity of the signal, are probably due to the non-uniform distribution of Pd NPs on the support surface resulting in their local accumulation, as evidenced by EDX maps (Figure 5).

Table 5. Effect of support modification on the contribution of different electronic states of Pd and Au on the catalyst surface calculated according to XPS.

Catalyst	$Pd^{(0, 2+ or 4+)}$ Relative Content, %			$Au^{(0 or 1+)}$ Relative Content, %	
	Pd^0 (335.9–336.1 eV)	Pd^{2+} (337.7–337.8 eV)	Pd^{4+} (338.7 eV)	Au^0 (84.1–84.2 eV)	Au^+ (85.2–85.4 eV)
Pd/Cp	59	23	18	-	-
Pd-Au/Cp	84	9	7	93	7
Pd/Cp-COOH	78	14	8	-	-
Pd-Au/Cp-COOH	80	12	8	90	10
Pd/Cp-butyl	83	11	6	-	-
Pd-Au/Cp-butyl	83	12	5	93	7
Pd/Cp-NH_2	89	8	3	-	-
Pd-Au/Cp-NH_2	91	7	2	92	8

Table 6. Surface concentration of elements on the catalyst surface (at.%) determined by XPS.

Catalyst	Element				
	C1s	O1s	N1s	Pd3d	Au4f
Pd/Cp	95.9	3.8	0.07	0.2	-
Pd-Au/Cp	95.3	2.9	0.06	0.1	0.3
Pd/Cp-COOH	91.8	7.1	0.20	1.0	-
Pd-Au/Cp-COOH	93.0	5.9	n.d.	0.8	0.2
Pd/Cp-butyl	94.1	4.5	0.02	1.3	-
Pd-Au/Cp-butyl	94.2	4.2	0.17	1.1	0.4
Pd/Cp-NH_2	77.1	16.8	1.69	2.7	-
Pd-Au/Cp-NH_2	81.5	13.1	1.62	2.5	1.23

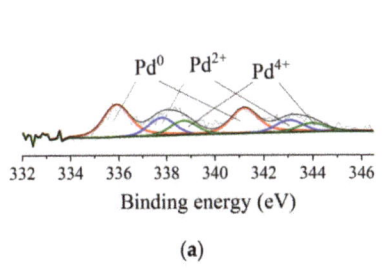

(a)

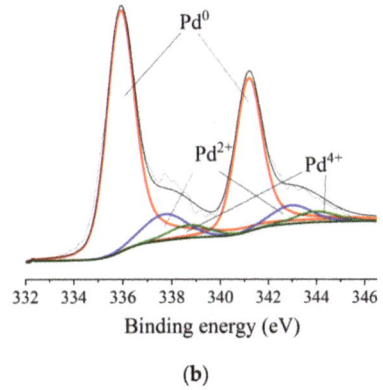

(b)

Figure 6. *Cont.*

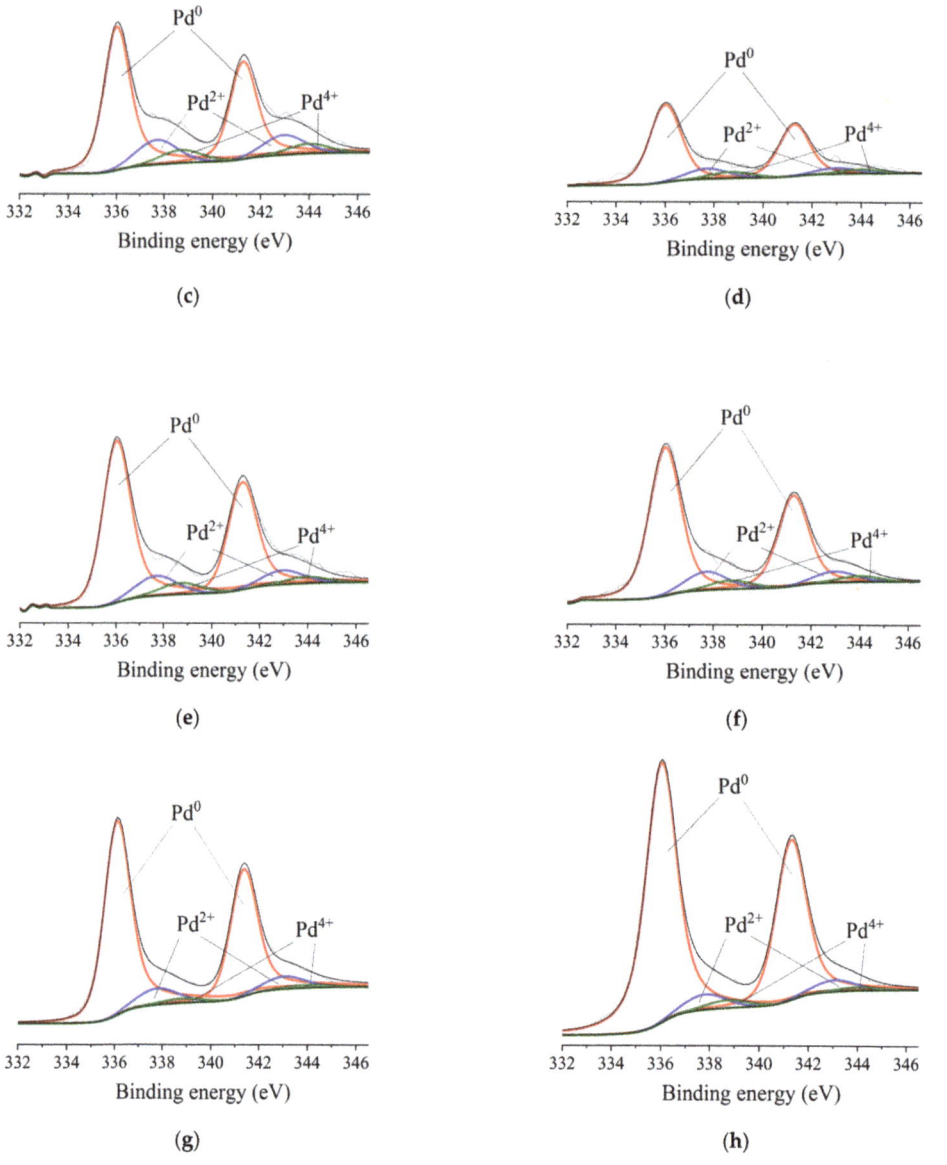

Figure 6. Pd3d XPS spectra: (**a**) Pd/Cp; (**b**) Pd-Au/Cp; (**c**) Pd/Cp-COOH; (**d**) Pd-Au/Cp-COOH; (**e**) Pd/Cp-butyl; (**f**) Pd-Au/Cp-butyl; (**g**) Pd/Cp-NH$_2$; (**h**) Pd-Au/Cp-NH$_2$.

Figure 7 shows the Au4f XPS spectra. According to the presented data, gold is in the following two states on the surface of all bimetallic catalysts: Au^0 and Au$^+$, with bonding energies (Au4f$_{7/2}$) 84.1–84.2 and 85.2–85.4 eV, respectively. The ratio between these states changes insignificantly when the surface of Sibunit is modified by butyl, carboxyl and amino groups (Table 5). However, the surface concentration of gold, for the studied catalysts, varies in the range of 0.2–1.23 at.% (Table 6) and increases as follows: Pd-Au/Cp-COOH < Pd-Au/Cp < Pd-Au/Cp-butyl < Pd-Au/Cp-NH$_2$.

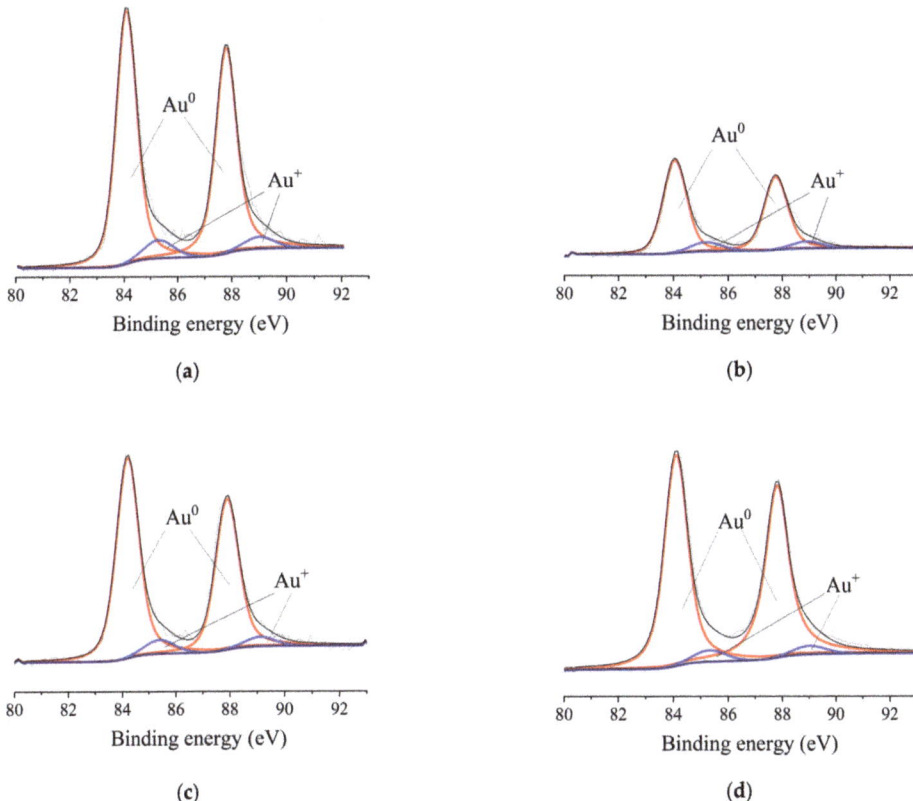

Figure 7. Au4f XPS spectra of (**a**) Pd-Au/Cp, (**b**) Pd-Au/Cp-COOH, (**c**) Pd-Au/Cp-butyl and (**d**) Pd-Au/Cp-NH$_2$.

For all studied catalysts, the O1s peak deconvolved into four states (Figure S1, Supplementary Materials) related to oxygen atoms within carbonyl groups with BE (O1s) = 531.5–531. 6 eV (C=O); oxygen atoms bound in single bonds to carbon atoms with BE O1s) = 532.5–532.7 eV (C-O); oxygen atoms in hydroxyl groups with BE (O1s) = 533.7–533.9 eV (C-OH) and in carboxyl groups and/or adsorbed water with BE (O1s) = 535.0–535.1 eV (O in H$_2$O or COOH) [52–54]. The relative contribution of each oxygen state is presented in Table 7. The main contribution is made by oxygen bound in single bonds with carbon, with the content in the catalyst varying from 46% (Pd/Cp-NH$_2$) to 65% (Pd/Cp). The fraction of oxygen bound to hydrogen and carbon in the hydroxyl group varies from 16% (Pd/Cp) to 32% (Pd/Cp-COOH and Pd/Cp-NH$_2$). For the samples modified by carboxyl and amino groups, the fraction of oxygen as part of the C-OH groups is less for bimetallic catalysts; for unmodified samples, this fraction is smaller for the monometallic sample. In the case of butyl containing mono- and bimetallic catalysts, the fraction of oxygen in the composition of C-OH groups is almost the same. The relative content of oxygen bound by a double bond with carbon (C=O) is the highest for the Pd-Au/Cp-COOH catalyst (19%), and smallest for Pd/Cp-COOH (9%). For all other catalysts, the oxygen content of the C=O group is approximately the same and varies from 12 to 16%. The amount of oxygen in the form of adsorbed water and/or carboxyl groups (O in H$_2$O or COOH) does not exceed 9% for all the samples. It is worth noting separately that the highest surface oxygen concentration was found for Pd/Cp-NH$_2$ (16.8 at.%) and Pd-Au/Cp-NH$_2$ (13.1 at.%) catalysts, followed by Pd/Cp-COOH (7.1 at.%) and Pd-Au/Cp-COOH (5.9 at.%), for Pd/Cp-butyl (4.5 at.%)

and Pd-Au/Cp-butyl (4.2 at.%); unmodified mono- and bimetallic catalysts have the lowest oxygen concentration of 3.8 and 2.9 at.%, respectively (Table 6).

Table 7. Catalytic results of the hydrogenation of furfural.

Entry	Catalyst	d_m, nm	Conversion for 5 h, %	Selectivity at 70% of Conversion, %					
				FA	THFA	2-MF	2-MTHF	IPFE	Others
1	Pd/Cp	4.2	80	33	3	2	0	51	11
2	Pd/Cp-butyl	4.6	93	55	18	3	0	22	2
3	Pd/Cp-COOH	4.2	96	27	4	2	0	27	41
4	Pd/Cp-NH$_2$	4.4	66	1	0	0	0	62	37
5	Pd-Au/Cp	3.9	97	26	2	2	0	69	1
6	Pd-Au/Cp-butyl	3.2	75	74	5	3	0	14	4
7	Pd-Au/Cp-COOH	3.6	97	54	7	6	0	32	1
8	Pd-Au/Cp-NH$_2$	3.4	49	31	3	1	0	16	49

FA—furfuryl alcohol; THFA—tetrahydrofurfuryl alcohol; 2-MF—2-methylfuran; 2-MTHF—2 methyltetrahydrofuran; IPFE—isopropyl furfuryl ether; others—furan, tetrahydrofuran, 1,2—pentanediol, etc. Reaction conditions: 0.3 M furfural (0.2882 g) in 10 mL 2-propanol, amount of catalyst 0.0645 g (Pd/furfural = 1:500 mol/mol; Pd-Au/furfural = 1:580 mol/mol), T = 50 °C, pH2 = 5 bar, t = 5 h, stirring 1000 rpm.

The C1s XPS spectra of the studied catalysts are shown in Figure S2, Supplementary Materials. The C1s peaks were deconvolved into five components characterizing the carbon states in C-C (284.8 eV), C-O (285.5–285.6 eV), C=O (286.7–286.9 eV), O-C=O (288.7–288.9 eV) and π-π* (291.0–291.2 eV) [54–58]. Based on the analysis of the contributions of the different carbon states (Table S3), the following conclusions can be drawn: the main contribution is made by C-C, the relative carbon content in this state varies from 64 to 70%; the carbon content of the oxygen-containing functional groups varies from 25–31%, and 3–5% in the π-π* bonds. It is important to note that the C=O value for the NH$_2$-modified samples is the highest, compared to the other samples, due to the possible overlapping peaks of the C=O and N-C=O functional groups, which are approximately in the same range of the binding energies [59]. In general, modifying the Sibunit surface or introducing gold along with palladium has little effect on changing the contribution of the different carbon states.

3.8. Hydrogenation of Furfural

The catalytic behavior of the supported mono- and bimetallic unmodified and modified catalysts was evaluated in the reaction of liquid-phase hydrogenation of furfural at 50 °C and 5 bar H$_2$ (Table 7, Scheme 1, Figures 8 and 9).

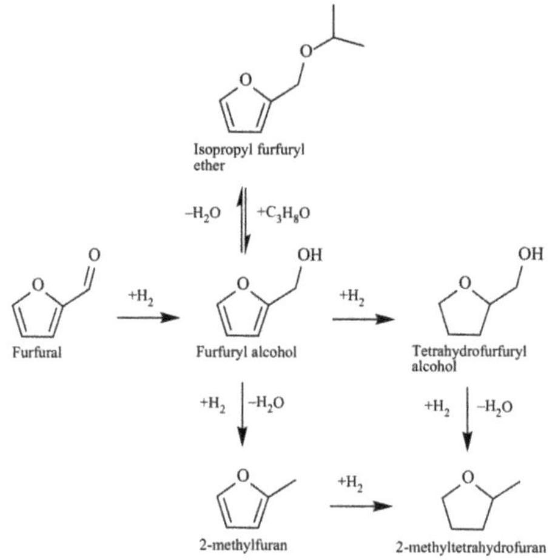

Scheme 1. Possible products of the furfural hydrogenation.

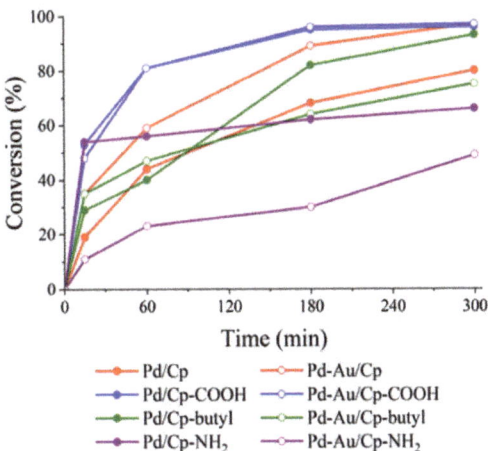

Figure 8. Conversion of furfural in hydrogenation processes over different catalysts. Reaction conditions: 0.3 M furfural (0.2882 g) in 10 mL 2-propanol, amount of catalyst 0.0645 g (Pd/furfural = 1:500 mol/mol; Pd-Au/furfural = 1:580 mol/mol), T = 50 °C, p (H2) = 5 bar, t = 5 h, stirring 1000 rpm.

The highest furfural conversion (96–97%) was achieved on Pd/Cp-COOH, Pd-Au/Cp-COOH and Pd-Au/Cp (Table 7, entries 3, 5 and 7, Figure 8). However, despite the high activity (furfural conversion), the selectivity for the desired products (furfuryl alcohol, tetrahydrofurfuryl alcohol, 2-methylfuran, 2-methyltetrahydrofuran) using these catalysts did not exceed 67% (Table 7, entries 3, 5 and 7, Figure 9). It is worth noting that, for Pd/Cp-COOH, more than 40% of the reaction products were unidentified (Table 7, entry 3, Figure 9) and for Pd-Au/Cp, the main hydrogenation product was isopropyl furfuryl ether, formed as a result of the interaction between the reagent (furfuryl) and the solvent (isopropyl alcohol) (Table 7, entry 5, Figure 9). For Pd/Cp, the conversion of furfural was 80% (Table 7, entry 1, Figure 9) and the distribution of reaction products for this sample was similar to Pd-Au/Cp (Table 7, entry 5, Figure 9). In the case of the monometallic sample modified by butyl groups, the conversion of furfural was 93% with a wide distribution of reaction products, among which, furfural and tetrahydrofurfural accounted for 55% and 18%, respectively (Table 7, entry 2, Figure 9). When gold was introduced into this catalyst, along with palladium, the selectivity changed dramatically and furfuryl alcohol was the major product of the reaction (Table 7, entry 6, Figure 9). Overall, by evaluating the reaction product distribution at the same conversion level (70%), this catalyst was the most selective for the desired products compared to the other studied samples. At the same time, it should be noted that this sample had the lowest conversion of furfural (75%) among the studied catalysts. The least active and selective were catalysts on the basis of Pd and Pd-Au NPs supported on Sibunit modified with amino groups (Table 7, Figures 4 and 9). The highest conversion achieved for these samples was 66%. The bimetallic catalyst is more selective for the main products but less active than the corresponding monometallic catalyst.

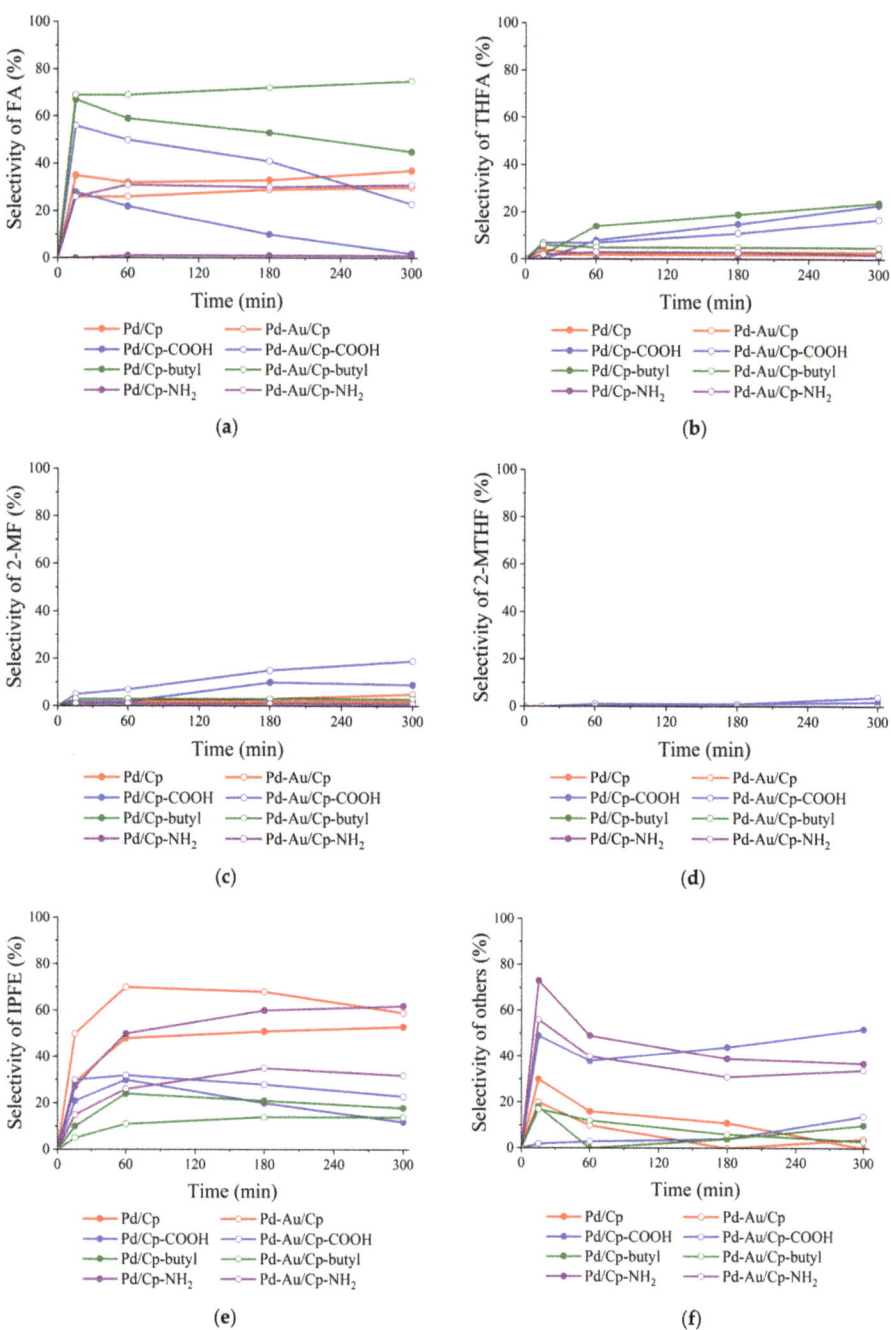

Figure 9. Time evolution of catalytic hydrogenation of furfural: (**a**) furfuryl alcohol; (**b**) tetrahydrofurfuryl alcohol; (**c**) 2-methylfuran; (**d**) 2-methyltetrahydrofuran; (**e**) isopropyl furfuryl ether; (**f**) others. Reaction conditions: 0.3 M furfural (0.2882 g) in 10 mL 2-propanol, amount of catalyst 0.0645 g (Pd/furfural = 1:500 mol/mol; Pd-Au/furfural = 1:580 mol/mol), T = 50 °C, pH$_2$ = 5 bar, t = 5 h, stirring 1000 rpm.

4. Conclusions

The core idea of the present study is to show that the catalytic behavior (activity and selectivity) of furfural hydrogenation catalysts can be influenced not only by changing the reaction parameters [60] or the content of supported metals [61], but also by changing the surface chemistry of the support, if all other conditions are kept equal. By analyzing the above results of the physicochemical and catalytic studies, it can be concluded that the most significant variation in the physicochemical properties of catalysts, after modification of the support, is observed in a change in the average size of metal particles and its distribution, which in turn affects the catalytic behavior of the studied materials. At the same time, it should be taken into account that the nature of the functional groups (butyl, carboxyl or amino group) plays an equally important role alongside the particle size and its distribution. The most striking example, in this case, is a comparison of the catalytic characteristics for Pd-Au/Cp-butyl and Pd-Au/Cp-NH_2 with close NP size Pd-Au NPs, but very different catalytic performances (furfural conversion and selectivity). It is worth noting separately that no correlation between the structural (Figure 4), textural (Table 3), acid–base (Table 4), electronic (Table 5, Tables S2 and S3; Figures 6 and 7, Figures S1 and S2) and catalytic properties (Table 7, Figures 8 and 9) of the studied materials was found. The structural and electronic properties of the catalysts were the least affected by the modification. The analysis of the kinetic curves revealed the considerable effect of surface functional groups on activity and selectivity, which cannot be explained by the identified minor changes in the physicochemical properties of the studied catalysts. Thus, we hypothesize that the selectivity can be achieved by specific interactions of intermediates with functional groups associated with electrostatic and/or hydrophobic/hydrophilic binding. At the same time, the in-depth evaluation of the mechanism obviously requires a comprehensive study, which is outside the scope of this paper. Nevertheless, the main conclusion of this study is the possibility to finely tune the performance of a catalyst, including the selectivity, by appropriate modification of the carbon support.

Supplementary Materials: The following supporting information can be downloaded at: https://www.mdpi.com/article/10.3390/ma15134695/s1, Figure S1: O1s XPS spectra of (**a**) Pd/Cp; (**b**) Pd-Au/Cp; (**c**) Pd/Cp-COOH; (**d**) Pd-Au/Cp-COOH; (**e**) Pd/Cp-butyl; (**f**) Pd-Au/Cp-butyl; (**g**) Pd/Cp-NH_2 and (**h**) Pd-Au/Cp-NH_2; Figure S2: C1s XPS spectra of (**a**) Pd/Cp; (**b**) Pd-Au/Cp; (**c**) Pd/Cp-COOH; (**d**) Pd-Au/Cp-COOH; (**e**) Pd/Cp-butyl; (**f**) Pd-Au/Cp-butyl; (**g**) Pd/Cp-NH_2 and (**h**) Pd-Au/Cp-NH_2; Figure S3: O1s XPS spectra of (**a**) Cp; (**b**) Cp-COOH; (**c**) Cp-butyl and (**d**) Cp-NH_2; Figure S4: C1s XPS spectra of (**a**) Cp; (**b**) Cp-COOH; (**c**) Cp-butyl and (**d**) Cp-NH_2; Table S1: Effect of support modification on contribution of different electronic states of oxygen on the support surface calculated according to XPS; Table S2: Effect of support modification on contribution of different electronic states of carbon on the support surface calculated according to XPS; Table S3: Surface concentration of elements on the support surface (at.%) determined by XPS; Table S4: Effect of support modification on contribution of different electronic states of oxygen on the support surface calculated according to XPS; Table S5: Effect of support modification on contribution of different electronic states of carbon on the support surface calculated according to XPS; Table S6: Surface concentration of elements on the support surface (at.%) determined by XPS.

Author Contributions: D.G. was responsible for the preparation of the supported Pd and Pd-Au catalysts, performed catalytic tests on these materials and the Hammett indicator method, interpreted XPS, XRD and TEM results and contributed to the writing; E.K. and E.P. were responsible for supervising the catalytic experiments, participated in the conceptualization and methodology of most of the characterization methods and contributed to the writing; S.A.C.C. was responsible for the XPS and BET analyses; E.S. was responsible for the modifying of Sibunit by tosylate salts and contributed to the writing; S.P. was responsible for the XRD analyses; S.A. was responsible for the methodology of the catalytic tests with supported Pd and Pd-Au on Sibunit materials; A.V. and L.P. dealt with the methodology of the preliminary catalytic tests; E.K., S.A.C.C., A.V., L.P., P.P., N.B. and A.P. provided the means for the realization of this work and contributed to the supervision and paper revision. All authors have read and agreed to the published version of the manuscript.

Funding: The research is funded by the Ministry of Education and Science of the Russian Federation Program № 075-03-2021-287/6, Sevastopol State University Research grant 42-01-09/169/2021-4 (Russia).

Institutional Review Board Statement: Not applicable.

Informed Consent Statement: Not applicable.

Data Availability Statement: Data available upon request.

Acknowledgments: The TEM and EDX analyses were carried out at the Innovation Centre for Nanomaterials and Nanotechnologies of Tomsk Polytechnic University. ICP-AES was performed using the core facilities of TPU's "Physics and Chemical Methods of Analysis". XPS analyses were carried out in CEMUP—Center of Materials of the University of Porto, Portugal. Fundação para a Ciência e a Tecnologia for Scientific Employment Stimulus Institutional Call (CEECINST/00102/2018), UIDB/50006/2020 and UIDP/50006/2020 (LAQV), UIDB/00100/2020 and UIDP/00100/2020 (Centro de Química Estrutural).

Conflicts of Interest: The authors declare no conflict of interest.

References

1. Inagaki, M.; Kang, F.; Toyoda, M.; Konno, H. *Advanced Materials Science and Engineering of Carbon*; Butterworth-Heinemann: Waltham, MA, USA, 2013; ISBN 9780124077898.
2. Kubozono, Y. *Physics and Chemistry of Carbon-Based Materials: Basics and Applications*; Springer Nature Singapore Pte. Ltd.: Singapore, 2019; ISBN 9789811334177.
3. Gawande, M.B.; Fornasiero, P.; Zbořil, R. Carbon-based single-atom catalysts for advanced applications. *ACS Catal.* **2020**, *10*, 2231–2259. [CrossRef]
4. Inagaki, M. Structure and texture of carbon materials. In *Carbons for Electrochemical Energy Storage and Conversion Systems*; Beguin, F., Frackowiak, E., Eds.; CRC Press: Boca Raton, FL, USA, 2009; pp. 37–76.
5. Gopinath, K.P.; Vo, D.V.N.; Gnana Prakash, D.; Adithya Joseph, A.; Viswanathan, S.; Arun, J. Environmental applications of carbon-based materials: A Review. *Environ. Chem. Lett.* **2021**, *19*, 557–582. [CrossRef]
6. Cazorla-Amorós, D. Grand Challenges in carbon-based materials research. *Front. Mater.* **2014**, *1*, 6. [CrossRef]
7. Auer, E.; Freund, A.; Pietsch, J.; Tacke, T. Carbons as supports for industrial precious metal catalysts. *Appl. Catal. A Gen.* **1998**, *173*, 259–271. [CrossRef]
8. Doesburg, E.B.M.; De Jong, K.P.; Van Hooff, J.H.C. Preparation of catalyst supports, zeolites and mesoporous materials. In *Studies in Surface Science and Catalysis*; Moulijn, J.A., van Leeuwen, P.W.N.M., van Santen, R.A., Averill, B.A., Eds.; Elsevier: Amsterdam, The Netherlands, 1999; Volume 123, pp. 433–457.
9. Doesburg, E.B.M.; De Jong, K.P.; Van Hooff, J.H.C. Preparation of supported catalysts. In *Studies in Surface Science and Catalysis*; Moulijn, J.A., van Leeuwen, P.W.N.M., van Santen, R.A., Averill, B.A., Eds.; Elsevier: Amsterdam, The Netherlands, 1999; Volume 123, pp. 459–485.
10. Surovikin, V.F.; Surovikin, Y.V.; Tsekhanovich, M.S. New fields in the technology for manufacturing carbon-carbon materials. Application of carbon-carbon materials. *Russ. J. Gen. Chem.* **2007**, *77*, 2301–2310. [CrossRef]
11. Rodríguez-Reinoso, F.; Sepulveda-Escribano, A. Carbon as catalyst support. In *Carbon Materials for Catalysis*; John Wiley & Sons: Hoboken, NJ, USA, 2008; pp. 131–155. [CrossRef]
12. Rehman, A.; Park, M.; Park, S.J. Current progress on the surface chemical modification of carbonaceous materials. *Coatings* **2019**, *9*, 103. [CrossRef]
13. Mahouche-Chergui, S.; Gam-Derouich, S.; Mangeney, C.; Chehimi, M.M. Aryl diazonium salts: A new class of coupling agents for bonding polymers, biomacromolecules and nanoparticles to surfaces. *Chem. Soc. Rev.* **2011**, *40*, 4143–4166. [CrossRef] [PubMed]
14. Ahmad, R.; Boubekeur-Lecaque, L.; Nguyen, M.; Lau-Truong, S.; Lamouri, A.; Decorse, P.; Galtayries, A.; Pinson, J.; Felidj, N.; Mangeney, C. Tailoring the surface chemistry of gold nanorods through Au-C/Ag-C covalent bonds using aryl diazonium salts. *J. Phys. Chem. C* **2014**, *118*, 19098–19105. [CrossRef]
15. Assresahegn, B.D.; Brousse, T.; Bélanger, D. Advances on the use of diazonium chemistry for functionalization of materials used in energy storage systems. *Carbon* **2015**, *92*, 362–381. [CrossRef]
16. Filimonov, V.D.; Trusova, M.; Postnikov, P.; Krasnokutskaya, E.A.; Lee, Y.M.; Hwang, H.Y.; Kim, H.; Chi, K.W. Unusually stable, versatile, and pure arene diazonium tosylates: Their preparation, structures, and synthetic applicability. *Org. Lett.* **2008**, *10*, 3961–3964. [CrossRef]
17. Ahmad, A.A.L.; Marutheri Parambath, J.B.; Postnikov, P.S.; Guselnikova, O.; Chehimi, M.M.; Bruce, M.R.M.; Bruce, A.E.; Mohamed, A.A. Conceptual developments of aryldiazonium salts as modifiers for gold colloids and surfaces. *Langmuir* **2021**, *37*, 8897–8907. [CrossRef]
18. Bensghaïer, A.; Mousli, F.; Lamouri, A.; Postnikov, P.S.; Chehimi, M.M. *The Molecular and Macromolecular Level of Carbon Nanotube Modification Via Diazonium Chemistry: Emphasis on the 2010s Years*; Springer International Publishing: Berlin/Heidelberg, Germany, 2020; Volume 3, ISBN 0123456789.

19. German, D.; Pakrieva, E.; Kolobova, E.; Carabineiro, S.A.C.; Stucchi, M.; Villa, A.; Prati, L.; Bogdanchikova, N.; Corberán, V.C.; Pestryakov, A. Oxidation of 5-hydroxymethylfurfural on supported Ag, Au, Pd and bimetallic Pd-Au catalysts: Effect of the support. *Catalysts* **2021**, *11*, 115. [CrossRef]
20. Amadou, J.; Chizari, K.; Houllé, M.; Janowska, I.; Ersen, O.; Bégin, D.; Pham-Huu, C. N-doped carbon nanotubes for liquid-phase C=C bond hydrogenation. *Catal. Today* **2008**, *138*, 62–68. [CrossRef]
21. Liang, W.; Dong, J.; Yao, M.; Fu, J.; Chen, H.; Zhang, X. Enhancing the selectivity of Pd/C catalysts for the direct synthesis of H_2O_2 by HNO_3 pretreatment. *New J. Chem.* **2020**, *44*, 18579–18587. [CrossRef]
22. Bianchi, C.L.; Biella, S.; Gervasini, A.; Prati, L.; Rossi, M. Gold on carbon: Influence of support properties on catalyst activity in liquid-phase oxidation. *Catal. Lett.* **2003**, *85*, 91–96. [CrossRef]
23. Matsagar, B.M.; Hsu, C.; Chen, S.S.; Ahamad, T.; Alshehri, S.M.; Tsang, D.C.W.; Wu, K.C. Sustainable energy & fuels selective hydrogenation of furfural to tetrahydrofurfuryl alcohol over a rh-loaded carbon catalyst in aqueous solution under mild conditions. *Sustain. Energy Fuels* **2020**, *4*, 293–301. [CrossRef]
24. Gilkey, M.J.; Panagiotopoulou, P.; Mironenko, A.V.; Jenness, G.R.; Vlachos, D.G.; Xu, B. Mechanistic insights into metal lewis acid-mediated catalytic transfer hydrogenation of furfural to 2-Methylfuran. *ACS Catal.* **2015**, *5*, 3988–3994. [CrossRef]
25. Mariscal, R.; Ojeda, M. Environmental science molecule for the synthesis of chemicals and fuels. *Energy Environ. Sci.* **2016**, *9*, 1144–1189. [CrossRef]
26. Chen, S.; Wojcieszak, R.; Dumeignil, F.; Marceau, E. How catalysts and experimental conditions determine the selective hydroconversion of furfural and 5-Hydroxymethylfurfural. *Chem. Rev.* **2018**, *118*, 11023–11117. [CrossRef] [PubMed]
27. Aldosari, O.F.; Iqbal, S.; Miedziak, P.J.; Brett, G.L.; Jones, D.R.; Liu, X.; Edwards, J.K.; Morgan, D.J.; Knight, D.K.; Hutchings, G.J. Pd-Ru/TiO_2 catalyst—An active and selective catalyst for furfural hydrogenation. *Catal. Sci. Technol.* **2016**, *6*, 234–242. [CrossRef]
28. Yu, W.; Tang, Y.; Mo, L.; Chen, P.; Lou, H.; Zheng, X. One-step hydrogenation-esterification of furfural and acetic acid over bifunctional Pd catalysts for bio-oil upgrading. *Bioresour. Technol.* **2011**, *102*, 8241–8246. [CrossRef] [PubMed]
29. Lee, J.; Woo, J.; Nguyen-Huy, C.; Lee, M.S.; Joo, S.H.; An, K. Highly dispersed Pd catalysts supported on various carbons for furfural hydrogenation. *Catal. Today* **2020**, *350*, 71–79. [CrossRef]
30. Du, J.; Zhang, J.; Sun, Y.; Jia, W.; Si, Z.; Gao, H.; Tang, X.; Zeng, X.; Lei, T.; Liu, S.; et al. Catalytic transfer hydrogenation of biomass-derived furfural to furfuryl alcohol over in-situ prepared nano Cu-Pd/C catalyst using formic acid as hydrogen source. *J. Catal.* **2018**, *368*, 69–78. [CrossRef]
31. Huang, R.; Cui, Q.; Yuan, Q.; Wu, H.; Guan, Y.; Wu, P. Total hydrogenation of furfural over Pd/Al_2O_3 and Ru/ZrO_2 mixture under mild conditions: Essential role of tetrahydrofurfural as an intermediate and support effect. *ACS Sustain. Chem. Eng.* **2018**, *6*, 6957–6964. [CrossRef]
32. Mironenko, R.M.; Talsi, V.P.; Gulyaeva, T.I.; Trenikhin, M.V.; Belskaya, O.B. Aqueous-phase hydrogenation of furfural over supported palladium catalysts: Effect of the support on the reaction routes. *React. Kinet. Mech. Catal.* **2019**, *126*, 811–827. [CrossRef]
33. Guselnikova, O.; Kalachyova, Y.; Hrobonova, K.; Trusova, M.; Barek, J.; Postnikov, P.; Svorcik, V.; Lyutakov, O. SERS platform for detection of lipids and disease markers prepared using modification of plasmonic-active gold gratings by lipophilic moieties. *Sens. Actuators B Chem.* **2018**, *265*, 182–192. [CrossRef]
34. Guselnikova, O.; Postnikov, P.; Elashnikov, R.; Trusova, M.; Kalachyova, Y.; Libansky, M.; Barek, J.; Kolska, Z.; Švorčík, V.; Lyutakov, O. Surface modification of Au and Ag Plasmonic thin films via diazonium chemistry: Evaluation of structure and properties. *Colloids Surf. A Physicochem. Eng. Asp.* **2017**, *516*, 274–285. [CrossRef]
35. Campisi, S.; Ferri, D.; Villa, A.; Wang, W.; Wang, D.; Kröcher, O.; Prati, L. Selectivity control in palladium-catalyzed alcohol oxidation through selective blocking of active sites. *J. Phys. Chem. C* **2016**, *120*, 14027–14033. [CrossRef]
36. Cornelio, B.; Saunders, A.R.; Solomonsz, W.A.; Laronze-Cochard, M.; Fontana, A.; Sapi, J.; Khlobystov, A.N.; Rance, G.A. Palladium nanoparticles in catalytic carbon nanoreactors: The effect of confinement on Suzuki-Miyaura reactions. *J. Mater. Chem. A* **2015**, *3*, 3918–3927. [CrossRef]
37. Taran, O.P.; Descorme, C.; Polyanskaya, E.M.; Ayusheyev, A.B.; Besson, M.; Parmon, V.N. Catalysts based on carbon material "sibunit" for the deep oxidation of organic toxicants in water solutions. Aerobic oxidation of phenol in the presence of oxidized carbon and Ru/C catalysts. *Katal. V Promyshlennosti* **2013**, *1*, 40–50.
38. Selen, V.; Güler, Ö.; Özer, D.; Evin, E. Synthesized multi-walled carbon nanotubes as a potential adsorbent for the removal of methylene blue dye: Kinetics, isotherms, and thermodynamics. *Desalin. Water Treat.* **2016**, *57*, 8826–8838. [CrossRef]
39. Delaporte, N.; Belanger, R.L.; Lajoie, G.; Trudeau, M.; Zaghib, K. Multi-carbonyl molecules immobilized on high surface area carbon by diazonium chemistry for energy storage applications. *Electrochim. Acta* **2019**, *308*, 99–114. [CrossRef]
40. Toupin, M.; Bélanger, D. Spontaneous functionalization of carbon black by reaction with 4-Nitrophenyldiazonium cations. *Langmuir* **2008**, *24*, 1910–1917. [CrossRef]
41. Lyskawa, J.; Grondein, A.; Bélanger, D. Chemical modifications of carbon powders with aminophenyl and cyanophenyl groups and a study of their reactivity. *Carbon* **2010**, *48*, 1271–1278. [CrossRef]
42. Briggs, B.D.; Bedford, N.M.; Seifert, S.; Koerner, H.; Ramezani-Dakhel, H.; Heinz, H.; Naik, R.R.; Frenkel, A.I.; Knecht, M.R. Atomic-scale identification of Pd leaching in nanoparticle catalyzed C-C coupling: Effects of particle surface disorder. *Chem. Sci.* **2015**, *6*, 6413–6419. [CrossRef]

43. Matczak, P. Computational study of the adsorption of molecular hydrogen on PdAg, PdAu, PtAg, and PtAu dimers. *React. Kinet. Mech. Catal.* **2011**, *102*, 1–20. [CrossRef]
44. Boronin, A.I.; Slavinskaya, E.M.; Danilova, I.G.; Gulyaev, R.V.; Amosov, Y.I.; Kuznetsov, P.A.; Polukhina, I.A.; Koscheev, S.V.; Zaikovskii, V.I.; Noskov, A.S. Investigation of palladium interaction with cerium oxide and its state in catalysts for low-temperature CO oxidation. *Catal. Today* **2009**, *144*, 201–211. [CrossRef]
45. Mirkelamoglu, B.; Karakas, G. The role of alkali-metal promotion on CO oxidation over PdO/SnO_2 catalysts. *Appl. Catal. A Gen.* **2006**, *299*, 84–94. [CrossRef]
46. Mucalo, M.R.; Cooney, R.P.; Metson, J.B. Platinum and palladium hydrosols: Characterisation by X-ray photoelectron spectroscopy and transmission electron microscopy. *Colloids Surf.* **1991**, *60*, 175–197. [CrossRef]
47. Kibis, L.S.; Titkov, A.I.; Stadnichenko, A.I.; Koscheev, S.V.; Boronin, A.I. X-ray photoelectron spectroscopy study of Pd oxidation by RF discharge in oxygen. *Appl. Surf. Sci.* **2009**, *255*, 9248–9254. [CrossRef]
48. Díez, N.; Śliwak, A.; Gryglewicz, S.; Grzyb, B.; Gryglewicz, G. Enhanced reduction of graphene oxide by high-pressure hydrothermal treatment. *RSC Adv.* **2015**, *5*, 81831–81837. [CrossRef]
49. Ivanova, A.S.; Korneeva, E.V.; Slavinskaya, E.M.; Zyuzin, D.A.; Moroz, E.M.; Danilova, I.G.; Gulyaev, R.V.; Boronin, A.I.; Stonkus, O.A.; Zaikovskii, V.I. Role of the support in the formation of the properties of a Pd/Al_2O_3 catalyst for the low-temperature oxidation of carbon monoxide. *Kinet. Catal.* **2014**, *55*, 748–762. [CrossRef]
50. Wertheim, G.K. Core-electron binding energies in free and supported metal clusters. *Zeitschrift für Phys. B Condens. Matter* **1987**, *66*, 53–63. [CrossRef]
51. Yu, W.; Hou, H.; Xin, Z.; Niu, S.; Xie, Y.; Ji, X.; Shao, L. Nanosizing Pd on 3D porous carbon frameworks as effective catalysts for selective phenylacetylene hydrogenation. *RSC Adv.* **2017**, *7*, 15309–15314. [CrossRef]
52. Zhou, S.; Hao, G.; Zhou, X.; Jiang, W.; Wang, T.; Zhang, N.; Yu, L. One-pot synthesis of robust superhydrophobic, functionalized graphene/polyurethane sponge for effective continuous oil-water separation. *Chem. Eng. J.* **2016**, *302*, 155–162. [CrossRef]
53. Zhou, J.H.; Sui, Z.J.; Zhu, J.; Li, P.; Chen, D.; Dai, Y.C.; Yuan, W.K. Characterization of surface oxygen complexes on carbon nanofibers by TPD, XPS and FT-IR. *Carbon* **2007**, *45*, 785–796. [CrossRef]
54. Zhu, J.; Xiong, Z.; Zheng, J.; Luo, Z.; Zhu, G.; Xiao, C.; Meng, Z.; Li, Y.; Luo, K. Nitrogen-doped graphite encapsulated Fe/Fe_3C nanoparticles and carbon black for enhanced performance towards oxygen reduction. *J. Mater. Sci. Technol.* **2019**, *35*, 2543–2551. [CrossRef]
55. Ye, W.; Li, X.; Zhu, H.; Wang, X.; Wang, S.; Wang, H.; Sun, R. Green fabrication of cellulose/graphene composite in ionic liquid and its electrochemical and photothermal properties. *Chem. Eng. J.* **2016**, *299*, 45–55. [CrossRef]
56. Zhang, L.; Li, Y.; Zhang, L.; Li, D.W.; Karpuzov, D.; Long, Y.T. Electrocatalytic oxidation of NADH on graphene oxide and reduced graphene oxide modified screen-printed electrode. *Int. J. Electrochem. Sci.* **2011**, *6*, 819–829.
57. Chen, X.; Wang, X.; Fang, D. A Review on C1s XPS-spectra for some kinds of carbon materials. *Fuller. Nanotub. Carbon Nanostructures* **2020**, *28*, 1048–1058. [CrossRef]
58. Bourlier, Y.; Bouttemy, M.; Patard, O.; Gamarra, P.; Piotrowicz, S.; Vigneron, J.; Aubry, R.; Delage, S.; Etcheberry, A. Investigation of InAlN layers surface reactivity after thermal annealings: A complete XPS study for HEMT. *ECS J. Solid State Sci. Technol.* **2018**, *7*, P329–P338. [CrossRef]
59. Ayiania, M.; Smith, M.; Hensley, A.J.R.; Scudiero, L.; McEwen, J.S.; Garcia-Perez, M. Deconvoluting the XPS spectra for nitrogen-doped chars: An analysis from first principles. *Carbon* **2020**, *162*, 528–544. [CrossRef]
60. Mironenko, R.M.; Belskaya, O.B. Effect of the conditions for the aqueous-phase hydrogenation of furfural over Pd/C catalysts on the reaction routes. *AIP Conf. Proc.* **2019**, *2141*, 020010. [CrossRef]
61. Salnikova, K.E.; Matveeva, V.G.; Larichev, Y.V.; Bykov, A.V.; Demidenko, G.N.; Shkileva, I.P.; Sulman, M.G. The liquid phase catalytic hydrogenation of furfural to furfuryl alcohol. *Catal. Today* **2019**, *329*, 142–148. [CrossRef]

Article

CVD-Synthesis of N-CNT Using Propane and Ammonia

Valery Skudin [1], Tatiana Andreeva [1], Maria Myachina [2,*] and Natalia Gavrilova [2]

[1] Department of Chemical Technology of Carbon Materials, Mendeleev University of Chemical Technology of Russia, Miusskaya Sq., 9, 125047 Moscow, Russia; skudin.v.v@muctr.ru (V.S.); kpolyashova@yandex.ru (T.A.)
[2] Department of Colloid Chemistry, Mendeleev University of Chemical Technology of Russia, Miusskaya sq., 9, 125047 Moscow, Russia; gavrilova.n.n@muctr.ru
* Correspondence: miachina.m.a@muctr.ru

Abstract: N-CNT is a promising material for various applications, including catalysis, electronics, etc., whose widespread use is limited by the significant cost of production. CVD-synthesis using a propane–ammonia mixture is one of the cost-effective processes for obtaining carbon nanomaterials. In this work, the CVD-synthesis of N-CNT was conducted in a traditional bed reactor using catalyst: $(Al_{0.4}Fe_{0.48}Co_{0.12})_2O_3 + 3\% MoO_3$. The synthesized material was characterized by XPS spectroscopy, ASAP, TEM and SEM-microscopy. It is shown that the carbon material contains various morphological structures, including multiwalled carbon nanotubes (MWCNT), bamboo-like structures, spherical and irregular sections. The content of structures (bamboo-like and spherical structure) caused by the incorporation of nitrogen into the carbon nanotube structure depends on the synthesis temperature and the ammonia content in the reaction mixture. The optimal conditions for CVD-synthesis were determined: the temperature range (650–700 °C), the composition ($C_3H_8/NH_3 = 50/50\%$) and flow rate of the ammonia-propane mixture (200 mL/min).

Keywords: N-CNT; CVD-synthesis; carbon nanotube; XPS; TEM; nitrogen doping by ammonia

Citation: Skudin, V.; Andreeva, T.; Myachina, M.; Gavrilova, N. CVD-Synthesis of N-CNT Using Propane and Ammonia. *Materials* **2022**, *15*, 2241. https://doi.org/10.3390/ma15062241

Academic Editor: Ilya Mishakov

Received: 15 February 2022
Accepted: 15 March 2022
Published: 18 March 2022

Publisher's Note: MDPI stays neutral with regard to jurisdictional claims in published maps and institutional affiliations.

Copyright: © 2022 by the authors. Licensee MDPI, Basel, Switzerland. This article is an open access article distributed under the terms and conditions of the Creative Commons Attribution (CC BY) license (https://creativecommons.org/licenses/by/4.0/).

1. Introduction

Steady interest in carbon nanotubes (CNTs) is due to their unique properties; already today, they find: practical applications—for the creation of fire-retardant materials, fuel cell electrodes, in catalysis—as catalyst supports, in nanoelectronics—for the creation of one-dimensional conductors, nanosized transistors, supercapacitors, in technology—as additives to polymer and inorganic composites to increase mechanical strength, electrical conductivity and heat resistance [1–8].

A new approach to changing the chemical and electrical properties of CNTs is the modification of the carbon structure by a heteroatom, nitrogen. Currently, to obtain nitrogen-doped carbon nanotubes (N-CNTs), methods and approaches based on the direct formation of material from a nitrogen-containing carbon precursor or on the thermal treatment of undoped CNTs in a nitrogen-containing atmosphere are being developed [9–14].

The development of new economical methods for the synthesis of carbon nanomaterials in large quantities is a very urgent task because their widespread use is currently constrained by their high cost, which does not allow the use of N-CNTs on an industrial scale. Admittedly, the most flexible, providing a variety of possible synthesis modes, is the chemical vapor deposition (CVD) method [2].

It is known that the qualitative and quantitative composition of carbon nanotubes obtained by this method depends on the temperature, duration of synthesis, catalyst and gas mixture compositions [15–21].

It is known that the growth mechanism of N-CNTs is different from the CNT mechanism only by fact that the destruction of nitrogen precursor leads to the formation of nitrogen which diffuses like carbon to the catalyst volume [21].

N-CNT can form bamboo-like or spherical section structures. There are four stages reported in the literature model:

(1) the catalyst reacts with carbon, forming carbide particle;
(2) carbon is forming graphite layer on the surface of carbide particle;
(3) new layers of graphite are formed with a cup-shaped structure;
(4) the cup slides leaving a gap at the tip of the particle [22].

The choice of the catalyst composition for the synthesis of N-CNT by the method of direct incorporation of nitrogen into the carbon structure was made on the basis of preliminary experiments and literature analysis [23–25]. The choice of catalyst composition was justified by the following considerations. We proceeded from the fact that the catalyst should have the ability to:

- form metastable carbides;
- accelerate the reactions of dissociation of hydrocarbons with the formation of carbon;
- form metastable nitrides;
- to accelerate the reactions of dissociation of nitrogen-containing compounds, with the formation of nitrogen.

These conditions are satisfied by elements of groups VI and VIII, which can be found in the compositions of catalysts for dehydrogenation, oxidative conversion, and dissociation of hydrocarbons. Most of these catalysts containing these elements simultaneously exhibit catalytic activity in the reaction of synthesis-dissociation of ammonia. The most accessible and cheap, at the same time, are compounds of Fe, Co, Ni, Mo, etc. [1,26].

Compounds of groups III, IV and VI can be considered as promoters that ensure the stability of the catalyst structure under the reaction conditions (Al, Cu, Mo) or as a support (Si, Al). In this case, compounds of VI and VIII groups can participate as a catalyst in both reactions [2].

The choice of the composition of the initial gas mixture is determined by availability and price. It is clear that CH_4 and NH_3 are the most accessible for these purposes, but despite this, only a few publications were devoted to the N-CNT synthesis using ammonia and propane [27–29]. These compounds are considered among the most stable when heated, but the dissociation reactions of methane and ammonia on iron and cobalt start at very different temperatures. Approximate decomposition temperature ranges for methane: 600–800 °C; propane: 400–700 °C; ammonia: 300–500 °C [29,30]. At high temperatures in the presence of catalysts in a methane-hydrogen mixture, it is difficult to maintain a high concentration of ammonia, since its dissociation begins at a noticeable rate already at temperatures of 300–350 °C. As a result, the concentration of nitrides on the surface of the catalysts and the content of ammonia in the gas mixture at N-CNT synthesis temperatures will be extremely low; this will not allow getting products with a high nitrogen content in them. In this work, these considerations substantiated the choice of propane (C_3H_8), which exhibits high reactivity even at a temperature of 600 °C. Therefore, a mixture of propane with ammonia in various ratios was used as the initial mixture in this work. The novelty of this work is due to the possibility of obtaining nitrogen-doped carbon nanomaterials using inexpensive precursors and at a relatively low temperature of 650 °C. In the future, the development of a method for obtaining a material with a certain morphological structure and a certain nitrogen content will make it possible to evaluate the mechanism of the formation of carbon nanotubes through the stage of formation of nitrides.

The aim of this work was to establish the dependence between the nitrogen content in the N-CNT, the composition of the initial mixture of propane and ammonia and synthesis temperature, and also established the morphological composition of the synthesized carbon materials.

2. Materials and Methods

2.1. Materials

For the synthesis of N-CNTs by chemical vapor deposition, the following gases were used: nitrogen—N_2 (99.99%), liquefied propane C_3H_8 (CH_4-0.3%, C_2H_6-4, 7%, C_3H_8-95%), anhydrous liquefied ammonia NH_3 (99.9%). All gases were purchased by NII KM (Moscow, Russia).

To obtain the catalyst, iron (III) nitrate, $Fe(NO_3)_3 \cdot 9H_2O$ nonahydrate; cobalt (II) nitrate $Co(NO_3)_2$ $6H_2O$ hexahydrate; aluminum nitrate nonahydrate $Al(NO_3)_3 \cdot 9H_2O$; aminoacetic acid (glycine) H_2NCH_2COOH; ammonium paramolybdate $(NH_4)_6Mo_7O_{24}$ $4H_2O$. All reagents were of the reagent grade and were purchased by CT Lantan (Moscow, Russia).

2.2. Synthesis of Catalyst $(Al_{0.4}Fe_{0.48}Co_{0.12})_2O_3 + 3\%\ MoO_3$

In a beaker, weighed portions of crystalline hydrates of iron (III) nitrates (2.852 g), cobalt (0.493 g), aluminum (2.119 g) and glycine (1.711 g) are placed. Pure water (2 mL) is added to a sample of crystalline ammonium paramolybdate and heated to 40 °C, stirring until the salt is completely dissolved. Then, the resulting solution of ammonium paramolybdate is transferred into a beaker with weighed portions of salts, the mixture is stirred until complete dissolution. The mixture is heated and stirred until a clear solution of intense red-brown color is formed for 1 h at a temperature of 40 °C. The resulting solution is transferred to a porcelain cup, which is placed in a muffle furnace preheated to 550 °C for 10 min, then removed, cooled to room temperature.

2.3. Synthesis of Nitrogen-Doped Carbon Nanotubes

Synthesis of N-CNTs was carried out in a horizontal steel reactor with a fixed catalyst bed with a constant catalyst mass 0.05 g $(Al_{0.4}Fe_{0.48}Co_{0.12})_2O_3 + 3\%\ MoO_3)$. The synthesis was carried out in the temperature range of 650–800 °C for 1 h at a constant total flow rate of 200 mL/min. Samples were obtained at various ratios of propane/ammonia (C_3H_8/NH_3, vol%): 100; 25/75; 50/50; 75/25; 90/10.

The flow rate of propane and ammonia was set using RRG-12 flow regulators (Eltochpribor, Zelenograd, Russia). The temperature regime for the synthesis of carbon nanotubes was set using a temperature controller TERMODAT-17E6 (PP "Control Systems", Perm, Russia). The flow rate of the product mixture was determined with an ADM G6691A flow meter (Agilent Tech., Santa-Clara, CA, USA). The ammonia flow rate at the outlet of the reactor was determined by gas titration.

2.4. Characterization of Nitrogen-Doped Carbon Nanotubes

The sizes of the carbon nanotubes were determined by LEO 912AM Omega (Carl Zeiss, Oberkochen, Germany) transmission electron microscope. Images were acquired at 100 kV accelerating voltage. The analysis of the microphotographs and the calculation of particle sizes were carried out using the Image Tool V.3.00 (Image Tool Software, UTHSCSA, San Antonio, TX, USA). At least 100 particles per sample were processed. The outer and inner diameters of the nanotubes were determined.

The morphology of carbon materials was studied using a scanning electron microscope (SEM). The micrographs of the samples were taken on JSM 6510 LV + SSD X-MAX microscopes (JEOL, Tokyo, Japan) at an accelerating voltage of 20 kV.

The XPS spectra were recorded using ESCA X-ray photoelectron spectrometer (OMICRON Nanotechnology GmbH, Taunusstein, Germany). The samples of N-CNT investigated by XPS spectroscopy contained catalyst.

The parameters of the porous structure of the samples were calculated based on the isotherm of low-temperature nitrogen adsorption. The studies were carried out on a Gemini VII analyzer (Micromeritics, Norcross, GA, USA) at the Center for Shared Use. DI. Mendeleev. The specific surface area was determined by the BET method. The total pore volume was found from the maximum value of the relative pressure, equal to 0.995. The predominant pore diameter was calculated using the BJH method.

3. Results

3.1. Influence of Flow Rate and Composition of the Initial Gas Mixture

It is known that ammonia on iron catalysts can dissociate at atmospheric pressure already at temperatures of 300–350 °C. Hydrocarbon gases under these conditions remain practically inert and begin to dissociate at a noticeable rate at temperatures above 600 °C; therefore, it is proposed to use propane as a carbon-containing precursor.

In order to maintain the ammonia concentration under the synthesis conditions at a sufficiently high level, it is necessary to reduce the residence time of the reaction medium in the reactor by increasing its flow rate and lowering the temperature in the reactor.

As can be seen from the results in Table 1, at a temperature in the reactor of 800 °C and a change in the mole flow of pure ammonia at the inlet to the reactor from 18.7 to 37.3 kmole/s, the mole flow of ammonia at the outlet remains practically unchanged. The result obtained indicates that in the studied range of mole flow rates in the outgoing mixture, an equilibrium concentration of unreacted ammonia is established, the level of which is determined by the temperature in the reactor. As the temperature in the reactor decreases, the flow rate and the equilibrium concentration of ammonia under stationary conditions increase.

Table 1. Dependence of the ammonia flow rate at the outlet of the reactor on its flow rate at the inlet and the temperature in the reactor.

Temperature, °C	Initial Ammonia Mole Flow Rate, kmol/s	Outlet Ammonia Mole Flow Rate, kmol/s
800	18.7	1.4
	22.4	1.4
	26.1	1.4
	29.9	1.4
	37.3	1.5
750	26.1	2.0
700	26.1	3.3
650	26.1	5.1

We concluded that it is not advisable to increase the ammonia concentration at the reactor outlet by reducing the contact time (or, which is also an increase in the linear velocity), since this can lead to catalyst carryover from the reactor and may be accompanied by an unsustainable increase in the consumption of raw materials. Secondly, as can be seen from the table, it is more rational to reduce the temperature in the reactor.

For the synthesis of carbon nanotubes, the value of 26/1 kmol/s (200 mL/min) was chosen as the optimal flow rate of the initial mixture of propane and ammonia, while the synthesis of the carbon material was carried out at different ammonia contents in the initial mixture (C_3H_8/NH_3, vol%): 100; 25:75; 50:50; 75:25; 90:10. The initial temperature for synthesis was chosen as 650 °C.

In order to establish the dependence of the amount of nitrogen introduced into CNTs on the different compositions of the initial gas mixture, as well as to determine the electronic state of atoms on the surface of the material under study, the obtained samples were investigated by X-ray photoelectron spectroscopy.

In Figure 1, the typical XPS spectra of CNT, obtained from propane are shown. For carbon, the line shape of the spectrum has a maximum with E = 284.6 eV, which is typical for sp^2 hybridization carbon structures. A characteristic peak for nitrogen is observed; in a pyridine-like (398.8 eV) state, other forms of nitrogen are absent. The formation of a product containing nitrogen in its structure could occur only from a mixture of gases that was formed when nitrogen was supplied to the reactor during its heating or cooling. Therefore, N_2 can also be a nitrogen precursor gas, under the given synthesis conditions.

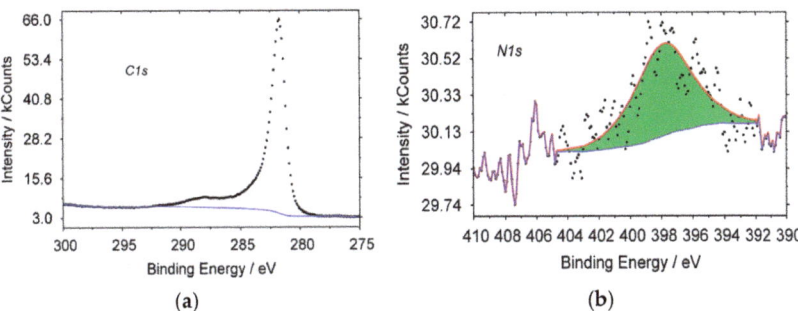

Figure 1. Typical XPS spectra for CNT, synthesized from propane: (**a**) carbon, (**b**) nitrogen.

XPS spectra (N_{1s}) of samples with different ammonia content in the initial gas mixture are shown in Figure 2. Characteristic peaks of different state of nitrogen are observed in all investigated samples: in the pyridine-like (N_{Py}, 398.8 eV), graphite-like (N_Q 401.7 eV) states and oxidized forms of nitrogen (N_{Ox} 405.6 eV).

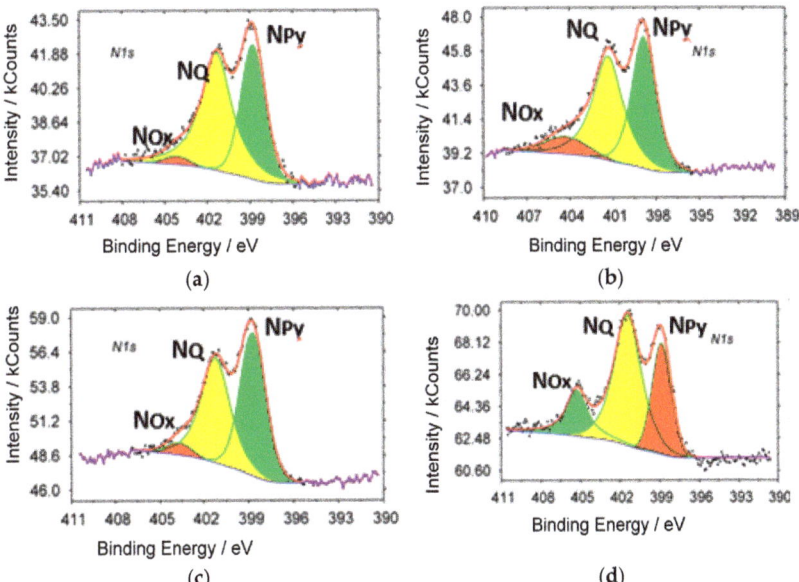

Figure 2. XPS spectra of the samples with different ammonia content in initial mixture: (**a**) 75%; (**b**) 50%; (**c**) 25%; (**d**) 10%.

The content of the different states of nitrogen, carbon and oxygen in synthesized samples are presented in Table 2.

Figures 3 and 4 show the dependences of the content of total nitrogen and various forms of nitrogen on the initial content of ammonia in the reaction mixture.

Table 2. Percentage of elements in N-CNT at 650 °C with different ammonia content in initial mixture.

NH_3 Content, % vol.	-	10	25	50	75
C_{1s}, % 284.6 eV	98.5	94.8	88.2	93.4	83.6
$(N)_{total}$, %	0.4	3.2	5.2	5.5	4.7
$(N_{1s}$ 398.8 eV) N_{Py}, %	0.4	0.6	2.6	2.4	1.9
$(N_{1s}$ 401.7 eV) N_Q, %	-	1.7	2.4	2.6	2.6
$(N_{1s}$ 405.6 eV) N_{Ox}, %	-	0.9	0.2	0.5	0.1
$(O)_{total}$, %	1.2	1.4	4.4	1.0	1.2

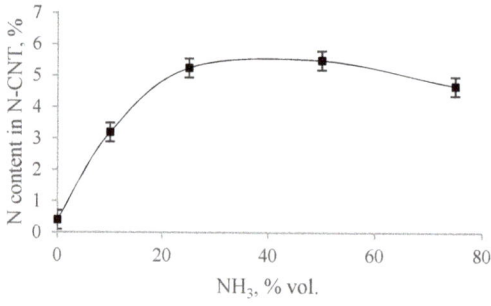

Figure 3. Dependence of N_{total} content in samples on NH_3 content in the mixture.

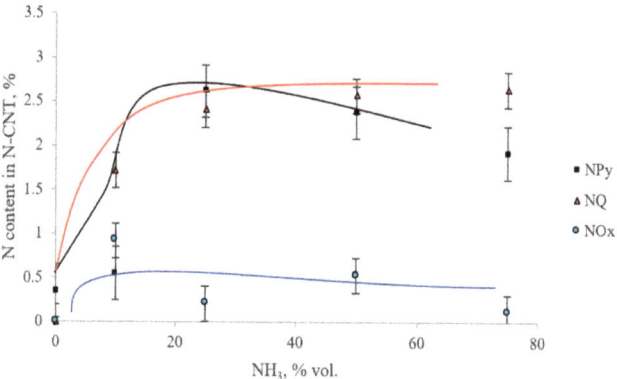

Figure 4. Dependence of the content of various forms of nitrogen on the content of NH_3 in the mixture.

From Figures 3 and 4 it can be seen that with an increase in the ammonia content in the initial gas mixture, the total nitrogen content in the samples first increases to the maximum nitrogen content in the sample, synthesized from a mixture of C_3H_8/NH_3 = (50/50%), and then decreases. In this case, the maximum content of the pyridine-like form of nitrogen is observed in the sample, which is synthesized from a mixture of C_3H_8/NH_3 (75/25%). A graphite-like form is seen in the sample C_3H_8/NH_3 (25/75%). The smallest content of oxidized forms of nitrogen contains the N-CNT, synthesized using C_3H_8/NH_3 (25/75%).

According to the literature, the maximum nitrogen content in a doped carbon material can reach about 10%; however, a high nitrogen content is not always justified from the point of view of further use, including as catalyst supports [6,8,31]. Significant incorporation of nitrogen into the structure of carbon nanotubes occurs at a high level of ammonia; however, at a concentration of more than 50%, the process of nitrogen doping slows down, and such an effect of the ammonia content may be due to the different mechanism of formation of nitrogen-doped structures [8]. To establish the mechanism, first of all, it is required to trace which phase transformations the catalyst undergoes. We assume that nitrides or carbonitrides are formed during CVD-synthesis, and the ammonia concentration has a direct effect on this process.

Figure 5 shows TEM-images of samples of synthesized N-CNT using a gas mixture with a different ratio of C_3H_8/NH_3 at a temperature of 650 °C.

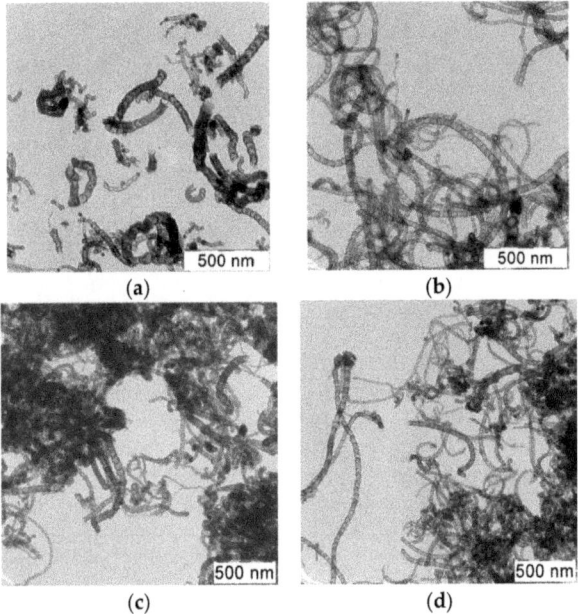

Figure 5. (**a**,**b**) TEM—images of N-CNT obtained using a gas mixture of C_3H_8/NH_3 (50/50%); (**c**) 25/75%; (**d**) 75/25%; at a temperature of 650 °C.

As can be seen from the images, the synthesized product is presented by different structures CNTs. There are several types of morphological structures found in the obtained products of nitrogen-doped carbon materials, including multiwalled carbon nanotubes (MWCNT), bamboo-like structures, spherical and irregular sections [13,25]. These types of structures are presented on Figure 6.

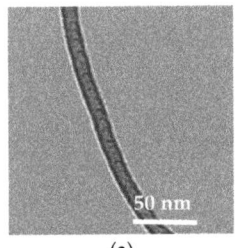 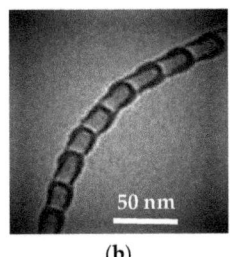 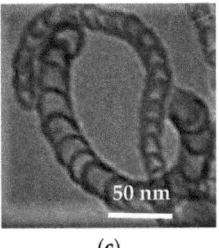 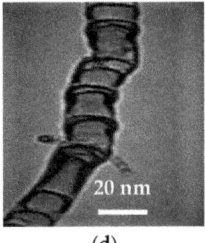

(**a**) (**b**) (**c**) (**d**)

Figure 6. Morphological structures of CNT: (**a**) MWCNT; (**b**) bamboo-like structures; (**c**) spherical sections; (**d**) irregular structures.

From a review of the literature, we can conclude that the nanotubes with bamboo-like, spherical and irregular structures (Figure 6b–d) are most likely nitrogen-doped carbon materials [13,25].

Based on the results of transmission electron microscopy of the samples, Table 3 contains the percentage of different type structures in samples, synthesized in the presence of NH_3 content, and the values of predominant inner and outer diameters.

Table 3. Percentage of different morphological structures in the samples obtained at a temperature of 650 °C.

NH_3 Content, % vol.	10	25	50	75
Morphological Structure		Content, % *		
MWCNT	21.1	5.7	9.1	20.8
bamboo-like structure	21.2	11.4	22.7	43.4
spherical section structure	15.2	68.7	59.1	26.4
irregular structure	51.5	14.2	9.1	9.4
Diameter				
Predominant inner, nm	26	31	24	32
Predominant outer, nm	16	24	15	21

* Percentage of particles of a given morphological structure in the sample according to TEM.

It can be seen that the composition of the initial gas mixture affects the morphology, at various ratios of initial gases, certain morphological structures prevail in the product.

With an increase in the content of ammonia in the mixture, there is an increase in the content of bamboo-like structures and a decrease in the content of irregular structures. For the sample synthesized from C_3H_8/NH_3 (50/50%) mixture, the quantitative ratio of all the observed structures occupies an intermediate position with respect to other samples. It also corresponds to the highest total content of fibers with bamboo-like and spherical sections (81.8%).

The values of the predominant outer and inner diameters lie in the range from 31 to 24 and from 24 to 15, respectively. Despite the change in the fractional composition of the carbon material with a change in the initial content of ammonia, no significant difference in the sizes of nanotubes is observed.

3.2. Temperature Effect

The choice of synthesis temperature is based on the dependence of the content of nitrogen atoms embedded in the CNT structure. For this purpose, several CNT samples were synthesized in an ammonium-propane mixture at several temperatures at a constant ratio of ammonia and propane in the initial gas mixture (50/50%). The Table 4 shows the values of product yield and residual catalyst content.

Table 4. Product yield of the synthesized samples at the different initial content of NH_3 and different temperature.

Temperature, °C	Product Yield, $g_{CNT}/g_{catalyst}$		
	NH_3 = 25% vol.	NH_3 = 50% vol.	NH_3 = 75% vol.
650	3.7	1.6	0.7
700	-	3.4	-
750	-	4.8	-
800	-	20.1	-

An important parameter in the synthesis of CNTs is their yield, which is determined by the ratio of the mass of the formed product to the mass of the initial catalyst. As can be seen from the presented data, with an increase in the synthesis temperature, an increase in the yield of N-CNTs is observed; moreover, the increase in initial ammonia content leads to the decrease of the N-CNT yield.

It is known that the amount of nitrogen contained in CNTs is affected by the conditions of synthesis [22,31]. Table 5 contains the results of XPS-analysis for samples, synthesized at different temperature and constant gas mixture (50/50%).

Table 5. Percentage of elements in N-CNT at different temperature (NH_3 content in reagent mixture of 50% vol.).

Temperature, °C	650	700	750	800
C_{1s}, % 284.6 eV	93.4	92.7	94.3	96.0
$(N)_{total}$, %	5.5	4.6	3.8	2.3
(N_{1s} 398.8 eV) N_{Py}, %	2.4	2.3	1.3	0.5
(N_{1s} 401.7 eV) N_Q, %	2.6	1.3	1.7	1.2
(N_{1s} 405.6 eV) N_{Ox}, %	0.5	0.6	0.7	0.7
$(O)_{total}$, %	1.0	2.7	1.9	1.7

From the data presented in Table 5, it can be seen that with an increase in the process temperature from 650 °C to 800 °C, a general decrease in the nitrogen content in N-CNTs is observed. This result may be due to a decrease in the concentration of ammonia in the gas phase.

It can be also seen that a decrease in temperature promotes the synthesis of materials containing non-oxidized forms of nitrogen incorporated into the carbon structure. An increase in the synthesis temperature leads to the appearance of oxidized forms of nitrogen. At temperatures of 700 °C and 750 °C, nitrogen in the tubes is in two states: pyridine-like and graphite-like, while the content of nitrogen in the graphite-like state is higher than in the pyridine-like state at any synthesis temperature. With an increase in temperature to 800 °C, a third form of nitrogen appears: oxidized. In addition, oxygen was found in the product, the content of which decreases with increasing synthesis temperature. One of the reasons for the appearance of oxygen in the samples, is residual catalyst in N-CNT, which components can be only partially reduced to a metallic state.

Figure 7 shows TEM and SEM-images of samples of synthesized N-CNT using a C_3H_8/NH_3 (50/50%) gas mixture at a temperature 800 °C.

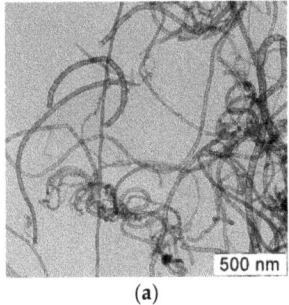
(a)

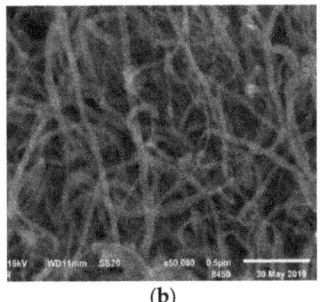
(b)

Figure 7. TEM (a) and SEM-image (b) of N-CNT obtained using a gas mixture of C_3H_8-NH_3 (50/50%) at a temperature of 800 °C.

As can be seen from the presented figures, the carbon nanomaterial is presented by nanotubes of various morphologies, including multiwalled carbon nanotubes (MWCNT), bamboo-like structures, spherical and irregular sections, similar to the material synthesized at 650 °C.

Based on the obtained TEM-images, the predominant inner and outer diameters and percentage of different morphological structures of nitrogen-doped CNTs were calculated. These results are presented in the Table 6.

Table 6. Percentage of different morphological structures in the samples obtained at different temperature (NH_3 content in reagent mixture—50% vol.).

Temperature, °C	650	700	750	800
Morphological Structures	\multicolumn{4}{c}{Content, % *}			
MWCNT	9.1	7.3	24.1	6.5
bamboo-like structure	22.7	26.8	31.5	31.4
spherical section structure	59.1	48.8	18.5	39.3
irregular structure	9.1	17.1	25.9	22.8
Diameter				
Predominant inner, nm	26	25	20	25
Predominant outer, nm	16	12	8	15

* Percentage of particles of a given morphological structure in the sample according to TEM.

It can be seen that the synthesis temperature affects the fractional composition of the resulting carbon material. With an increase in temperature, the content of bamboo-like and irregular structures increases, and the content of structures with spherical sections decreases. Based on the fact that only bamboo-like and spherical structures can be attributed to nitrogen-doped carbon nanotubes, the optimal temperature range for synthesis is 650 °C.

The values of the predominant outer and inner diameters lie in the range from 26 to 20 and from 16 to 8, respectively. There is no significant change in the particle size with an increase in the synthesis temperature.

For investigating of doped N-CNT porous structure samples o with a total nitrogen content $(N)_{total}$ of 4.6 and 2.3%, obtained at temperatures of 700 and 800 °C, were studied using low-temperature nitrogen adsorption, which are presented on the Figure 8.

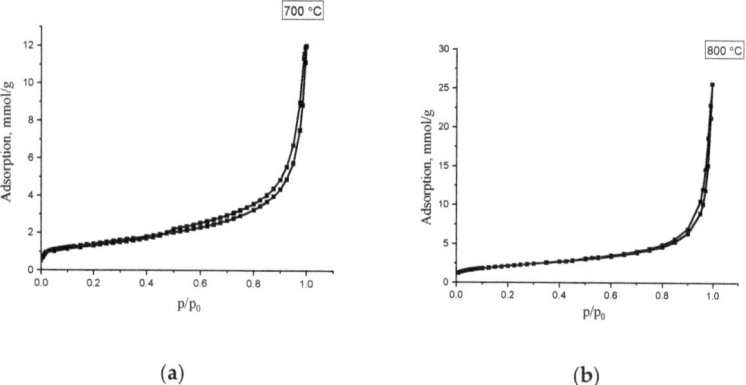

(a) (b)

Figure 8. Low-temperature nitrogen adsorption isotherms for N-CNT, synthesized at the temperature of 700 °C (**a**) and 800 °C (**b**).

The isotherms refer to Type II according to the De-Boer classification, and indicates the occurrence of polymolecular adsorption. Based on the obtained isotherms, the main characteristics of the porous structure of were calculated: specific surface area, pore volume, and predominant pore diameter. The obtained values are shown in Table 7.

Table 7. Porous structure of the synthesized samples at the initial content of $NH_3 = 50\%$ vol. and different temperature.

Parameter	Temperature, °C	
	700	800
Surface area, m^2/g	113	215
Pore volume, cm^3/g	0.4	1.2
Predominant pore diameter, nm	4	3

As can be seen from the presented data, the studied samples of carbon material have mesoporous structure, the specific surface of which is 113 and 215 m^2/g. The pore volume of the samples is 0.4 and 1.2 cm^3/g, the main contribution to which is made by mesopores, which are formed due to the interparticle volume between carbon nanotubes. Mesopore size distribution calculated BJH method (Barrett–Joyner–Halenda) shows that the predominant pore size is about 3–4 nm.

An increase in the surface area and pore volume may be associated with an increase in the content of amorphous carbon with an increase in the synthesis temperature.

4. Discussion

The main purpose of this work is to choose the conditions for a simple and cost-effective process for obtaining N-CNTs, in which the incorporation of the N atom into the structure of nanotubes will be possible. The carbon material was obtained by embedding a carbon material into the crystal lattice during synthesis using a $(Al_{0.4}Fe_{0.48}Co_{0.12})_2O_3$ catalyst doped with MoO_3 (3%) and a propane–ammonia gas mixture. The choice of ammonia and propane as a precursor of the doped carbon material was due to the low cost and the possibility of obtaining the material already at a temperature of 650 °C. An analysis of the literature data showed that this temperature of N-CNT synthesis is relatively low compared to other variants of CVD-synthesis using light hydrocarbons [22].

The content of nitrogen embedded in the crystalline structure of carbon varied due to changes in the ammonia content in the reaction mixture and the synthesis temperature. A total nitrogen content of about 5% has already been achieved with an ammonia content of 25% in the mixture. A further increase in the ammonia content to 90% did not lead to a significant change in the content of total nitrogen; however, the propane/ammonia ratio affects the content of a certain form of nitrogen (graphite-like, pyridine-like and oxidized forms of nitrogen), as well as the fractional composition of N-CNT. According to the TEM results, it was found that the synthesized carbon material contains various morphological structures, including multiwalled carbon nanotubes (MWCNT), bamboo-like structures, spherical and irregular section structure [26]. In accordance with the literature data, only bamboo-like and spherical section structures can be classified as structures in which nitrogen is embedded in the crystalline structure of carbon [13,25]. It was found that the maximum number of such structures is formed when using an equimolar propane/ammonia mixture.

The next part of the work was devoted to establishing the influence of the synthesis temperature on the N content and the morphological composition of the carbon material. A series of experiments was carried out at a constant propane/ammonia equimolar ratio; it was found that with an increase in temperature from 650 to 800 °C, the product yield increases, but the total nitrogen content decreases from 5.5 to 2.3%. In addition, the analysis of TEM and SEM images showed that there is a change in the content of various morphological structures; there is a decrease in the content of the spherical section structure and irregular structures begin to accumulate. An increase in the product yield, accumulation of irregular structures, and an increase in the specific surface area of the final product indicate the accumulation of amorphous carbon, which in turn leads to a deterioration in the quality of the resulting product. Based on this, it is advisable to carry out the synthesis of N-CNT at temperatures not exceeding 700 °C.

The temperature dependence of the growth rate of N-CNT can be related to several factors such as concentration, diffusion rate, and growth rate at the interface between the catalyst and the formed nanotube. With an increase in temperature, the rate of diffusion of carbon and nitrogen atoms increases significantly, and the growth rate of N-CNT will correspondingly increase [12].

As a result of this work, the optimal conditions for obtaining doped carbon material (at least 80% N-CNT, total nitrogen content at least 5–6%) from readily available precursors were established. Further studies of CVD-synthesis of N-CNT using propane–ammonia mixture can be directed to: detailed consideration of the growth process of carbon nanotubes and phase transformations of the catalyst during growth; an increase in the activity of catalysts based on Al, Co, Fe oxides and a search for promoters, including molybdenum oxides.

5. Conclusions

In this work, nitrogen-doped carbon materials were synthesized using propane and ammonia by the CVD method. It was shown that the total nitrogen content, as well as the content of a certain form of nitrogen (graphite-like, pyridine-like and oxidized forms of nitrogen) depend on the initial ammonia content in the reaction mixture and the synthesis temperature.

The optimal conditions for N-CNT synthesis were chosen as follows: mixture flow rate, 200 mL/min; the composition of the reaction mixture—C_3H_8/NH_3 (50/50%); temperature range—650–700 °C. Under these conditions, a mesoporous carbon material is formed with a total nitrogen content of about 5% and with a content of nitrogen-doped structures (bamboo-like and spherical) of at least 80%.

Author Contributions: Conceptualization, V.S.; methodology, T.A. and V.S.; investigation, T.A., N.G. and M.M.; data curation, M.M. and V.S.; writing—original draft preparation, M.M.; writing—review and editing, M.M., N.G., V.S.; supervision V.S. All authors have read and agreed to the published version of the manuscript.

Funding: The study was funded by a Russian Science Foundation, grant number 21-73-00303.

Institutional Review Board Statement: Not applicable.

Informed Consent Statement: Not applicable.

Data Availability Statement: The data presented in this study are available on request from the corresponding author.

Conflicts of Interest: The authors declare no conflict of interest.

References

1. Wang, X.-D.; Vinodgopal, K.; Dai, G.-P. *Synthesis of Carbon Nanotubes by Catalytic Chemical Vapor Deposition. Perspective of Carbon Nanotubes*, 1st ed.; IntechOpen: London, UK, 2019; pp. 13–33.
2. Eatemadi, A.; Daraee, H.; Karimkhanloo, H.; Kouhi, M.; Zarghami, N.; Akbarzadeh, A. Carbon nanotubes: Properties, synthesis, purification, and medical applications. *Nanoscale Res. Lett.* **2014**, *9*, 393–406. [CrossRef] [PubMed]
3. Feng, L.; Xie, N.; Zhong, J. Carbon Nanofibers and Their Composites: A Review of Synthesizing, Properties and Applications. *Materials* **2014**, *7*, 3919–3945. [CrossRef]
4. Bikiaris, D. Microstructure and properties of polypropylene/carbon nanotube nanocomposites. *Materials* **2010**, *3*, 2884–2946. [CrossRef]
5. Poudel, R.Y.; Li, W. Synthesis, properties, and applications of carbon nanotubes filled with foreign materials: A review. *Mater. Today Phys.* **2018**, *7*, 7–34. [CrossRef]
6. Kanygin, M.A.; Sedelnikova, O.V.; Asanov, I.P.; Bulusheva, L.G.; Okotrub, A.V.; Kuzhir, P.P.; Plyushch, A.O.; Maksimenko, S.A.; Lapko, K.N.; Sokol, A.A.; et al. Effect of nitrogen doping on the electromagnetic properties of carbon nanotube-based composites. *J. Appl. Phys.* **2013**, *113*, 144315–144318. [CrossRef]
7. Vikkisk, M.; Kruusenberg, I.; Joost, U.; Shulga, E. Electrocatalysis of oxygen reduction on nitrogen-containing multi-walled carbon nanotube modified glassy carbon electrodes. *Electrochim. Acta* **2013**, *87*, 709–716. [CrossRef]
8. Sharifi, T.; Hu, G.; Jia, X.E.; Wagberg, T. Formation of active sites for oxygen reduction reactions by transformation of nitrogen functionalities in nitrogen-doped carbon nanotubes. *ACS Nano* **2012**, *10*, 8904–8912. [CrossRef]

9. Gong, K.; Du, F.; Xia, Z.; Durstock, M.; Dai, L. Nitrogen-doped carbon nanotube arrays with high electrocatalytic activity for oxygen reduction. *Science* **2009**, *323*, 760–764. [CrossRef]
10. Zhang, D.; Hao, Y.; Zheng, L.; Ma, Y.; Feng, H.; Luo, H. Nitrogen and sulfur co-doped ordered mesoporous carbon with enhanced electrochemical capacitance performance. *J. Mater. Chem. A.* **2013**, *1*, 7584–7591. [CrossRef]
11. Panchakarla, L.S.; Govindaraj, A.; Rao, C.N.R. Nitrogen- and Boron-Doped Double-Walled Carbon Nanotubes. *ACS Nano* **2007**, *1*, 494–500. [CrossRef]
12. Zhong, Y.; Jaidann, M.; Zhang, Y.; Zhang, G.; Liu, H.; Ionescu, I.; Lussier, L.-S. Synthesis of high nitrogen doping of carbon nanotubes and modeling the stabilization of filled DAATO@CNTs (10,10) for nanoenergetic materials. *J. Phys. Chem. Solids* **2010**, *71*, 134–139. [CrossRef]
13. Ayala, P.; Arenal, R.; Rümmeli, M.; Rubio, A.; Pichler, T. The doping of carbon nanotubes with nitrogen and their potential applications. *Carbon* **2010**, *48*, 575–586. [CrossRef]
14. Gourari, D.E.; Razafinimanana, M.; Monthioux, M.; Arenal, R.; Valensi, F.; Joulie, S.; Serin, V. Synthesis of (B-C-N) Nanomaterials by Arc Discharge Using Heterogeneous Anodes. *Plasma Sci. Technol.* **2016**, *18*, 465–468. [CrossRef]
15. Manawi, Y.; Ihsanullah, I.; Samara, A.; Al-Ansari, T.; Atieh, M. A Review of Carbon Nanomaterials' Synthesis via the Chemical Vapor Deposition (CVD) Method. *Materials* **2018**, *11*, 822. [CrossRef] [PubMed]
16. Shukrullah, S.; Mohamed, N.M.; Shaharun, M.S.; Naz, M.Y. Mass Production of Carbon Nanotubes Using Fluidized Bed Reactor: A Short Review. *Trends Appl. Sci. Res.* **2014**, *9*, 121–131. [CrossRef]
17. Ibrahim, E.M.M.; Khavrus, V.O.; Leonhardt, A.; Hampel, S.; Oswald, S.; Rummeli, M.H. Synthesis, characterization, and electrical properties of nitrogen-doped single-walled carbon nanotubes with different nitrogen content. *Diam. Relat. Mater.* **2010**, *19*, 1199–1206. [CrossRef]
18. Paul, R.; Du, F.; Dai, L.; Ding, Y.; Wang, Z.L.; Wei, F.; Roy, A. 3D Heteroatom-Doped Carbon Nanomaterials as Multifunctional Metal-Free Catalysts for Integrated Energy Devices. *Adv. Mater.* **2019**, *31*, 1805598. [CrossRef]
19. Hou, P.X.; Orikasa, H.; Yamazaki, T.; Matsuoka, K.; Tomita, A.; Setoyama, N.; Fukushima, Y.; Kyotani, T. Synthesis of nitrogen-containing microporous carbon with a highly ordered structure and effect of nitrogen doping on H_2O adsorption. *Chem. Mater.* **2005**, *17*, 5187–5193. [CrossRef]
20. Szymanski, G.S.; Grzybek, T.; Papp, H. Influence of nitrogen surface functionalities on the catalytic activity of activated carbon in low temperature SCR of NOx with NH3. *Catal. Today.* **2004**, *90*, 51–59. [CrossRef]
21. Van Dommele, S.; Romero-Izquirdo, A.; Brydson, R.; de Jong, K.P.; Bitter, J.H. Tuning nitrogen functionalities in catalytically grown nitrogen-containing carbon nanotubes. *Carbon* **2008**, *46*, 138–148. [CrossRef]
22. Podyacheva, O.Y.; Ismagilov, Z.R. Nitrogen-doped carbon nanomaterials: To the mechanism of growth, electrical conductivity and application in catalysis. *Catalysis Today* **2015**, *249*, 12–22. [CrossRef]
23. An, B.; Xu, S.; Li, L.; Tao, J.; Huang, F.; Geng, X. Carbon nanotubes coated with a nitrogen-doped carbon layer and its enhanced electrochemical capacitance. *J. Mater. Chem. A.* **2013**, *1*, 7222–7228. [CrossRef]
24. Arjmand, M.; Chizari, K.; Krause, B.; Potschke, P.; Sundararaj, U. Effect of synthesis catalyst on structure of nitrogen-doped carbon nanotubes and electrical conductivity and electromagnetic interference shielding of their polymeric nanocomposites. *Carbon* **2016**, *98*, 358–372. [CrossRef]
25. Chizari, K.; Vena, A.; Laurentius, L.; Sundararaj, U. The effect of temperature on the morphology and chemical surface properties of nitrogen-doped carbon nanotubes. *Carbon* **2014**, *68*, 369–379. [CrossRef]
26. Nxumalo, E.N.; Coville, N.J. Nitrogen Doped Carbon Nanotubes from Organometallic Compounds: A Review. *Materials* **2010**, *3*, 2141–2171. [CrossRef]
27. Jafarpour, S.M.; Kini, M.; Schulz, S.E. Effects of catalyst configurations and process conditions on the formation of catalyst nanoparticles and growth of single-walled carbon nanotubes. *Microelectron. Eng.* **2017**, *167*, 95–104. [CrossRef]
28. Chun, K.-Y.; Lee, H.S.; Lee, C.J. Nitrogen doping effects on the structure behavior and the field emission performance of doublewalled carbon nanotubes. *Carbon* **2009**, *47*, 169–177. [CrossRef]
29. Mishakov, I.V.; Bauman, Y.I.; Streltsov, I.A.; Korneev, D.V.; Vinokurova, O.B.; Vedyagin, A.A. The regularities of the formation of carbon nanostructures from hydrocarbons based on the composition of the reaction mixture. *Resour. Effic. Technol.* **2016**, *2*, 61–67. [CrossRef]
30. Zhang, X.; Lu, Z.; Ma, M.; Yang, Z. Adsorption and dissociation of ammonia on small iron clusters. *Int. J. Hydrogen Energy* **2015**, *40*, 346–352. [CrossRef]
31. Mabena, L.F.; Ray, S.S.; Mhlanga, S.D.; Coville, N.J. Nitrogen-doped carbon nanotubes as a metal catalyst support. *Appl. Nanosci.* **2011**, *1*, 67–77. [CrossRef]

Article

Modification of Gold Zeolitic Supports for Catalytic Oxidation of Gluconic Acid

Adrian Walkowiak *, Joanna Wolska *, Anna Wojtaszek-Gurdak, Izabela Sobczak, Lukasz Wolski and Maria Ziolek

Faculty of Chemistry, Adam Mickiewicz University, Uniwersytetu Poznańskiego 8, 61–614 Poznań, Poland; anna.gurdak@amu.edu.pl (A.W.-G.); sobiza@amu.edu.pl (I.S.); wolski.lukasz@amu.edu.pl (L.W.); ziolek@amu.edu.pl (M.Z.)
* Correspondence: adrian.walkowiak@amu.edu.pl (A.W.); j.wolska@amu.edu.pl (J.W.);
  Tel.: +48-618-291-794 (A.W.)

Abstract: Activity of gold supported catalysts strongly depends on the type and composition of support, which determine the size of Au nanoparticles (Au NPs), gold-support interaction influencing gold properties, interaction with the reactants and, in this way, the reaction pathway. The aim of this study was to use two types of zeolites: the three dimensional HBeta and the layered two-dimensional MCM-36 as supports for gold, and modification of their properties towards the achievement of different properties in oxidation of glucose to gluconic acid with molecular oxygen and hydrogen peroxide. Such an approach allowed establishment of relationships between the activity of gold catalysts and different parameters such as Au NPs size, electronic properties of gold, structure and acidity of the supports. The zeolites were modified with (3-aminopropyl)-trimethoxysilane (APMS), which affected the support features and Au NPs properties. Moreover, the modification of the zeolite lattice with boron was applied to change the strength of the zeolite acidity. All modifications resulted in changes in glucose conversion, while maintaining high selectivity to gluconic acid. The most important findings include the differences in the reaction steps limiting the reaction rate depending on the nature of the oxidant applied (oxygen vs. H_2O_2), the important role of porosity of the zeolite supports, and accumulation of negative charge on Au NPs in catalytic oxidation of glucose.

Keywords: gold zeolites; amino-organosilane modifier; boron modifier; selective glucose oxidation with O_2 and H_2O_2; microwave-assisted oxidation; base-free oxidation

Citation: Walkowiak, A.; Wolska, J.; Wojtaszek-Gurdak, A.; Sobczak, I.; Wolski, L.; Ziolek, M. Modification of Gold Zeolitic Supports for Catalytic Oxidation of Glucose to Gluconic Acid. *Materials* 2021, 14, 5250. https://doi.org/10.3390/ma14185250

Academic Editor: Ilya V. Mishakov

Received: 10 July 2021
Accepted: 7 September 2021
Published: 13 September 2021

Publisher's Note: MDPI stays neutral with regard to jurisdictional claims in published maps and institutional affiliations.

Copyright: © 2021 by the authors. Licensee MDPI, Basel, Switzerland. This article is an open access article distributed under the terms and conditions of the Creative Commons Attribution (CC BY) license (https://creativecommons.org/licenses/by/4.0/).

1. Introduction

Over 30 years of gold catalysis, since the Haruta's pioneering discovery [1] of unique activity of Au nanoparticles (NPs), has brought a huge number of scientific publications concerning the application of Au NPs, usually supported on porous matrices, in different oxidation processes. The mechanisms of many reactions performed successfully on gold catalysts have not been fully disclosed yet. Recently, Hutchings [2] has pointed out that work on improvement of catalyst performance or designing new ones must be based on deep understanding of the reaction mechanism. In the last few years, Au NPs have been deposited on several different supports (e.g., mesoporous silica [3], metal oxides [4–6] and carbons [7,8]) and applied as catalysts in glucose oxidation. However, glucose oxidation carried out over gold catalysts supported on zeolites belongs to the reactions whose pathways have not been fully solved [9]. Attempts at unravelling the reaction mechanism can be undertaken on the basis of some knowledge about the structure and type of active sites. The aim of this study was to provide the necessary information on the active sites and the interaction between modifiers in the zeolites' supports and active gold species, which is known to affect the catalytic activity.

For the catalysts based on zeolites as supports and gold NPs as the main active centers it is relatively easy to determine the structure and type of active sites (e.g., Brønsted acid sites (BAS) in zeolites). Moreover, zeolites can be easily modified by isomorphous substitution in their skeleton (e.g., by incorporation of boron instead of aluminum in the aluminosilica framework [10,11]) or by post-synthesis modification (e.g., by functionalization of the zeolite surface with amino-organosilanes [12,13]). Changes in the zeolite composition imply changes in the surface (acid-base) properties and thus modify the interaction between the support and gold NPs. Furthermore, a large number of zeolite structures [14] opens an opportunity to study the effect of the support structure on the activity and selectivity of gold catalysts. Moreover, the acid-base active centers in zeolites can act as sites of chemisorption of organic compounds applied as reactants and, therefore, take part in the mechanism of catalytic oxidation on Au NPs. Acid-base properties of non-reducible supports (such as zeolites) have been recognized as important in oxidation processes [15]. As concerns sugars oxidation, the BAS on the catalyst surface are able to protonate glucose molecules making them more susceptible to oxidation [16]. Recently [17], we have postulated that the first step of the base-free glucose oxidation on gold supported on Beta zeolite is the protonation of the carbonyl oxygen (at carbon atom C1) in the glucose molecule, in which the BAS of the zeolite support take part. It has been indicated that the chemisorption of glucose on BAS is not the rate limiting step of the reaction, but the strength of BAS determines the selectivity of the reaction. Oxygen chemisorption on Au NPs and/or its interaction with chemisorbed glucose has been postulated as limiting the total reaction rate. The role of the Au NPs size in glucose oxidation has been stressed in many papers and it has been indicated that the optimal gold particle size (to ensure the highest glucose conversion) is ca 7 nm [18] or 9 nm [9,19] for gold supported on carbons and on zeolites [17]. The Au NPs of this size have also been identified as the most active in glucose oxidation on AuPd supported on titanate nanotubes [20]. The question is, if this size of AuNPs would ensure the best performance of all gold supported catalysts applied in glucose oxidation. To answer this question two different zeolite structures (3D Beta and 2D MCM-36) were used in this study as supports of AuNPs of the size of ca 7 nm, and the obtained gold catalysts were applied in base-free glucose oxidation.

To get a deeper insight into the role of BAS in the glucose oxidation pathway, modification of the 2D zeolite support (MCM-36) with boron to change the zeolite acidity was applied. Moreover, amine groups ((3-aminopropyl)trimethoxysilane) were introduced into the zeolite supports for their modification and tailor properties of gold phase. Another problem that we would like to address in this study concerns the role of the oxidant type (molecular oxygen vs. hydrogen peroxide) on the activity of zeolite-based gold catalysts studied.

2. Materials and Methods

2.1. Materials/Compounds

The chemicals used in this work were: glucose (Sigma Aldrich, Saint Louis, MO, USA, 99.5%), chloroauric acid ($HAuCl_4 \cdot xH_2O$, Sigma Aldrich, Saint Louis, MO, USA, 99.995%), (3-aminopropyl)trimethoxysilane (APMS, Sigma Aldrich, Saint Louis, MO, USA, 97%), sodium borohydride (Sigma Aldrich, >98%), boric acid (H_3BO_3, Sigma Aldrich, Saint Louis, MO, USA, 99.97%), toluene (Sigma Aldrich, Saint Louis, MO, USA, HPLC grade), acetonitrile (Sigma Aldrich, Saint Louis, MO, USA, HPLC grade), SiO_2 (Ultrasil, Wesseling, Germany, 3VN, Degussa), sodium aluminate (Riedel-de Haen, Seelze, Germany, 53% of Al_2O_3 and 42.5% of Na_2O), hexamethyleneimine (HMI; Sigma Aldrich, Saint Louis, MO, USA, 99%), cetyltrimethylammonium chloride aqueous solution (CTMACl; 25 wt.%, Sigma Aldrich), tetrapropylammonium hydroxide aqueous solution (TPAOH; 40%, Merck, Darmstadt, Germany), tetraethylorthosilicate (TEOS; Sigma Aldrich, Saint Louis, MO, USA, 98%), sodium hydroxide (NaOH, POCH, Gliwice, Poland, analytical grade), nitric acid (HNO_3, CHEMPUR, Piekary Śląskie, Poland, aqueous solution (65%), analytical grade), hydrogen peroxide (30 wt.%, StanLab, Lublin, Poland), pyridine (C_5H_5N, Sigma Aldrich, Saint Louis, MO, USA, 99.8%), and deionized water.

Proton form of Beta zeolite (HBeta) was obtained via calcination of the commercial ammonium form of Beta (Alfa Aesar, Haverhill, MA, USA, Si/Al = 19). Calcination conditions: 15 h in 550 °C (heating rate 1 °C min^{-1}).

2.2. Synthesis of MCM-36 Zeolite

The MCM-36 zeolite was obtained by the modification of the layered precursor MCM-22 in the two-step procedures reported in [21]. The MCM-22 zeolite was prepared according to the procedure described in [22]. The following reagents were used for the synthesis of MCM-22 zeolite: SiO$_2$ sodium aluminate, deionized water, sodium hydroxide, and hexamethyleneimine (HMI) as the template. The synthesis gel had the following composition: Si/Al = 20, H$_2$O/SiO$_2$ = 49, Na/SiO$_2$ = 0.23, HMI/SiO$_2$ = 0.51. The synthesis mixture was stirred for 30 min at r.t. (room temperature) and then loaded into a Teflon-lined Parr reactor (300 mL). Hydrothermal synthesis was carried out at 150 °C for 120 h upon continuous stirring. The product was recovered by filtration, washed with deionized water, dried at 110 °C. Dried MCM-22 zeolite (with template) was mixed with CTMACl and TPAOH aqueous solutions at a relative weight ratio of 1:4.6:1.16. The mixture was continuously stirred at 90 °C for 24 h. The resulting swollen material was filtered and washed with a small amount of deionized water and dried at 80 °C overnight. The next step was pillaring of dried swollen MWW zeolite. The excess of TEOS was added as a pillaring agent to the swollen MWW zeolite. The mixture was stirred and heated under reflux at 95 °C overnight. The solid was isolated by centrifugation, hydrolyzed in the centrifugation tube by adding 10–20 mL of water and stirring overnight. It was then centrifuged and dried at 60 °C overnight. Final calcination was carried out at 550 °C for 5 h.

2.3. Modification of MCM-36 Zeolites with Boron

B/MCM-36 catalyst was prepared by the impregnation method using an aqueous solution of H$_3$BO$_3$ (assumed boron loading: 1 wt.%). The heated catalyst was placed in a flask connected to a rotary vacuum evaporator and evaporated for 1 h. Next, the proper amount of boric acid solution was added and the wet material was rotated (100 rpm) at 80 °C. After evaporation, the material was dried in an oven at 80 °C for 12 h and then calcined at 550 °C for 5 h.

2.4. Modification with (3-Aminopropyl)trimethoxysilane (APMS)

A given zeolite (HBeta or MCM-36 or B/MCM-36) was dispersed in toluene and to this dispersion APMS was added (2.5 mL of aminosilane per 1 g of zeolite). The as-prepared mixture was refluxed for 18 h and then filtered and washed successively with toluene, deionized water and acetonitrile. The product was dried overnight at 100 °C. The so obtained materials are labelled as NH$_2$/**X**, where **X** stands for the type of zeolite support (HBeta or MCM-36 or B/MCM-36).

2.5. Gold Catalysts Preparation

The aminosilane-modified zeolite support was dispersed in water. Then, a certain amount of the aqueous solution of chloroauric acid was added in order to obtain ca 2 wt.% of gold loading in the final product. The amount of gold precursor required was calculated relative to the weight of the zeolite support without APMS modifier. The suspension was stirred vigorously for 1 h. After stirring, the yellowish solid with incorporated gold species was separated from the colorless solution via vacuum filtration. The resulted yellowish solid was dispersed in a small amount of water and, after 15 min of vigorous stirring, the aqueous solution of sodium borohydride was added (0.1 M; molar ratio NaBH$_4$:Au = 10:1). The mixture immediately turned purple and was stirred for another 20 min. The colored product was separated by vacuum filtration and next dried in 80 °C overnight. Last of all, the samples were calcined in order to remove the aminosilane modifier (calcination conditions: 4 h in 500 °C, heating rate: 2 °C min^{-1}). The as-prepared samples were denoted as Au/**X**, where **X** stands for the type of zeolite support (HBeta or MCM-36 or B/MCM-36) and made the group of calcined gold catalysts.

An independent series of amine-containing materials was prepared. This time the final products were not calcined (the synthesis procedure ended with the drying step) and hence the amino-organosilane modifier was not removed. For this series of materials the amount of gold precursor required to obtain ca 2 wt.% of gold loading in the final product was calculated relative to the weight of the zeolite support containing APMS modifier. The as-prepared materials were labeled as Au/NH$_2$/**X**, where **X** stands for the type of zeolite support (HBeta or MCM-36 or B/MCM-36) and made the group of non-calcined gold catalysts.

The Au/HBeta and Au/MCM-36 samples were furthermore modified with APMS following the procedure described in Section 2.4. The as-prepared materials were denoted as NH$_2$/Au/**X**, where **X** stands for HBeta or MCM-36 and made the group of calcined gold catalysts modified with amino-organosilane.

2.6. Characterization of the Materials

X-ray powder diffraction (XRD) patterns of zeolites were obtained at r.t. on a Bruker AXS D8 Advance (Billerica, MA, USA) apparatus using CuKα radiation (λ = 0.154 nm), with a step of 0.05° in the wide-angle range (2θ = 6–60°).

The N$_2$ adsorption/desorption analysis was performed at −196 °C using a Micromeritics ASAP2020 Physisorption Analyzer (Norcross, GA, USA). The samples were pre-treated in situ under vacuum at 200 °C. The surface area calculated by the BET method, external surface area and micropore volume was estimated according to the t-plot, whereas the total pore volume was calculated at relative pressure (p/p$_0$) of 0.98.

Transmission electron microscopy (TEM) investigation was carried out by means of a Tecnai Osiris instrument (FEI/Thermo Fisher, Waltham, MS, USA) operating at an accelerating voltage of 200 kV. The samples for TEM characterization were placed on a lacey-carbon film supported on a Cu TEM grid. The gold particle sizes were measured with the use of ImageJ software (1.50e version). The mean Au NPs size of each sample was calculated on the basis of 200 counts.

X-ray Photoelectron Spectra (XPS) were obtained on a spectrometer equipped with monochromatic Al-Kα source emitting photons of energy of 1486.71 eV (150 W) and a hemispherical analyzer (PHOIBOS 150 MCD NAP, SPECS, Berlin, Germany) set to the pass energy of 60 eV and 20 eV for survey and regions, respectively. XPS measurement was performed in an ultrahigh vacuum (UHV) under a pressure $< 5 \times 10^{-9}$ mbar and the flood gun was on. A studied sample was deposited on a sample holder using a double-side carbon adhesive tape. Any charging that occurred during the measurements (due to incomplete neutralization of ejected surface electrons) was accounted for by rigidly shifting the entire spectrum by a distance needed to set the binding energy of the C1s, assigned to adventitious carbon, to the assumed value of 284.8 eV.

UV-vis spectra of the samples were recorded using a Varian-Cary 300 Scan UV–vis spectrophotometer (Candela, Warszawa, Poland) equipped with a diffuse-reflectance accessory. Powdered samples were placed in a cell equipped with a quartz window. The spectra were recorded in the range from 190 to 800 nm (the wavelength resolution was 0.4 nm). Spectralon was used as a reference material.

The actual gold loading in zeolites was measured by inductively coupled plasma atomic emission spectroscopy with an ICP-OES SPECTRO BLUE TI spectrometer (Kleve, Germany). The gold phase was extracted from the sample by means of aqua regia.

The amount of nitrogen in the samples was examined by elemental analysis (Elemental Analyser Vario EL III, Elementar Corporation, Langenselbold, Germany).

^{11}B MAS NMR studies were performed on a Bruker NMR 500 MHz spectrometer (11.4 T) (Billerica, MA, USA) at a resonance frequency of 160.46 MHz using a pulse of 0.4 µs (π/11). The chemical shift was determined using boric acid as a reference. All samples were hydrated prior to measurement by placing them overnight in a desiccator with a saturated solution of magnesium nitrate.

Infrared spectra combined with pyridine adsorption were recorded using a Bruker Invenio FTIR spectrometer (Billerica, MA, USA) with an in situ vacuum cell (resolution 4 cm^{-1}, number of scans = 64). Solids in the form of thin wafers (ca. 10 mg cm^{-2}.) were placed inside the cell and then evacuated at 350 °C for 2 h. After that, pyridine was admitted at 150 °C. After saturation with pyridine, the solids were degassed at 150, 200, 250 and 300 °C in vacuum for 30 min at each temperature. The spectrum without adsorbed pyridine (after sample activation) was subtracted from all recorded spectra. The numbers of Lewis and Brønsted acidic sites were calculated assuming the extinction coefficient ε for the band at ca 1545 cm^{-1} = 0.044 cm^2 μmol^{-1} (Brønsted acidic sites) and 1450 cm^{-1} = 0.165 cm^2 μmol^{-1} (Lewis acid sites) [23].

The attenuated total reflectance–Fourier transform infrared (ATR–FTIR) spectra measurements were performed on a Bruker Vertex 70 spectrometer (Billerica, MA, USA) equipped with an ATR attachment with a diamond crystal plate. The catalysts in portions of 15 mg were treated with 0.5 mL of 0.2 M glucose solution and heated at 80 °C, then placed over the diamond crystal, and force was applied to the sample by rotation of the pressure clamp to its click-stop release. The spectra were recorded in the range from 4000 to 400 cm^{-1} (resolution 4 cm^{-1}, number of scans = 64). A background air spectrum was subtracted from each spectrum of the solid sample.

2.7. Glucose Oxidation

Glucose oxidation reactions with molecular oxygen as an oxidant were carried out in a pressure batch reactor from Parr (Moline, IL, USA), equipped with temperature and pressure controllers. In a typical experiment, 20 mL of 0.2 M aqueous glucose solution and the corresponding zeolite catalyst in the amount to achieve the 1970/1 glucose/Au molar ratio, were added to the reactor. The reaction was performed without base addition. After closing, the reactor was purged several times with oxygen before reaction, then it was heated up to 110 °C (a heating rate 2 °C min^{-1}) and pressured at 0.5 MPa with molecular oxygen. The reactions were conducted for 60 min. After the reaction, oxygen was released and the reactor was cooled down to room temperature. Afterwards, the reaction mixtures were withdrawn using a syringe, filtered through a Millipore filter (0.2 μm, PTFE, Maidstone, Great Britain) and analyzed by ultrahigh performance liquid chromatography on a UPLC Acquity Arc Waters instrument (Milford, MA, USA).

To investigate time-dependent glucose conversion, the reactions were performed over selected zeolite catalysts (Au/MCM-36 and Au/NH$_2$/MCM-36) for 15, 30, 60, 120 and 180 min (reaction conditions: 20 mL of 0.2 M glucose solution, glucose/Au (molar ratio) = 1970/1, T = 110 °C, p O$_2$: 0.5 MPa, stirring rate: 600 rpm). The kinetics of glucose oxidation over the selected catalysts was estimated on the basis of pseudo-first order model expressed by the following equation: $ln(C/C_0) = -kt$ where C_0 is the initial concentration of glucose, C is the glucose concentration after given reaction time, k is the reaction rate constant and t is the time of the reaction. According to this model, the reaction rate constant can be determined by drawing a plot of $-ln(C/C_0)$ versus t, whose slope is equal to the value of k. R^2 coefficients were determined to assess the goodness of fit to the pseudo-first order model.

Microwave-assisted oxidation of glucose with hydrogen peroxide as an oxidant. The reaction tests were performed in a microwave synthesis platform MicroSYNTH (from Milestone, Sorisole, Italy) equipped with a contactless temperature infrared controller. Glucose aqueous solution in a portion of 7.64 mL (0.3243 g of glucose in deionized water) was introduced into a 10 mL glass tube. Then, an appropriate amount of a given catalyst (Au/glucose molar ratio: 1970/1) was added. Finally, 0.36 mL of a hydrogen peroxide solution was introduced. After the oxidant introduction, the glucose concentration was of 0.2 M, and the H$_2$O$_2$ to glucose molar ratio was equal to 2.2. The reactions were conducted for 10 min at 110 °C (stirring rate: 80% of the maximum setting). The post-reaction mixture processing and analysis were the same as in the case of glucose oxidation with the use of molecular oxygen.

2.8. Analysis of Reactant and Products

The quantitative analyses of the reaction mixtures were performed using an ultrahigh performance liquid chromatograph UPLC Acquity Arc Waters (Milford, MA, USA) and the products of the reaction were analyzed by two detectors, refractive index (RI 2414) and a photodiode array (PDA 2998). The reactant and the products were separated on a Shodex sugar column SH1011, heated at 30 °C. The eluent was an aqueous solution of H_2SO_4 (0.005 M) and its flow rate was set at 0.6 mL min^{-1}. The samples to be analyzed were collected at the end of the reaction: 1 mL of the reactant/products solution was diluted in 50 mL of deionized water.

3. Results and Discussion
3.1. Composition of Catalysts and Their Textural/Structural Properties

The textural and structural parameters of the prepared catalysts were characterized by XRD and low temperature adsorption/desorption of nitrogen. Figure 1 presents the diffractograms of MCM-36 and HBeta zeolites before and after modifications with boron, amino-organosilane species and gold. All zeolites showed well-defined crystal structures. The XRD patterns present reflections typical of MCM-36 and HBeta zeolite structures. It is worth noting that additional reflections characteristic of metallic gold can be observed in the XRD patterns of the calcined gold catalysts. Modification of MCM-36 zeolite with boron did not change the zeolite structure. The XRD pattern of B/MCM-36 (Figure 1) did not exhibit the presence of boron oxide or boron acid crystal phases. To examine in detail the localization of boron species in MCM-36 zeolites, the ^{11}B MAS NMR studies were performed. The ^{11}B MAS NMR spectra of boron containing MCM-36 (see Supplementary Data; Figure S1) showed two signals in the region characteristic of boron species in tetrahedral environments. The most intensive signal centered at around 2 ppm according to literature [24,25] is assigned to the framework-related tetrahedral deformed zeolite [HO-B-(OSi)$_3$]$^-$. This result indicates that the applied modification method led to partial replacement of aluminum in the zeolite skeleton with boron.

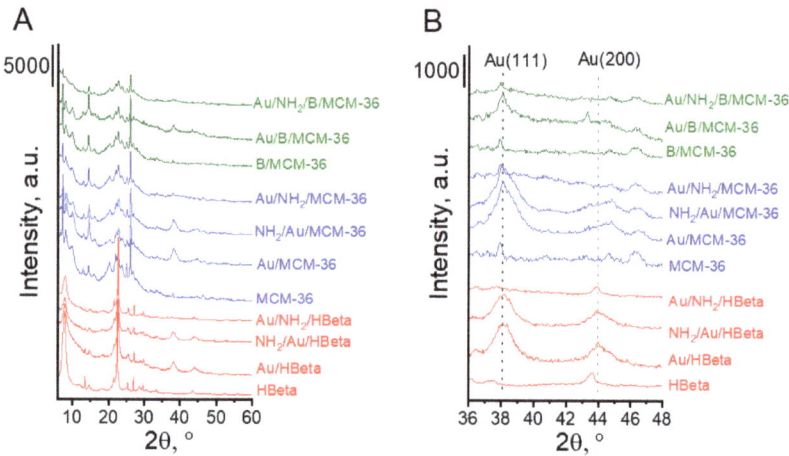

Figure 1. XRD patterns of investigated catalysts. (**A**) The range of 6–60° 2θ; (**B**) the range of XRD peaks typical of Au(111) and Au(200) planes.

Summarizing the modification with boron, amine species and gold do not adversely affect the structural properties of the zeolites but have a significant influence on the textural properties. In Table 1, the texture parameters of all prepared materials are collected. The addition of gold or especially amino-organosilane (APMS) to each type of support (HBeta, MCM-36, B/MCM-36) caused a decrease in the BET surface area, external surface area and pore volume. Significant decrease in the textural parameters of zeolites after APMS modification indicates that the amine species partially block the zeolite pores. This effect was the most pronounced for $NH_2/Au/Beta$ in which the zeolite was modified with amino-organosilane after gold loading and calcination.

Table 1. Composition, mean Au NPs sizes and textural properties of prepared catalysts.

Entry	Catalyst	%wt. Au [a]	%wt. N [b]	Au NPs Size [c], nm	BET Surface Area, $m^2 g^{-1}$	t-Plot External Surface Area $m^2 g^{-1}$	Single Point Total Pore Volume [d] $cm^3 g^{-1}$	t-Plot Micropore Volume, $cm^3 g^{-1}$
1.	HBeta	–	traces	–	526	149	0.34	0.20
2.	Au/HBeta	1.87	traces	7.0	338	82	0.24	0.14
3.	Au/NH$_2$/HBeta	2.37	1.56	2.5	132	78	0.13	0.03
4.	NH$_2$/Au/HBeta	1.78	1.74	6.6	57	25	0.04	0.02
5.	MCM-36	–	traces	–	534	361	0.54	0.10
6.	Au/MCM-36	1.84	traces	7.0	413	191	0.43	0.10
7.	Au/NH$_2$/MCM-36	2.33	1.50	2.8	217	154	0.32	0.03
8.	NH$_2$/Au/MCM-36	1.91	1.36	6.7	210	106	0.24	0.06
9.	B/MCM-36	–	traces	–	546	401	0.53	0.08
10.	Au/B/MCM-36	1.58	traces	5.3	401	241	0.41	0.09
11.	Au/NH$_2$/B/MCM-36	1.55	1.87	3.2	108	38	0.21	0.02

[a] Gold loadings were determined by ICP-OES. [b] The amount of nitrogen in the samples was examined by elemental analysis. [c] The mean gold particle sizes were calculated from TEM images. The calculations were performed on the basis of 200 particles. [d] The total pore volumes were evaluated at a relative pressure of 0.98.

Catalyst composition estimated from ICP-OES for gold and elemental analyzes for nitrogen is shown in Table 1. The catalysts based on aluminosilicate HBeta and MCM-36 zeolites have gold content in the range of 1.78–2.37 wt.%. The boron containing zeolite showed a lower efficiency of gold introduction (1.55 and 1.58 wt.%). Moreover, the amount of gold is similar for the non-calcined gold zeolite containing amino-organosilane (Au/NH$_2$/B/MCM-36) and calcined Au/B/MCM-36. A different tendency was observed for the gold catalyst based on aluminosilicate zeolites (HBeta, MCM-36). The applied method of modification with amino-organosilane affects the final content of gold in the catalysts. The highest content of gold was observed in the non-calcined gold zeolites containing amino-organosilane. The results of elemental analysis indicate insignificant difference in nitrogen content in the non-calcined gold zeolites containing amino-organosilane (Au/NH$_2$/zeolite) and the calcined gold zeolites modified with amino-organosilane after gold loading (NH$_2$/Au/zeolite), based on the same type of zeolites.

3.2. Surface Properties of Catalysts
3.2.1. Acidity of Selected Catalysts

The acidic properties of zeolite supports and calcined gold zeolites were characterized by FTIR spectroscopy with pyridine adsorption (Figure S2). The studies were not performed for the zeolites containing amino-organosilane due to a relatively low decomposition temperature of the modifier. It is known that pyridine is a suitable probe molecule to detect the presence of both Brønsted (the bands at 1545 and 1620 cm^{-1} from protonated pyridine) and Lewis (the bands at 1455 and 1610 cm^{-1} from coordinative bonded pyridine) acidity [26–28]. In the spectra of all zeolites (before and after gold introduction) after pyridine adsorption and evacuation at 200, 250 and 300 °C (it means when the bands at 1447 and 1597 cm^{-1} from pyridine hydrogen bonded to silanol groups disappeared), the bands from Brønsted acid sites (BAS) and Lewis acid sites (LAS) are well visible (Figure S2). However, these bands are more intense for the supports than for the gold-containing zeolites. The number of BAS for all zeolites studied was estimated on the basis of pyridine chemisorbed after evacuation at 150, 200, 250 and 300 °C, whereas the number of LAS–after evacuation at 300 °C (Table 2). Moreover, the ratio of pyridine chemisorbed on BAS/LAS is shown. As follows, the BAS/LAS ratio is much higher for all zeolites of MCM-36 type (between ca 4–6) than for the HBeta samples (ca. 1.5) and it is higher for the supports than for gold-zeolites. It indicates that the number of BAS is higher on MCM-36 than on HBeta samples in relation to the number of LAS and that the number of BAS decreased after modification of both supports with gold. Interestingly, the introduction of boron into the framework of MCM-36 does not significantly change the BAS/LAS ratio (5.73 for MCM-36 and 6.31 for B/MCM-36), the numbers of BAS and LAS in both zeolites are comparable. Moreover, the gold loading on MCM-36 and B/MCM-36 only slightly decreased the number of BAS and therefore, their numbers are almost the same on Au/MCM-36 and Au/B/MCM-36 (3.77 and 4.07, respectively). The strength of BAS was estimated as the percentage of the amount of pyridine desorbed from BAS after evacuation at 300 °C relative to that desorbed at 200 °C. The higher the percentage of pyridine desorbed, the lower the strength of BAS. The data shown in Table 2 indicate that the strength of BAS is very high and comparable for MCM-36 and Au/MCM-36 (22.4 and 24.7% of desorbed pyridine, respectively). The modification of MCM-36 with boron significantly increased the strength of BAS on B/MCM-36 (5.3% of desorbed pyridine), but the modification of this support with gold again led to a decrease in the BAS strength (35.2% of desorbed pyridine for Au/B/MCM-36). The Brønsted acid sites present on HBeta and Au/HBeta are characterized by much lower strength than those on the corresponding MCM-36 and Au/MCM-36 zeolites.

Table 2. Content of Brønsted (BAS) and Lewis (LAS) acid sites occupied by pyridine after desorption at different temperatures. The calculation based on intensity of IR bands (1545 cm^{-1} for BAS and 1455 cm^{-1} for LAS) and extinction coefficients of 0.044 cm^2 µmol^{-1} for BAS and 0.165 cm^2 µmol^{-1} for LAS from [23].

Entry	Catalyst	Evacuation Temp., °C	Number of BAS Occupied by Pyridine, after Evacuation, µmol g^{-1}	Number of LAS Occupied by Pyridine after Evacuation at 300 °C, µmol g^{-1}	Pyridine Desorbed at 300 °C from BAS, % [a]	BAS/LAS Ratio After Evacuation at 300 °C
1.	HBeta	150 200 250 300	434 452 360 297	– – – 193	34.3	1.54
2.	Au/HBeta	150 200 250 300	344 348 244 192	– – – 129	44.8	1.49
3.	MCM-36	150 200 250 300	425 406 373 315	– – – 55	22.4	5.73
4.	Au/MCM-36	150 200 250 300	188 150 148 113	– – – 30	24.7	3.77
5.	Au/B/MCM-36	150 200 250 300	198 176 150 114	– – – 28	35.2	4.07
6.	B/MCM-36	150 200 250 300	482 360 328 341	– – – 54	5.3	6.31

[a] Related to the amount of pyridine chemisorbed after evacuation at 200 °C.

3.2.2. Particle Size and Oxidation State of Gold

The oxidation state of gold in zeolites was determined on the basis of complementary measurements of XRD patterns, UV–vis spectra, HR-TEM images and XPS study. Moreover, the particle size of gold was measured on the basis of HR-TEM images.

The presence of metallic gold in the samples studied was confirmed by all techniques used. The diffraction peaks at $2\theta = 8.2°$ and $44.4°$, typical of metallic gold [29], were found in the XRD patterns of gold-containing zeolites after calcination (Au/HBeta, Au/MCM-36, Au/B/MCM-36) and the calcined samples further modified with the amino-organosilane (NH_2/Au/HBeta and NH_2/Au/MCM-36) (Figure 1B). In the XRD patterns of gold non-calcined zeolites functionalized with APMS and modified with gold (Au/NH_2/HBeta, Au/NH_2/MCM-36, Au/NH_2/B/MCM-36) the reflections corresponding to metallic gold were not detected, suggesting the presence of very small gold crystallites.

To confirm the above suggestions, HR-TEM images were taken. They allowed us to get the information about the shape of gold crystallites, average gold particle size and distribution of Au NPs on the surface of zeolites, depending on the preparation conditions. Spherical metallic gold nanoparticles (NPs) were well visible in the HR-TEM images of all gold catalysts (Figure 2). In the MCM-36 structure, gold was localized in mesopores between zeolites layers (Figure 3). The average size of gold particles is shown in Table 1 and the Au NPs distributions histograms, plotted on the basis of HR-TEM images, are shown in Figure 4. The influence of calcination process after functionalization of zeolite with APMS and gold introduction is clearly seen. Gold particles on the surface of calcined catalysts (Au/HBeta, Au/MCM-36, Au/B/MCM-36) were much larger (d_{av} = 5–7 nm) than the particles on non-calcined gold-catalysts containing APMS (Au/NH_2/HBeta, Au/NH_2/MCM-36, Au/NH_2/B/MCM-36) (d_{av} = 2.5–3.5 nm) indicating that Au agglomeration took place during calcination. The difference between both kinds of samples is well visible in histograms. They show a wide Au NPs size distribution for the calcined samples (the particles diameter oscillates between 3 and 12 nm) and clearly narrower particle size distribution for the non-calcined zeolites prepared without calcination (between 1–6 nm). The results obtained are consistent with the literature data [30] which demonstrated a similar narrow particle size distribution for Pd NPs loaded on silica functionalized with APMS. The treatment of calcined samples with APMS did not change significantly the size of gold particles.

UV-vis spectroscopy was used as the next complementary technique for gold characterization. The UV-vis spectra of all Au-containing samples show the characteristic surface plasmon resonance (SPR) band at ca 500 nm, typical of metallic gold [31] (Figure 5). However, the differences in intensity of the SPR band are observed, depending on gold-zeolite composition and preparation procedure. It is known from literature [32,33] that the intensity of SPR band is related to the diameter of gold nanoparticles. The intensity of the band increases with increasing size of metal particles. The small intensity of the band at around 512–520 nm, observed in the spectra of non-calcined catalysts in comparison to that of the analogous band in the spectra of calcined samples, indicates the presence of very small gold crystallites in Au/NH_2/HBeta and Au/NH_2/MCM-36. Similar intensities of the band at around 520 nm in the UV-vis spectra of calcined gold-zeolites and that in the spectra of calcined gold-zeolites further treated with APMS, indicate that the introduction of amino-organosilane after sample calcination did not influence the size of metal particles. These results are in line with the results of XRD study and TEM measurements. However, it is important to note that the presence of organosilane (APMS) in zeolites leads to the appearance of new bands in the UV-vis spectra. Two bands at ca 200 and 260 nm are characteristic of amine species [34–36]. Moreover, an additional band at 448 nm is visible in the spectrum of NH_2/Au/HBeta. The appearance of a secondary peak in the UV-vis spectrum at a shorter wavelength relative to the typical gold SPR band position was observed earlier for gold core–mesoporous silica shell nanoparticles (Au-MMS-rods) [37]. On the other hand, it has been documented that gold nanospheres show one absorption peak in the UV-vis spectrum, while nanorods show two characteristic absorption peaks due to transverse plasmon resonance and longitudinal plasmon resonance [37]. That is why, in the case of zeolites with spherical particles, the presence of the secondary band cannot be related to transverse/longitudinal SPR and should be a result of amino-organosilane localization in the vicinity of gold. The SPR band is sensitive to dielectric properties of the environment in the vicinity of the illuminated nanoparticle. Changes in the environment of the nanoparticle cause a shift in the maximum of the plasmon absorption band. Taking into account that the secondary band in the shorter wavelength region is not present in the spectrum of $NH_2/Au/MCM-36$, one can suggest the presence of gold particles in two different surroundings only on the surface of $NH_2/Au/HBeta$. Most probably on the surface of $NH_2/Au/HBeta$ zeolite, APMS anchors to the hydroxyls which are close to gold and that is why amino-organosilanes are in very close proximity to gold changing dielectric properties of gold environment (the SPR band at 448 nm). In the layered and pillared MCM-36, gold is preferentially localized in mesopores (between zeolite layers) and APMS is anchored to hydroxyl groups relatively away from the gold. As a result, only one Au SPR band at ca 520 nm is visible in the spectra of $NH_2/Au/MCM-36$.

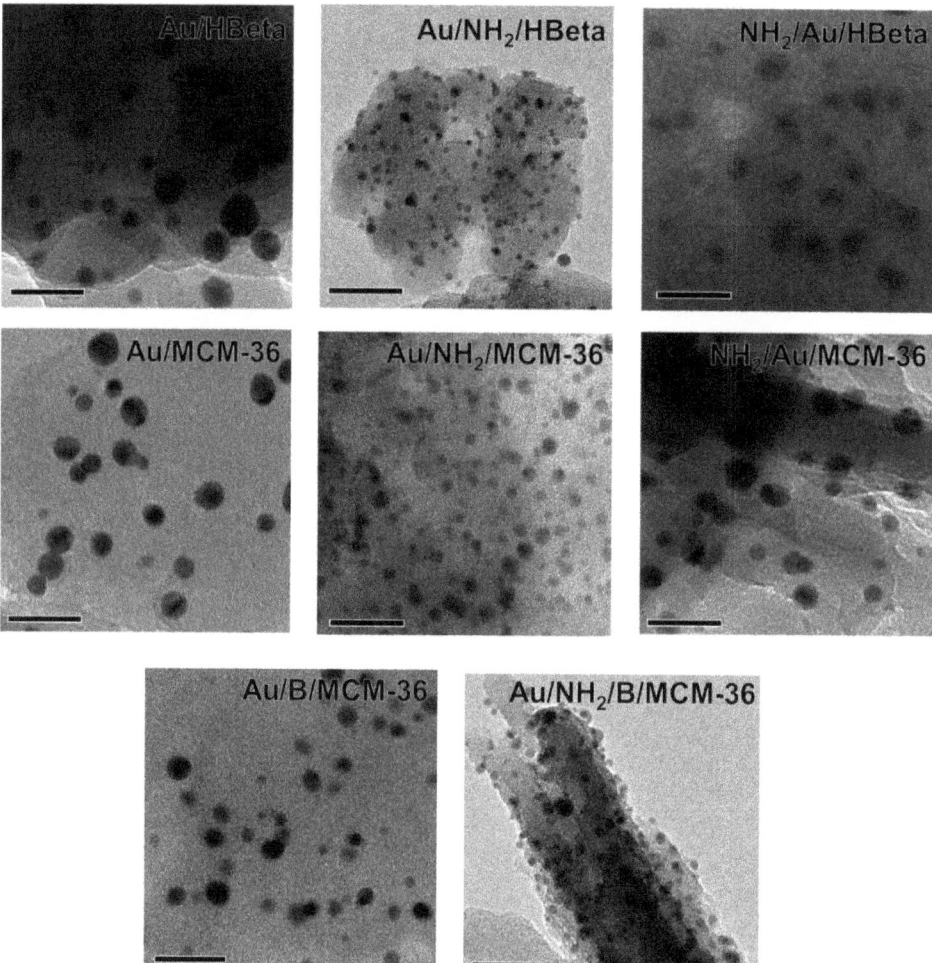

Figure 2. HR-TEM images of investigated gold catalysts. Scale bars are equal to 20 nm in each case.

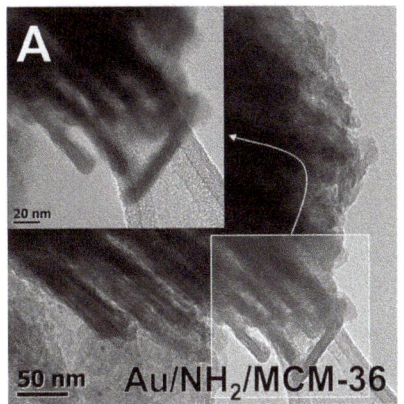
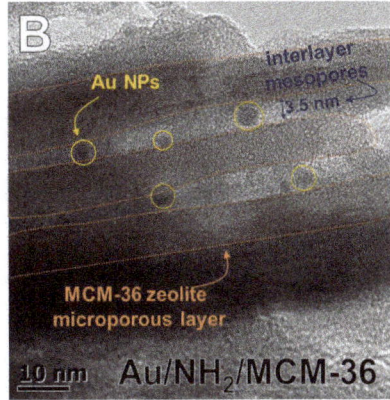

Figure 3. (**A**) An exemplary HR-TEM image indicating layered structure of MCM-36 support; (**B**) HR-TEM image presenting Au NPs location in interlayer spaces of MCM-36 support.

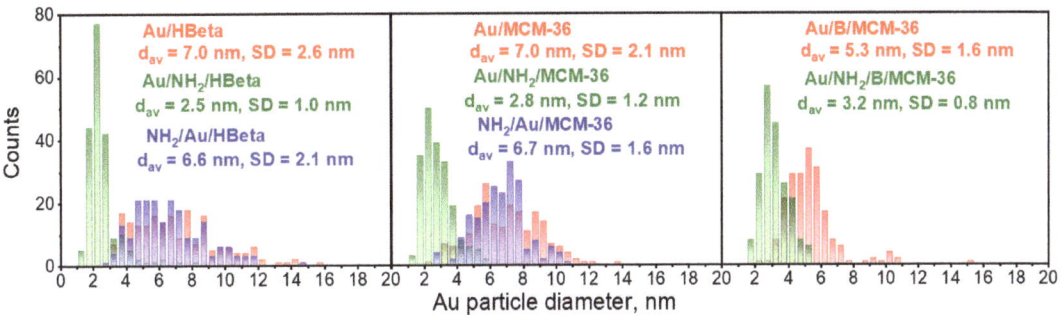

Figure 4. Gold particle size distribution histograms of the catalysts. The values of mean Au NPs diameter (d_{av}) and standard deviation (SD) were included. The number of particles measured was 200 in each case.

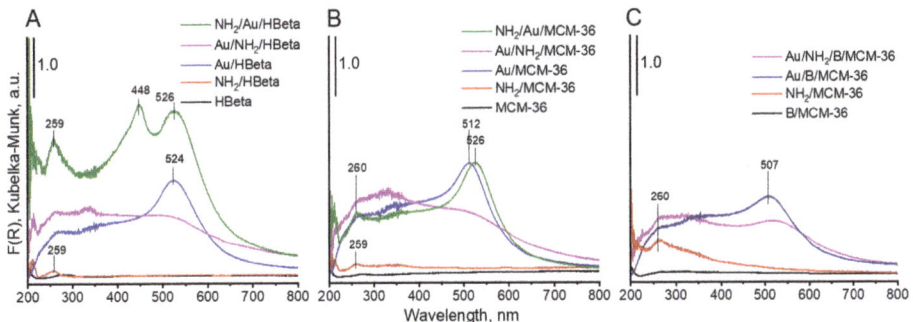

Figure 5. DR-UV-vis spectra of prepared materials.

The presence of metallic gold particles on the surface of all catalysts was further confirmed by X-ray photoelectron spectroscopy. The XPS spectra of gold zeolites based on HBeta and MCM-36 are compared in Figure 6. In all Au 4f spectra, two components separated by 3.7 eV, namely Au $4f_{7/2}$ and Au $4f_{5/2}$, were identified. According to literature [38], the Au $4f_{7/2}$ peak of metallic gold species corresponds to the binding energy value of 84.0 eV. Interestingly, for all materials prepared in this study, the binding energy corresponding to the Au $4f_{7/2}$ peak was lower than 84.0 eV, indicating that in all the samples the gold species were present in metallic form on whose surface negative charge was accumulated [39]. The higher the BE shift towards lower binding energy value of Au $4f_{7/2}$, the higher the negative charge accumulated on the surface of metallic gold species [39]. From among all calcined gold catalysts, the lowest BE of Au NPs was observed for Au/MCM-36 catalyst (BE = 83.3 eV; Figure 6). For Au/B/MCM-36 and Au/HBeta the BE of Au NPs was slightly higher and took the values of 83.6 and 83.7 eV, respectively. As far as the electronic properties of gold species are concerned, it is important to underline that Au NPs in the non-calcined gold catalysts were characterized by similar binding energy values as those determined for the calcined samples. Thus, in view of this observation, one can conclude that the presence of amine group has no significant impact on the electronic properties of gold species. Interestingly, in the calcined gold catalysts modified with amino-organosilane, the binding energy of gold species was the lowest from among all materials synthesized in this work, suggesting the presence of strong electronic interaction between the amino-organosilane and Au NPs. Thus, these results show the difference between the amine species on gold zeolites calcined prior to modification with APMS and those on the non-calcined (containing amino-organosilane) samples.

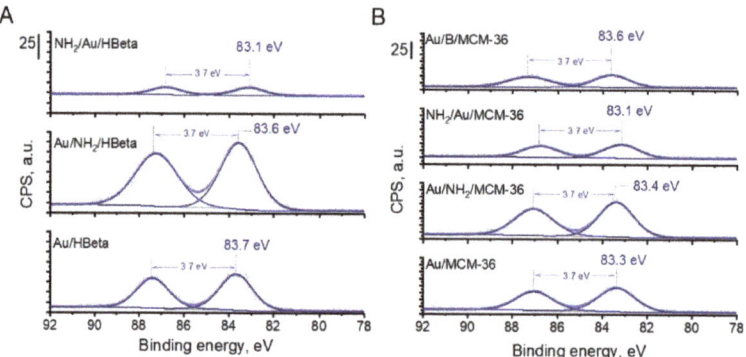

Figure 6. Au 4f region of XP spectra of gold-containing zeolites from (**A**) HBeta; (**B**) MCM-36 and B/MCM-36 series. The values of binding energy corresponding to the Au $4f_{7/2}$ peaks are marked.

3.3. Glucose Oxidation with Molecular Oxygen

3.3.1. Influence of Zeolite Support on the Activity of Gold Catalysts

Table 3 shows the results of glucose oxidation after 1 h of the reaction performed over all catalysts studied in this work. Independently of the glucose conversion, the selectivity to gluconic acid over gold catalysts was very high (ca. 99%). All zeolite supports, i.e., HBeta, MCM-36 and B/MCM-36, exhibited low glucose conversion. Deposition of gold species on the zeolite supports grafted with APMS and subjected to calcination (the gold calcined zeolites) resulted in a significant increase in the catalysts activities. In view of these results, one can conclude that zeolite supports were almost inactive in glucose oxidation and the main active components responsible for efficient glucose oxidation to gluconic acid were gold nanoparticles. From among all calcined gold zeolites, the highest glucose conversion of 65.2% was observed for Au/MCM-36 (Table 3, Entry 6). Interestingly, in the case of Au/HBeta, which had the same average gold particle size as that of Au/MCM-36, glucose conversion was significantly lower and was found to be of 50.6% (Table 3, Entry 2). In view of these results, one can conclude that the type of zeolite support had a significant impact on the activity of gold particles. The gold catalysts supported on MCM-36, characterized by larger pore size and stronger Brønsted acidity (see Tables 1 and 2), were found to be more active than those supported on HBeta. It shows that the porosity of the support and the strength of its BAS can have significant impact on the activity of gold nanoparticles in glucose oxidation. Interestingly, the lowest glucose conversion of

45.8% was observed for Au/B/MCM-36. This catalyst had slightly smaller gold particle size than Au/MCM-36 (5.3 vs. 7.0 nm, respectively), the same pore structure, almost the same number of BAS but a weaker Bronsted acidity (see Table 2). Since smaller gold particles should result, in general, in higher glucose conversion than that observed for the catalysts with larger gold particles [15], one can expect that a lower glucose conversion in the presence of Au/B/MCM-36 containing smaller Au NPs can be assigned to the lower strength of acid sites. According to previous reports [17,40,41], BAS can play an important role in transformation of glucose molecule to geminal 1,1-diol (hydrate), which is then oxidized by active oxygen species adsorbed on gold particles. The stronger the BAS, the more labile the protons, and thus, the more efficient the protonation of glucose molecule to form *gem*-diol. As far as differences in the activity of calcined gold zeolites are concerned, it is important to underline that Au/MCM-36, Au/B/MCM-36 and Au/HBeta exhibited different electronic properties of gold species. As described in the XPS section, Au NPs supported on Au/MCM-36 sample were characterized by significantly lower binding energy than that of the gold species in Au/B/MCM-36 and Au/HBeta (83.3 vs. 83.6 and 83.7 eV, respectively; Figure 6). According to literature [39], a shift of BE of gold species towards binding energy values lower than 84.0 eV may indicate accumulation of partial negative charge on the surface of metallic gold species. The stronger the BE shift, the stronger the partial negative charge accumulated on Au NPs. Formation of metallic gold species with partial negative charge on their surface has been previously reported by Xu et al. [39]. These authors have established that accumulation of negative charge on the surface of Au NPs facilitated activation of molecular oxygen and increased the activity of gold catalysts in oxidation reactions. Thus, in view of previous literature reports, we hypothesize that the highest activity of Au/MCM-36 may result not only from different porosity and acidity of the supports in comparison to Au/HBeta, but also from different electronic properties of the active phase, i.e., the gold species. Since Au/B/MCM-36 and Au/MCM-36 exhibited the same pore structure, but different electronic properties of gold species, one can expect that this hypothesis is very probable. The role of partial negative charge on the surface of gold species in oxidation of glucose will be discussed below.

Table 3. Results of base-free oxidation of glucose with molecular oxygen over selected catalysts.

Entry	Catalyst	Glucose Conv. [a], %	Selectivity, %		TOF [b], h^{-1}
			A	B	
1.	HBeta	1.2	82.0	18.0	–
2.	Au/HBeta	50.6	98.8	1.2	24,200
3.	Au/NH$_2$/HBeta	60.7	99.2	0.8	10,400
4.	NH$_2$/Au/HBeta	12.0	>99.9	traces	5410
5.	MCM-36	6.2	89.0	11.0	–
6.	Au/MCM-36	65.2	99.2	0.8	31,300
7.	Au/NH$_2$/MCM-36	77.5	99.3	0.7	14,800
8.	NH$_2$/Au/MCM-36	61.0	99.1	0.9	28,000
9.	B/MCM-36	<1	>99.9	traces	–
10.	Au/B/MCM-36	45.8	98.9	1.1	16,600
11.	Au/NH$_2$/B/MCM-36	68.1	99.1	0.9	14,900

[a] Reaction conditions: 20 mL of 0.2 M glucose solution, glucose/Au (molar ratio) = 1970/1, mixing rate = 600 rpm, T = 110 °C, pO$_2$ = 0.5 MPa, time = 1 h; A–gluconic acid, B–glucuronic acid. [b] The number of moles of gold atoms localized on the external surface of spherical Au particles was taken into account in the TOF calculations: (the number of moles of glucose converted after 1 h) × (the number of moles of gold atoms localized on the external surface of the Au NPs in a given mass of the catalyst)$^{-1}$ × h^{-1}; based on Au NPs size calculated from TEM.

As described in Section 3.2.2., non-calcined gold catalysts contained significantly smaller gold particles than the calcined catalysts. Thus, one can expect that smaller gold particle size should result in a higher catalytic activity. Indeed, it was revealed that the activity of all non-calcined gold catalysts after 1 h of the reaction was much higher than that observed for calcined samples. From among all non-calcined samples, the highest glucose conversion of 77.5% was observed for Au/NH$_2$/MCM-36 catalyst. It is worth noting that amine-grafted zeolite supports exhibited almost the same activity as parent zeolites (results not shown). Thus, the amine groups themselves were not able to take part in oxidation of glucose to gluconic acid. The possible synergistic interaction between amine groups and Au NPs, which could influence the gold activity, has been excluded. As described in the XPS section, the presence of amine group of APMS in the neighborhood of gold species has no impact on the electronic properties of gold species which were similar in Au/MCM-36 and Au/NH$_2$/MCM-36.

To shed more light on the role of amine groups in glucose oxidation over gold catalysts, additional samples, in which APMS was grafted on the surface of calcined samples, were synthesized (NH$_2$/Au/HBeta and NH$_2$/Au/MCM-36) and had similar average gold particle size as the parent samples before anchoring of APMS (see Table 1 and Figure 4). As implied by Table 3 data, the influence of APMS grafting on the activity of calcined gold catalysts was strongly affected by the type of zeolite support. In the case of the catalysts supported on HBeta, modification with APMS significantly diminished the catalytic activity of the parent sample. Interestingly, in the case of NH$_2$/Au/MCM-36, modification with APMS had negligible influence on glucose conversion. In view of these results, one can conclude that the type of zeolite support had a significant impact on the role of amine groups, and thus, on the overall catalytic performance of the catalysts. This hypothesis was further confirmed by different optical properties of gold species in NH$_2$/Au/HBeta and NH$_2$/Au/MCM-36 (see Figure 5). As described in the TEM section, the most important difference between Au-HBeta and Au-MCM-36 was localization of gold species. In Au/HBeta, gold catalysts were localized mainly on the external surface of the support (see Figure S3), while in Au/MCM-36 they were localized mainly between the zeolite layers (see Figure 3B). Taking into account the results of UV-vis measurements, it is very likely that gold particles in NH$_2$/Au/HBeta were covered by APMS, while in NH$_2$/Au/MCM-36 the distance between the Au NPs localized inside the zeolite layers and APMS modifier was much larger, preventing deactivation of the gold species. Since glucose conversion in the presence of NH$_2$/Au/MCM-36 was only slightly lower than that in the presence of Au/MCM-36, we inferred that deactivation of NH$_2$/Au/HBeta resulted from covering of gold particles with the amino-organosilane, but not from the strong interaction between the amine group of APMS and the gold species. Moreover, a higher activity of non-calcined gold catalysts (Au/NH$_2$/zeolite) than that of calcined NH$_2$/Au/zeolite materials indicated that the role of amino-organosilane in both types of materials (the calcined catalysts modified with amino-organosilane and non-calcined ones) was different.

Differences in activities of non-calcined gold zeolites containing amino-organosilane (Au/NH$_2$/MCM-36 and Au/NH$_2$/HBeta) and calcined gold zeolites modified with amino-organosilane after gold loading (NH$_2$/Au/MCM-36 and NH$_2$/Au/HBeta) can be explained on the basis of ATR spectra of zeolites after adsorption of glucose solution and drying at 80 °C. As can be seen in Figure S4, the FTIR spectra of the zeolites and glucose show the main bands in the wavenumber range 1300–700 cm^{-1}. The characteristic bands assigned to glucose have maxima at 1222 and 1200 (δCH + δOH in plane), 1148, 1050 and 994 (νCO + νCC), 1106 and 1021 (νCO), 915 (νCO + νCCH + γ_{as} ring of pyranose), 838 (δOH) and 772 cm^{-1} (δCCO + δCCH) [42]. Some of these bands (1200, 1148, 1106, 1021, 994, 915 and 772 cm^{-1}) are also visible in the spectra of the zeolites treated with glucose, but they are shifted to the lower wavenumbers (1190, 1145, 1077, 1017, 985, 900 and 770 cm^{-1}, respectively). The highest blue-shift is observed for the bands that are overlapped by those assigned to the T-O-T (T = Si or Al) vibrations in the zeolites (at 1220, 1087 and 797 cm^{-1}). The latter are also shifted in the spectra of the zeolites after glucose adsorption (to ca 1212, 1048 and 793 cm^{-1}, respectively). It is important to note that the spectra of all zeolites studied after glucose adsorption are similar in the range 1300-700 cm^{-1}.

Significant differences in the ATR spectra are observed in the C=O vibration region (1800–1600 cm^{-1}). Two bands at ca 1737 and 1777 cm^{-1} are present only after adsorption of glucose on Au/NH$_2$/MCM-36 and Au/NH$_2$/HBeta. The same bands were previously observed in the spectra of gold-zeolites after glucose oxidation and were assigned to C=O stretching vibrations in the reaction product—gluconic acid [17]. In the ATR study in this work, the oxidation of glucose cannot be considered, but the above-mentioned bands can be assigned to chemisorption of glucose. Such a chemisorption with the participation of protons and water had been proposed earlier [17] and the coordination of glucose-water-proton complex to the zeolite surface had been shown. The glucose-water-proton complex contains carboxyl groups whose C=O vibrations appear in the IR spectrum as the two, above-mentioned bands. The question is why such a chemisorption was observed only on two catalysts. The chemisorption of this type cannot be explained only as due to the presence of amine groups on the zeolite surface as the two bands were not observed in the spectra of NH_2/Au/zeolite samples. Therefore, we hypothesize that the amine groups in Au/NH_2/zeolite are protonated during the procedure of gold anchoring (from auric acid precursor). It is known that in the process of gold anchoring on APMS functionalized samples in the first step amine groups are protonated and $AuCl_4^-$ anion is connected to the NH_3^+ moiety [43,44]. The hypothesis about partial protonation of the amine groups of APMS in non-calcined gold catalysts was confirmed by XPS. According to literature [45], the amine groups of APMS on functionalized surfaces are characterized by the binding energy of ca 399.1–399.6 eV, while the protonated amine groups are characterized by the binding energy of ca 400.9–401.7 eV. As can be seen from Figure S5, the N 1s region of the non-calcined gold catalysts and the calcined gold catalysts modified with amino-organosilane differed substantially. For the non-calcined samples, one can observe a more pronounced peak with the binding energy typical of protonated amine groups (region highlighted in yellow). Interestingly, the highest relative contribution of protonated amine groups was observed for Au/NH_2/HBeta catalyst, indicating different efficiency of amine protonation in MCM-36 and HBeta catalysts. In view of these results, we hypothesize that in the non-calcined samples (Au/NH_2/zeolite) the protonated amine groups take part in the chemisorption of glucose via aldehyde groups. It is not the case for the catalysts in which APMS was added after gold loading followed by calcination.

Taking into account the above described results one can conclude that the enhancement in glucose conversion in the presence of Au/NH_2/zeolite is caused, to some extent, by the increase in the glucose chemisorption rate attributed to the presence of protonated amine groups.

Normalization of catalysts activity in glucose oxidation to the exposed surface gold atoms expressed in turnover frequency, TOF, provides more detailed insight into the reactivity of Au NPs on the surface of the catalysts (see Table 3). It was found that the highest reactivity of surface gold atoms was observed for calcined gold catalysts characterized by gold particle size in the range from 5.3 to 7.0 nm. It is important to emphasize that for all calcined samples, TOF values are strongly correlated with electronic properties of gold species. The lower the BE of the gold species, the higher the TOF values. These results clearly show that accumulation of partial negative charge on the surface of metallic gold species is one of the most important factors determining the activity of surface gold atoms in glucose oxidation. For non-calcined samples, in which gold particles were much smaller (ca. 2.5–3.2 nm in diameter; Table 1), TOF values were significantly lower. Thus, the results obtained in this study are in agreement with the previous literature reports [17], in which the highest reactivity of surface gold atoms was observed for gold particles with diameter of ca 30 nm, and the TOF value was reduced as the gold particle size decreased.

In order to evaluate the catalytic activity of the Au/MCM-36 (i.e., the catalyst with the highest TOF value) in base-free glucose oxidation with molecular oxygen, several recent reports on this topic were analyzed. Table 4 contains a comparison of selected parameters characterizing gold catalysts (on various supports such as zeolites (MCM-36 and HBeta) [17], titania (TH) [4], mesoporous carbon (CMK) [46], mesoporous silica (SBA-15) [47], hydroxyapatite-layered double hydroxide composite (HAP-LDH) [48]) and their catalytic performance in base-free glucose oxidation. The most active catalyst from each work was selected for this comparison. It is noteworthy that Au/MCM-36 is characterized by the highest TOF value from all the corresponding catalysts, even though it contains the largest Au NPs.

Table 4. Comparison of TOF values of different gold catalysts in base-free glucose oxidation with molecular oxygen as oxidant.

Entry	Catalyst Symbol	%wt. Au [a]	Au NPs Size, nm	Reaction Conditions	TOF, h^{-1}	Reference
1.	Au/MCM-36	1.84	7.0	20 mL of 0.2 M glucose solution, glucose/Au molar ratio = 1970, T = 110 °C, p(O$_2$) = 0.5 MPa	31,300 [a]	This work
2.	Au/TH-150	0.5	1.2	10 mL of 0.1 M glucose solution, glucose/Au molar ratio = 1000, T = 110 °C, p(O$_2$) = 1 MPa	1920 [b]	[4]
3.	Au/CMK-3	0.94	3.0	20 mL of 0.1 M glucose solution, glucose/Au molar ratio = 1000, T = 110 °C, p(O$_2$) = 0.3 MPa	17,700 [c]	[46]
4.	Au-HBeta(AP)	1.4	6.0	20 mL of 0.2 M glucose solution, glucose/Au molar ratio = 1970, T = 110 °C, p(O$_2$) = 0.5 MPa	26,500 [d]	[17]
5.	Au/SBA-15	2.1	4.9	10 mL of 0.2 M glucose solution glucose/Au molar ratio = 1970, T = 110 °C, p(O$_2$) = 0.5 MPa	16,900 [d]	[47]
6.	Au/HAP-LDH	0.22	6.6	12 mL of 0.167 M glucose solution glucose/Au molar ratio = 1000, T = 110 °C, p(O$_2$) = 0.5 MPa	20,200 [e]	[48]

Calculated after: [a] 60 min of the reaction; [b] 15 min of the reaction; [c] 5 min of the reaction; [d] 120 min of the reaction; [e] calculated for low conversions (<30%).

3.3.2. Time Dependence of Glucose Conversion

To get a deeper insight into glucose oxidation over the most active gold catalysts supported on MCM-36 zeolite, kinetic measurements were performed. Figure 7A shows glucose conversion over Au/MCM-36 and Au/NH$_2$/MCM-36 samples as a function of reaction time. It was found that the non-calcined catalyst, irrespectively of the reaction time, exhibited significantly higher glucose conversion than that observed for the calcined one. As can be seen from Figure 7A, Au/NH$_2$/MCM-36 reached the reaction equilibrium after 120 min, while the same glucose conversion over Au/MCM-36 was observed 60 min later, i.e., after 180 min of the reaction. To shed more light onto the kinetics of glucose oxidation over gold catalysts supported on MCM-36, the experimental data were fitted to pseudo-first order kinetic model (see Figure 7B). It was found that the experimental data obtained for the catalysts fit in within this model. For the reactions with the use of Au/MCM-36 and Au/NH$_2$/MCM-36, the R^2 values were higher than 0.95 (see Figure 7B). The highest reaction rate constant of 0.0742 min^{-1} was observed for Au/NH$_2$/MCM-36 sample, and was ca 1.5 times higher than that estimated for Au/MCM-36.

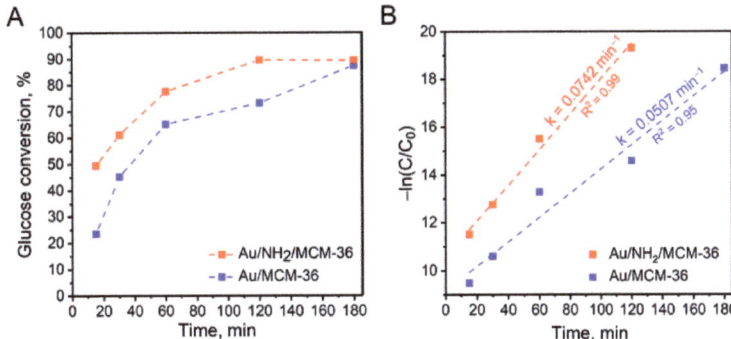

Figure 7. (**A**) Effect of the reaction time on the activity of Au/MCM-36 and Au/NH$_2$/MCM-36 in base-free glucose oxidation with molecular oxygen; (**B**) pseudo-first order plot for determination of glucose oxidation rate. The reaction rate constants (k) were marked in the graph. *Reaction conditions:* 20 mL of 0.2 M glucose solution, glucose/Au (molar ratio) = 1970/1, mixing rate = 600 rpm, T = 110 °C, p O$_2$ = 0.5 MPa, time = 1 h; A—gluconic acid, B—glucuronic acid.

3.4. Glucose Oxidation with Hydrogen Peroxide

Table 5 summarizes the results of the microwave-assisted oxidation of glucose with the use of hydrogen peroxide as an oxidant. The glucose conversion in the presence of zeolite supports was relatively low and did not exceed 15%. Only gold catalysts showed significant activities. The selectivity to gluconic acid for the gold-containing zeolites was high and reached >98% in each case.

Table 5. Results of microwave-assisted oxidation of glucose with hydrogen peroxide over selected catalysts.

Entry	Catalyst	Glucose Conv. [a], %	Selectivity, % A	Selectivity, % B	TOF [b], h^{-1}
1.	HBeta	13.3	93.9	6.1	–
2.	Au/HBeta	49.1	99.0	1.0	158,000
3.	Au/NH$_2$/HBeta	44.4	98.9	1.1	50,900
4.	NH$_2$/Au/HBeta	12.2	95.2	4.8	37,100
5.	MCM-36	3.8	83.6	16.4	–
6.	Au/MCM-36	59.2	99.2	0.8	191,000
7.	Au/NH$_2$/MCM-36	59.5	99.2	0.8	76,500
8.	NH$_2$/Au/MCM-36	45.9	99.0	1.0	142,000
9.	B/MCM-36	7.2	72.2	27.7	–
10.	Au/B/MCM-36	49.5	99.0	1.0	121,000
11.	Au/NH$_2$/B/MCM-36	56.6	99.1	0.9	83,800

[a] Reaction conditions: glucose/Au (molar ratio) = 1970/1, 8 mL of 0.2 M glucose solution, T = 110 °C, 2.2 equiv. H$_2$O$_2$, time = 10 min.; A–gluconic acid, B–glucuronic acid. [b] The number of moles of gold atoms localized on the external surface of spherical Au particles was taken into account in the TOF calculations: (the number of moles of glucose converted after 1 h) × (the number of moles of gold atoms localized on the external surface of the Au NPs in a given mass of the catalyst)$^{-1}$ × h^{-1}; based on Au NPs size calculated from TEM.

The dependence of glucose conversion on the structural/textural properties of calcined gold zeolites in the microwave-assisted glucose oxidation with the use of hydrogen peroxide appeared to be similar to that reported above for glucose oxidation with the use of molecular oxygen. The activities of the Au/HBeta and Au/MCM-36 gold zeolites differed even though their gold particle size was the same (ca. 7 nm). The glucose conversion higher by ca 10 percentage points was observed for Au/MCM-36 catalyst. This difference can be attributed to the larger pore size in the latter sample. However, the impact of electronic properties of Au NPs and acidic centers on the zeolite supports on the activity in glucose oxidation with H$_2$O$_2$ should also be taken into account. According to the literature data [49], hydrogen peroxide is chemisorbed on Au surface active species. The chemisorption is accompanied by decomposition of H_2O_2. One of the proposed mechanisms of chemisorption is the redox mechanism [49] in which electron transfer from gold species to H_2O_2 occurs, which leads to the formation of OH^- ions and $HO^\bullet$ radicals. The OH^- ions react with free H_2O_2 to yield HO_2^- and H_2O. Finally, in the reaction between HO_2^- and $HO^\bullet$ radicals, O_2 and H_2O are generated. The higher the electron density on the surface of Au NPs, the faster the electron transfer in the first step of this redox pathway. As the electron density (concluded from XPS results) on the calcined and non-calcined gold zeolites based on HBeta was lower than that on the catalysts based on MCM-36 and the catalysts of the latter series were more active, one can conclude that the rate of hydrogen peroxide decomposition is crucial for the activity of gold zeolites in glucose oxidation with H_2O_2. The acidity of the support for gold seems not to play an important role in this reaction because Au/HBeta with a higher concentration of BAS showed a lower activity than Au/MCM-36 containing a lower number of BAS. Moreover, the presence of protonated amine groups grafted on the zeolitic surface in the non-calcined gold catalysts did not result in reaching distinctly superior glucose conversion when compared to that of the calcined ones, as reported for the oxidation with molecular oxygen. A slight activity enhancement was observed only for $Au/NH_2/B/MCM$-36 when compared that of Au/B/MCM-36 (56.6% vs. 49.5%, respectively). It suggests that the rate limiting reaction step in the oxidation with H_2O_2 is different from that in the oxidation with molecular oxygen.

Modification of the calcined Au/zeolites by APMS led to different activities depending on the zeolite structure. As mentioned before, it is expected that in $NH_2/Au/HBeta$ sample, Au NPs are covered with the amine modifier (as concluded from UV-vis results) and therefore, the activity significantly decreased in both the oxidation with H_2O_2 as well as with O_2. It is not the case for $NH_2/Au/MCM$-36 and, interestingly, a decrease in $NH_2/Au/MCM$-36 activity after the amino-organosilane modifier introduction is relatively low.

Similarly to the above-mentioned results for glucose oxidation with O_2, the highest TOF values (Table 5) were observed for the calcined gold zeolites containing the largest Au NPs, i.e., those of ca 7 nm in diameter (Au/MCM-36 and Au/HBeta). Slightly lower TOF number were obtained for the amino-organosilane-grafted $NH_2/Au/MCM$-36 catalysts. In contrast, the lowest activity expressed by the TOF number was found to show $NH_2/Au/HBeta$.

The glucose conversions for all the examined gold catalysts did not exceed 60%. One can suppose that it was due to the oxidant depletion during the reaction run. However, the presence of H_2O_2 remaining in the post-reaction solutions was confirmed by a simple qualitative test (by applying potassium iodide and starch mixture). Nevertheless, to make sure that the initial amount of H_2O_2 was not the limiting factor affecting the catalytic performance, a test with the use of a larger amount of H_2O_2 (3.0:1.0 oxidant to glucose molar ratio) for a chosen catalyst (Au/HBeta) was performed. It was established that the increase in H_2O_2/glucose molar ratio from 2.2 to 3.0 did not result in any significant enhancement of glucose conversion (49.1 vs. 50.8%, respectively). It was therefore decided to keep the H_2O_2/glucose molar ratio of 2.2 in all the experiments.

The mechanism of microwave-assisted glucose oxidation in the presence of hydrogen peroxide has not been yet clearly elucidated and current literature does not provide exhaustive answers to the questions about possible reaction pathways in the glucose oxidation reaction. Rautiainen et al. [50] have shown a correlation between the rate of oxygen formation and the activity of gold catalysts. The most active Au/Al_2O_3 exhibited the highest rate of H_2O_2 decomposition towards O_2 formation. Bearing in mind that the differences in glucose conversion between the calcined and non-calcined gold zeolites were negligible when using hydrogen peroxide as an oxidative agent, it is supposed that the molecular oxygen evolution resulting from hydrogen peroxide decomposition on Au NPs—rather than the main oxidation step of carbonyl group of a glucose molecule—could be a possible rate-limiting step in the overall process. Microwave-assisted oxidation of glucose with H_2O_2 is undoubtedly a promising alternative to the conventional oxidation of glucose with oxygen, but further research should be carried out to shed more light on the mechanism of hydrogen peroxide activation.

4. Conclusions

Different combinations in modification of the zeolite supports (APMS grafting on calcined and non-calcined gold zeolites as well as boron impregnation) for gold and the use of zeolites of different structures (two-dimensional MCM-36 and three-dimensional HBeta) allowed us to establish some important correlations between various parameters and the activity of gold catalysts in glucose oxidation with molecular oxygen and hydrogen peroxide. Much higher glucose conversion in oxidation with both, molecular oxygen and hydrogen peroxide, obtained in the presence of Au/MCM-36 and Au/NH$_2$/MCM-36 than that got for Au/HBeta and Au/NH$_2$/HBeta, respectively, indicated the role of the support porosity/structure/texture properties. The more open the structure (larger pores), the higher the glucose conversion. For the supports of the same structural/textural features (Au/B/MCM-36 and Au/MCM-36), the level of negative charge accumulation on metallic gold particles (a lower for the boron containing catalyst) determined chemisorption of molecular oxygen and decomposition of hydrogen peroxide, and in this way influenced glucose conversion (higher for the reaction performed in the presence of Au/MCM-36 with a higher electron density on gold particles). The role of gold NPs with partial negative charge on their surface was also confirmed by comparison of the activity of a single surface gold atom (expressed by TOF) for two catalysts containing Au NPs of the same average size of 7 nm, namely Au/HBeta (lower partial negative charge on gold particles) and Au/MCM-36 zeolites (higher partial negative charge on Au NPs). The first catalyst showed a much lower TOF value.

Non-calcined Au/NH$_2$/zeolites contained the smallest Au NPs. The presence of APMS in non-calcined gold zeolites, independently of the support structure and composition, led to a significant growth of the glucose conversion in the oxidation with molecular oxygen and almost did not change their activity in the oxidation with hydrogen peroxide, although the gold particles were well dispersed. The amine groups, protonated during the procedure of gold anchoring, took part in the chemisorption of glucose, which enhanced the catalytic activity in glucose oxidation with oxygen. The presence of APMS in calcined gold zeolites (NH$_2$/Au/zeolites) did not enhance their activity, which emphasizes the role of the protonated amine groups. As glucose chemisorption on the protonated amine species did not increase the glucose conversion in the oxidation with H$_2$O$_2$, one can conclude that in this case the reaction rate was not limited by sugar chemisorption but by decomposition of hydrogen peroxide on gold particles.

Supplementary Materials: The following are available online at https://www.mdpi.com/article/10.3390/ma14185250/s1, Figure S1: ^{11}B MAS NMR spectrum of B/MCM-36 sample, Figure S2: FTIR spectra after (a) adsorption of pyridine at 150 °C and desorption at (b) 150 °C, (c) 200 °C, (d) 250 °C, (e) 300 °C. All the spectra were obtained by subtraction the spectrum after activation and normalized to the density of a wafer of ca 10 mg cm^{-2}, Figure S3: Representative TEM image of Au/HBeta showing Au NPs localization on the external surface of the zeolite, Figure S4: ATR-FTIR spectra of selected materials before and after glucose solution treatment and drying at 80 °C, Figure S5: N 1s region of XP spectra of gold-containing zeolites for (A) HBeta; (B) MCM-36 series. The regions of BE values typical to the protonated and non-protonated amine groups are marked in yellow and blue, respectively.

Author Contributions: Conceptualization, A.W., J.W., A.W.-G., I.S., L.W. and M.Z.; methodology, A.W., J.W., A.W.-G., I.S., L.W. and M.Z.; validation, A.W., J.W., A.W.-G., I.S. and L.W.; formal analysis, A.W., J.W., A.W.-G., I.S. and L.W.; investigation, A.W., J.W., A.W.-G., I.S.; resources, A.W., J.W., A.W.-G. and M.Z.; data curation, A.W., J.W., A.W.-G., I.S. and L.W.; writing—original draft preparation, A.W., J.W., A.W.-G., I.S. and L.W.; writing—review and editing, A.W., J.W., A.W.-G., I.S., L.W. and M.Z.; visualization, A.W. and J.W.; supervision, L.W. and M.Z.; funding acquisition, A.W. and M.Z. All authors have read and agreed to the published version of the manuscript.

Funding: Financial support from the National Science Centre in Poland (Grant No. 2018/29/B/ST5/00137) and Polish Ministry of Education and Science ("Diamentowy Grant" program, Grant No. DI2018 002248) is gratefully acknowledged.

Institutional Review Board Statement: Not applicable.

Informed Consent Statement: Not applicable.

Data Availability Statement: The data presented in this study are available on request from the corresponding authors via e-mail: adrian.walkowiak@amu.edu.pl (A.W.), j.wolska@amu.edu.pl (J.W.).

Conflicts of Interest: The authors declare no conflict of interest.

References

1. Haruta, M. Gold catalysts prepared by coprecipitation for low-temperature oxidation of hydrogen and of carbon monoxide. *J. Catal.* **1989**, *115*, 301–309. [CrossRef]
2. Hutchings, G.J. Spiers Memorial Lecture: Understanding reaction mechanisms in heterogeneously catalysed reactions. *Faraday Discuss.* **2021**, *229*, 9–34. [CrossRef] [PubMed]
3. Ortega-Liebana, M.C.; Bonet-Aleta, J.; Hueso, J.L.; Santamaria, J. Gold-Based Nanoparticles on Amino-Functionalized Mesoporous Silica Supports as Nanozymes for Glucose Oxidation. *Catalysts* **2020**, *10*, 333. [CrossRef]
4. Guo, S.; Fang, Q.; Li, Z.; Zhang, J.; Zhang, J.; Li, G. Efficient base-free direct oxidation of glucose to gluconic acid over TiO2-supported gold clusters. *Nanoscale* **2018**, *11*, 1326–1334. [CrossRef]
5. da Silva, A.G.; Rodrigues, T.S.; Candido, E.G.; de Freitas, I.C.; da Silva, A.H.; Fajardo, H.V.; Balzer, R.; Gomes, J.F.; Assaf, J.M.; de Oliveira, D.C.; et al. Combining active phase and support optimization in MnO2-Au nanoflowers: Enabling high activities towards green oxidations. *J. Colloid Interface Sci.* **2018**, *530*, 282–291. [CrossRef]
6. Megías-Sayago, C.; Reina, T.R.; Ivanova, S.; Odriozola, J.A. Au/CeO2-ZnO/Al2O3 as Versatile Catalysts for Oxidation Reactions: Application in Gas/Liquid Environmental Processes. *Front. Chem.* **2019**, *7*, 504. [CrossRef]
7. Franz, S.; Shcherban, N.D.; Bezverkhyy, I.; Sergiienko, S.A.; Simakova, I.L.; Salmi, T.; Murzin, D.Y. Catalytic activity of gold nanoparticles deposited on N-doped carbon-based supports in oxidation of glucose and arabinose mixtures. *Res. Chem. Intermed.* **2021**, 1–15. [CrossRef]
8. Perovic, M.; Zeininger, L.; Oschatz, M. Immobilization of Gold-on-Carbon Catalysts Onto Perfluorocarbon Emulsion Droplets to Promote Oxygen Delivery in Aqueous Phase D-Glucose Oxidation. *ChemCatChem* **2020**, *13*, 196–201. [CrossRef]
9. Wojcieszak, R.; Ferraz, C.P.; Sha, J.; Houda, S.; Rossi, L.M.; Paul, S. Advances in Base-Free Oxidation of Bio-Based Compounds on Supported Gold Catalysts. *Catalysts* **2017**, *7*, 352. [CrossRef]
10. Mihályi, M.R.; Kollár, M.; Klébert, S.; Mavrodinova, V. Transformation of ethylbenzene-m-xylene feed over MCM-22 zeolites with different acidities. *Appl. Catal. A Gen.* **2014**, *476*, 19–25. [CrossRef]
11. Shvets, O.; Shamzhy, M.; Yaremov, P.S.; Musilová, Z.; Procházková, D.; Čejka, J. Isomorphous Introduction of Boron in Germanosilicate Zeolites with UTL Topology. *Chem. Mater.* **2011**, *23*, 2573–2585. [CrossRef]
12. Zhang, X.; Lai, E.S.M.; Martin-Aranda, R.; Yeung, K.L. An investigation of Knoevenagel condensation reaction in microreactors using a new zeolite catalyst. *Appl. Catal. A Gen.* **2004**, *261*, 109–118. [CrossRef]
13. Chatti, R.; Bansiwal, A.K.; Thote, J.A.; Kumar, V.; Jadhav, P.; Lokhande, S.K.; Biniwale, R.B.; Labhsetwar, N.K.; Rayalu, S.S. Amine loaded zeolites for carbon dioxide capture: Amine loading and adsorption studies. *Microporous Mesoporous Mater.* **2009**, *121*, 84–89. [CrossRef]
14. Baerlocher, C.; McCusker, L. Database of Zeolite Structures. Framework Type LTA (Material: Linde Type A, Zeolite A). 2020. Available online: http://www.iza-structure.org/databases/ (accessed on 10 July 2021).
15. Ishida, T.; Murayama, T.; Taketoshi, A.; Haruta, M. Importance of Size and Contact Structure of Gold Nanoparticles for the Genesis of Unique Catalytic Processes. *Chem. Rev.* **2019**, *120*, 464–525. [CrossRef]
16. Ouellette, R.J.; Rawn, J.D. Aldehydes and Ketones: Nucleophilic Addition Reactions. In *Organic Chemistry: Structure, Mechanism, Synthesis*, 2nd ed.; Elsevier: Oxford, UK, 2018; pp. 595–623. [CrossRef]
17. Wolska, J.; Walkowiak, A.; Sobczak, I.; Wolski, L.; Ziolek, M. Gold-containing Beta zeolite in base-free glucose oxidation—The role of Au deposition procedure and zeolite dopants. *Catal. Today* **2021**. [CrossRef]
18. Prati, L.; Villa, A.; Lupini, A.R.; Veith, G.M. Gold on carbon: One billion catalysts under a single label. *Phys. Chem. Chem. Phys.* **2012**, *14*, 2969–2978. [CrossRef] [PubMed]
19. Megías-Sayago, C.; Santos, J.L.; Ammari, F.; Chenouf, M.; Ivanova, S.; Centeno, M.; Odriozola, J. Influence of gold particle size in Au/C catalysts for base-free oxidation of glucose. *Catal. Today* **2018**, *306*, 183–190. [CrossRef]
20. Khawaji, M.; Zhang, Y.; Loh, M.; Graça, I.; Ware, E.; Chadwick, D. Composition dependent selectivity of bimetallic Au-Pd NPs immobilised on titanate nanotubes in catalytic oxidation of glucose. *Appl. Catal. B Environ.* **2019**, *256*, 117799. [CrossRef]
21. Roth, W.; Chlubná, P.; Kubů, M.; Vitvarová, D. Swelling of MCM-56 and MCM-22P with a new medium—surfactant–tetramethylammonium hydroxide mixtures. *Catal. Today* **2013**, *204*, 8–14. [CrossRef]
22. Leonowicz, M.E.; Lawton, J.A.; Lawton, S.L.; Rubin, M.K. MCM-22: A Molecular Sieve with Two Independent Multidimensional Channel Systems. *Science* **1994**, *264*, 1910–1913. [CrossRef]

23. Gil, B.; Marszałek, B.; Micek-Ilnicka, A.; Olejniczak, Z. The Influence of Si/Al Ratio on the Distribution of OH Groups in Zeolites with MWW Topology. *Top. Catal.* **2010**, *53*, 1340–1348. [CrossRef]
24. Koranyi, T.I.; Nagy, J.B. Distribution of Aluminum and Boron in the Periodical Building Units of Boron-Containing β Zeolites. *J. Phys. Chem. B* **2006**, *110*, 14728–14735. [CrossRef] [PubMed]
25. Liu, H.; Ernst, H.; Freude, D.; Scheffler, F.; Schwieger, W. In situ 11B MAS NMR study of the synthesis of a boron-containing MFI type zeolite. *Microporous Mesoporous Mater.* **2002**, *54*, 319–330. [CrossRef]
26. Parry, E. An infrared study of pyridine adsorbed on acidic solids. Characterization of surface acidity. *J. Catal.* **1963**, *2*, 371–379. [CrossRef]
27. Busch, O.M.; Brijoux, W.; Thomson, S.J.; Schüth, F. Spatially resolving infrared spectroscopy for parallelized characterization of acid sites of catalysts via pyridine sorption: Possibilities and limitations. *J. Catal.* **2004**, *222*, 174–179. [CrossRef]
28. Emeis, C. Determination of Integrated Molar Extinction Coefficients for Infrared Absorption Bands of Pyridine Adsorbed on Solid Acid Catalysts. *J. Catal.* **1993**, *141*, 347–354. [CrossRef]
29. Sekhar, A.S.; Sivaranjani, K.; Gopinath, C.S.; Vinod, C. A simple one pot synthesis of nano gold–mesoporous silica and its oxidation catalysis. *Catal. Today* **2012**, *198*, 92–97. [CrossRef]
30. da Silva, F.P.; Fiorio, J.L.; Rossi, L.M. Tuning the Catalytic Activity and Selectivity of Pd Nanoparticles Using Ligand-Modified Supports and Surfaces. *ACS Omega* **2017**, *2*, 6014–6022. [CrossRef]
31. Feldheim, D.L.; Foss, C.A. *Metal Nanoparticles Synthesis: Characterization and Applications*; Marcel Dekker: New York, NY, USA, 2002.
32. Nakamura, T.; Herbani, Y.; Ursescu, D.; Banici, R.; Dabu, R.V.; Sato, S. Spectroscopic study of gold nanoparticle formation through high intensity laser irradiation of solution. *AIP Adv.* **2013**, *3*, 082101. [CrossRef]
33. Abdelhalim, M.A.K.; Mady, M.M.; Ghannam, M.M. Physical Properties of Different Gold Nanoparticles: Ultraviolet-Visible and Fluorescence Measurements. *J. Nanomed. Nanotechnol.* **2012**, *3*. [CrossRef]
34. Dyer, J.R. *Applications of Absorption Spectroscopy of Organic Compounds*; Prentice-Hall: Englewood Cliffs, NJ, USA, 1965.
35. Tiwari, I.; Singh, M.; Tripathi, V.S.; Lakshminarayana, G.; Nogami, M. An Amperometric Sensor for Nanomolar Detection of Hydrogen Peroxide Based on Encapsulation of Horseradish Peroxidase in Thymol Blue-Ormosil Composite. *Sens. Lett.* **2011**, *9*, 1323–1330. [CrossRef]
36. Sobczak, I.; Calvino-Casilda, V.; Wolski, L.; Siodła, T.; Martin-Aranda, R.; Ziolek, M. The role of gold dopant in AP-Nb/MCF and AP-MCF on the Knoevenagel condensation of ethyl cyanoacetate with benzaldehyde and 2,4-dichlorobenzaldehyde. *Catal. Today* **2018**, *325*, 81–88. [CrossRef]
37. Dias, D.R.; Moreira, A.F.; Correia, I.J. The effect of the shape of gold core–mesoporous silica shell nanoparticles on the cellular behavior and tumor spheroid penetration. *J. Mater. Chem. B* **2016**, *4*, 7630–7640. [CrossRef] [PubMed]
38. Zhang, P.; Sham, T.K. X-Ray Studies of the Structure and Electronic Behavior of Alkanethiolate-Capped Gold Nanoparticles: The Interplay of Size and Surface Effects. *Phys. Rev. Lett.* **2003**, *90*, 245502. [CrossRef] [PubMed]
39. Xu, Y.; Li, J.; Zhou, J.; Liu, Y.; Wei, Z.; Zhang, H. Layered double hydroxides supported atomically precise Aun nanoclusters for air oxidation of benzyl alcohol: Effects of size and active site structure. *J. Catal.* **2020**, *389*, 409–420. [CrossRef]
40. Besson, M.; Gallezot, P. Selective oxidation of alcohols and aldehydes on metal catalysts. *Catal. Today* **2000**, *57*, 127–141. [CrossRef]
41. Sharma, A.S.; Kaur, H.; Shah, D. Selective oxidation of alcohols by supported gold nanoparticles: Recent advances. *RSC Adv.* **2016**, *6*, 28688–28727. [CrossRef]
42. Ibrahim, M.; Alaam, M.; Elhaes, H.; Jalbout, A.F.; De Leon, A. Analysis of the structure and vibrational spectra of glucose and fructose. *Eclet. Quim.* **2006**, *31*, 15–21. [CrossRef]
43. Liu, X.; Wang, A.; Yang, X.; Zhang, T.; Mou, C.-Y.; Su, D.-S.; Li, J. Synthesis of Thermally Stable and Highly Active Bimetallic Au−Ag Nanoparticles on Inert Supports. *Chem. Mater.* **2008**, *21*, 410–418. [CrossRef]
44. Wang, A.; Liu, X.Y.; Mou, C.-Y.; Zhang, T. Understanding the synergistic effects of gold bimetallic catalysts. *J. Catal.* **2013**, *308*, 258–271. [CrossRef]
45. Graf, N.; Yegen, E.; Gross, T.; Lippitz, A.; Weigel, W.; Krakert, S.; Terfort, A.; Unger, W.E. XPS and NEXAFS studies of aliphatic and aromatic amine species on functionalized surfaces. *Surf. Sci.* **2009**, *603*, 2849–2860. [CrossRef]
46. Qi, P.; Chen, S.; Chen, J.; Zheng, J.; Zheng, X.; Yuan, Y. Catalysis and Reactivation of Ordered Mesoporous Carbon-Supported Gold Nanoparticles for the Base-Free Oxidation of Glucose to Gluconic Acid. *ACS Catal.* **2015**, *5*, 2659–2670. [CrossRef]
47. Wisniewska, J.; Sobczak, I.; Ziolek, M. Gold based on SBA-15 supports–Promising catalysts in base-free glucose oxidation. *Chem. Eng. J.* **2020**, *413*, 127548. [CrossRef]
48. Zhuge, Y.; Fan, G.; Lin, Y.; Yang, L.; Li, F. A hybrid composite of hydroxyapatite and Ca–Al layered double hydroxide supported Au nanoparticles for highly efficient base-free aerobic oxidation of glucose. *Dalton Trans.* **2019**, *48*, 9161–9172. [CrossRef] [PubMed]
49. Kiyonaga, T.; Jin, Q.; Kobayashi, H.; Tada, H. Size-Dependence of Catalytic Activity of Gold Nanoparticles Loaded on Titanium (IV) Dioxide for Hydrogen Peroxide Decomposition. *ChemPhysChem* **2009**, *10*, 2935–2938. [CrossRef] [PubMed]
50. Rautiainen, S.; Lehtinen, P.; Vehkamäki, M.; Niemelä, K.; Kemell, M.; Heikkilä, M.; Repo, T. Microwave-assisted base-free oxidation of glucose on gold nanoparticle catalysts. *Catal. Commun.* **2016**, *74*, 115–118. [CrossRef]

Review

Recent Advances in Structured Catalytic Materials Development for Conversion of Liquid Hydrocarbons into Synthesis Gas for Fuel Cell Power Generators

Vladislav Shilov [1,2], Dmitriy Potemkin [1,2], Vladimir Rogozhnikov [1] and Pavel Snytnikov [1,*]

1. Boreskov Institute of Catalysis SB RAS, Pr. Lavrentieva 5, 630090 Novosibirsk, Russia
2. Faculty of Natural Science, Novosibirsk State University, Pirogova St., 2, 630090 Novosibirsk, Russia
* Correspondence: pvsnyt@catalysis.ru

Abstract: The paper considers the current state of research and development of composite structured catalysts for the oxidative conversion of liquid hydrocarbons into synthesis gas for fuel cell feeding and gives more detailed information about recent advances in the Boreskov Institute of Catalysis. The main factors affecting the progress of the target reaction and side reactions leading to catalyst deactivation are discussed. The properties of the $Rh/Ce_{0.75}Zr_{0.25}O_2/Al_2O_3/FeCrAl$ composite multifunctional catalyst for the conversion of diesel fuel into synthesis gas are described. The results of the catalyst testing and mathematical modeling of the process of diesel fuel steam–air conversion into synthesis gas are reported.

Keywords: fuel cell; reforming; fuel processing; liquid fuel; diesel fuel; hydrogen; catalyst; syngas

1. Introduction

It was expected in most of the research work carried out over the past 20–25 years that by the time of mass use of fuel cells, the infrastructure for their supply with fuel (hydrogen) would already be created, including efficient logistics, a fully developed hydrogen refueling network and sufficient hydrogen long-time storing capacities. However, these forecasts appeared overly optimistic. Even regarding widely used natural gas, the existing and actively developing infrastructure for its transportation and consumption turns to be insufficient to cover completely the current level of mass demand.

That is why all the world's research centers involved in R&D of fuel-cell-based power units show interest in using fuel of common types, such as natural gas, liquefied petroleum gas, gasoline, aviation kerosene, diesel fuel, methanol, ethanol, etc. Besides, compared to other currently available hydrogen storage technologies, hydrocarbon fuels demonstrate the highest hydrogen content per unit volume [1].

These fuels can hardly be oxidized directly in the anode space of a fuel cell (FC), since they are inert at low temperatures, and can initiate electrode coking and FC failure at high temperatures. Therefore, the fuel is first converted into a hydrogen-rich gas, which is then oxidized in a fuel cell. Depending on the FC type, hydrogen for FC fueling can contain some amounts of CO and CO_2. Figure 1 presents a typical scheme of catalytic processes for the production and cleanup of hydrogen-rich gas mixtures from various fuels for fuel cell feeding. Note that sulfur-containing compounds are poisonous for FC of all types; therefore, any fuel must be desulfurized either at the stage of production or prior to using it for hydrogen production. The processes for liquid and gaseous fuels desulfurization are well-developed and widely used in industry, but remain beyond consideration in this review.

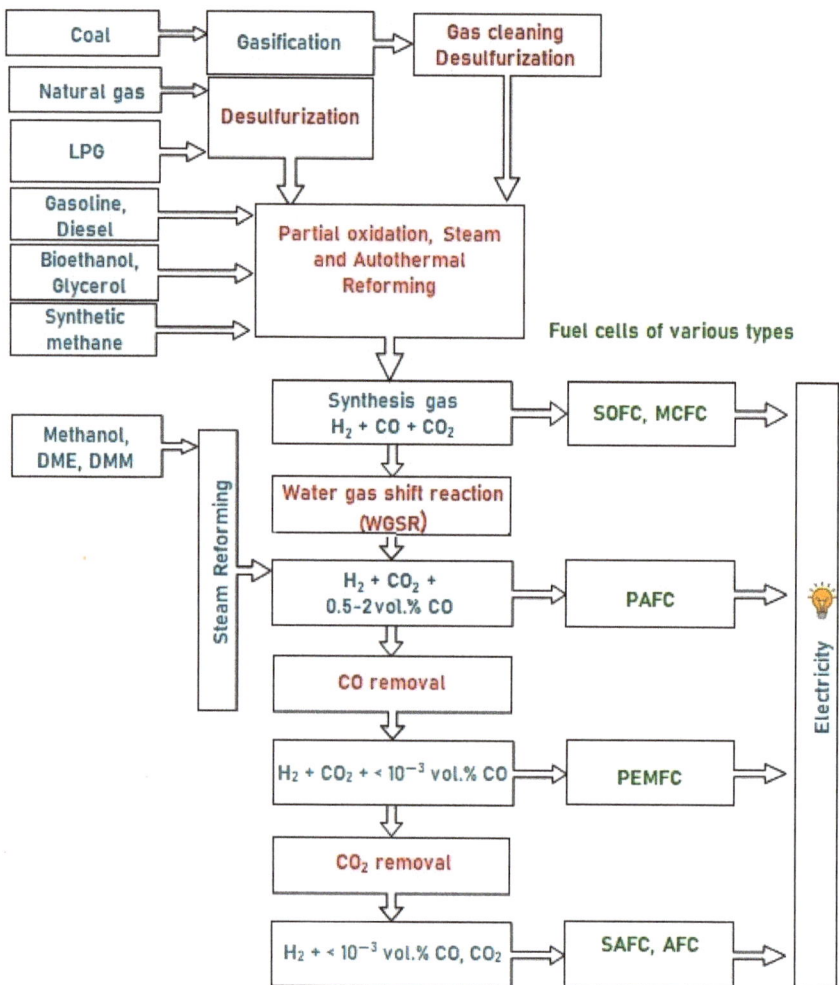

Figure 1. A typical scheme of hydrogen production from various fuels for feeding fuel cells of different types.

Synthesis gas is produced by the processes of partial oxidation (PO) (1), steam reforming (SR) (2) and autothermal reforming (ATR) (3). The most efficient process is the SR of hydrocarbons or alcohols, which, depending on the fuel type, is performed at temperatures of 300–900 °C, with excess water (molar ratio H_2O/C = 1.1–5), and provides the highest hydrogen yield. However, since SR is an endothermic process and therefore requires a significant heat supply, a steam-generating system and water conditioning, it is usually realized in combination with stationary power plants based on high-power FCs.

$$C_nH_m + \frac{n}{2}O_2 \rightarrow nCO + \frac{m}{2}H_2 \quad \Delta H < 0 \tag{1}$$

$$C_nH_m + nH_2O \rightarrow nCO + [\frac{m}{2}+n]H_2 \quad \Delta H > 0 \tag{2}$$

$$C_nH_m + \frac{n}{4}O_2 + \frac{n}{2}H_2O \rightarrow nCO + [\frac{m}{2}+\frac{n}{2}]H_2 \quad \Delta H \leq 0 \tag{3}$$

For mobile decentralized power systems based on low-power fuel cells, the hydrogen-rich gas generating unit must be compact, highly productive and water-independent, and operate at moderate temperatures. In this regard, the PO and ATR processes are of particular interest.

Besides the R&D studies on the conversion of natural gas and propane–butane mixtures, considerable interest is focused on the use of the main logistics fuels (gasoline, aviation kerosene and diesel fuel), based on the concept of hybrid systems, when fuel cells serve as an auxiliary power unit supplementary to the main power system—a truck diesel engine, aircraft turbines, etc. This approach promotes the efficiency of fuel consumption in modes when the main engine is either turned off or idling during stops, loading/unloading operations in cargo terminals, airport parking, etc. [2–6].

Oxidative conversion (SR, ATR and PO) of liquid hydrocarbons into synthesis gas e is based on the process of steam reforming of naphtha, which has been used since the 1960s in the petrochemical and ammonia industries for the production of synthesis gas and hydrogen [7]. The low sulfur content ($\leq$0.001 wt.%) in diesel fuel of modern grades greatly promoted the development of fuel processors [3,8–10]. Indeed, over the past decade, the number of R&D works aimed at finding and studying catalysts for the conversion of kerosene [3,4,11–13], diesel [14–17], renewable natural raw materials—biodiesel [18,19] and glycerin (a by-product of biomass processing) [20]—as well as fuel processor designing [5,6,13], has increased significantly [2,4–6,11,13,19–31].

In contrast to monofuels (methanol, dimethyl ether, ethanol, etc.), which are also considered as promising raw materials for the production of hydrogen (hydrogen-rich gas) [32], the middle distillates—a product of oil refining (kerosene and diesel fuel)—are multi-component mixtures, mainly containing saturated and aromatic hydrocarbons, that seriously complicate both the study of the reforming processes and comparative analysis of the results reported by various research teams. It should be taken into account that the fuel composition can vary greatly depending on the initial oil feedstock and oil refining technologies, as well as on the individual characteristics of fuel production (including seasonal ones) [33,34]. To exclude this ambiguity, laboratory studies are mainly carried out with the use of model mixtures simulating diesel fuel or aviation kerosene chemical characteristics and the important physical properties affecting the heat and mass transfer processes. This approach simplifies the experiments, promotes the study of how individual fuel components affect the conversion process and provides comparative analysis of the catalysts' activity.

Dodecane [35–41], tetradecane [42–45] or hexadecane [46–53] were often used as model substances in the studies of SR, PO and ATR processes. The effect of chemical additives of other classes (for example, toluene, alkylbenzenes, decalin, tetralin, naphthalene, 1-methylnaphthalene, simulating cycloalkanes, aromatics and polyaromatics) was studied as well [36,37,43,44,46,54]. The tests showed that, compared to paraffins and cycloparaffins, aromatic compounds are much harder to convert to synthesis gas and facilitate carbon formation [52,55–59]. In addition, in the presence of aromatic compounds, the rate of heterogeneous reactions of paraffin conversion decreases; that facilitates non-catalytic homogeneous processes, releasing ethylene, which also accelerates the catalyst coking [60].

In a number of works and experiments were carried out using synthetic commercial fuels and mixtures—for example, Norpar13 (ExxonMobil, Spring, TX, USA) [61], consisting of saturated hydrocarbons with an average number of carbon atoms equal to 13, NExBTL biodiesel (Neste Oil, Espoo, Finland) [14,62], GTL diesel (Shell MDS, Bintulu, Malaysia) [4], EcoPar diesel (EcoPar, Gothenburg, Sweden) [4] and GTL kerosene (Shell MDS, Bintulu, Malaysia) [4]. All these fuels are characterized by extremely low content of aromatic and sulfurous compounds—less than 0.1 and 0.0001 wt.%, respectively. So, it is no wonder that the best reported results were achieved with the use of exactly these fuels: experiments in the presence of Rh-based catalysts under appropriate reaction conditions demonstrated a 100% fuel conversion, high yield of the main reaction products (H_2 and CO), negligible content of C_{2+} byproducts and stable operation of the catalyst for several thousand hours. The results of experiments with the fuels produced at oil refineries by traditional methods—MK1 diesel (Shell, Stockholm, Sweden) [14,17,62–64], SD10 (Preem, Stockholm, Sweden) [27,65], Ultimate diesel (ARAL, Hamburg, Germany) [4,17,63,64,66], automotive diesel [27] or Jet-A aviation kerosene (after additional desulphurization) [4,27]—were less impressive. The sulfur content in these fuels was a few ppm, the content of aromatics—5 (MK1), 13–18 (SD10, Ultimate diesel, Jet-A) and 24 wt.% (automotive diesel). The presence of aromatics significantly complicated the process of fuel conversion and worsened its characteristics: the fuel converted incompletely, oily residue was observed on the surface of the aqueous condensate (as the reaction is always carried out in a significant excess of steam against stoichiometry), the content of C_{2+} byproducts in the gas phase was high and carbon deposits were formed both on the catalyst and on the structural elements of the fuel reformer. To prevent undesirable reaction routes, engineering solutions were proposed, including the use of various types of nozzles, ultrasonic sprayers, specially designed evaporation and mixing chambers for homogenization of the reaction mixture before supplying to the catalyst [4,63,67,68].

Significant attention was paid to the development of structured catalysts for oxidative conversion (SR, ATR, and PO) of liquid hydrocarbons. The approaches were mainly based on the experience of creating catalytic afterburners for automobile exhausts. Cordierite blocks were used for supporting an active catalytic layer. A large number of various catalytic systems have been studied, mainly based on Rh, Ru, Pd, Pt, Ni and Co, and their bimetallic compositions in combination with various supports comprised of individual or mixed oxides of Zr, Ce, Gd, La, Y, Pr and Al, doped with alkali and alkali-earth metals; the systems supported on perovskite and pyrochlore were studied [14–17,39,42,43,45,49,52–54,65,69–83].

Although Ni-based catalysts are widely used in industrial reactors for SR of natural gas into synthesis gas, their use for the conversion of diesel fuel and aviation kerosene seems to be rather problematic. The authors of review [83] compared and analyzed the results of studies of a large number of nickel catalysts in the reactions of partial oxidation and carbon dioxide conversion of methane, and concluded that although the use of a $Ce_{1-x}Zr_xO_2$ mixed oxide, possessing high oxygen mobility, as a support, the introduction of perovskite $BaTiO_3$, characterized by a large number of oxygen vacancies, into the catalyst composition and the catalyst doping with K, Ca, Y, La and Pr oxides all contributed to an increase in catalyst activity compared to Ni/Al_2O_3, these factors appeared unable to impede completely the processes of carbon formation, and the catalysts suffered rapid coking. Nevertheless, many research teams around the world persistently undertake attempts to create active and stable catalysts for the pre-reforming and steam reforming of diesel and kerosene fuels [16,54,73,84–88].

Comparative studies of noble-metal-containing catalysts showed that the Pt-, Ru- and Pd-based systems stood behind the Rh-based ones in catalytic activity, stability and coking resistance [39,40,49,52,53], and therefore these metals were considered mainly as doping additives. It should be noted that, even in laboratory experiments, the catalyst depositing (coating) on substrates with high thermal conductivity is a necessary trick for preventing undesirable hot spot formation in the catalyst, which can accelerate coking processes. Laboratory studies showed that the Rh- and Rh-Pt-based ceramic block catalysts exceed in activity, selectivity and stability other catalytic systems at diesel ATR [55–57]. Quite naturally, cordierite-supported $RhPt/Al_2O_3$-CeO_2 commercial catalysts (Umicore AG&Co. KG, Hanau, Germany) were used in pilot tests of more than a dozen fuel reformer modifications performed by the research team from Forschungszentrum Jülich (Jülich, Germany) [4–6]. To reduce temperature heterogeneity at diesel ATR, induced by the high exothermic effect of oxidation reactions, proceeding predominantly in the frontal zone of the catalytic block, and the high endothermic effect of the reactions of steam and carbon dioxide reforming, which take place in its tail section, it seems reasonable to use, instead of cordierite ceramics, a metal support composed of FeCrAl wire mesh, which has a high thermal conductivity [89]. Since the coefficient of thermal expansion (CTE) of high-temperature metal alloys twice exceeds that of oxide coatings, the latter often suffer cracking and destruction during heating. This problem was addressed in recent works on the development of a catalyst for diesel fuel ATR, carried out at the Boreskov Institute of Catalysis, SB RAS [18,48–51,90–99]. An elegant approach has been proposed based on supporting a needle-shaped coating instead of a continuous catalytic layer [90]. In such a "flexible" coating, individual elements of the oxide layer can move relative to each other during heating/cooling and respective expansion/compression of the metal substrate, thus preventing degradation of the catalytic layer.

2. Active Component of Catalysts for the Conversion of Diesel Fuel into Synthesis Gas

At present, the main amount of commercial diesel fuel (DF) produced at refineries by the process of diesel fraction hydrotreating meets the Euro 5 standard in terms of fractional composition and the content of polycyclic aromatics and sulfur. Table 1 presents a typical average composition of diesel fuel.

Table 1. A typical average composition of commercial diesel fuel.

Component	Content (vol.%)
n-Paraffins	20
iso-paraffins	15–20
Cycloparaffins	35
Alkylbenzenes	20–23
Diaromatic hydrocarbons	5
Polycyclic aromatics	<2
Sulfur compounds	0.0005–0.0008 (5–8 ppm)

As noted above, a key factor for providing efficient catalytic conversion of diesel fuel into synthesis gas is to ensure the stable operation and coking resistance of the catalyst. The most active and stable catalysts for diesel fuel conversion are Rh- and other precious metal systems supported on oxide carriers containing mobile lattice oxygen, mainly zirconium and cerium oxides. The mobile lattice oxygen participates in the oxidation of incipient carbon deposits and thus significantly improves the catalyst stability. The most active carrier in this regard is cerium oxide, but at temperatures above 600 °C it is not strongly sintered and therefore cannot be used in its pure form. So, mixed oxides of composition $Ce_xZr_{1-x}O_{2-\delta}$ were chosen as the support, as they possess both high mobility of lattice oxygen and thermal stability.

The main efforts were aimed at developing a procedure for depositing nanoparticles of platinum group metals (Ru, Rh, Pd and Pt) onto oxide supports, which would provide a high particle dispersion and adhesion to the support and be quite simple and adaptable for coating the structured substrates.

As a result of the research, a method of sorption–hydrolytic precipitation was proposed [49]. The method is based on the slow kinetics of ligand exchange in alkaline solutions of chloride complexes of platinum metals. This approach allowed the selection of appropriate concentrations of metal chlorides and precipitant (Na_2CO_3) and a temperature to obtain a metastable solution, in which homogeneous precipitation of platinum metal hydroxides is impeded for kinetic reasons. After immersing the carrier into the solution, the precipitation of metal hydroxide particles in its pores proceeds by the heterogeneous nucleation mechanism.

According to CO chemisorption data, the average size of the Rh, Ru and Pt particles deposited by the sorption–hydrolytic procedure on a commercial support of composition $Ce_{0.75}Zr_{0.25}O_2$ (hereinafter CZ) was 1.1, 1.2 and 1.8 nm, respectively [49]. Transmission electron microscopy (TEM) data showed that Rh particles exist on the support surface predominantly in the form of 1–2 nm clusters. (Figure 2a). After HD ATR experiments, the support crystallites increased in size to 20–30 nm, while the aggregate size remained the same (Figure 2b). Additionally, carbon species in the form of 1–2 graphite layers appeared occasionally on the surface (Figure 2c). The size of Rh particles increased slightly to 2–4 nm. After oxidative treatment of the catalyst, the carbon deposits disappeared, while the size of support crystallites and Rh particles remained unchanged.

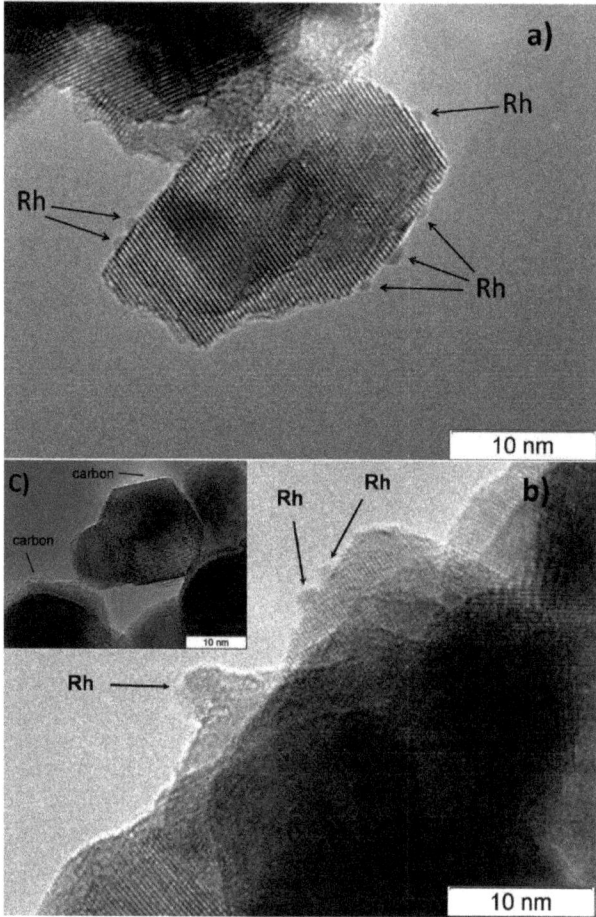

Figure 2. TEM images of 1 wt.% Rh/CZ: (**a**) as-prepared and (**b,c**) used in HD ATR [49].

Note that the DF ATR catalysts must be highly active and stable under DF SR conditions, since oxygen is rapidly consumed under ATR conditions and, in fact, most of the catalyst layer operates under SR conditions. The resulting Rh/CZ, Ru/CZ and Pt/CZ catalysts were studied in SR of n-hexadecane (HD), which served as a model DF compound. The fuel conversion was evaluated gravimetrically by collecting unreacted fuel into a condensate vessel and calculated using the following equation:

$$X\,(\%) = \frac{V_0 * t - m}{V_0 * t} * 100,$$

where X (%) is fuel conversion, V_0—fuel flow rate (g/h), t—sampling time (h) and m—sample weight (g).

Figure 3 shows the time dependences of HD conversion, and product distribution for the Pt/CZ, Ru/CZ and Rh/CZ catalysts at HD SR. It is seen that Pt/CZ showed the worst catalytic properties: at 550 °C, it failed to achieve complete HD conversion and demonstrated its decrease from 59 to 27% within 3 h. The Ru/CZ catalyst rapidly lost activity after 5 h on stream and respective HD conversion decreased to 46%. The Rh/CZ catalyst demonstrated stable operation at 550 °C for 8 h, provided 100% HD conversion and the product concentrations (vol.%) of 54 H_2, 18 CO_2, 5 CO and ~6% CH_4, which were close to the thermodynamically equilibrium ones. Then the catalyst was regenerated with hydrogen and retested at 650 °C. The HD conversion was 100%. The reaction product distribution was similar to the thermodynamically equilibrium one calculated for a temperature of 650 °C.

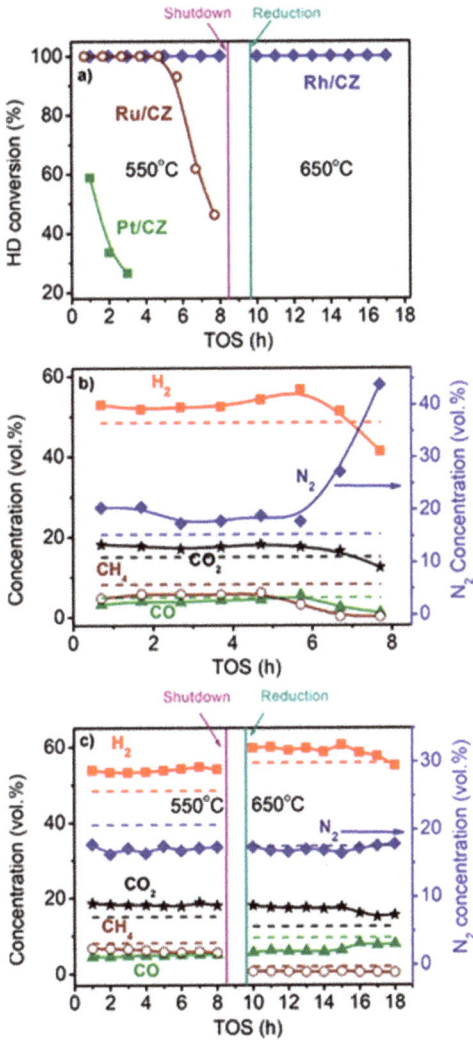

Figure 3. The HD conversion (**a**) and product distribution on dry basis (**b**,**c**) over 1.9 wt.% Pt/CZ (**a**), 1 wt.% Ru/CZ (**a**,**b**) and 1 wt.% Rh/CZ (**a**,**c**) in the HD SR as a function of time on stream at H_2O/C = 3.0, T = 550–650 °C and GHSV = 23,000 h^{-1}. Points—experiment, dashed line—equilibrium [49].

Thus, the activity and stability of the prepared noble-metal-based catalysts decreased in the following order: Rh/CZ > Ru/CZ >> Pt/CZ [49]—in good agreement with the results of other studies discussed above.

Note that catalysts supported on mixed cerium–zirconium oxides are usually inappropriate for practical use in granular form owing to insufficient mechanical strength and poor formability. Therefore, the feasibility of using a mixed Al_2O_3-$Ce_{0.75}Zr_{0.25}O_2$ oxide as a support (commercial product 50 wt.% Al_2O_3 and 50 wt.% $Ce_{0.75}Zr_{0.25}O_2$, hereinafter CZA) and alumina as a binding agent was investigated. The alumina additive is also intended to improve the catalyst thermal stability.

Besides Rh/CZ and Rh/CZA, Figure 4 presents the test results for Rh/CZA doped with MgO (Rh/CZA-Mg) and Rh-based catalyst supported on CZ containing 20 wt.% pseudoboehmite (Rh/CZB) as a structural promoter. Clearly, the Rh/CZ and Rh/CZB catalysts had the highest activity and coking resistance. The higher the aluminum content in the catalysts, the more rapidly they lost activity owing to acceleration of the side process of carbon deposition. Most likely, the catalyst deactivation is associated with the presence of acid sites on the alumina surface, while support doping with basic oxide MgO facilitated a significant increase in catalyst stability.

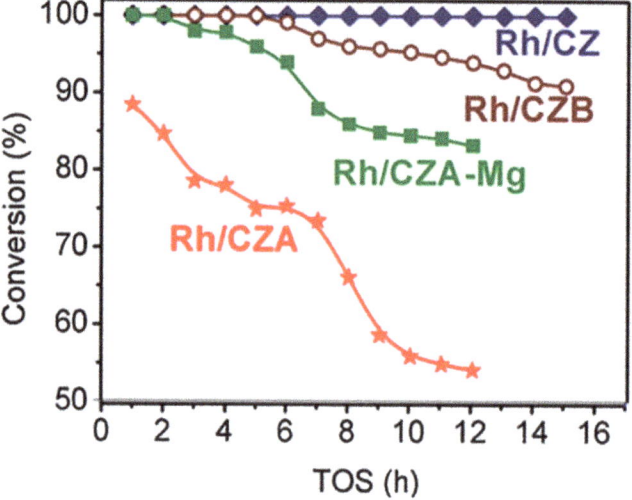

Figure 4. The time dependence of HD conversion under the following SR conditions: GHSV = 23,000 h^{-1}, H_2O/C = 3, T = 550 °C [50].

Rh/CZA was the most susceptible to coking: it accumulated 3.7 wt.% of carbon in 12 h of HD SR. Since the size of Rh particles in Rh/CZA after annealing remained unchanged (Figure 5f), rapid deactivation of the catalyst is explained by coke formation. As known, γ-Al_2O_3 contains acid sites, which are responsible for coke formation [13,27]. As proved by the results of Rh/CZA-Mg catalytic activity tests (Figure 4) and TPO data [50], the blocking of acid sites by Mg cations causes a threefold decrease in the amount of carbon formed. Compared to Rh/CZA-Mg and Rh/CZA, the use of 20 wt.% pseudoboehmite as a binder in Rh/CZB improved the catalyst performance in HD SR (Figure 4), but the carbon productivity exceeded that of Rh/CZA-Mg. The same amount of carbon was observed in Rh/CZB after HD ATR. Among the studied catalysts, Rh/CZ accumulated the lowest amount of carbon: 1.2 wt.% for 15 h of HD SR and 1.5 wt.% for 12 h of HD ATR.

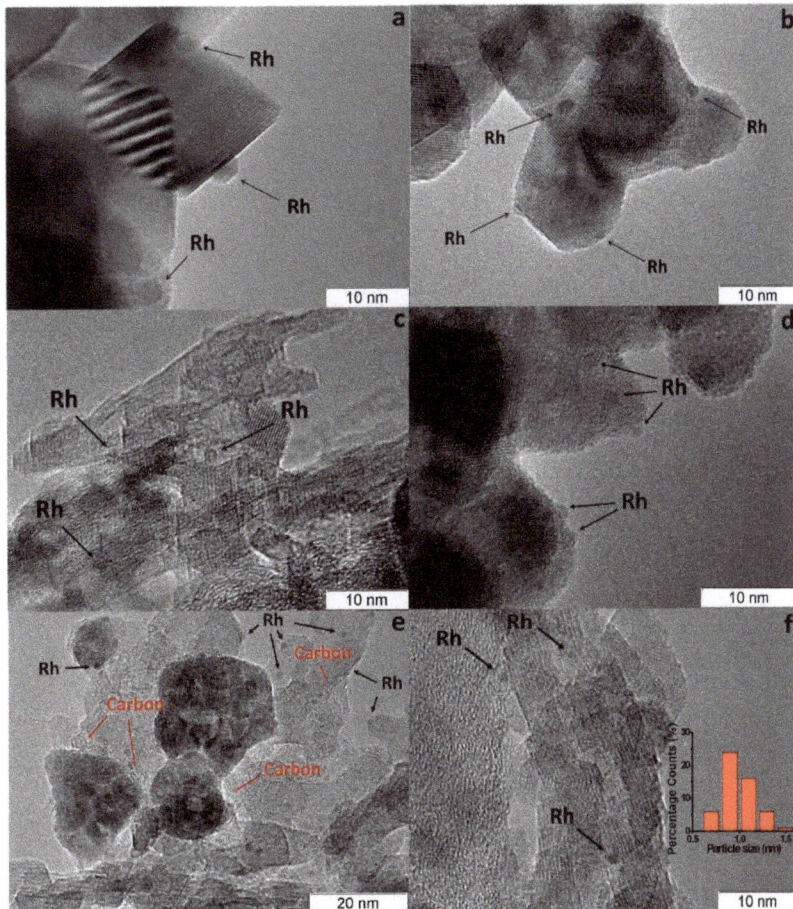

Figure 5. TEM images of as-prepared Rh/CZ (**a**), Rh/CZB (**c**) and used in HD SR Rh/CZ (**b**), Rh/CZB (**d**), Rh/CZA (**e**) and annealed Rh/CZA (**f**) catalysts [50].

Thus, to ensure the stable operation in DF SR and ATR of Rh-based catalysts supported on alumina-containing carriers, it is necessary to "neutralize" completely the acidity of the alumina surface [50].

3. Structured Composite Catalysts Supported on FeCrAl Alloy Wire Mesh

The DF ATR process is characterized by a combination of exo- and endothermic reactions. As discussed below in Section 8, modern understanding of the process mechanism assumes that fast complete oxidation reactions occur in the frontal part of the reactor with the release of heat, which is then consumed along the catalyst bed during the endothermic processes of steam and carbon-dioxide reforming of hydrocarbons. Therefore, when carrying out the ATR process, to prevent overheating in the frontal section and overcooling in the tail section, the structure of the catalyst bed must ensure efficient heat transfer between these zones. Conventional granular ceramic catalysts have low thermal conductivity and are hardly appropriate for this purpose. Besides, to reduce the pressure drop in the reactor, it is necessary to use large-size catalyst grains (1 cm and larger), which inevitably leads to a low utilization factor of the catalyst grain under conditions of fast ATR reactions.

For ensuring efficient implementation of the ATR process, it was proposed to use composite catalytic systems of the "metal nanoparticles/active oxide nanoparticles/structural oxide component/structured metal substrate" type. A structured metal substrate made of heat-resistant FeCrAl alloy facilitates fast heat removal/supply for exo-/endothermic reactions, has sufficient hydrodynamic characteristics, allows manufacturing products of various geometric shape and easy process scaling. The structural oxide component (alumina) provides thermal stability and high specific surface area and increases mechanical strength for the supported catalytic coating. The active oxide component (cerium oxide and mixed cerium–zirconium oxides with a fluorite structure) participates in the activation of water and oxygen molecules, improves the coke resistance because of high oxygen mobility and keeps the active component in a fine dispersed state owing to strong metal–carrier interaction. Metal nanoparticles of 1–2 nm size are involved in the activation of hydrocarbon molecules.

Based on this concept, the structured catalysts of composition 0.24 wt.% Rh/Ce$_{0.75}$Zr$_{0.25}$O$_2$/Al$_2$O$_3$/FeCrAl (Rh/CZB/FCA) were prepared and tested. The structured metal substrate was made of FeCrAl alloy wire mesh (wire thickness of 0.25 mm, cell size of 0.5 × 0.5 mm). Structural oxide component, a layer of η-Al$_2$O$_3$ in the amount of 6 wt.%, was supported on the metal substrate to ensure reliable adhesion of CZ active oxide nanoparticles (Figure 6). By calcining in air, α-Al$_2$O$_3$ layer was pre-formed on FeCrAl wire mesh; then a modified Bayer method (using aluminum hydroxide) was used to deposit an η-Al$_2$O$_3$ coating with a flexible ("breathing") needle-shaped morphology [90]. According to SEM data (Figure 6), aluminum oxide consisted of tubular or acicular crystals (5–10 μm thick, up to 50 μm long).

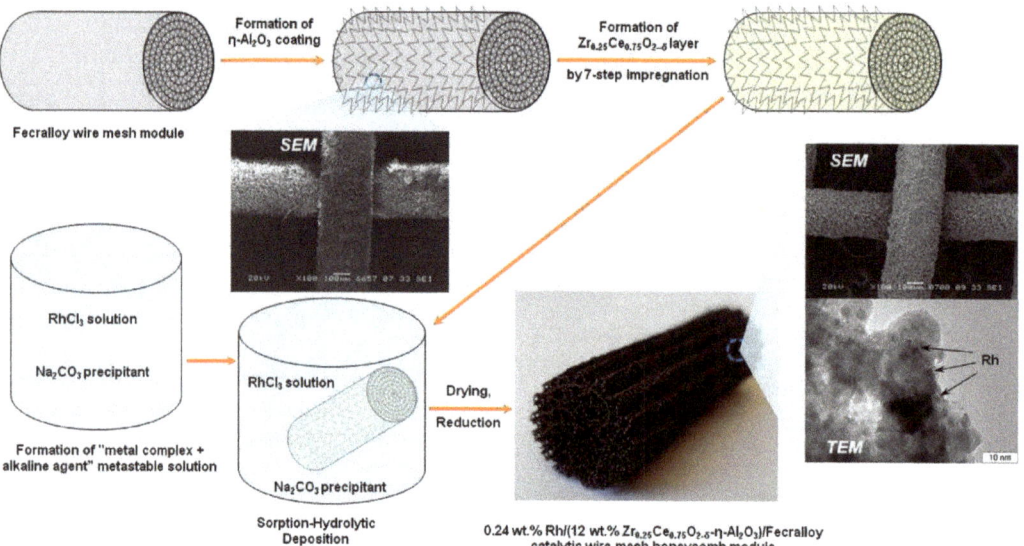

Figure 6. Schematic preparation procedure of Rh/CZB/FCA catalytic block, its general view, SEM and TEM images [51].

The obtained η-Al$_2$O$_3$/FeCrAl sample was repeatedly impregnated with a solution of Ce(NO$_3$)$_3$·6H$_2$O and ZrO(NO$_3$)$_2$·2H$_2$O (Ce/Zr = 3) and calcined at 800 °C [51]. Thus, the 12 wt.% Ce$_{0.75}$Zr$_{0.25}$O$_2$/Al$_2$O$_3$/FeCrAl (CZB/FCA) composite support was obtained. Rh nanoparticles in the amount of 0.24 wt.% were deposited on the CZB/FCA by sorption–hydrolytic precipitation method. The RhCl$_3$ solution was mixed with Na$_2$CO$_3$ in the ratio Na/Cl = 1. At room temperature, the resulting solution is metastable with regard to the homogeneous precipitation of rhodium hydroxide. The solution was brought into contact with CZB/FCA at T = 75 °C to initiate the hydrolysis that facilitated uniform deposition of rhodium particles throughout the structured support surface. At the final stage, the structured Rh/CZB/FCA catalyst was dried in air and reduced in hydrogen flow at 250 °C for 30 min.

The TEM images of as-prepared Rh/CZB/FCA catalyst (Figure 7) [50] show that the support consisted of ~1 μm-sized Al_2O_3 particles containing 5–20 nm $Ce_{1-x}Zr_xO_{2-\delta}$ crystallites on their surface and aggregated into large porous species. Rh particles on the support surface were predominantly in the form of 1–2 nm clusters, though 3–4 nm particles were observed as well (Figure 7b).

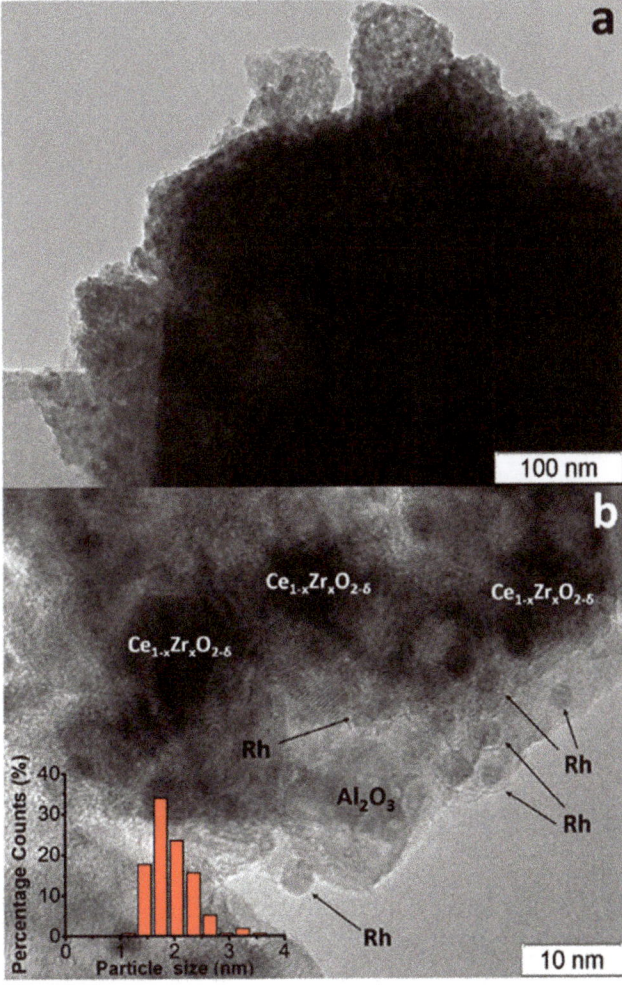

Figure 7. TEM images of as-prepared (**a**,**b**) Rh/$Ce_{0.75}Zr_{0.25}O_2$-Al_2O_3/FeCrAl and Rh particles size distribution (**b**) [50].

The SEM micrographs of Al_2O_3/FeCrAl (Figure 8a) and Rh/CZB/FCA (Figure 8b,c) clearly show that the η-Al_2O_3 layer consisting of tubular or acicular crystals (5–10 μm thick, up to 50 μm long) evenly covers the surface of the FeCrAl wire mesh. The thickness of the η-Al_2O_3 layer was 30–50 μm. After depositing Rh/$Ce_{1-x}Zr_xO_{2-\delta}$ onto the η-Al_2O_3 layer, the surface microstructure changed (Figure 8b): the crystal surface became rougher, though the thickness of the final layer remained the same (30–50 μm).

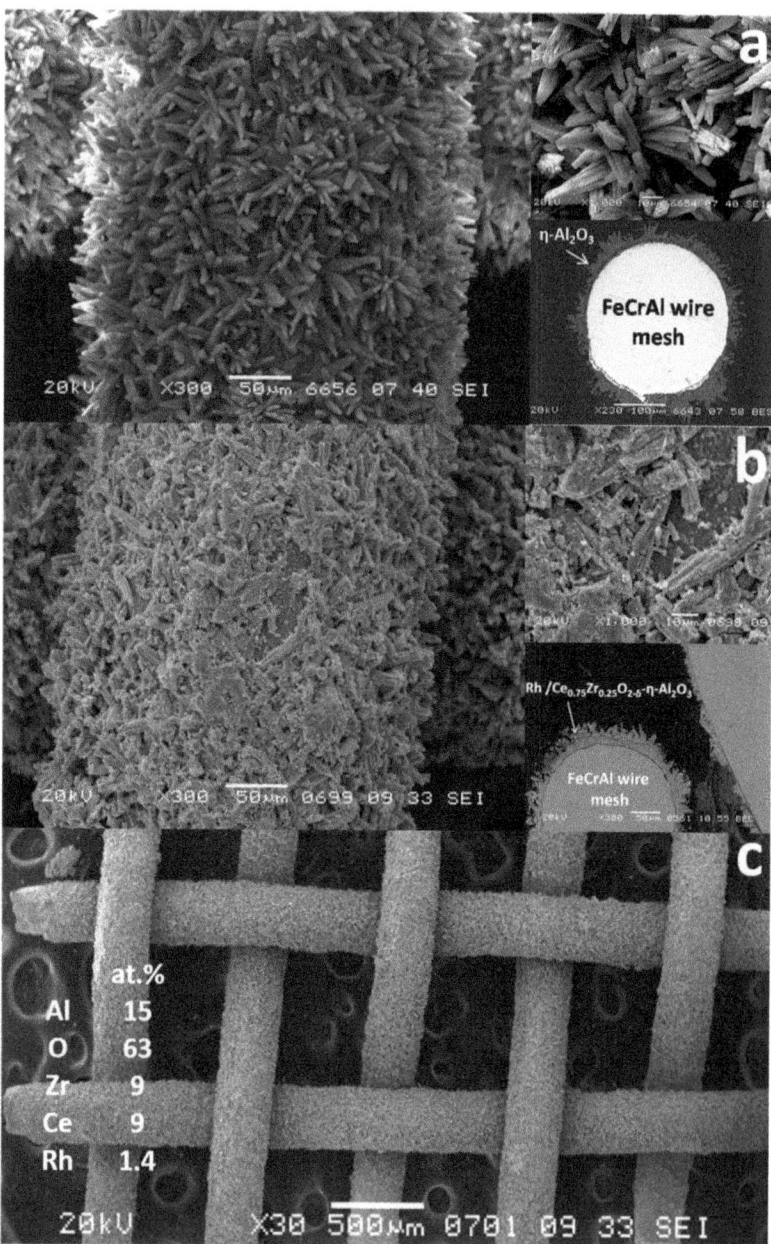

Figure 8. SEM images of η-Al$_2$O$_3$/FeCrAl (**a**) and as-prepared Rh/Ce$_{0.75}$Zr$_{0.25}$O$_2$-Al$_2$O$_3$/FeCrAl (**b**,**c**) catalyst [50].

According to EDX data for several 100 × 100 μm areas, the surface of the Rh/CZB/FCA catalyst contained Rh, Ce, Zr, Al and O. The element concentrations in all regions were the same, which proves a uniform distribution of Rh and Ce$_{1-x}$Zr$_x$O$_{2-\delta}$ over the surface of alumina crystals and agrees well with the TEM data.

Thus, it was shown [50] that the structured catalyst $Rh/Ce_{0.75}Zr_{0.25}O_2$-$Al_2O_3/FeCrAl$ is a composite in which aluminum oxide, chemically bonded to the metal substrate, provides the mechanical strength of the catalytic layer and keeps Rh and $Ce_{1-x}Zr_xO_{2-\delta}$ particles in a highly dispersed state.

In comparative studies of the catalytic properties of granular Rh/CZB and composite Rh/CZB/FCA in HD ATR, the Rh/CZB catalyst showed stable operation for 6 h on stream and product distribution close to the thermodynamically equilibrium one. Furthermore, the HD conversion and concentrations of the main products decreased, while the outlet concentrations of C_2–C_5 components increased (Figure 9a). Rh/CZB/FCA under HD ATR conditions demonstrated a 100% HD conversion for 12 h on stream (Figure 9b) even at a higher space velocity compared to that in the experiment with the granular catalyst. The outlet product concentrations were close to the thermodynamically equilibrium values. Thus, the structured Rh/CZB/FCA catalyst showed high activity in HD ATR and provided hydrogen productivity of 2.5 $kg_{H_2} kg_{cat}^{-1} h^{-1}$; it obviously possesses a high potential for the ATR of commercial diesel fuel.

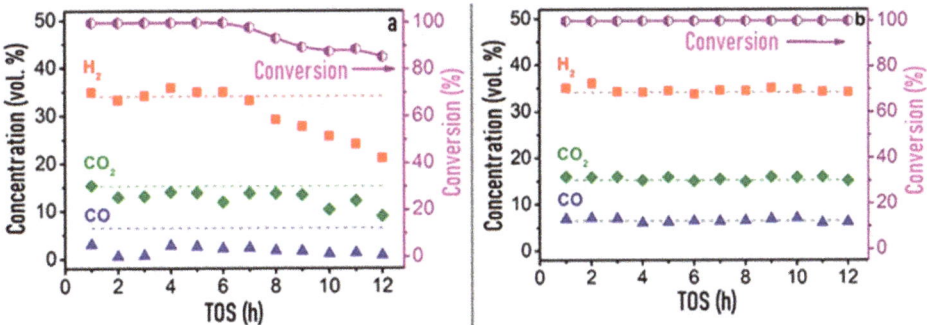

Figure 9. The time-on-steam dependence of HD conversion under ATR conditions: WHSV 30,000 $cm^3 g^{-1} h^{-1}$ (a) and 90,000 $cm^3 g^{-1} h^{-1}$ (b), T = 650 °C, H_2O/C = 2.5, O_2/C = 0.5 over Rh/CZB (a) and Rh/CZB/FCA (b) catalysts. Adapted from [50].

4. Testing of Structured Catalysts

Catalytic modules of the same composition ($Rh/Ce_{0.75}Zr_{0.25}O_2$-$Al_2O_3/FeCrAl$), but different lengths—10, 20 and 60 mm (hereinafter referred to as Rh10, Rh20 and Rh60, respectively)—were used in several series of experiments on HD SR and HD ATR to obtain detailed information on the outlet product distribution and temperature profile along the length of the catalytic block [96]. All tests were carried out at constant temperature (T = 750 °C) and inlet molar ratios H_2O/C = 2.6 and O_2/C = 0.4 and with variable space velocity of the reaction mixture.

It was found [96] that the HD conversion in SR experiments increased with the length of the catalytic block and amounted to 77, 80 and 92% for Rh10, Rh20 and Rh60, respectively; the outlet concentrations of H_2, CO and CO_2 increased as well and approached thermodynamically equilibrium values for the Rh60 catalyst. The outlet C_1–C_5 byproduct concentrations decreased (especially in the case of ethylene—from 7.5 to ~2 vol.%) with increasing catalytic block length.

The HD conversion byproducts (C_2–C_5 hydrocarbons) mainly contained 1-alkenes. Based on the product distribution data, it was assumed that during the SR process, thermal cracking of HD takes place to form 1-alkenes and hydrogen; CO and CO_2 are formed by the reaction of steam with carbon-containing intermediates on the catalyst surface. It also cannot be excluded that the hydrogen released in these processes participates in the hydrocracking reaction with the formation of light alkanes.

Another series of experiments with catalytic blocks of different lengths was performed under HD ATR conditions [96]. It demonstrated significantly different data on the HD conversion, target products distribution and composition of byproducts and intermediates (C_2–C_5 hydrocarbons), compared to respective HD SR data.

It should be noted that none of the experiments demonstrated unreacted oxygen at the reactor outlet. HD ATR in the presence of oxygen and steam proceeds very quickly: the required contact time of the reaction mixture with the catalyst did not exceed 0.03 s (i.e., even a 10 mm long catalytic block was sufficient). A sixfold longer contact time—0.18 s (block of 60 mm length)—was needed only to reach thermodynamic equilibrium between the target reaction products (H_2, CO and CO_2), and to reduce the content of reaction byproducts (primarily ethylene). For Rh60, the outlet concentrations of CH_4, C_2H_4, C_3H_6 and C_4H_8 were 200, 400, 90 and 20 ppm, respectively.

Based on the results obtained, the catalytic block under HD ATR conditions can be conditionally divided into two zones (Figure 10): zone 1, where fast reactions of deep oxidation and cracking occur, and zone 2, which involves the slower processes of steam reforming, dehydrogenation and methanation, proceeding along the entire length of the catalytic layer and being responsible for the formation of the final reaction products [96].

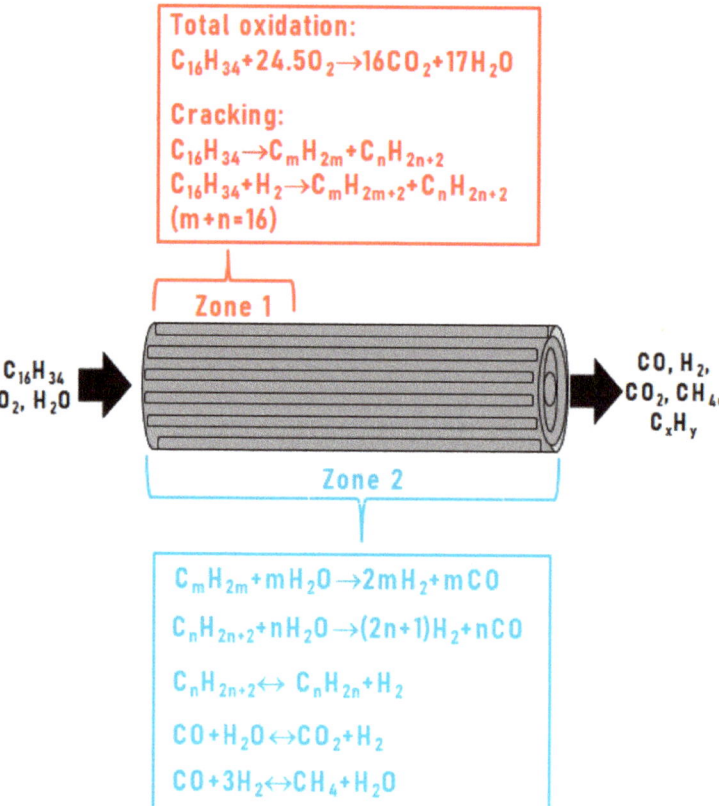

Figure 10. Scheme of autothermal reforming of hexadecane [96].

The studies of the DF ATR process (Figure 11) revealed that at a temperature of 750 °C and a space velocity of 30,000 h^{-1}, the DF conversion decreased from 100 to 97.8% in 4 h and the catalyst suffered coking. The oily residue collected from the surface of the aqueous condensate at the reactor outlet contained 78 wt.% of mono-, di- and polyaromatic hydrocarbons (Table 2), which are hardly convertible to synthesis gas and, most likely, contribute to the formation of carbon on the catalyst surface.

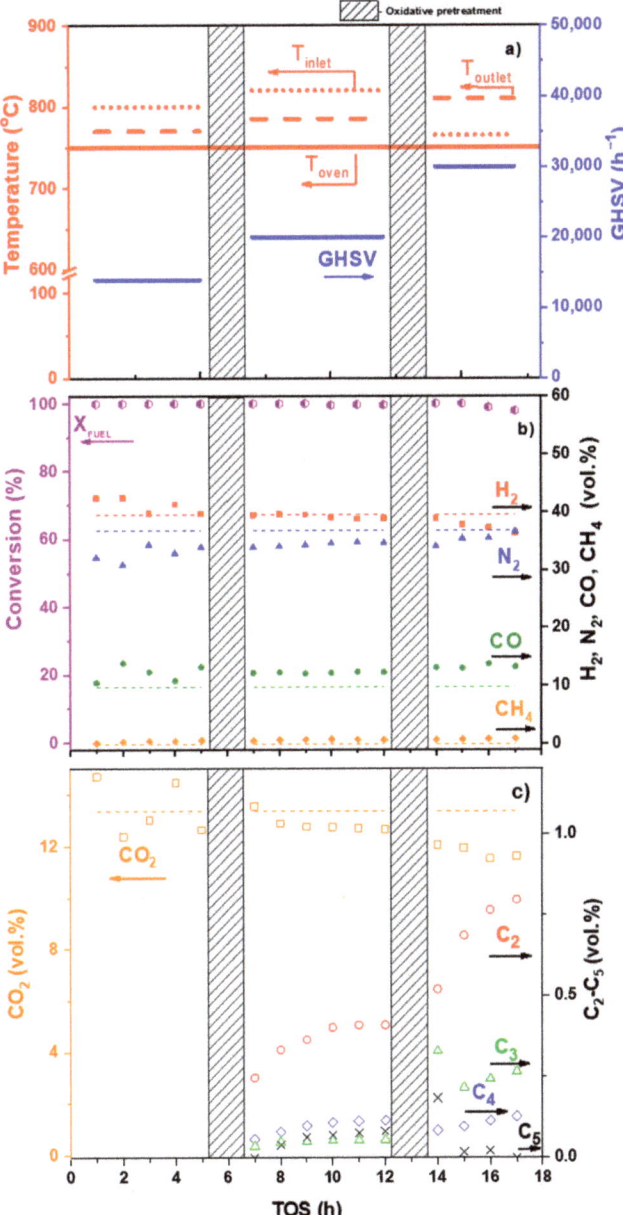

Figure 11. DF ATR over Rh/CZB/FCA under the following conditions: GHSV 14,000–30,000 h^{-1}, T_{oven} = 750 °C, H_2O/C = 2.5, O_2/C = 0.4. Experimental conditions (**a**); diesel fuel conversion (%) and H_2, N_2, CO, CH_4 concentrations (vol. %) (dry basis) (**b**); CO_2 and by-products concentrations (vol. %) (dry basis) (**c**). Points—experiment; dashed lines—equilibrium concentrations. Adapted from [91].

Table 2. Qualitative and quantitative analysis of winter diesel fuel and oily residue [91].

Composition	Winter DF	Oily Residue	Dimension
Monoaromatics	25	23	wt.%
Diaromatics	5	46	wt.%
Polyaromatics	1	9	wt.%
Sulfur	8	-	ppm
H/C	1.94	-	-

To check the assumption about the key effect of di- and polyaromatic compounds on the fuel reforming process, comparative experiments were carried out using catalytic blocks of different lengths in the ATR of DF model blend containing various classes of organic compounds: 75 wt.% hexadecane + 20 wt.% o-xylene + 5 wt.% naphthalene [97].

Conversion of hydrocarbons at ATR of the DF model blend increased with the increasing lengths of the catalytic blocks (decreasing GHSV) as follows: 97.5% for Rh10, 98.7% for Rh20 and 99.1% for Rh60. As Figure 12 shows, the lower the GHSV, the higher the conversion of both aliphatic and mono/diaromatic hydrocarbons. However, in the presence of aromatic compounds, the conversion of n-hexadecane decreases considerably (without aromatics, the HD conversion exceeds 99% even on Rh10). The set of organic compounds in the oily residue was identified by GC-MS analysis. These compounds represent byproducts and intermediates of the reactions of alkylation, dealkylation, isomerization, condensation, cracking and dehydrogenation accompanying the ATR process [97].

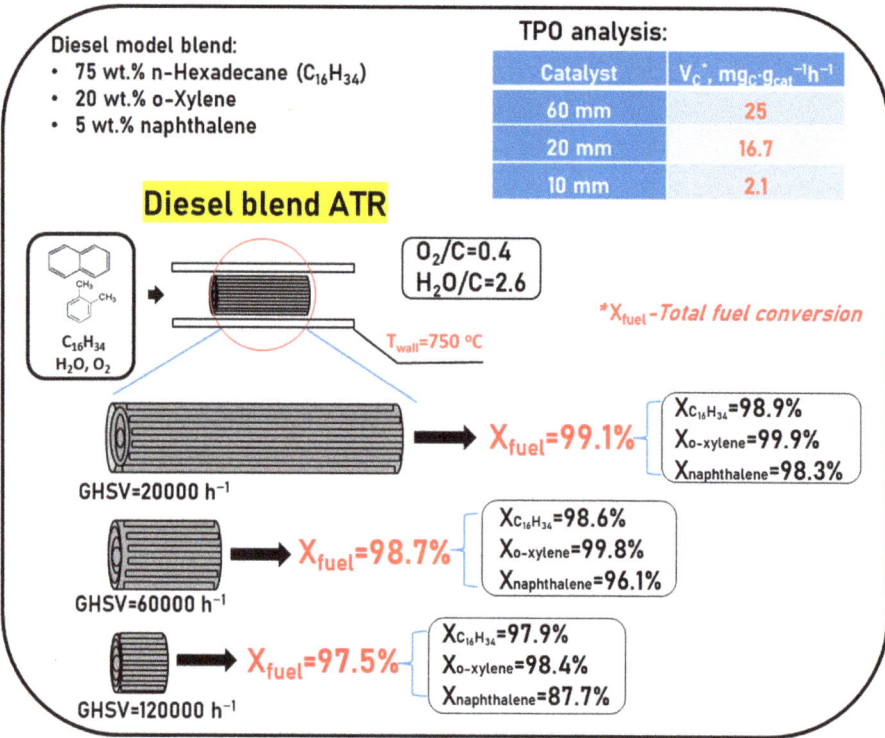

Figure 12. Autothermal reforming of DF model blend over Rh10, Rh20 and Rh60 at furnace temperature 750 °C. Adapted from [97].

TPO experiments show (Figure 12) that the most profound increase (around an order of magnitude) of the average specific velocity of carbon formation is observed when changing from module Rh10 to Rh20, and less significant—between modules Rh20 and Rh60. These data mean that the reaction of complete oxidation, which proceeds in the frontal catalyst section at the higher temperature, prevents carbon accumulation even in the presence of large amounts of aromatic hydrocarbons in the feed mixture. In the next catalyst sections, where the endothermic reactions of hydrocarbon conversion proceed and the temperature becomes lower, considerable catalyst coking is observed. Carbon formation is promoted both by the condensation reactions of low reactive aromatic compounds and by high gas-phase concentration of ethylene—a known precursor of carbon deposits on the catalyst surface [97].

As mentioned above, all oxygen at diesel fuel ATR is consumed in the narrow catalyst layer at the frontal section of the block. Most likely, the reaction of complete oxidation involves predominantly aliphatic hydrocarbons that have a lower C–C bond energy. Aromatic compounds react with oxygen much more slowly owing to their lower activity caused by strong carbon–carbon bonds in the aromatic ring. To compare the reactivity of aliphatic, mono- and diaromatic hydrocarbons under steam reforming conditions, a set of experiments with individual compounds—HD, o-xylene and 1-methylnaphthalene—were performed using catalytic block Rh60 (Figure 13).

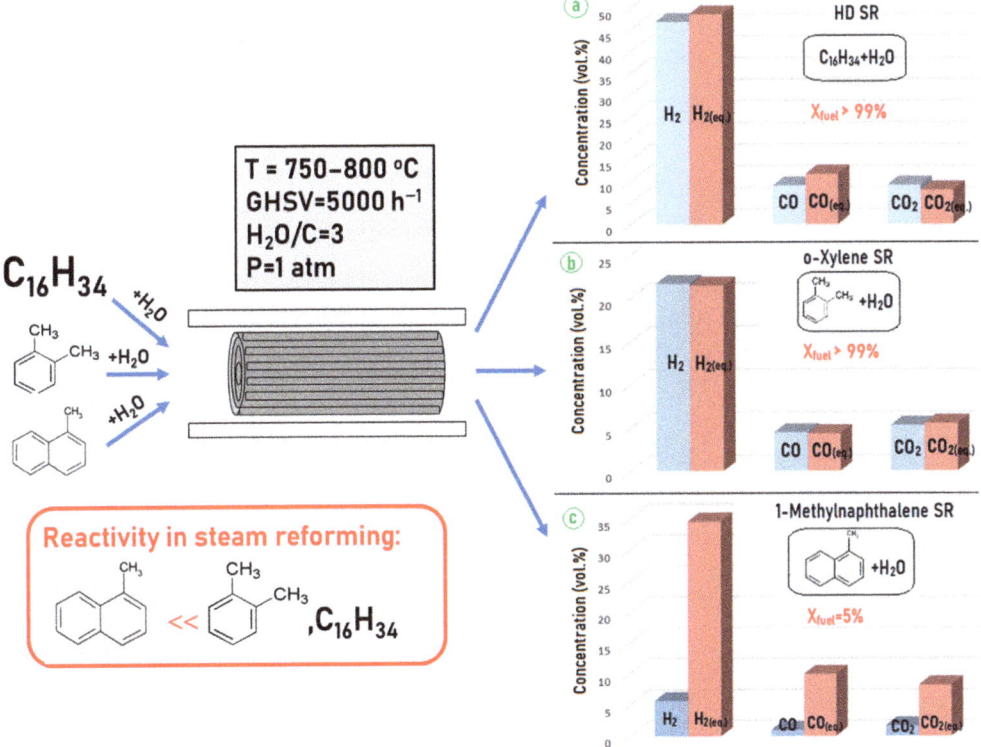

Figure 13. Steam reforming of HD (**a**), o-xylene (**b**) and 1-methylnaphthalene (**c**) over Rh60. Comparison of products concentrations with equilibrium values (red columns). Adapted from [97].

It is seen that the Rh60 catalyst is efficient in the SR of aliphatic and monoaromatic compounds, but demonstrates a relatively low efficiency in the reforming of di- and polyaromatic compounds. In fact, the situation occurs when aliphatic hydrocarbons are converted in the frontal part of the catalytic block at high temperature, whereas less reactive compounds undergo SR in the tail part of the block in the presence of synthesis gas components. Therefore, the key factor for ensuring stable operation of DF ATR catalysts is their activity and stability in SR of di- and polyaromatic compounds [97].

5. Catalyst Coking and Regeneration

To study the process of catalyst coking and predominant carbon localization in the catalyst structure, a series of experiments was carried out with a sample of composition 0.24 wt.% $Rh/6\%Ce_{0.75}Zr_{0.25}O_2/6\%Al_2O_3/FeCrAl$.

The catalytic block was tested in the SR and ATR of various hydrocarbon fuels in a quartz fixed-bed flow reactor for about 250 h (Figure 14) under the following operating conditions: atmospheric pressure, temperature range 550–800 °C, GHSV 1000–20,000 h^{-1}, molar ratios H_2O/C = 2–3, O_2/C = 0.28–0.5. The experiment was stopped when the catalyst turned to a deactivated state after DF ATP. The number of regeneration procedures in air flow at 600 °C exceeded 30 [93].

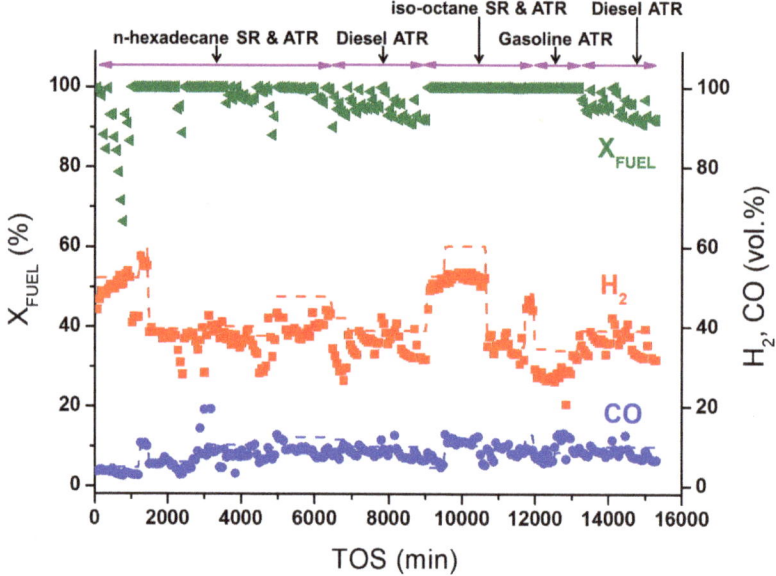

Figure 14. Fuel conversion (X_{FUEL}), H_2 and CO concentrations (dry basis) as a function of time on stream during SR and ATR of different fuels. Points—experimental data, dashed lines—equilibrium concentrations [93].

After the experiments, various parts of the catalytic block were examined. The outer layer of the catalyst was removed mechanically, the cylindrical structure was untwisted, and the sections located in the front and end zones of the catalytic block were cut out from the mesh for SEM analysis. Besides, a part of the catalytic coating (from the front and end zones of the catalytic block) was removed from the surface of the wire mesh and examined by TEM method. For comparison, the samples from the front and end parts of the as-prepared catalytic block (reference sample) were examined using the same methods. The samples for analysis were collected in this manner because autothermal reforming combines exothermic total oxidation reactions and endothermic reactions of steam reforming. As noted earlier, the process of complete oxidation of hydrocarbons proceeds quickly and is localized in the front part of the catalytic block, while steam reforming is a slower process and covers almost the entire catalytic block length [93].

The SEM images (Figure 15) of the as-prepared and used catalytic blocks showed that the catalytic coating stayed undamaged (kept integrity) after SR and ATR of hydrocarbon fuels. The Rh/Ce$_{0.75}$Zr$_{0.25}$O$_2$/Al$_2$O$_3$ coating remained dense and uniform both at the front and at the end parts of the catalytic block. This result is particularly significant for the front part of the catalytic block since it is subjected to rapid temperature fluctuations during the start-up and shutdown procedures, and exactly in this part, the highest temperature is reached during stationary operation.

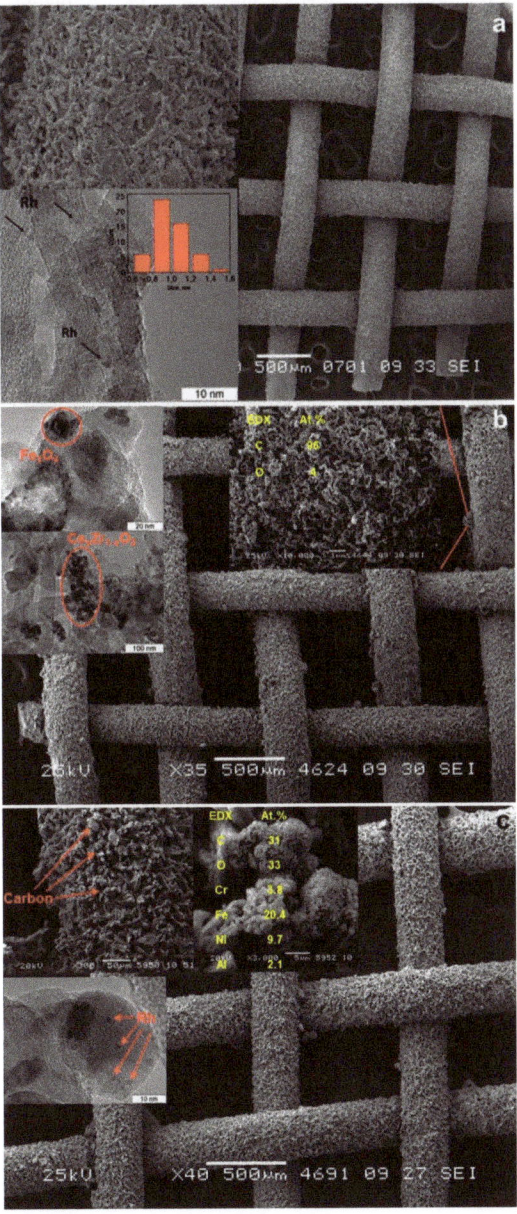

Figure 15. The TEM and SEM images and respective EDX patterns for the as-prepared (**a**) and the used catalytic block (250 h on stream): the frontal part (**b**) and the end part (**c**) [93].

The formation of carbon "knobs" 5–50 μm in size was observed on the surface of both the front and end parts of the block after 250 h on stream. According to TEM and EDX data, the "knobs" were formed from carbon nanofibers. The TEM and EDX analyses revealed also the presence of iron particles near to these carbon species. Probably, coke formation on the catalyst surface is promoted by nano-sized iron impurities. This assumption seems to be quite reasonable because iron is known as one of the best catalysts for the growth of carbon nanofibers and nanotubes. Iron on the catalyst surface can appear both from the FeCrAl structural support during high-temperature transformations and from the reactor material. Another potential source of iron is the thermocouple jacket material, which has contact with the catalytic block and undergoes slow corrosion in the presence of steam and oxygen at temperatures of 750–800 °C at the block inlet. The problem of how Fe nanoparticles appear on the catalyst surface requires further study. This example clearly shows that under extreme reaction conditions of DF ATR, undesirable side processes can be initiated by materials other than those of the structured catalyst. In this regard, the choice of appropriate materials for the manufacture of various parts of the reformer and auxiliary devices seems to be of key importance for providing long-term and stable catalyst operation.

The TEM images of the used catalytic block show that the Rh nanoparticles kept a highly dispersed state and shape. This observation proves the high stability of the catalyst microstructure, at least in the end part of the catalytic block. Unfortunately, attempts to get acceptable-quality images of Rh particles in the front part of the used catalytic block appeared unsuccessful.

Microscopic studies of the catalytic block confirmed the formation of carbon nanofibers on the catalyst surface at DF ATR. It should be noted that the process of coke formation occurs on the surface of the catalytic coating and causes no destruction of it. Most likely, the formation of carbon fibers is caused by contamination of the catalytic coating with iron nanoparticles. The location of carbon deposits on the catalyst surface allows their simple oxidation and removal during regeneration. Thus, it was shown that the prepared catalytic block is stable and can be regenerated under the conditions of SR and ATR of hydrocarbons; no morphological violations and microstructure degradations were observed in either the frontal or tail parts of the catalytic block.

Additionally, the processes of carbon removal from the catalyst surface in the process of oxidation by steam or oxygen were studied. For this purpose, the Rh/CZB/FCA catalyst was subjected to preliminary coking under the DF ATR conditions. The catalyst regeneration by oxygen or steam was performed in a flow of composition 20 vol.% O_2 and 80 vol.% Ar or 75 vol.% H_2O and 25 vol.% Ar, respectively, at the furnace heating rate of 10 °C/min from 350 °C to 750 °C. The product distribution was determined using a Stanford Research QMS 200 mass spectrometer in real time.

It was found (Figure 16) that the reaction of carbon deposit oxidation by oxygen begins to proceed actively at a temperature of 450 °C and releases a large amount of CO_2. The main part of the carbon is oxidized long before 750 °C, which is the operating temperature at DF ATR. After the catalyst was regenerated, its activity in DF ATR turned to the initial level without degradation of the catalytic coating.

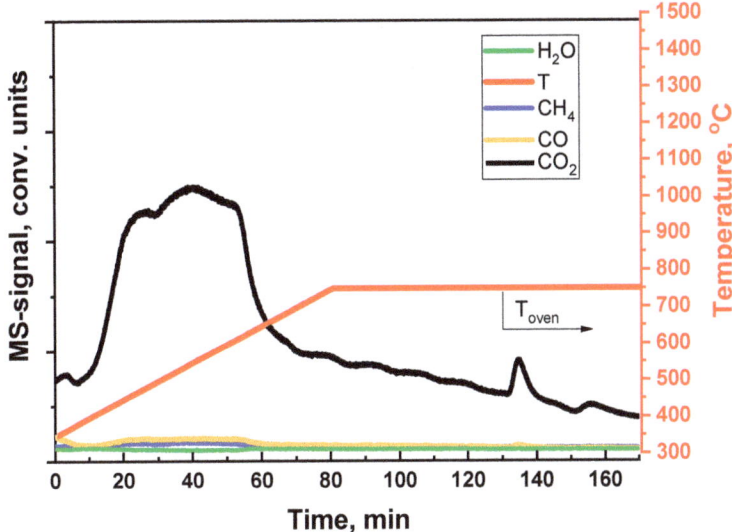

Figure 16. Removal of carbon, deposited on the Rh/CZB/FCA surface at DF ATR, by oxygen.

In experiments on catalyst regeneration by steam (Figure 17), the reaction of carbon deposit oxidation proceeds actively starting from the temperature of 550 °C and is accompanied by the formation of hydrogen (Figure 17). The high measured values of hydrogen concentration, in comparison with those of other reaction products, are associated with the high sensitivity of the device to hydrogen. Probably, H_2 is formed not only by the whisker carbon oxidation reaction, but also by the steam reforming of gum carbon, which contains hydrogen in its composition. Traces of carbon were observed even after two hours of catalyst regeneration by steam at the furnace temperature of 750 °C.

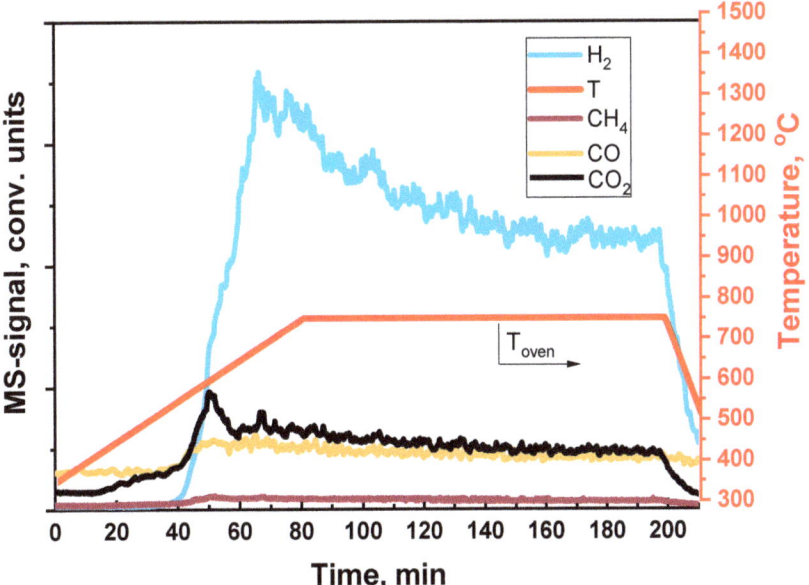

Figure 17. Removal of carbon, deposited on the Rh/CZB/FCA surface at DF ATR, by steam.

The performed studies showed that oxidative regeneration with air is the most effective way to remove carbon deposited on the catalyst during DF ATR. At the same time, water vapor is also able to oxidize the carbon deposits, but at a lower rate. The rate of carbon deposit oxidation by steam at the DF ATR operating temperatures dictates the period of catalyst life between regenerations. This parameter is one of the most important properties of the catalyst: upon reaching a certain (perfect) rate, the need for regeneration can be avoided completely. Note that water vapor oxidizes carbon only at high temperatures; therefore, to ensure stable operation of the catalyst with real diesel fuel containing di- and polyaromatics, it is recommended to increase the temperature at the end part of the catalytic block. This purpose can be achieved, for example, by increasing the O_2:C input ratio.

6. Testing of Diesel Reformer

To carry up-scale tests, a model diesel fuel reformer was developed, which included a diesel fuel burner, a steam generator and a superheater, a gas-liquid nozzle for DF evaporation by hot steam, a zone for mixing the steam-fuel blend with air and the 0.24 wt.% $Rh/6\%Ce_{0.75}Zr_{0.25}O_2/6\%Al_2O_3$/FeCrAl-structured catalytic block with a diameter of 42 mm and a length of 120 mm (Figure 18).

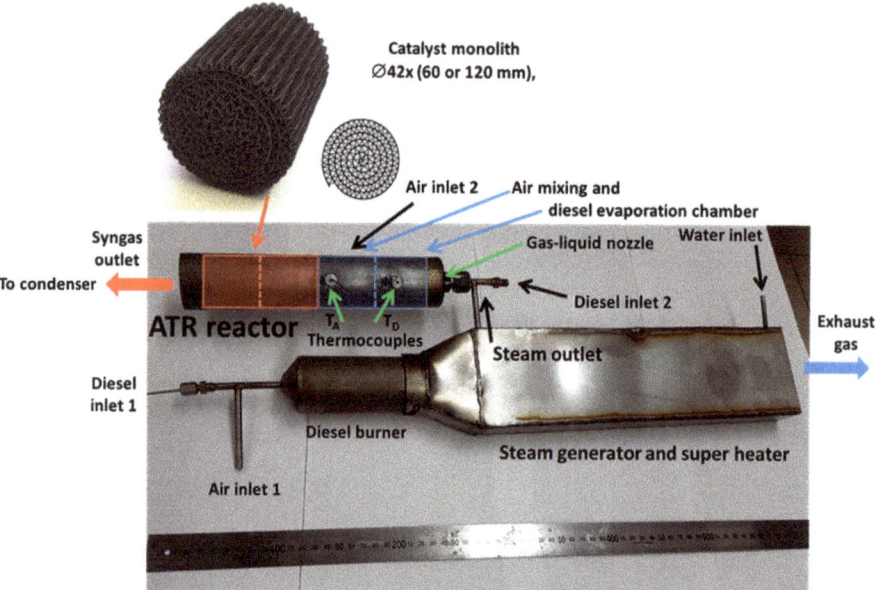

Figure 18. A general view of the ATR diesel reactor together with a starting diesel burner, an evaporator and a steam superheater [91].

The studies were carried out at high molar ratios O_2/C = 0.6 and 0.7 to reduce the coking processes; the reaction mixture space velocity was 6750 and 7500 h^{-1}, respectively. Six thermocouples were inserted into the catalytic block to record the temperature in certain points. Figure 19a demonstrates the thermocouple locations and a typical temperature profile inside the catalytic block during experiments after reaching a steady state. Clearly, the temperature in different parts of the catalytic block was constant in time.

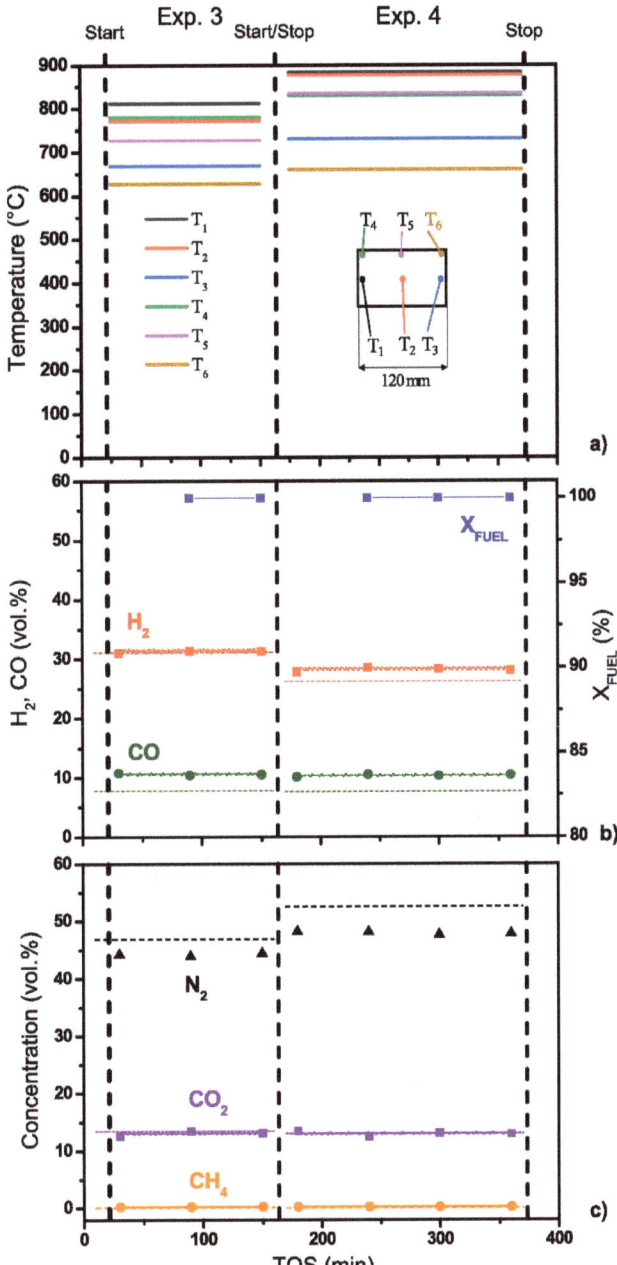

Figure 19. The temperatures in six points of Rh/CZ-42-120 catalytic block (**a**), diesel conversion, H_2 and CO (**b**), N_2, CO_2 and CH_4 concentrations (dry basis) (**c**) as a function of time on stream during DF ATR experiments. Points and solid lines—experimental data, dashed lines—equilibrium concentrations. Experimental conditions: $O_2:C = 0.6$; GHSV = 6750 h^{-1}; $O_2:C = 0.7$, GHSV = 7500 h^{-1} [91].

Under these reaction conditions, the synthesis gas productivity of the reformer was about 0.5 m^3/h. Figure 19b,c show the synthesis gas composition on a dry gas basis.

Despite the high total aromatics content in DF (Table 2), the condensate at the reformer outlet consisted of an almost clear aqueous solution with a faint odor. HPLC analysis revealed no noticeable amounts of unreacted hydrocarbons in the solution.

Thus, upscaling the catalyst and ATR process of diesel fuel with a high content of aromatic hydrocarbons was proved feasible. The performance of the model reformer is sufficient to feed a 0.5 kW high-temperature SOFC. The complete conversion of the di-aromatic components of diesel fuel is most likely associated with the high temperature of the block owing to using gas mixtures with a high O_2:C molar ratio [91].

7. Development of a Mathematical Model for the Diesel Fuel Conversion

For further process upscaling, a mathematical model of the DF ATR over a 0.24 wt.% Rh/6%$Ce_{0.75}Zr_{0.25}O_2$/6%Al_2O_3/FeCrAl-structured catalyst was developed. HD was used as a model fuel in the calculations. To build a stationary model using Comsol Multiphysics software (version 6.0), the catalytic block was represented by a homogeneous porous medium, and the computational domain was defined in a two-dimensional axisymmetric geometry [94–96,98].

For the modeling, a simplified set of reactions was used, which described the main stages of the reforming:

$$C_{16}H_{34} + 24.5O_2 \rightarrow 16CO_2 + 17H_2O \quad \Delta H_{298} = -9800 \text{ kJ/mol} \quad (4)$$

$$C_{16}H_{34} + 16H_2O \rightarrow 33H_2 + 16CO \quad \Delta H_{298} = 2500 \text{ kJ/mol} \quad (5)$$

$$CO + H_2O \leftrightarrow CO_2 + H_2 \quad \Delta H_{298} = -41 \text{ kJ/mol} \quad (6)$$

$$CO + 3H_2 \leftrightarrow CH_4 + H_2O \quad \Delta H_{298} = -206 \text{ kJ/mol} \quad (7)$$

$$CO + 0.5O_2 \rightarrow CO_2 \quad \Delta H_{298} = -283 \text{ kJ/mol} \quad (8)$$

$$H_2 + 0.5O_2 \rightarrow H_2O \quad \Delta H_{298} = -286 \text{ kJ/mol} \quad (9)$$

Reactions were also introduced in the model to account for the formation of light hydrocarbons, namely, alkanes and alkenes C_2–C_5. Since most of them are represented by C_2 hydrocarbons, only ethane and ethylene were considered in the model. Possible routes for the formation of C_{2+} compounds are the hydrogenolysis of hexadecane and the interaction of CO with hydrogen, similar to the methanation Reaction (7). Thermodynamic analysis showed that the first route is the most probable, so the final set of assumed reactions for the formation and transformation of C_{2+} compounds was represented by the following reactions:

$$C_{16}H_{34} + 7H_2 \rightarrow 8C_2H_6 \quad \Delta H_{298} = -305 \text{ kJ/mol} \quad (10)$$

$$C_2H_6 \leftrightarrow C_2H_4 + H_2 \quad \Delta H_{298} = 137 \text{ kJ/mol} \quad (11)$$

$$C_2H_6 + H_2 \rightarrow 2CH_4 \quad \Delta H_{298} = -65 \text{ kJ/mol} \quad (12)$$

$$C_2H_6 + 2H_2O \rightarrow 5H_2 + 2CO \quad \Delta H_{298} = 348 \text{ kJ/mol} \quad (13)$$

$$C_2H_4 + 2H_2O \rightarrow 4H_2 + 2CO \quad \Delta H_{298} = 315 \text{ kJ/mol} \quad (14)$$

$$C_2H_6 + 3.5O_2 \rightarrow 2CO_2 + 3H_2O \quad \Delta H_{298} = -1428 \text{ kJ/mol} \quad (15)$$

$$C_2H_4 + 3O_2 \rightarrow 2CO_2 + 2H_2O \quad \Delta H_{298} = -1323 \text{ kJ/mol} \quad (16)$$

In real conditions, Reactions (4), (5) and (10) proceed through several stages with successive decomposition of HD and the formation of various intermediates. However, none of the intermediates were found in the outlet gas mixture; therefore, the presented total process scheme was used.

Another important issue is related to the probable cracking of hexadecane without the participation of hydrogen. However, when the reaction of this type was added to the scheme, no improvement in the description of the experimental data was reached, and it was decided to exclude this stage from consideration.

The mathematical model included the processes of mass transfer and heat transfer, accounting for the changes in the velocity and pressure fields in the system [94,95].

When solving the minimization task for various operating temperatures, the following set of kinetic equations and kinetic parameter values was obtained that provided the best matching between the experimental and calculated data (Table 3):

Table 3. The optimum kinetic equations and kinetic parameters of Reactions (4)–(6) [95].

No.	Kinetic Equation *	Pre-exponential Factor k_i^0	Activation Energy E_i, kJ/mol
1.	$W_1 = k_1 c_{C_{16}H_{34}} c_{O_2}$	1.9×10^6	47.3
2.	$W_2 = k_2 c_{C_{16}H_{34}} c_{H_2O}$	5×10^{12}	208
3.	$W_3 = \frac{k_3}{c_{H_2}} \left(c_{CO} c_{H_2O} - \frac{c_{CO_2} c_{H_2}}{\exp\left(\frac{4577.8}{T} - 4.33\right)} \right)$	3×10^8	125
4.	$W_4 = k_4 \left(c_{CO} c_{H_2}^3 - \frac{c_{CH_4} c_{H_2O}}{\exp\left(\frac{26800}{T} - 29.8\right)} \right)$	2×10^2	10
5.	$W_5 = k_5 c_{CO} c_{O_2}$	8×10^8	80
6.	$W_6 = k_6 c_{H_2} c_{O_2}$	2×10^5	40
7.	$W_7 = k_7 c_{C_{16}H_{34}} c_{H_2}^{0.5}$	3×10^{10}	124.7
8.	$W_8 = k_8 \left(c_{C_2H_6} - \frac{c_{C_2H_4} c_{H_2}}{\exp\left(\frac{-16850}{T} + 16.5\right)} \right)$	7×10^4	8.4
9.	$W_9 = k_9 c_{C_2H_6} c_{H_2}^{0.1}$	1011	149
10.	$W_{10} = k_{10} c_{C_2H_6} c_{H_2O}^2$	7×10^{13}	208
11.	$W_{11} = k_{11} c_{C_2H_4} c_{H_2O}^2$	1.5×10^{13}	208
12.	$W_{12} = k_{12} c_{C_2H_6} c_{O_2}^{0.5}$	2×10^6	53
13.	$W_{13} = k_{13} c_{C_2H_4} c_{O_2}^{0.5}$	10^6	47

* $k_i = k_i^0 \exp(-E_i/(R_g \cdot T))$.

At mathematical modeling, the 3D geometry was simplified to a two-dimensional axisymmetric one assuming a homogeneous porous medium in the catalytic block region (Figure 20). The geometry included also the region of the gas flow (10 mm long) before and after the catalytic block (cylinder with a radius of 9 mm and a length of 20 mm) with insulating material (a layer of 1 mm thick mineral wool along the entire outer surface—orange area in Figure 20) [95].

The inlet temperature of the gas flow was set according to the experimental readings of a thermocouple located before the catalytic block inlet. The temperature at the outer surface of the insulating material was set equal to the furnace temperature. An infinitely fast heat exchange was set at the boundary between the catalyst and mineral wool, i.e., the temperatures of both sides were equal.

In the ATR experiments, the inlet gas mixture contained 8.8 vol.% oxygen, 1.35 vol.% hexadecane and 56.8 vol.% water; the rest is nitrogen. Figures 21–23 present the simulation results. It is seen that oxygen (Figure 21, right) and most of the hexadecane (Figure 22, left) are consumed in a narrow region at the catalyst inlet owing to the HD oxidation reaction. This reaction is highly exothermic and maintains a high temperature in the catalyst block, peaking at around 890 °C at its frontal part. Then the temperature maximum dissipates both in the axial and radial direction both because of heat losses to the environment and the progress of the less rapid endothermic reactions. In the SR process, the central part of the block is colder than its walls because the process is supported by the reactor furnace, while the catalyst heating is limited by its radial thermal conductivity. As for the ATR process, it is supported by the heat of the reactions (mainly the HD oxidation) over the entire cross-section of the block. If we pay attention to the radial temperature gradients, they are more profound in the ATR than in the SR owing to stronger (higher) heat fluxes. Therefore, in further studies, adequate attention should be paid to an increase in the thermal conductivity of the catalytic block for ATR processes.

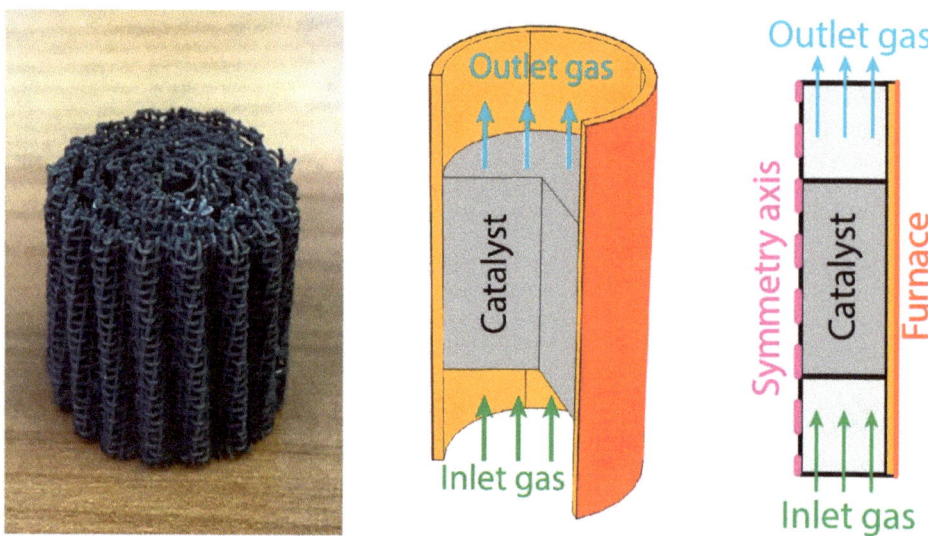

Figure 20. The catalytic block (**left**), considered 3D process geometry (**center**) and 2D axisymmetric computational domain (**right**) [95].

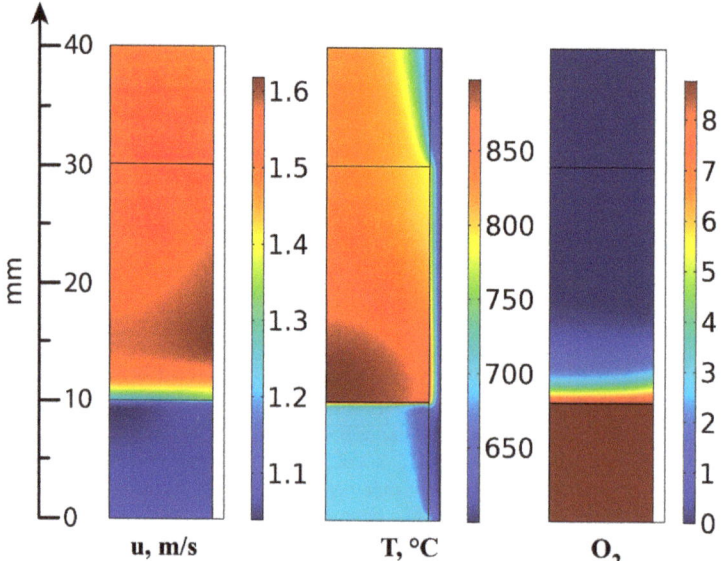

Figure 21. Gas velocity (**left**), temperature distribution (**center**) and oxygen concentration (**right**; vol.%) in ATR mode [95].

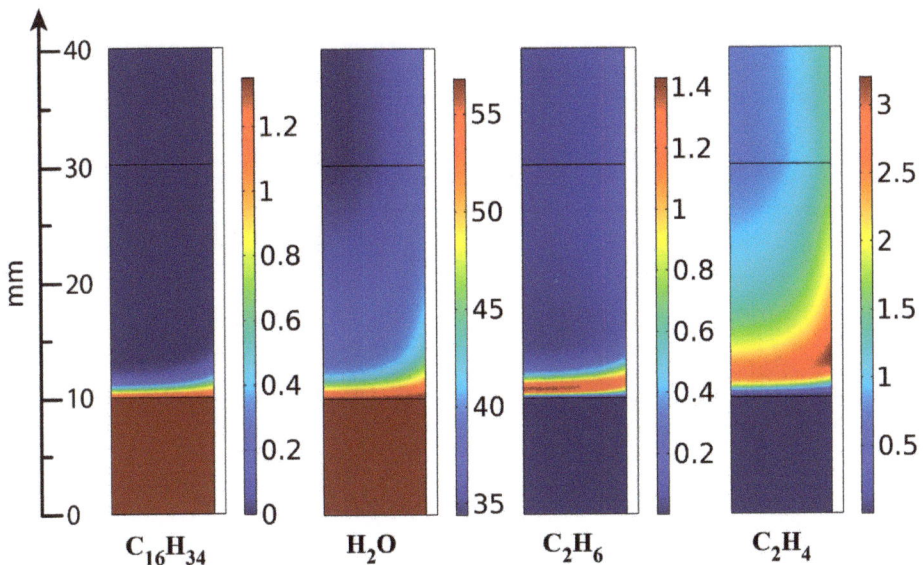

Figure 22. $C_{16}H_{34}$, water, C_2H_6 and C_2H_4 concentrations (vol.%) distribution in ATR [95].

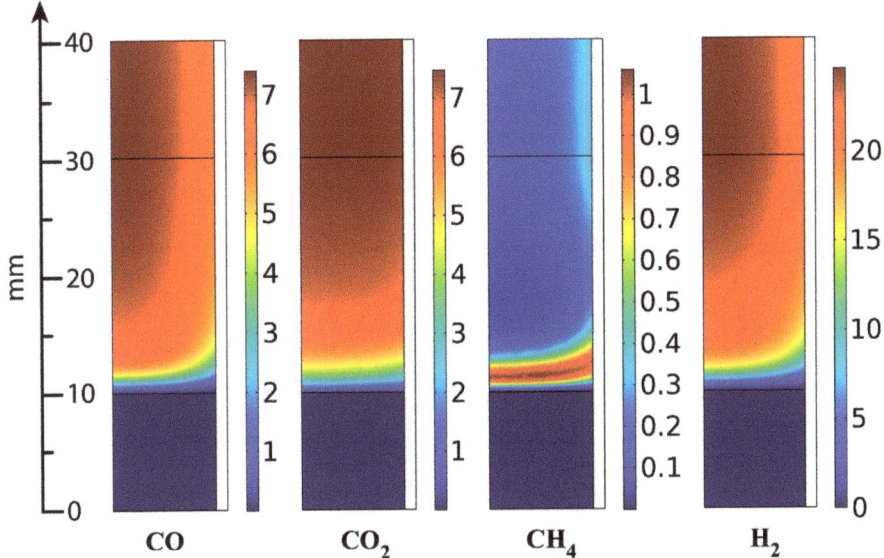

Figure 23. CO, CO_2, CH_4 and H_2 concentrations (vol.%) distribution in ATR [95].

The C_{2+} fraction is formed in the narrow frontal part of the catalytic block (Figure 22) because of the rapid hydrogenolysis of hexadecane; then its concentration passes through a maximum and decreases owing to SR Reactions (13) and (14), facilitated by high temperatures. Here, the total yield of C_{2+} compounds is significantly lower than in the SR experiments.

Similarly, methane is formed in the narrow frontal part of the block (Figure 23) and then consumed during the reactions; its outlet concentration in ATR is lower than in the SR process. This observation correlates also with the higher temperature, which, according to reaction thermodynamics, shifts the equilibrium toward methane SR and facilitates an increase in CO and hydrogen concentrations.

Table 4 presents the calculated and experimental results of the outlet product distribution. The experimental results were averaged for the three experiments since they showed similar values of the outlet product concentrations.

Table 4. The outlet product distributions in the model and experimental ATP processes [95].

	HD Conversion, %	Outlet Product Distribution (vol.%)						
		CO	CO_2	CH_4	H_2	N_2	H_2O	C_{2+}
Model	99.9	6.7	7.4	0.23	23	25.5	35.9	1.1
Experiment (averaged data)	99.2	5.3	8.5	0.25	23	25.8	36.1	1.2

Clearly, the experimental and calculated results agree well, and the equilibrium state can be achieved with a catalytic block length exceeding 300 mm.

Thus, experimental studies of diesel fuel reforming on the structured $Rh/Ce_{0.75}Zr_{0.25}O_2$-Al_2O_3/FeCrAl catalytic block in the ATR and SR processes allowed the development of kinetic and mathematical models that provide a qualitatively adequate and quantitatively accurate description of experimental data [94–96,98].

8. Conclusions

In this review, a method was described for the synthesis of $Rh/Ce_{0.75}Zr_{0.25}O_2/Al_2O_3/$FeCrAl catalyst for the conversion of diesel fuel into synthesis gas. Each structural component of this catalyst has its specific function. The structured metal substrate made of FeCrAl alloy provides fast heat removal/supply for exo-/endothermic reactions, possesses sufficient hydrodynamic characteristics, facilitates the manufacturing of the blocks of various geometric shapes and allows easy process upscaling. The structural oxide component (alumina) provides thermal stability and high specific surface area, strengthens mechanically the supported catalytic coating and protects the metal substrate. The active oxide component (mixed cerium–zirconium oxide with a fluorite structure) participates in the activation of water and oxygen molecules, improves the catalyst's coke resistance through high oxygen mobility and keeps the active component in a fine dispersed state owing to strong metal–carrier interaction. The Rh nanoparticles of 1–2 nm size are involved in the activation of hydrocarbon molecules.

The synthesized catalysts were active and stable in the ATR of diesel fuel of various types, including those with a high content of aromatic hydrocarbons. The catalysts demonstrated high efficiency in the conversion of other liquid hydrocarbon fuels (gasoline and biodiesel) into synthesis gas [18,92]. The oxidative regeneration of catalysts was proved feasible; the conditions for stable catalyst operation were determined. Both the catalyst and the process are easily scalable.

A mathematical model of the process is proposed. It provides a qualitatively adequate and quantitatively accurate description of the experimental results.

It is reasonable to focus subsequent studies on further improving the catalyst's coke resistance and optimizing the system for high-boiling diesel fuel evaporating and mixing the steam–fuel blend with air.

Author Contributions: Conceptualization, V.S., D.P. and P.S.; methodology, V.S. and V.R.; software, validation, V.S.; formal analysis, investigation, D.P., V.R. and V.S.; resources, data curation, writing—original draft preparation, V.S., D.P. and P.S.; supervision, P.S. All authors have read and agreed to the published version of the manuscript.

Funding: This research was funded by the Russian Science Foundation (Project № 19-19-00257).

Institutional Review Board Statement: Not applicable.

Informed Consent Statement: Not applicable.

Data Availability Statement: Not applicable.

Acknowledgments: The authors are grateful to their colleagues A.N. Zagoruiko, S.V. Zazhigalov, P.A. Simonov, V.D. Belyaev, N.V. Ruban, A.V. Kulikov, E.Yu. Gerasimov, V.P. Pakharukova, K.I. Shefer, A.V. Ischenko, T.B. Shoinkhorova and V.A. Sobyanin for the comprehensive assistance in performing the research and the fruitful discussion of the results.

Conflicts of Interest: The authors declare that they have no known competing financial interest or personal relationships that could have appeared to influence the work reported in this paper.

Abbreviations

ATR—autothermal reforming; CFD—computational fluid dynamics; GHSV—gas hourly space velocity; LPG—liquefied petroleum gas; PO—partial oxidation; SR—steam reforming; SOFC—solid oxide fuel cell; MCFC—molten carbonate fuel cell; PAFC—phosphoric acid fuel cell; PEMFC—proton-exchange membrane fuel cell; DME dimethyl ether; DMM—dimethoxy methane; HD—hexadecane; DF—diesel fuel.

References

1. Speight, J.G. Fuels for Fuel Cells. In *Fuel Cells: Technologies for Fuel Processing*; Elsevier: Amsterdam, The Netherlands, 2011; pp. 29–48.
2. Peters, R.; Pasel, J.; Samsun, R.C.; Scharf, F.; Tschauder, A.; Stolten, D. Heat Exchanger Design for Autothermal Reforming of Diesel. *Int. J. Hydrogen Energy* **2018**, *43*, 11830–11846. [CrossRef]
3. Samsun, R.C.; Pasel, J.; Peters, R.; Stolten, D. Fuel Cell Systems with Reforming of Petroleum-Based and Synthetic-Based Diesel and Kerosene Fuels for APU Applications. *Int. J. Hydrogen Energy* **2015**, *40*, 6405–6421. [CrossRef]
4. Pasel, J.; Samsun, R.C.; Peters, R.; Stolten, D. Fuel Processing of Diesel and Kerosene for Auxiliary Power Unit Applications. *Energy Fuels* **2013**, *27*, 4386–4394. [CrossRef]
5. Samsun, R.C.; Prawitz, M.; Tschauder, A.; Pasel, J.; Pfeifer, P.; Peters, R.; Stolten, D. An Integrated Diesel Fuel Processing System with Thermal Start-up for Fuel Cells. *Appl. Energy* **2018**, *226*, 145–159. [CrossRef]
6. Samsun, R.C.; Krekel, D.; Pasel, J.; Prawitz, M.; Peters, R.; Stolten, D. A Diesel Fuel Processor for Fuel-Cell-Based Auxiliary Power Unit Applications. *J. Power Sources* **2017**, *355*, 44–52. [CrossRef]
7. Rostrup-Nielsen, J.R.; Christensen, T.S.; Dybkjaer, I. Steam Reforming of Liquid Hydrocarbons. In *Studies in Surface Science and Catalysis*; Elsevier: Amsterdam, The Netherlands, 1998; pp. 81–95.
8. Bae, J.; Lee, S.; Kim, S.; Oh, J.; Choi, S.; Bae, M.; Kang, I.; Katikaneni, S.P. Liquid Fuel Processing for Hydrogen Production: A Review. *Int. J. Hydrogen Energy* **2016**, *41*, 19990–20022. [CrossRef]
9. Xu, X.; Li, P.; Shen, Y. Small-Scale Reforming of Diesel and Jet Fuels to Make Hydrogen and Syngas for Fuel Cells: A Review. *Appl. Energy* **2013**, *108*, 202–217. [CrossRef]
10. Abatzoglou, N.; Fauteux-Lefebvre, C. Review of Catalytic Syngas Production through Steam or Dry Reforming and Partial Oxidation of Studied Liquid Compounds. *Wiley Interdiscip Rev. Energy Environ.* **2016**, *5*, 169–187. [CrossRef]
11. Pasel, J.; Samsun, R.C.; Peters, R.; Thiele, B.; Stolten, D. Long-Term Stability at Fuel Processing of Diesel and Kerosene. *Int. J. Hydrogen Energy* **2014**, *39*, 18027–18036. [CrossRef]
12. Yoon, S.; Bae, J.; Kim, S.; Yoo, Y.S. Self-Sustained Operation of a KWe-Class Kerosene-Reforming Processor for Solid Oxide Fuel Cells. *J. Power Sources* **2009**, *192*, 360–366. [CrossRef]
13. Samsun, R.C.; Pasel, J.; Janßen, H.; Lehnert, W.; Peters, R.; Stolten, D. Design and Test of a 5 KW High-Temperature Polymer Electrolyte Fuel Cell System Operated with Diesel and Kerosene. *Appl. Energy* **2014**, *114*, 238–249. [CrossRef]
14. Granlund, M.Z.; Jansson, K.; Nilsson, M.; Dawody, J.; Pettersson, L.J. Evaluation of Co, La, and Mn Promoted Rh Catalysts for Autothermal Reforming of Commercial Diesel: Aging and Characterization. *Appl. Catal. B* **2015**, *172–173*, 145–153. [CrossRef]
15. Cheekatamarla, P.K.; Lane, A.M. Efficient Bimetallic Catalysts for Hydrogen Generation from Diesel Fuel. *Int. J. Hydrogen Energy* **2005**, *30*, 1277–1285. [CrossRef]
16. Lee, S.; Bae, M.; Bae, J.; Katikaneni, S.P. Ni–Me/Ce0.9Gd0.1O2−x (Me: Rh, Pt and Ru) Catalysts for Diesel Pre-Reforming. *Int J. Hydrogen Energy* **2015**, *40*, 3207–3216. [CrossRef]
17. Karatzas, X.; Dawody, J.; Grant, A.; Svensson, E.E.; Pettersson, L.J. Zone-Coated Rh-Based Monolithic Catalyst for Autothermal Reforming of Diesel. *Appl. Catal. B* **2011**, *101*, 226–238. [CrossRef]
18. Shilov, V.A.; Rogozhnikov, V.N.; Ruban, N.V.; Potemkin, D.I.; Simonov, P.A.; Shashkov, M.V.; Sobyanin, V.A.; Snytnikov, P.V. Biodiesel and Hydrodeoxygenated Biodiesel Autothermal Reforming over Rh-Containing Structured Catalyst. *Catal. Today* **2021**, *379*, 42–49. [CrossRef]
19. Martin, S.; Kraaij, G.; Ascher, T.; Baltzopoulou, P.; Karagiannakis, G.; Wails, D.; Wörner, A. Direct Steam Reforming of Diesel and Diesel–Biodiesel Blends for Distributed Hydrogen Generation. *Int. J. Hydrogen Energy* **2015**, *40*, 75–84. [CrossRef]

20. Schwengber, C.A.; Alves, H.J.; Schaffner, R.A.; da Silva, F.A.; Sequinel, R.; Bach, V.R.; Ferracin, R.J. Overview of Glycerol Reforming for Hydrogen Production. *Renew. Sustain. Energy Rev.* **2016**, *58*, 259–266. [CrossRef]
21. Lindermeir, A.; Kah, S.; Kavurucu, S.; Mühlner, M. On-Board Diesel Fuel Processing for an SOFC–APU—Technical Challenges for Catalysis and Reactor Design. *Appl. Catal. B* **2007**, *70*, 488–497. [CrossRef]
22. Zhang, S.; Wang, X.; Xu, X.; Li, P. Hydrogen Production via Catalytic Autothermal Reforming of Desulfurized Jet-A Fuel. *Int. J. Hydrogen Energy* **2017**, *42*, 1932–1941. [CrossRef]
23. Xu, X.; Zhang, S.; Li, P. Autothermal Reforming of N-Dodecane and Desulfurized Jet-A Fuel for Producing Hydrogen-Rich Syngas. *Int. J. Hydrogen Energy* **2014**, *39*, 19593–19602. [CrossRef]
24. Fabiano, C.; Italiano, C.; Vita, A.; Pino, L.; Laganà, M.; Recupero, V. Performance of 1.5 Nm3/h Hydrogen Generator by Steam Reforming of n-Dodecane for Naval Applications. *Int. J. Hydrogen Energy* **2016**, *41*, 19475–19483. [CrossRef]
25. Xu, X.; Zhang, S.; Wang, X.; Li, P. Fuel Adaptability Study of a Lab-Scale 2.5 KWth Autothermal Reformer. *Int. J. Hydrogen Energy* **2015**, *40*, 6798–6808. [CrossRef]
26. Pasel, J.; Samsun, R.C.; Tschauder, A.; Peters, R.; Stolten, D. Advances in Autothermal Reformer Design. *Appl. Energy* **2017**, *198*, 88–98. [CrossRef]
27. Pasel, J.; Samsun, R.C.; Tschauder, A.; Peters, R.; Stolten, D. A Novel Reactor Type for Autothermal Reforming of Diesel Fuel and Kerosene. *Appl. Energy* **2015**, *150*, 176–184. [CrossRef]
28. Krekel, D.; Samsun, R.C.; Pasel, J.; Prawitz, M.; Peters, R.; Stolten, D. Operating Strategies for Fuel Processing Systems with a Focus on Water–Gas Shift Reactor Stability. *Appl. Energy* **2016**, *164*, 540–552. [CrossRef]
29. Pasel, J.; Samsun, R.C.; Tschauder, A.; Peters, R.; Stolten, D. Water-Gas Shift Reactor for Fuel Cell Systems: Stable Operation for 5000 Hours. *Int. J. Hydrogen Energy* **2018**, *43*, 19222–19230. [CrossRef]
30. Lindström, B.; Karlsson, J.A.J.; Ekdunge, P.; de Verdier, L.; Häggendal, B.; Dawody, J.; Nilsson, M.; Pettersson, L.J. Diesel Fuel Reformer for Automotive Fuel Cell Applications. *Int. J. Hydrogen Energy* **2009**, *34*, 3367–3381. [CrossRef]
31. Karatzas, X.; Nilsson, M.; Dawody, J.; Lindström, B.; Pettersson, L.J. Characterization and Optimization of an Autothermal Diesel and Jet Fuel Reformer for 5kWe Mobile Fuel Cell Applications. *Chem. Eng. J.* **2010**, *156*, 366–379. [CrossRef]
32. Li, D.; Li, X.; Gong, J. Catalytic Reforming of Oxygenates: State of the Art and Future Prospects. *Chem. Rev.* **2016**, *116*, 11529–11653. [CrossRef]
33. Shafer, L.; Striebich, R.; Gomach, J.; Edwards, T. Chemical Class Composition of Commercial Jet Fuels and other Specialty Kerosene Fuels. In Proceedings of the 14th AIAA/AHI Space Planes and Hypersonic Systems and Technologies Conference, American Institute of Aeronautics and Astronautics, Reston, Viriginia, 6 November 2006.
34. Edwards, T.; Maurice, L.Q. Surrogate Mixtures to Represent Complex Aviation and Rocket Fuels. *J. Propuls. Power* **2001**, *17*, 461–466. [CrossRef]
35. Guggilla, V.S.; Akyurtlu, J.; Akyurtlu, A.; Blankson, I. Steam Reforming of n-Dodecane over Ru−Ni-Based Catalysts. *Ind Eng Chem Res* **2010**, *49*, 8164–8173. [CrossRef]
36. Lee, S.; Bae, J.; Katikaneni, S.P. $La_{0.8}Sr_{0.2}Cr_{0.95}Ru_{0.05}O_3-$ and $Sm_{0.8}Ba_{0.2}Cr_{0.95}Ru_{0.05}O_3-$ as Partial Oxidation Catalysts for Diesel. *Int J Hydrogen Energy* **2014**, *39*, 4938–4946. [CrossRef]
37. Zheng, J.; Strohm, J.J.; Song, C. Steam Reforming of Liquid Hydrocarbon Fuels for Micro-Fuel Cells. Pre-Reforming of Model Jet Fuels over Supported Metal Catalysts. *Fuel Process. Technol.* **2008**, *89*, 440–448. [CrossRef]
38. Zheng, Q.; Janke, C.; Farrauto, R. Steam Reforming of Sulfur-Containing Dodecane on a Rh–Pt Catalyst: Influence of Process Parameters on Catalyst Stability and Coke Structure. *Appl. Catal. B* **2014**, *160–161*, 525–533. [CrossRef]
39. Vita, A.; Italiano, C.; Fabiano, C.; Pino, L.; Laganà, M.; Recupero, V. Hydrogen-Rich Gas Production by Steam Reforming of n-Dodecane. *Appl. Catal. B* **2016**, *199*, 350–360. [CrossRef]
40. Vita, A.; Italiano, C.; Pino, L.; Laganà, M.; Recupero, V. Hydrogen-Rich Gas Production by Steam Reforming of n-Dodecane. Part II: Stability, Regenerability and Sulfur Poisoning of Low Loading Rh-Based Catalyst. *Appl. Catal. B* **2017**, *218*, 317–326. [CrossRef]
41. Jung, S.Y.; Ju, D.G.; Lim, E.J.; Lee, S.C.; Hwang, B.W.; Kim, J.C. Study of Sulfur-Resistant Ni–Al-Based Catalysts for Autothermal Reforming of Dodecane. *Int. J. Hydrogen Energy* **2015**, *40*, 13412–13422. [CrossRef]
42. Haynes, D.J.; Berry, D.A.; Shekhawat, D.; Spivey, J.J. Catalytic Partial Oxidation of N-Tetradecane Using Rh and Sr Substituted Pyrochlores: Effects of Sulfur. *Catal. Today* **2009**, *145*, 121–126. [CrossRef]
43. Haynes, D.J.; Campos, A.; Berry, D.A.; Shekhawat, D.; Roy, A.; Spivey, J.J. Catalytic Partial Oxidation of a Diesel Surrogate Fuel Using an Ru-Substituted Pyrochlore. *Catal. Today* **2010**, *155*, 84–91. [CrossRef]
44. Shekhawat, D.; Gardner, T.H.; Berry, D.A.; Salazar, M.; Haynes, D.J.; Spivey, J.J. Catalytic Partial Oxidation of N-Tetradecane in the Presence of Sulfur or Polynuclear Aromatics: Effects of Support and Metal. *Appl. Catal. A Gen* **2006**, *311*, 8–16. [CrossRef]
45. Haynes, D.J.; Berry, D.A.; Shekhawat, D.; Spivey, J.J. Catalytic Partial Oxidation of N-Tetradecane Using Pyrochlores: Effect of Rh and Sr Substitution. *Catal. Today* **2008**, *136*, 206–213. [CrossRef]
46. Fauteux-Lefebvre, C.; Abatzoglou, N.; Blanchard, J.; Gitzhofer, F. Steam Reforming of Liquid Hydrocarbons over a Nickel–Alumina Spinel Catalyst. *J. Power Sources* **2010**, *195*, 3275–3283. [CrossRef]
47. Lakhapatri, S.L.; Abraham, M.A. Deactivation Due to Sulfur Poisoning and Carbon Deposition on Rh-Ni/Al_2O_3 Catalyst during Steam Reforming of Sulfur-Doped n-Hexadecane. *Appl. Catal. A Gen.* **2009**, *364*, 113–121. [CrossRef]

48. Shoynkhorova, T.B.; Rogozhnikov, V.N.; Simonov, P.A.; Snytnikov, P.V.; Salanov, A.N.; Kulikov, A.V.; Gerasimov, E.Y.; Belyaev, V.D.; Potemkin, D.I.; Sobyanin, V.A. Highly Dispersed Rh/Ce$_{0.75}$Zr$_{0.25}$O$_2$-δ-η-Al$_2$O$_3$/FeCrAl Wire Mesh Catalyst for Autothermal n-Hexadecane Reforming. *Mater Lett* **2018**, *214*, 290–292. [CrossRef]
49. Shoynkhorova, T.B.; Simonov, P.A.; Potemkin, D.I.; Snytnikov, P.V.; Belyaev, V.D.; Ishchenko, A.V.; Svintsitskiy, D.A.; Sobyanin, V.A. Highly Dispersed Rh-, Pt-, Ru/Ce$_{0.75}$Zr$_{0.25}$O$_2$-δ Catalysts Prepared by Sorption-Hydrolytic Deposition for Diesel Fuel Reforming to Syngas. *Appl. Catal. B* **2018**, *237*, 237–244. [CrossRef]
50. Shoynkhorova, T.B.; Snytnikov, P.V.; Simonov, P.A.; Potemkin, D.I.; Rogozhnikov, V.N.; Gerasimov, E.Y.; Salanov, A.N.; Belyaev, V.D.; Sobyanin, V.A. From Alumina Modified Rh/Ce$_{0.75}$Zr$_{0.25}$O$_2$-δ Catalyst towards Composite Rh/Ce$_{0.75}$Zr$_{0.25}$O$_2$-δ-η-Al$_2$O$_3$/FeCrAl Catalytic System for Diesel Conversion to Syngas. *Appl Catal B* **2019**, *245*, 40–48. [CrossRef]
51. Shoynkhorova, T.B.; Rogozhnikov, V.N.; Ruban, N.V.; Shilov, V.A.; Potemkin, D.I.; Simonov, P.A.; Belyaev, V.D.; Snytnikov, P.V.; Sobyanin, V.A. Composite Rh/Ce$_{0.75}$Zr$_{0.25}$O$_2$-δ-η-Al$_2$O$_3$/Fecralloy Wire Mesh Honeycomb Module for Natural Gas, LPG and Diesel Catalytic Conversion to Syngas. *Int. J. Hydrogen Energy* **2019**, *44*, 9941–9948. [CrossRef]
52. Alvarez-Galvan, M.C.; Navarro, R.M.; Rosa, F.; Briceño, Y.; Gordillo Alvarez, F.; Fierro, J.L.G. Performance of La,Ce-Modified Alumina-Supported Pt and Ni Catalysts for the Oxidative Reforming of Diesel Hydrocarbons. *Int. J. Hydrogen Energy* **2008**, *33*, 652–663. [CrossRef]
53. Goud, S.K.; Whittenberger, W.A.; Chattopadhyay, S.; Abraham, M.A. Steam Reforming of N-Hexadecane Using a Pd/ZrO$_2$ Catalyst: Kinetics of Catalyst Deactivation. *Int. J. Hydrogen Energy* **2007**, *32*, 2868–2874. [CrossRef]
54. Liu, L.; Hong, L. Ni/Ce1−xMx Catalyst Generated from Metallo-Organic Network for Autothermal Reforming of Diesel Surrogate. *Appl. Catal. A Gen.* **2013**, *459*, 89–96. [CrossRef]
55. Kang, I.; Bae, J.; Bae, G. Performance Comparison of Autothermal Reforming for Liquid Hydrocarbons, Gasoline and Diesel for Fuel Cell Applications. *J. Power Sources* **2006**, *163*, 538–546. [CrossRef]
56. Kopasz, J.P.; Applegate, D.; Miller, L.; Liao, H.K.; Ahmed, S. Unraveling the Maze: Understanding of Diesel Reforming through the Use of Simplified Fuel Blends. *Int. J. Hydrogen Energy* **2005**, *30*, 1243–1250. [CrossRef]
57. Cheekatamarla, P.K.; Finnerty, C.M. Synthesis Gas Production via Catalytic Partial Oxidation Reforming of Liquid Fuels. *Int. J. Hydrogen Energy* **2008**, *33*, 5012–5019. [CrossRef]
58. Shamsi, A.; Baltrus, J.P.; Spivey, J.J. Characterization of Coke Deposited on Pt/Alumina Catalyst during Reforming of Liquid Hydrocarbons. *Appl Catal A Gen* **2005**, *293*, 145–152. [CrossRef]
59. Qi, A.; Wang, S.; Ni, C.; Wu, D. Autothermal Reforming of Gasoline on Rh-Based Monolithic Catalysts. *Int. J. Hydrogen Energy* **2007**, *32*, 981–991. [CrossRef]
60. Yoon, S.; Kang, I.; Bae, J. Suppression of Ethylene-Induced Carbon Deposition in Diesel Autothermal Reforming. *Int. J. Hydrogen Energy* **2009**, *34*, 1844–1851. [CrossRef]
61. Xie, C.; Chen, Y.; Li, Y.; Wang, X.; Song, C. Sulfur Poisoning of CeO$_2$–Al$_2$O$_3$-Supported Mono- and Bi-Metallic Ni and Rh Catalysts in Steam Reforming of Liquid Hydrocarbons at Low and High Temperatures. *Appl. Catal. A Gen.* **2010**, *390*, 210–218. [CrossRef]
62. Granlund, M.Z.; Jansson, K.; Nilsson, M.; Dawody, J.; Pettersson, L.J. Evaluation of Co, La, and Mn Promoted Rh Catalysts for Autothermal Reforming of Commercial Diesel. *Appl. Catal. B* **2014**, *154–155*, 386–394. [CrossRef]
63. Meißner, J.; Pasel, J.; Peters, R.; Samsun, R.C.; Tschauder, A.; Stolten, D. Elimination of By-Products of Autothermal Diesel Reforming. *Chem. Eng. J.* **2016**, *306*, 107–116. [CrossRef]
64. Karatzas, X.; Jansson, K.; Dawody, J.; Lanza, R.; Pettersson, L.J. Microemulsion and Incipient Wetness Prepared Rh-Based Catalyst for Diesel Reforming. *Catal. Today* **2011**, *175*, 515–523. [CrossRef]
65. Karatzas, X.; Creaser, D.; Grant, A.; Dawody, J.; Pettersson, L.J. Hydrogen Generation from N-Tetradecane, Low-Sulfur and Fischer–Tropsch Diesel over Rh Supported on Alumina Doped with Ceria/Lanthana. *Catal. Today* **2011**, *164*, 190–197. [CrossRef]
66. Peters, R.; Pasel, J.; Samsun, R.C.; Scharf, F.; Tschauder, A.; Müller, M.; Müller, A.; Beer, M.; Stolten, D. Spray Formation of Middle Distillates for Autothermal Reforming. *Int. J. Hydrogen Energy* **2017**, *42*, 16946–16960. [CrossRef]
67. Porš, Z.; Pasel, J.; Tschauder, A.; Dahl, R.; Peters, R.; Stolten, D. Optimised Mixture Formation for Diesel Fuel Processing. *Fuel Cells* **2008**, *8*, 129–137. [CrossRef]
68. Kang, I.; Bae, J.; Yoon, S.; Yoo, Y. Performance Improvement of Diesel Autothermal Reformer by Applying Ultrasonic Injector for Effective Fuel Delivery. *J. Power Sources* **2007**, *172*, 845–852. [CrossRef]
69. Kim, M.-Y.; Kyriakidou, E.A.; Choi, J.-S.; Toops, T.J.; Binder, A.J.; Thomas, C.; Parks, J.E.; Schwartz, V.; Chen, J.; Hensley, D.K. Enhancing Low-Temperature Activity and Durability of Pd-Based Diesel Oxidation Catalysts Using ZrO2 Supports. *Appl. Catal. B* **2016**, *187*, 181–194. [CrossRef]
70. Wong, A.P.; Kyriakidou, E.A.; Toops, T.J.; Regalbuto, J.R. The Catalytic Behavior of Precisely Synthesized Pt–Pd Bimetallic Catalysts for Use as Diesel Oxidation Catalysts. *Catal. Today* **2016**, *267*, 145–156. [CrossRef]
71. Xiong, H.; Peterson, E.; Qi, G.; Datye, A.K. Trapping Mobile Pt Species by PdO in Diesel Oxidation Catalysts: Smaller Is Better. *Catal. Today* **2016**, *272*, 80–86. [CrossRef]
72. Achouri, I.E.; Abatzoglou, N.; Fauteux-Lefebvre, C.; Braidy, N. Diesel Steam Reforming: Comparison of Two Nickel Aluminate Catalysts Prepared by Wet-Impregnation and Co-Precipitation. *Catal. Today* **2013**, *207*, 13–20. [CrossRef]
73. Koo, K.Y.; Park, M.G.; Jung, U.H.; Kim, S.H.; Yoon, W.L. Diesel Pre-Reforming over Highly Dispersed Nano-Sized Ni Catalysts Supported on MgO–Al$_2$O$_3$ Mixed Oxides. *Int. J. Hydrogen Energy* **2014**, *39*, 10941–10950. [CrossRef]

74. Fauteux-Lefebvre, C.; Abatzoglou, N.; Braidy, N.; Achouri, I.E. Diesel Steam Reforming with a Nickel–Alumina Spinel Catalyst for Solid Oxide Fuel Cell Application. *J. Power Sources* **2011**, *196*, 7673–7680. [CrossRef]
75. Krumpelt, M. Fuel Processing for Fuel Cell Systems in Transportation and Portable Power Applications. *Catal Today* **2002**, *77*, 3–16. [CrossRef]
76. Smith, M.W.; Shekhawat, D.; Berry, D.A.; Haynes, D.J.; Floyd, D.L.; Spivey, J.J.; Ranasingha, O. Carbon Formation on Rh-Substituted Pyrochlore Catalysts during Partial Oxidation of Liquid Hydrocarbons. *Appl. Catal. A Gen.* **2015**, *502*, 96–104. [CrossRef]
77. Villoria, J.A.; Alvarez-Galvan, M.C.; Al-Zahrani, S.M.; Palmisano, P.; Specchia, S.; Specchia, V.; Fierro, J.L.G.; Navarro, R.M. Oxidative Reforming of Diesel Fuel over $LaCoO_3$ Perovskite Derived Catalysts: Influence of Perovskite Synthesis Method on Catalyst Properties and Performance. *Appl. Catal. B* **2011**, *105*, 276–288. [CrossRef]
78. Mota, N.; Álvarez-Galván, M.C.; Al-Zahrani, S.M.; Navarro, R.M.; Fierro, J.L.G. Diesel Fuel Reforming over Catalysts Derived from $LaCo1-xRuxO_3$ Perovskites with High Ru Loading. *Int. J. Hydrogen Energy* **2012**, *37*, 7056–7066. [CrossRef]
79. Kondakindi, R.R.; Kundu, A.; Karan, K.; Peppley, B.A.; Qi, A.; Thurgood, C.; Schurer, P. Characterization and Activity of Perovskite Catalysts for Autothermal Reforming of Dodecane. *Appl. Catal. A Gen.* **2010**, *390*, 271–280. [CrossRef]
80. Navarro Yerga, R.M.; Álvarez-Galván, M.C.; Mota, N.; Villoria de la Mano, J.A.; Al-Zahrani, S.M.; Fierro, J.L.G. Catalysts for Hydrogen Production from Heavy Hydrocarbons. *ChemCatChem* **2011**, *3*, 440–457. [CrossRef]
81. Xue, Q.; Gao, L.; Lu, Y. Sulfur-Tolerant Pt/Gd_2O_3–CeO_2–Al_2O_3 Catalyst for High Efficiency H2 Production from Autothermal Reforming of Retail Gasoline. *Catal. Today* **2009**, *146*, 103–109. [CrossRef]
82. Mota, N.; Álvarez-Galván, M.C.; Villoria, J.A.; Rosa, F.; Fierro, J.L.G.; Navarro, R.M. Reforming of Diesel Fuel for Hydrogen Production over Catalysts Derived from $LaCo1-x M x O_3$ (M = Ru, Fe). *Top. Catal.* **2009**, *52*, 1995–2000. [CrossRef]
83. Xu, X.; Liu, X.; Xu, B. A Survey of Nickel-Based Catalysts and Monolithic Reformers of the Onboard Fuel Reforming System for Fuel Cell APU Applications. *Int. J. Energy Res* **2016**, *40*, 1157–1177. [CrossRef]
84. Xu, L.; Mi, W.; Su, Q. Hydrogen Production through Diesel Steam Reforming over Rare-Earth Promoted Ni/γ-Al_2O_3 Catalysts. *J. Nat. Gas Chem.* **2011**, *20*, 287–293. [CrossRef]
85. Sugisawa, M.; Takanabe, K.; Harada, M.; Kubota, J.; Domen, K. Effects of La Addition to Ni/Al_2O_3 Catalysts on Rates and Carbon Deposition during Steam Reforming of n-Dodecane. *Fuel Process. Technol.* **2011**, *92*, 21–25. [CrossRef]
86. Kim, T.; Song, K.H.; Yoon, H.; Chung, J.S. Steam Reforming of N-Dodecane over $K_2Ti_2O_5$-Added Ni-Alumina and Ni-Zirconia (YSZ) Catalysts. *Int. J. Hydrogen Energy* **2016**, *41*, 17922–17932. [CrossRef]
87. Liu, L.; Hong, L. Interactions between CeO_2 and Ni P for Enhancing Coking and Sulfur Resistance in Autothermal Reforming of Liquid Hydrocarbons. *Fuel* **2012**, *96*, 348–354. [CrossRef]
88. Pengpanich, S.; Meeyoo, V.; Rirksomboon, T.; Schwank, J. Iso-Octane Partial Oxidation over $Ni-Sn/Ce_{0.75}Zr_{0.25}O_2$ Catalysts. *Catal Today* **2008**, *136*, 214–221. [CrossRef]
89. Pauletto, G.; Vaccari, A.; Groppi, G.; Bricaud, L.; Benito, P.; Boffito, D.C.; Lercher, J.A.; Patience, G.S. FeCrAl as a Catalyst Support. *Chem. Rev.* **2020**, *120*, 7516–7550. [CrossRef]
90. Porsin, A.V.; Rogoznikov, V.N.; Kulikov, A.V.; Salanov, A.N.; Serkova, A.N. Crystallization of Aluminum Hydroxide in a Sodium Aluminate Solution on a Heterogeneous Surface. *Cryst. Growth Des.* **2017**, *17*, 4730–4738. [CrossRef]
91. Rogozhnikov, V.N.; Kuzin, N.A.; Snytnikov, P.V.; Potemkin, D.I.; Shoynkhorova, T.B.; Simonov, P.A.; Shilov, V.A.; Ruban, N.V.; Kulikov, A.V.; Sobyanin, V.A. Design, Scale-up, and Operation of a $Rh/Ce_{0.75}Zr_{0.25}O_2$-δ-η-Al_2O_3/FeCrAl Alloy Wire Mesh Honeycomb Catalytic Module in Diesel Autothermal Reforming. *Chem. Eng. J.* **2019**, *374*, 511–519. [CrossRef]
92. Potemkin, D.I.; Rogozhnikov, V.N.; Ruban, N.V.; Shilov, V.A.; Simonov, P.A.; Shashkov, M.V.; Sobyanin, V.A.; Snytnikov, P.V. Comparative Study of Gasoline, Diesel and Biodiesel Autothermal Reforming over Rh-Based FeCrAl-Supported Composite Catalyst. *Int. J. Hydrogen Energy* **2020**, *45*, 26197–26205. [CrossRef]
93. Rogozhnikov, V.N.; Potemkin, D.I.; Ruban, N.V.; Shilov, V.A.; Salanov, A.N.; Kulikov, A.V.; Simonov, P.A.; Gerasimov, E.Y.; Sobyanin, V.A.; Snytnikov, P.V. Post-Mortem Characterization of $Rh/Ce_{0.75}Zr_{0.25}O_2/Al_2O_3$/FeCrAl Wire Mesh Composite Catalyst for Diesel Autothermal Reforming. *Mater. Lett.* **2019**, *257*, 126715. [CrossRef]
94. Zazhigalov, S.V.; Rogozhnikov, V.N.; Snytnikov, P.V.; Potemkin, D.I.; Simonov, P.A.; Shilov, V.A.; Ruban, N.V.; Kulikov, A.V.; Zagoruiko, A.N.; Sobyanin, V.A. Simulation of Diesel Autothermal Reforming over $Rh/Ce_{0.75}Zr_{0.25}O_2$-δ-η-Al_2O_3/FeCrAl Wire Mesh Honeycomb Catalytic Module. *Chem. Eng. Process.-Process Intensif.* **2020**, *150*, 107876. [CrossRef]
95. Zazhigalov, S.V.; Shilov, V.A.; Rogozhnikov, V.N.; Potemkin, D.I.; Sobyanin, V.A.; Zagoruiko, A.N.; Snytnikov, P.V. Modeling of Hydrogen Production by Diesel Reforming over $Rh/Ce_{0.75}Zr_{0.25}O_2$-δ-η-Al_2O_3/FeCrAl Wire Mesh Honeycomb Catalytic Module. *Catal. Today* **2021**, *378*, 240–248. [CrossRef]
96. Shilov, V.A.; Rogozhnikov, V.N.; Zazhigalov, S.V.; Potemkin, D.I.; Belyaev, V.D.; Shashkov, M.V.; Zagoruiko, A.N.; Sobyanin, V.A.; Snytnikov, P.V. Operation of $Rh/Ce_{0.75}Zr_{0.25}O_2$-δ-η-Al_2O_3/FeCrAl Wire Mesh Honeycomb Catalytic Modules in Diesel Steam and Autothermal Reforming. *Int. J. Hydrogen Energy* **2021**, *46*, 35866–35876. [CrossRef]
97. Shilov, V.A.; Rogozhnikov, V.N.; Potemkin, D.I.; Belyaev, V.D.; Shashkov, M.V.; Sobyanin, V.A.; Snytnikov, P.V. The Influence of Aromatic Compounds on the Rh-Containing Structured Catalyst Performance in Steam and Autothermal Reforming of Diesel Fuel. *Int. J. Hydrogen Energy* **2022**, *47*, 11316–11325. [CrossRef]

98. Zazhigalov, S.V.; Shilov, V.A.; Rogozhnikov, V.N.; Potemkin, D.I.; Sobyanin, V.A.; Zagoruiko, A.N.; Snytnikov, P.V. Mathematical Modeling of Diesel Autothermal Reformer Geometry Modifications. *Chem. Eng. J.* **2022**, *442*, 136160. [CrossRef]
99. Ruban, N.; Rogozhnikov, V.; Zazhigalov, S.; Zagoruiko, A.; Emelyanov, V.; Snytnikov, P.; Sobyanin, V.; Potemkin, D. Composite Structured M/Ce$_{0.75}$Zr$_{0.25}$O$_2$/Al$_2$O$_3$/FeCrAl (M = Pt, Rh, and Ru) Catalysts for Propane and n-Butane Reforming to Syngas. *Materials* **2022**, *15*, 7336. [CrossRef] [PubMed]

Disclaimer/Publisher's Note: The statements, opinions and data contained in all publications are solely those of the individual author(s) and contributor(s) and not of MDPI and/or the editor(s). MDPI and/or the editor(s) disclaim responsibility for any injury to people or property resulting from any ideas, methods, instructions or products referred to in the content.

Article

Composite Structured M/Ce$_{0.75}$Zr$_{0.25}$O$_2$/Al$_2$O$_3$/FeCrAl (M = Pt, Rh, and Ru) Catalysts for Propane and n-Butane Reforming to Syngas

Natalia Ruban [1,2,*], Vladimir Rogozhnikov [1,3], Sergey Zazhigalov [1], Andrey Zagoruiko [1], Vyacheslav Emelyanov [2,4], Pavel Snytnikov [1,2], Vladimir Sobyanin [1] and Dmitriy Potemkin [1,2,5,*]

1. Boreskov Institute of Catalysis, Pr. Akademika Lavrentieva, 5, 630090 Novosibirsk, Russia
2. Faculty of Natural Science, Novosibirsk State University, Pirogova St., 2, 630090 Novosibirsk, Russia
3. Department of Physical and Colloid Chemistry, Faculty of Chemical Technology and Ecology, Gubkin Russian State University of Oil and Gas, Leninsky Pr., 65, 119991 Moscow, Russia
4. Nikolaev Institute of Inorganic Chemistry, Pr. Akademika Lavrentieva, 3, 630090 Novosibirsk, Russia
5. Department of Environmental Engineering, Novosibirsk State Technical University, Karl Marx Pr., 20, 630073 Novosibirsk, Russia
* Correspondence: natavruban@gmail.com (N.R.); potema@catalysis.ru (D.P.); Tel.: +7-913-932-46-20 (D.P.)

Abstract: Here, we report the preparation, characterization, and performance of reforming propane and n-butane into a syngas of composite structured M/Ce$_{0.75}$Zr$_{0.25}$O$_2$/Al$_2$O$_3$/FeCrAl (M = 0.46 wt.% Pt, 0.24 wt.% Rh, and 0.24 wt.% Ru) catalysts. The catalysts are composed of a high-heat-conducting FeCrAl block with preset geometry, with a surface nearly totally covered by θ-Al$_2$O$_3$. Afterwards, a layer of ceria–zirconia mixed oxide was deposited. The formed oxide coating was used as a support for 2–3 nm sized Pt, Rh, or Ru nanoparticles. The performance of the catalysts in propane steam reforming decreased in the order of Rh ≈ Ru > Pt. The reformates obtained in the propane steam reforming over Rh- and Ru/Ce$_{0.75}$Zr$_{0.25}$O$_2$/Al$_2$O$_3$/FeCrAl at 600 °C and GHSV = 8300 h^{-1} contained 65.2 and 62.4 vol.% of H$_2$, respectively, and can be used as a fuel for solid oxide fuel cells. In the oxidative steam reforming of propane at 700 °C and GHSV = 17,000 h^{-1}, the activities of the Rh- and Pt-based catalysts were similar and the compositions of the outlet gas mixtures were quite close to equilibrium in both cases. Increasing the reagent flow rate to 25,600 h^{-1} showed stability of the Rh/Ce$_{0.75}$Zr$_{0.25}$O$_2$/Al$_2$O$_3$/FeCrAl performance, whereas the Pt/Ce$_{0.75}$Zr$_{0.25}$O$_2$/Al$_2$O$_3$/FeCrAl activity decreased. A mathematical model considering the velocity field, mass balance, pressure, and temperature distribution, as well as the reaction kinetics, was suggested for the propane steam and oxidative steam reforming over the Pt- and Rh/Ce$_{0.75}$Zr$_{0.25}$O$_2$/Al$_2$O$_3$/FeCrAl catalysts. The model well described the experimental results.

Keywords: steam reforming; autothermal reforming; propane; butane; structured catalysts

Citation: Ruban, N.; Rogozhnikov, V.; Zazhigalov, S.; Zagoruiko, A.; Emelyanov, V.; Snytnikov, P.; Sobyanin, V.; Potemkin, D. Composite Structured M/Ce$_{0.75}$Zr$_{0.25}$O$_2$/Al$_2$O$_3$/FeCrAl (M = Pt, Rh, and Ru) Catalysts for Propane and n-Butane Reforming to Syngas. *Materials* **2022**, *15*, 7336. https://doi.org/10.3390/ma15207336

Academic Editors: Barbara Pawelec

Received: 20 September 2022
Accepted: 17 October 2022
Published: 20 October 2022

Publisher's Note: MDPI stays neutral with regard to jurisdictional claims in published maps and institutional affiliations.

Copyright: © 2022 by the authors. Licensee MDPI, Basel, Switzerland. This article is an open access article distributed under the terms and conditions of the Creative Commons Attribution (CC BY) license (https://creativecommons.org/licenses/by/4.0/).

1. Introduction

The most important global issues such as climate change and pollution of the environment focus the scientific society's attention on the development of ecology-friendly energy production. Although in some developed countries (i.e., Germany and Denmark), the share of energy generated by renewable sources is increasing in a stable tendency [1], it is predicted that fossil fuels' share in total energy supply will be significant for at least 20 more years. For these reasons, the development of more efficient and sustainable ways to generate energy from fossil fuels is an important and quite immediate task.

Fuel cells are considered as a perspective technology for efficient electricity generation directly from the energy of chemical bonds. The most convenient fuel for fuel cells is hydrogen. The use of hydrogen as a fuel for fuel cells leads to the stable working cycle of power generators and makes the technology environmentally friendly. On the other hand, at this moment, logistical problems associated with hydrogen are major restraining factors for fuel cell technologies. Nowadays, the main sources of hydrogen are fossil fuels such as natural gas and petrochemicals [2]. Therefore, hydrogen production from fossil fuels immediately before the working cycle of fuel cell-based power generators can solve the logistical problems and provide stable work for the fuel cells.

In the cases of logistics and energy density, the most attractive fuels are the liquid products of the oil industry: gasoline, diesel, or jet fuel. However, the composition of these types of fuel is complicated and strongly depends on the region and manufacturer, which can negatively impact the stability of power generation. Because of these reasons, liquified petroleum gases (LPGs), primarily consisting of propane and butane, may be considered as golden means. On the one hand, contrariwise liquid fuels in the composition of LPGs are minimally converted components, especially aromatic compounds. Further, LPGs are preferable to natural gas because of the liquefaction opportunity in quite soft conditions ($p > 10$ bar) [3]. In addition, synthetic LPGs can be easily obtained from "green" hydrogen and captured CO_2 via the hydrogenation reaction, closing the carbon-free energy cycle. Hence, the development of highly efficient catalysts for LPG reforming is an important task for fuel cell energy production.

The development of fuel cell technologies has created a special demand for mobile and local power stations. The current industrial technologies of hydrogen generation from hydrocarbons are not optimal at these types of stations [4]. First, commonly used steam reforming is an endothermal process; therefore, the technology requires a constant external heat input. A combination of endothermal steam and exothermal partial hydrocarbon oxidation, namely, oxidative steam reforming (OSR), can provide high hydrogen productivity and decrease heat input. For the effective use of OSR benefits, it is important to provide heat transfer from the short front zone of the catalyst, where exothermal oxidation takes place, to the zone of steam oxidation. Structured catalysts formed from metal alloys have a level high thermal conductivity [5]. Moreover, it has been shown that using structured forms of catalysts can improve the efficiency of hydrocarbon catalytic reforming and decrease the overall cost of the process [6–9].

In the reforming processes, Ni-based catalysts are often used because of their rather good efficiency and low cost [10]. However, the impact of the standard problem of coke formation on the Ni-based catalyst surfaces [11] increases in this condition, whereas regeneration is undesirable. For medium-scale local or mobile power stations applying expensive, but more stable and efficient, noble metal-based catalysts may be a reasonable solution. According to data collected by the scientific community, the most active catalysts of C_{2+}-hydrocarbon reforming are those based on Rh, Ru, Ir, and Pt [12–14]. Information about the compositions of efficient catalysts can be successfully used for the development of new catalytic system preparation methods [15], and in that work, we used information about the most active granular catalysts for designing structured catalysts.

Previously, we demonstrated a high-activity, Rh-based structured catalyst in the autothermal reforming of gasoline, diesel, and biodiesel [16,17]. It was found that the presence of aromatic compounds in the liquid fuel composition was associated with the decreasing efficiency of the catalytic reforming over the Rh-based structured catalyst. The composition of LPGs is more attractive because this type of fuel primarily consists of C_3-C_4 alkanes and does not contain aromatic compounds. In this work, we focused on the investigation of Rh-based structured catalyst activity in the steam and autothermal reforming of propane and n-butane as model compounds of LPGs. Ru- and Pt-based structured catalysts were prepared and tested with the fundamental goal of comparing the noble metal-based catalysts' molar activities in LPG reforming. In addition, this research was conducted for economic reasons: the extremely high price of Rh motivated us to search for less expensive catalysts with suitable activities.

2. Materials and Methods

2.1. Preparation of the Structured Catalysts

The preparation procedure of the structured catalysts started with the structured support formation. In the first step of the support preparation, a structured FeCrAl module (diameter = 18 mm and length = 60 mm) was formed from wire mesh of Fe, Cr, and Al alloy (NPO Souznichrom Inc., Moscow, Russia) and calcinated. In the next synthetic step, the structured FeCrAl module was covered by an θ-Al_2O_3 layer, according to reported technique in [18]. Then, on the θ-Al_2O_3/FeCrAl surface, a layer of Ce and Zr mixed oxide was formed. The preparation procedure included the 7-step impregnation of the θ-Al_2O_3/FeCrAl by a $Ce(NO_3)_3$ (AO Reahim, LLC, Moscow, Russia, 98% purity) and $ZrO(NO_3)_2$ (Interhim, LLC, Saint-Petersburg, Russia, 98% purity) mixed solution (the Ce:Zr molar ratio was 3:1). After each impregnation, the structured support was calcinated at 800 °C for 5 min and again for 30 min after the final impregnation. The structured support $Ce_{0.75}Zr_{0.25}O_2$/θ-Al_2O_3/FeCrAl was formed. The preparation procedure of the structured support was similar for the Rh-, Ru-, and Pt-based structured catalysts and is described schematically in Figure 1.

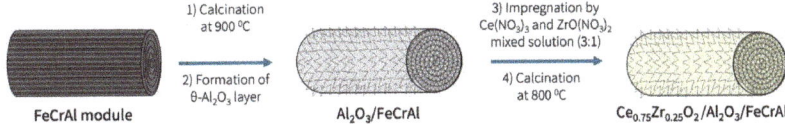

Figure 1. Scheme of the $Ce_{0.75}Zr_{0.25}O_2$/Al_2O_3/FeCrAl preparation.

The goal of this work was a comparison activity per 1 metal atom catalyst with an equimolar content of noble metal (M/CZA/FeCrAl, M = Rh, Pt, and Ru). For the Rh- and Pt-based structured catalysts, a sorption–hydrolytic preparation technique was used. According to the preparation procedure, a metal hydroxide was deposited on the structured support surface from the solution consisting of the noble metal precursor ($RhCl_3$·$4H_2O$ and K_2PtCl_4, respectively) and Na_2CO_3. The preparation procedure of the Ru/CZA/FeCrAl included impregnation of the structured support by a hot solution of the *fac*-[$RuNO(NH_3)_2(NO_3)_3$] precursor. The Ru-precursor was obtained from $Ru(OH)Cl_3$ (Krastsvetmet, JSC, Krasnoyarsk, Russia, 99% purity), according to the 3-step technique described in [19–21]. Afterward, the precursor fixation catalysts were regenerated in the N_2-H_2 gas mixture at 750 °C.

2.2. Catalysts Characterization

Pieces of the structured FeCrAl module and the Al_2O_3/FeCrAl were separated using shears and investigated by an XRD technique. The analysis was carried out on an Arl X'Tra (Thermo Fisher Scientific, Waltham, MA, USA). CuKα radiation was used. Measurements were carried out in the 2θ range from 20 to 70° (step = 0.05°). The signal accumulation time was 5 s. For the XRD analysis of the Ce and Zr mixed oxide, a small piece of support coating was mechanically separated from the structured support. The analysis was carried out on diffractometer Bruker D8 Advance (Bruker, Karlsruhe, Germany) in CuKα radiation mode (λ = 1.5418 Å). Bragg–Brentano focusing was used. The measurements were carried out in the 2θ range of 15–75° (step = 0.05°). The signal accumulation time was 3 s. The analysis of the diffraction pictures was carried out using the ICDD PDF-2 database.

The structured catalysts were studied using transmission electron microscopy. Samples of the catalytic coating were mechanically separated for the investigations. The studies of the samples were carried out using a Themis Z electron microscope (Thermo Fisher Scientific).

The Rh/CZA/FeCrAl catalyst was investigated by temperature-programmed oxidation (TPO) method after tests in the propane and n-butane steam reforming. A quartz tubular reactor was used for the TPO experiments. The oxidation processes were carried out in a mixture of O_2 and He (6 vol.% and 94 vol.%, respectively). A QMC-200 mass spectrometer was used for the registration of the CO_2 concentration. The sensitivity of the measurements was 0.05 mg of carbon.

2.3. Catalytic Activity Investigations

Investigations of the catalytic activity were carried out in fixed-bed flow reactor. The steam and oxidative steam reforming were carried out at atmospheric pressure. An external steam generator was used for superheating the water to 150 °C. In the oxidative steam reforming steam, the fuel and air were mixed immediately before the reaction zone. The reactor was preliminarily heated up to 600 °C and 700 °C in a nitrogen flow before the steam and oxidative steam reforming, respectively.

The steam reforming of propane and n-butane was carried out at $H_2O/C = 3$. In the oxidative steam reforming of the fuels, the H_2O/C and O_2/C ratios were fixed at 2.5 and 0.5, respectively. The reaction temperature was measured in the end zone of the catalyst.

Compositions of the obtained gas mixtures were analyzed using gas chromatography. For the analysis, a GC-1000 gas chromatograph (Chromos, Dzerzhinsk, Russia) was used. A thermal conductivity detector (TCD) and a flame ionization detector (FID) of the GC-1000 chromatograph allowed for analysis of the N_2, H_2, and carbon-containing components at the same time.

As experimental controls, an empty reactor and a reactor with a structured support $Ce_{0.75}Zr_{0.25}O_2/Al_2O_3/FeCrAl$ was heated up to 600 °C, and a propane–water mixture was added to the reaction zone. In both cases, the outlet gas mixture primarily consisted of propane (the water was separated before the analysis), whereas the hydrogen concentrations were lower than 2% in both cases.

3. Results

3.1. Characterization of the M/CZA/FeCrAl

The $Ce_{0.75}Zr_{0.25}O_2/Al_2O_3/FeCrAl$ structured supports were prepared and characterized using the XRD method after each synthetic step. As seen in Figure 2a, the FeCrAl surface was almost totally covered by θ-Al_2O_3 after the second synthetic step. The XRD data obtained for the sample of coating after the third synthetic stage (Figure 2b) proved that the support presented with a $Ce_{1-x}Zr_xO_2$ mixed oxide with a fluorite-type structure. According to preparation procedure, the Ce:Zr molar ratio in the precursor solution was 3:1 therefore the average composition of the support that can be described by $Ce_{0.75}Zr_{0.25}O_2$. As the preparation procedure of the $Ce_{1-x}Zr_xO_2$ mixed oxide was non-selective, mixed oxide was represented by the number of solid solutions with a variable composition. It was also confirmed by the XRD (Figure 2b) and HAADF-STEM (Figures 3c and 4c) analyses.

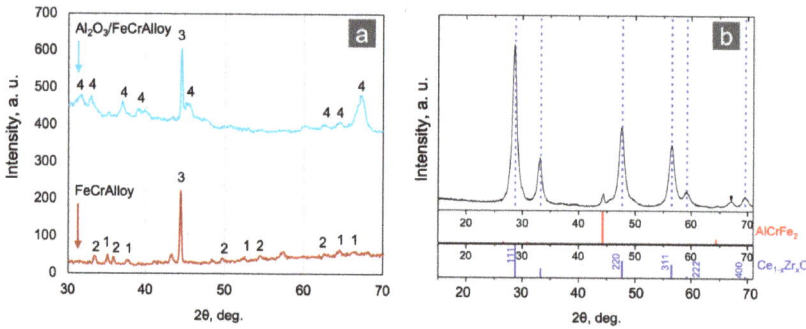

Figure 2. XRD patterns of the $Al_2O_3/FeCrAl$ and FeCrAl: (**a**) 1—reflexes of α-Al_2O_3, 2—Fe and Cr (oxides), 3—Fe, Cr (metals), and 4—θ-Al_2O_3; and (**b**) a sample of the supporting coating.

The support structure is stable in the condition of metal deposition, according to the literature data and our previous research [22]. A sample of the catalytic coating of the as-prepared Rh/CZA/FeCrAl was mechanically separated and studied using transmission electron microscopy (TEM). According to the TEM data (Figure 3a) and element distribution mapping (Figure 3c), the Rh particles were evenly distributed on the support surface. The average particle size was 2 nm (Figure 3b).

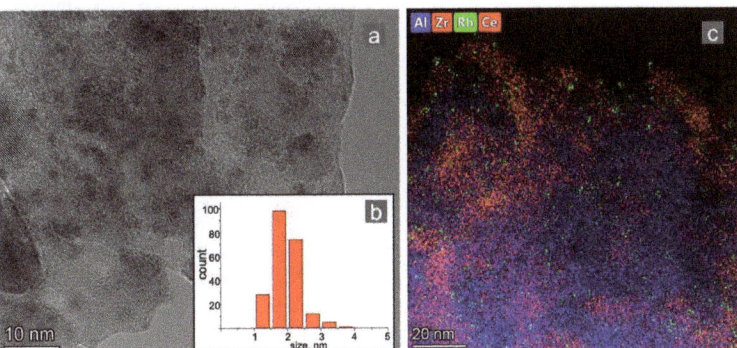

Figure 3. TEM images: (**a**) element distribution mapping of the (**c**) Rh/CZA/FeCrAl catalytic coating and (**b**) Rh particle size distribution.

For the preparation of the Ru/CZA/FeCrAl catalyst, an impregnation method was used. One of the common problems with this method is the agglomerates formation of active metal particles. In this work, the support was impregnated by the specific precursor of Ru with the goals of preventing agglomerates formation and simultaneously simplifying the structured catalyst preparation procedure. A sample of the Ru/CZA/FeCrAl catalytic coating was investigated by the TEM technique. As seen in Figure 4b, the average Ru particle size was 2.3 nm and only small quantities of 6–8 nm-sized particles were found.

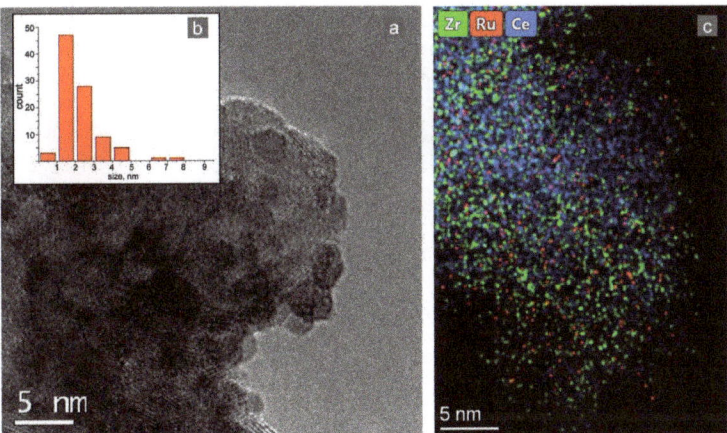

Figure 4. TEM images: (**a**) element distribution mapping of the (**c**) Ru/CZA/FeCrAl catalytic coating and (**b**) Ru particle size distribution.

3.2. Propane Steam Reforming over the M/CZA/FeCrAl

Oxidative steam reforming (OSR) is a combination of the partial and steam oxidation of hydrocarbons. Steam reforming proceeds more slowly than partial oxidation and the activity of the catalyst in this process is critically important. As the first stage of investigation, the propane steam reforming over the Ru-, Pt- and Rh/CZA/FeCrAl was studied. The results of the Ru- and Rh/CZA/FeCrAl catalytic activity investigations at 600 °C, with $H_2O/C = 3$ and a flow rate of 8300 h^{-1}, are presented in Figure 5, where compositions of the reaction products were compared with product distribution in thermodynamic equilibrium. As seen in Figure 5, the product distributions were quite close to an equilibrium condition in both cases, but the concentration of unconverted fuel was higher in the case of the Ru/CZA/FeCrAl. The obtained data for the Pt/CZA/FeCrAl was not presented in Figure 5 because of the low activity of the catalyst—the propane conversion was less than 10% in the investigated process.

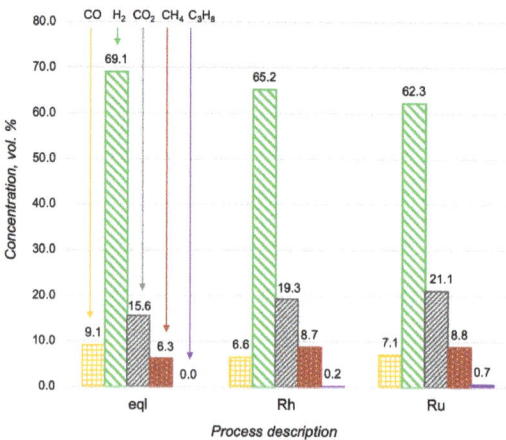

Figure 5. Product distribution in the propane steam reforming at 600 °C, with $H_2O/C = 3$ and a flow rate of 8300 h^{-1}, in a thermodynamic equilibrium (eql), over the Rh/CZA/FeCrAl (Rh), and over the Ru/CZA/FeCrAl (Ru).

Additional studies of the propane steam reforming at higher flow rates showed that the Rh/CZA/FeCrAl was more active than the Ru/CZA/FeCrAl (Figure 6). In particular, the concentration of the propane in the outlet gas mixture in the case of the Ru-based catalyst at 16,700 h^{-1} was approximately the same as the propane content in the reaction products over the Rh/CZA/FeCrAl at 25,000 h^{-1}. According to the obtained results, the Rh/CZA/FeCrAl was chosen for the investigations of the oxidative steam reforming of propane.

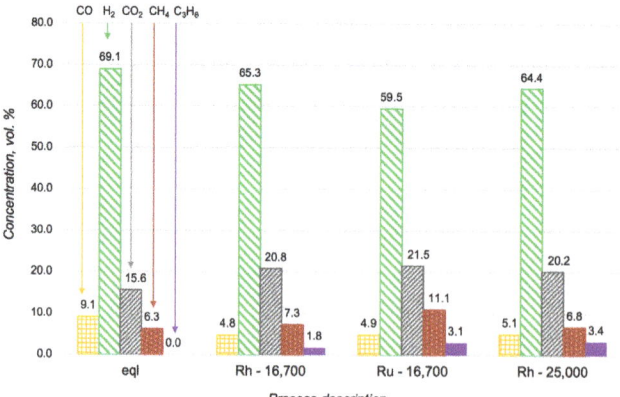

Figure 6. Product distribution in the propane steam reforming at 600 °C with $H_2O/C = 3$ in a thermodynamic equilibrium (eql), over the Rh/CZA/FeCrAl at GHSV = 16,700 h^{-1} (Rh—16,700), over the Ru/CZA/FeCrAl at GHSV = 16,700 h^{-1} (Ru—16,700), and over the Rh/CZA/FeCrAl at GHSV = 25,000 h^{-1} (Rh—25,000).

3.3. Propane Oxidative Steam Reforming over the Pt- and Rh/CZA/FeCrAl

The high activity of Pt-based catalysts in oxidative processes is well known [23,24]. It was the reason for the investigation of the Pt/CZA/FeCrAl in propane oxidative steam reforming, despite the obtained proof of its low activity in propane steam reforming. In Figure 7, the data of the Pt- and Rh/CZA/FeCrAl activities in propane oxidative steam reforming were compared with the product distribution in a thermodynamic equilibrium. The catalysts demonstrated high activities in propane oxidative steam reforming at 700 °C, with H_2O/C = 2.5, O_2/C = 0.5, and GHSV = 17,000 h^{-1}, and the product distributions were quite close to equilibrium in both cases. At higher reagent flow rates, the concentration of hydrogen in the outlet gas mixtures decreased simultaneously with the increasing C_{2+}-hydrocarbons and methane content (in the case of the Pt/CZA/FeCrAl). In general, the activity of the Pt/CZA/FeCrAl in propane oxidative steam reforming at GHSV = 25,600 h^{-1} was significantly lower than the Rh/CZA/FeCrAl activity.

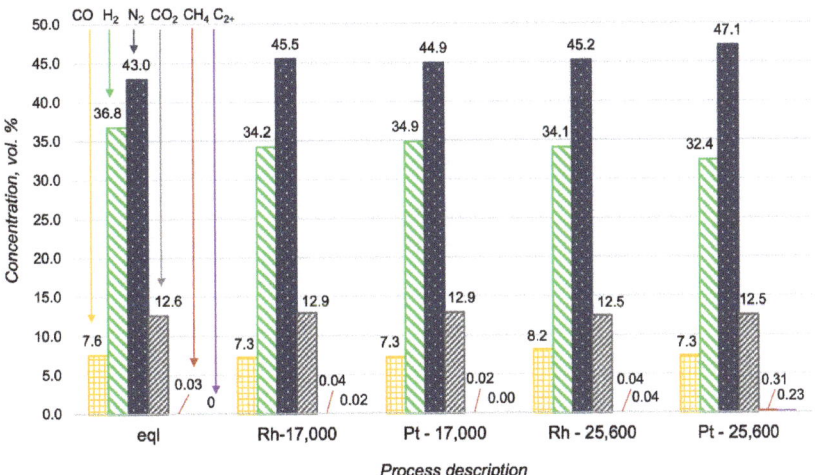

Figure 7. Product distribution in propane oxidative steam reforming at 700 °C with H_2O/C = 2.5 and O_2/C = 0.5 in a thermodynamic equilibrium (eql), over the Rh/CZA/FeCrAl and the Pt/CZA/FeCrAl at GHSV = 17,000 h^{-1} ("Rh—17,000" and "Pt—17,000", respectively), and over the Rh/CZA/FeCrAl and the Pt/CZA/FeCrAl at GHSV = 25,600 h^{-1} ("Rh—25,600" and "Pt—25,600", respectively).

More detailed information about hydrocarbon content in the reforming products is presented in Figure 8. Interestingly, C_{2+}-hydrocarbons were not observed in the outlet gas mixture obtained over the Pt/CZA/FeCrAl at 17,000 h^{-1}. This could be a sign of the high rate of propane oxidation in the initial zone of the Pt-based catalyst. The increasing of reagents flow rate led to decreases in the contact times between the reagents and the Pt active sites in the initial zone of the catalyst, and therefore, according to the low activity of Pt/CZA/FeCrAl in the steam reforming, some quantities of propane were not converted. The Rh/CZA/FeCrAl showed stable activity at 17,000 h^{-1}, which did not decrease at 25,600 h^{-1}.

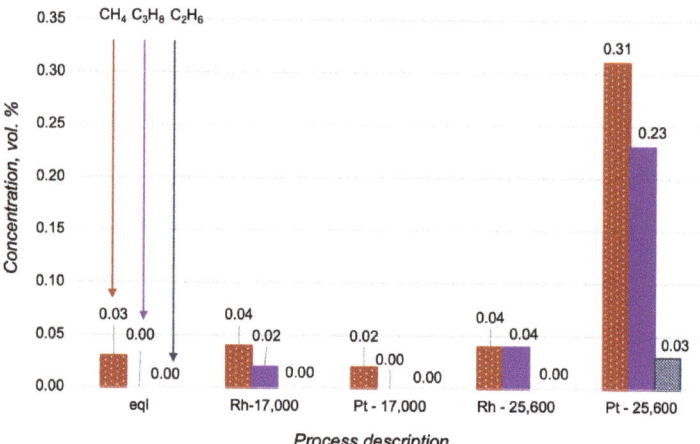

Figure 8. Hydrocarbon content in the products of the propane oxidative steam reforming at 700 °C, with $H_2O/C = 2.5$ $O_2/C = 0.5$ in a thermodynamic equilibrium (eql), over the Rh/CZA/FeCrAl and the Pt/CZA/FeCrAl at GHSV = 17,000 h^{-1} ("Rh—17,000" and "Pt—17,000", respectively), and over the Rh/CZA/FeCrAl and the Pt/CZA/FeCrAl at GHSV = 25,600 h^{-1} ("Rh—25,600" and "Pt—25,600", respectively).

3.4. n-Butane Reforming over the Rh/CZA/FeCrAl

The investigation of the propane steam and oxidative steam reforming showed that among investigated structured catalysts, the Rh/CZA/FeCrAl had the best efficiency. In commercial LPG compositions, there are hydrocarbons heavier than propane, such as butane. According to the literature data [25], the problem of carbon formation in hydrocarbons reforming can be associated with hydrocarbon chain length because of differences in the intermediate species structure and content. The next step of the study was to approve the Rh/CZA/FeCrAl properties in the n-butane reforming.

The process of the n-butane steam reforming was carried out at 600 °C, with $H_2O/C = 3$ and GHSV = 8125 h^{-1}. In Figure 9, the results of the Rh/CZA/FeCrAl activity investigation are presented. As seen in a comparison of Figures 5 and 9, the content of unconverted fuel in the products of the n-butane steam reforming was higher than that in the case of propane. In general, it can be concluded that the Rh/CZA/FeCrAl demonstrated a rather high activity in both propane and n-butane steam reforming.

The results of our previous study of liquid fuels reforming over a Rh-based structured catalyst showed that carbon fibers had formed on the catalyst surface [17,26]. It was concluded that carbon formation was associated with the content of aromatic compounds in the fuel composition. To improve on that conclusion, we investigated the Rh/CZA/FeCrAl after the propane and n-butane steam reforming using the TPO method. According to the obtained data, in both cases, no carbon fibers had formed on the catalyst surface.

The catalytic activity of the Rh/CZA/FeCrAl was measured in n-butane oxidative steam reforming at 700 °C, with $H_2O/C = 2.5$, $O_2/C = 0.5$, and GHSV = 16,600 h^{-1}. In Figure 10, the composition of the reforming products was compared with the product distribution in a thermodynamic equilibrium. According to the obtained data, in the catalytic reforming, the n-butane was nearly totally converted. In the case of the propane oxidative steam reforming, C_{2+}-hydrocarbons were not observed in the outlet gas mixture.

The measurements of the Rh/CZA/FeCrAl catalytic activity showed that the catalyst was highly efficient in the processes of propane and n-butane oxidative steam reforming, and in both cases, the fuel was converted to a hydrogen-rich gas mixture. It was especially important that in the steam and oxidative steam reforming of the fuels, carbon formation was not observed. It can be concluded that the Rh/CZA/FeCrAl is a perspective catalyst for LPG reforming with the goal of obtaining a syngas.

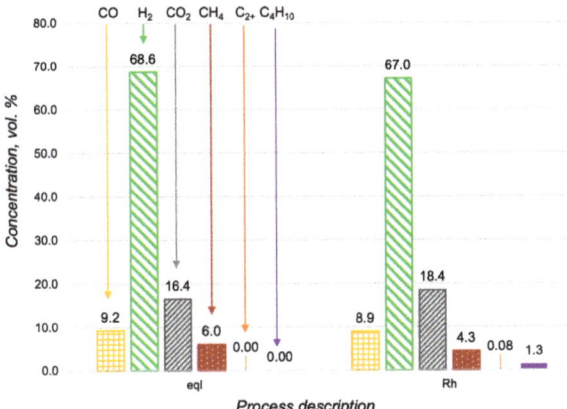

Figure 9. Hydrocarbon content in the products of the n-butane steam reforming at 600 °C, with $H_2O/C = 3$ and $O_2/C = 0.5$ in a thermodynamic equilibrium (eql) and over the Rh/CZA/FeCrAl at GHSV = 8125 h^{-1}.

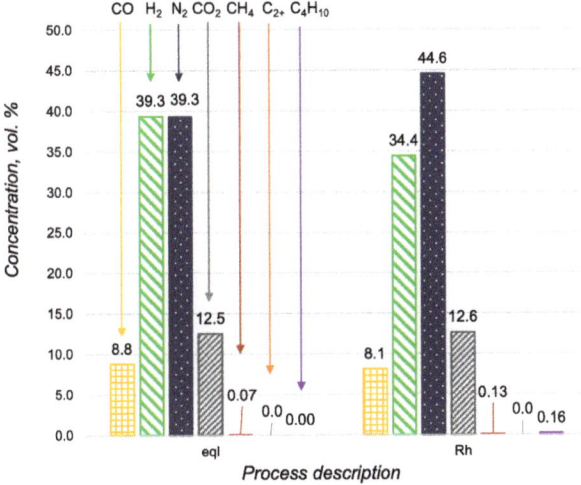

Figure 10. Product distribution in the n-butane oxidative steam reforming at 700 °C, with $H_2O/C = 2.5$ $O_2/C = 0.5$ in a thermodynamic equilibrium (eql) and over the Rh/CZA/FeCrAl at GHSV = 16,600 h^{-1}.

3.5. Mathematical Modelling

For processes such as the steam and oxidative steam reforming of propane, heat management plays a crucial role. The idea of material design was intended to improve the catalysts' ability to heat transfer for enhancing the slow endothermic reactions of the steam reforming. To quantify the effect, we performed the mathematical modelling of the propane steam and oxidative steam reforming over the Pt- and Rh-based catalysts.

We used the model in the 2D axisymmetric geometry developed earlier for the diesel ATR process over a Rh-based catalyst [27,28]. The catalytic module was considered as a porous medium. The model took into account the velocity field and pressure distribution (Navier–Stokes equations for the free medium and Brinkman equations for the porous medium of the catalytic module), mass balance (convective and diffusion transfers and reaction source), and temperature distribution for the gas and catalyst, as follows:

$$\nabla \cdot j_i + \rho(u \cdot \nabla)\omega_i = R_i, \tag{1}$$

$$j_i = -\left(\rho D_i^m \nabla \omega_i + \rho \omega_i D_i^m \frac{\nabla M_n}{M_n} - \rho \omega_i \sum_k \frac{M_i}{M_n} D_k^m \nabla x_k\right), \text{ and} \tag{2}$$

$$D_i^m = \frac{1 - \omega_i}{\sum_{k \neq i} \frac{c_k}{D_{ik}}}. \tag{3}$$

The velocity field (u) and pressure (p) were set by the "Brinkman Equations" interface shown in (4) and (5).

$$0 = \nabla \cdot \left[-pI + \mu \frac{1}{\varepsilon_p}\left(\nabla u + (\nabla u)^T\right) - \frac{2\mu}{3\varepsilon_p}(\nabla \cdot u)I\right] - u\left(\mu K^{-1} + \frac{Q_m}{\varepsilon_p^2}\right) \tag{4}$$

$$\nabla \cdot (\rho u) = Q_m \tag{5}$$

The heat transfer was assigned through the "Heat Transfer in Porous Media" for the gas and catalyst phases interface (i.e., it was supposed to be the quasi-homogeneous model, and so the gas and catalyst temperatures were described by the same variable).

$$\rho C_p u \cdot \nabla T - \nabla \cdot \left((1 - \varepsilon_p)\lambda_c + \varepsilon_p \lambda\right)\nabla T = Q \tag{6}$$

The chemical transformations were assumed to take place according to the following set of reactions:

$$CO + H_2O \leftrightarrow CO_2 + H_2 \quad \Delta H_{298} = -41 \text{ kJ/mol} \tag{7}$$

$$CH_4 + H_2O \leftrightarrow CO + 3H_2 \quad \Delta H_{298} = 206 \text{ kJ/mol} \tag{8}$$

$$CO + 0.5O_2 \rightarrow CO_2 \quad \Delta H_{298} = -283 \text{ kJ/mol} \tag{9}$$

$$H_2 + 0.5O_2 \rightarrow H_2O \quad \Delta H_{298} = -286 \text{ kJ/mol} \tag{10}$$

$$CH_4 + 2O_2 \rightarrow CO_2 + 2H_2O \quad \Delta H_{298} = -803 \text{ kJ/mol} \tag{11}$$

$$C_3H_8 + 5O_2 \rightarrow 3CO_2 + 4H_2O \quad \Delta H_{298} = -2045 \text{ kJ/mol} \tag{12}$$

$$C_3H_8 + 3H_2O \rightarrow 3CO + 7H_2 \quad \Delta H_{298} = 499 \text{ kJ/mol} \tag{13}$$

$$C_3H_8 + 2H_2 \rightarrow 3CH_4 \quad \Delta H_{298} = -121 \text{ kJ/mol} \tag{14}$$

The inlet gas mixture was supposed to be evenly mixed and have a composition similar to that reported for the experiments above. The occurrence of homogeneous reactions in a free medium before and after the catalyst was neglected. The inlet mixture temperature was set at 600 °C for the steam reforming and 700 °C for the oxidative steam reforming, and the catalyst was located inside the hot box (outer cartridge walls) with a constant set temperature of 700 °C. All the modification calculations were compared with the reference case—the cylindrical catalytic module with a 9 mm radius and a 60 mm length. The volumes of the catalyst in all the modifications were the same as in the reference case, unless otherwise stated. COMSOL Multiphysics was used for the simulation.

The experimental data were used to adjust the kinetic parameters. The set of kinetic equations and values of kinetic parameters used are presented in Table 1 (the kinetic constants were used in an Arrhenius form). The concentrations are given in mole fractions. For the steam reforming process, only the reactions without oxygen were used (as in Equations (7), (8), (13) and (14)).

Table 1. Kinetic parameters for Equations (7)–(14).

Catalyst	Reaction	Reaction Rate	k_0, mol/(m$^3\cdot$s)	E, kJ/mol
Pt/CZA/FeCrAl	7	$k_7\left(c_{CO}c_{H_2O} - \dfrac{c_{CO_2}c_{H_2}}{\exp\left(\frac{4577.8}{T}-4.33\right)}\right)$	1.3×10^6	41.5
	8	$k_8\left(c_{CH_4}c_{H_2O} - \dfrac{c_{CO}c_{H_2}^3}{\exp\left(-\frac{26800}{T}+29.8\right)}\right)$	5.3×10^5	50
	9	$k_9 c_{CO} c_{O_2}$	4.0×10^8	80
	10	$k_{10} c_{H_2} c_{O_2}$	2.0×10^6	40
	11	$k_{11} c_{CH_4} c_{O_2}^{0.5}$	8.0×10^6	100
	12	$k_{12} c_{C_3H_8} c_{O_2}$	4.9×10^8	100
	13	$k_{13} c_{C_3H_8} c_{H_2O}^{0.1}$	7.0×10^{11}	180
	14	$k_{14} c_{C_3H_8} c_{H_2}$	2.0×10^7	100
Rh/CZA/FeCrAl	7	$k_7\left(c_{CO}c_{H_2O} - \dfrac{c_{CO_2}c_{H_2}}{\exp\left(\frac{4577.8}{T}-4.33\right)}\right)$	1.0×10^6	41.5
	8	$k_8\left(c_{CH_4}c_{H_2O} - \dfrac{c_{CO}c_{H_2}^3}{\exp\left(-\frac{26800}{T}+29.8\right)}\right)$	3.6×10^3	10
	9	$k_9 c_{CO} c_{O_2}$	4.0×10^8	80
	10	$k_{10} c_{H_2} c_{O_2}$	2.0×10^6	40
	11	$k_{11} c_{CH_4} c_{O_2}^{0.5}$	8.0×10^6	100
	12	$k_{12} c_{C_3H_8} c_{O_2}$	1.0×10^8	100
	13	$k_{13} c_{C_3H_8} c_{H_2O}^3$	2.9×10^6	40
	14	$k_{14} c_{C_3H_8} c_{H_2}^2$	1.6×10^3	10

The modeling results of the propane SR (Table 2) and OSR (Table 3) are presented below. Generally, it is seen that the model can well describe both the SR and OSR of propane over the Pt/CZA/FeCrAl and the Rh/CZA/FeCrAl catalysts at various temperatures and flow rates. The high performance of both catalyst in the oxidative steam reforming (despite the low performance of the Pt/CZA/FeCrAl in the SR of propane) is associated with the high heat conductivity of the FeCrAl support and the effective heat transfer from the exothermal total oxidation of the propane in the front zone to the endothermal propane steam reforming in the end zone of the catalyst bed.

Table 2. Comparison of the modelling results and experiments for the propane SR.

Catalyst		Concentration on Dry Basis, vol.%				
		CO	CO$_2$	CH$_4$	H$_2$	C$_3$H$_8$
Pt/CZA/FeCrAl	Exp	0.6	11.2	0.2	38.7	49.2
	Model	0.7	10.9	0.3	39.1	47.8
Rh/CZA/FeCrAl	Exp	6.6	19.3	8.7	65.2	0.2
	Model	6.5	17.4	7.2	68.3	0.25

Table 3. Comparison of the modelling results and experiments for the propane OSR.

Catalyst		GHSV, h^{-1}	Concentration on Dry Basis, vol.%					
			CO	CO$_2$	CH$_4$	H$_2$	C$_3$H$_8$	N$_2$
Pt/CZA/FeCrAl	Exp	20,000	7.3	12.8	0.0	34.8	0.00	45.1
	Model	20,000	7.4	12.4	0.1	34.9	0.21	45.0
	Exp	30,000	7.4	12.5	0.3	32.2	0.28	47.3
	Model	30,000	7.8	12.2	0.1	34.9	0.17	44.9
Rh/CZA/FeCrAl	Exp	20,000	6.4	13.6	0.0	34.9	45.0	0.02
	Model	20,000	7.7	12.4	0.1	35.8	44.1	0.03
	Exp	30,000	6.7	13.5	0.0	34.6	45.2	0.05
	Model	30,000	8.2	12.0	0.1	35.3	44.5	0.06

4. Discussion

The efficiency of the studied catalysts was compared with the literature data on the activity of catalysts based on Rh and Pt. The properties of granular Pt/CeO_2 and Rh/Al_2O_3 catalysts in propane OSR were studied in [29] and [30]. The authors of [29] studied the properties of Pt/CeO_2 pellets and two structured Pt/CeO_2 catalysts: Pt/CeO_2/cordierite and a self-structured Pt/CeO_2 catalyst. A comparison of the results showed that the activity of both self-structured Pt/CeO_2 and Pt/CeO_2 deposited on a cordierite monolith in propane OSR were higher than the activity of a granular Pt/CeO_2 catalyst in the same conditions. In [30], the catalyst design was similar to the catalysts presented in this article, and Rh/Al_2O_3 deposited on a rectangular structural block of FeCrAl foil was studied. The Rh/CZA/FeCrAl and Pt/CZA/FeCrAl catalysts studied in this work outperformed the mentioned catalysts in syngas productivity.

Solid oxide fuel cells (SOFC) are one of the prospective types of power generators. The key benefits of SOFC are its wide range of working temperatures and flexibility in fuel composition. Fuels applicable for SOFC can contain inert components such as N_2 and CO_2 and up to 20% of CO [31], 10% of CH_4 [32], and 1.8% of C_{2+}-hydrocarbons [33]. In Table 4, the compositions of the syngases obtained over a M/CZA/FeCrAl in the investigated processes were compared with the requirements for the SOFC's fuel.

As seen in Table 4, in general, the syngases obtained over the Rh/CZA/FeCrAl in the propane and n-butane reforming processes are applicable for SOFCs. The Pt-based catalyst is also suitable for propane's conversion to syngas in oxidative steam reforming conditions.

Table 4. Comparison of the requirements for the SOFCs fuel and the compositions of the syngases obtained over the M/CZA/FeCrAl.

			Concentration on Dry Basis, vol.%			
			H_2	CO	CH_4	C_{2+}
	SOFC's Fuel Requirements		-	<20	<10	<1.8
Process	Catalyst	GHSV, h^{-1}	H_2	CO	CH_4	C_{2+}
C_3H_8 SR	Rh/CZA/FeCrAl	8300	65.2	6.6	8.7	0.2
		16,700	65.3	4.8	7.3	1.8
		25,000	64.4	5.1	6.8	3.4
	Ru/CZA/FeCrAl	8300	62.3	7.1	8.8	0.7
		16,700	59.5	4.9	11.1	3.2
C_3H_8 OSR	Rh/CZA/FeCrAl	17,000	34.2	7.3	0.04	0.04
		25,600	34.1	8.2	0.04	0.02
	Pt/CZA/FeCrAl	17,000	34.9	7.3	0.02	0.00
		25,600	32.4	7.3	0.31	0.23
n-C_4H_{10} SR	Rh/CZA/FeCrAl	8125	67.0	8.9	4.3	1.38
n-C_4H_{10} OSR		16,600	34.4	12.6	0.13	0.16

5. Conclusions

Highly dispersed composite structured $M/Ce_{0.75}Zr_{0.25}O_2/Al_2O_3$/FeCrAl (M = Pt, Rh, and Ru) catalysts were prepared and tested in the steam and oxidative steam reforming of propane. The preparation process was monitored by XRD and TEM at different stages. It was shown that the suggested preparation procedures allowed us to obtain noble metal particles with an average size of 2–3 nm.

$Rh/Ce_{0.75}Zr_{0.25}O_2/Al_2O_3$/FeCrAl outperformed other catalysts in propane reforming in terms of activity and syngas productivity. The syngases obtained in the propane steam reforming over Rh- and $Ru/Ce_{0.75}Zr_{0.25}O_2/Al_2O_3$/FeCrAl at 600 °C and GHSV = 8300 h^{-1} contained 65.2 and 62.4 vol.% of H_2, respectively, and can be used as fuel for solid oxide fuel cells. In the oxidative steam reforming of propane at 700 °C and GHSV= 17,000 h^{-1}, the activities of Rh- and Pt-based were similar and the compositions of the outlet gas mixtures were quite close to equilibrium in both cases. Increasing the reagent flow rate to 25,600 h^{-1}

showed the stability of the $Rh/Ce_{0.75}Zr_{0.25}O_2/Al_2O_3/FeCrAl$ performance, whereas the $Pt/Ce_{0.75}Zr_{0.25}O_2/Al_2O_3/FeCrAl$ activity decreased. The activity of the Rh-based catalyst was also proven in n-butane steam and oxidative steam reforming.

A mathematical model that included the velocity field, mass balance, pressure, and temperature distribution, as well as the reaction kinetics, was suggested for propane steam and oxidative steam reforming over the Pt- and $Rh/Ce_{0.75}Zr_{0.25}O_2/Al_2O_3/FeCrAl$ catalysts. The model well described the experimental results.

Author Contributions: Conceptualization, D.P. and P.S.; methodology, V.S.; software, S.Z. and A.Z.; validation, V.E.; investigation, N.R., V.R. and V.E.; data curation, V.S.; writing—original draft preparation, D.P. and N.R.; supervision, D.P.; project administration, D.P.; funding acquisition, D.P. All authors have read and agreed to the published version of the manuscript.

Funding: The work was funded by the Russian Science Foundation under the project N° 21-79-10377. D.I. Potemkin appreciates the financial support from the Russian Foundation for Basic Research under the project 19-33-60008.

Informed Consent Statement: Not applicable.

Data Availability Statement: Not applicable.

Acknowledgments: The experiments were carried out using the facilities of the shared research center "National center of investigation of catalysts" at the Boreskov Institute of Catalysis.

Conflicts of Interest: The authors declare that they have no known competing financial interests or personal relationships that could have appeared to influence the work reported in this paper.

References

1. IEA Energy and Carbon Tracker 2020. Available online: https://www.iea.org/fuels-and-technologies/renewables (accessed on 12 May 2022).
2. Pinsky, R.; Sabharwall, P.; Hartvigsen, J.; O'Brien, J. Comparative review of hydrogen production technologies for nuclear hybrid energy systems. *Prog. Nucl. Energy* **2020**, *123*, 103317. [CrossRef]
3. Antolini, E. Direct propane fuel cells. *Fuel* **2022**, *315*, 123152. [CrossRef]
4. Abe, J.O.; Popoola, A.P.I.; Ajenifuja, E.; Popoola, O.M. Hydrogen energy, economy and storage: Review and recommendation. *Int. J. Hydrogen Energy* **2019**, *44*, 15072–15086. [CrossRef]
5. Pauletto, G.; Vaccari, A.; Groppi, G.; Bricaud, L.; Benito, P.; Boffito, D.C.; Lercher, J.A.; Patience, G.S. FeCrAl as a Catalyst Support. *Chem. Rev.* **2020**, *120*, 7516–7550. [CrossRef] [PubMed]
6. Mundhwa, M.; Thurgood, C. Numerical study of methane steam reforming and methane combustion over the segmented and continuously coated layers of catalysts in a plate reactor. *Fuel Process. Technol.* **2017**, *158*, 57–72. [CrossRef]
7. Herdem, M.; Mundhwa, M.; Farhad, S.; Hamdullahpur, F. Catalyst layer design and arrangement to improve the performance of a microchannel methanol steam reformer. *Energy Convers. Manag.* **2019**, *180*, 149–161. [CrossRef]
8. Cherif, A.; Nebbali, R.; Lee, C.-J. Design and multiobjective optimization of membrane steam methane reformer: A computational fluid dynamic analysis. *Int. J. Energy Res.* **2022**, *46*, 8700–8715. [CrossRef]
9. Cherif, A.; Nebbali, R.; Lee, C.-J. Numerical analysis of steam methane reforming over a novel multi-concentric rings Ni/Al_2O_3 catalyst pattern. *Int. J. Energy Res.* **2021**, *45*, 18722–18734. [CrossRef]
10. Zhang, H.; Sun, Z.; Hu, Y.H. Steam reforming of methane: Current states of catalyst design and process upgrading. *Renew. Sustain. Energy Rev.* **2021**, *149*, 111330. [CrossRef]
11. Sehested, J. Four challenges for nickel steam-reforming catalysts. *Catal. Today* **2006**, *111*, 103–110. [CrossRef]
12. Kokka, A.; Katsoni, A.; Yentekakis, I.V.; Panagiotopoulou, P. Hydrogen production via steam reforming of propane over supported metal catalysts. *Int. J. Hydrogen Energy* **2020**, *45*, 14849–14866. [CrossRef]
13. Ramantani, T.; Evangeliou, V.; Kormentzas, G.; Kondarides, D.I. Hydrogen production by steam reforming of propane and LPG over supported metal catalysts. *Appl. Catal. B* **2022**, *306*, 121129. [CrossRef]
14. Schadel, B.T.; Duisberg, M.; Deutschmann, O. Steam reforming of methane, ethane, propane, butane, and natural gas over a rhodium-based catalyst. *Catal. Today* **2009**, *142*, 42–51. [CrossRef]
15. Lin, D.; Zhang, Q.; Qin, Z.; Li, Q.; Feng, X.; Song, Z.; Cai, Z.; Liu, Y.; Chen, X.; Chen, D.; et al. Reversing Titanium Oligomers Formation towards High-Efficiency and Green Synthesis of Titanium-containing Molecular Sieves. *Angew. Chem.* **2020**, *133*, 3485–3590. [CrossRef]
16. Potemkin, D.I.; Rogozhnikov, V.N.; Ruban, N.V.; Shilov, V.A.; Simonov, P.A.; Shashkov, M.V.; Sobyanin, V.A.; Snytnikov, P.V. Comparative study of gasoline, diesel and biodiesel autothermal reforming over Rh-based FeCrAl-supported composite catalyst. *Int. J. Hydrogen Energy* **2020**, *45*, 26197–26205. [CrossRef]

17. Ruban, N.V.; Potemkin, D.I.; Rogozhnikov, V.N.; Shefer, K.I.; Snytnikov, P.V.; Sobyanin, V.A. Rh- and Rh–Ni–MgO-based structured catalysts for on-board syngas production via gasoline processing. *Int. J. Hydrogen Energy* **2021**, *46*, 35840–35852. [CrossRef]
18. Porsin, A.V.; Kulikov, A.V.; Rogozhnikov, V.N.; Serkova, A.N.; Salanov, A.N.; Shefer, K.I. Structured reactors on a metal mesh catalyst for various applications. *Catal. Today* **2016**, *273*, 213–223. [CrossRef]
19. Kabin, E.V.; Emelyanov, V.A.; Vorobyev, V.A.; Alferova, N.I.; Tkachev, S.V.; Baidina, I.A. Reaction of trans-[RuNO(NH$_3$)$_4$OH]Cl$_2$ with nitric acid and synthesis of ammine(nitrato)nitrosoruthenium complexes. *Russ. J. Inorg. Chem.* **2012**, *57*, 1146–1153. [CrossRef]
20. Emelyanov, V.A.; Khranenko, S.P.; Belyaev, A.V. Nitrosation of ruthenium chloro complexes. *Russ. J. Inorg. Chem.* **2001**, *46*, 346–351.
21. Ilyin, M.A.; Emelanov, V.A.; Belyaev, A.V.; Makhinya, A.N.; Tkachev, S.V.; Alferova, N.I. New method for the synthesis of *trans*-hydroxotetraamminenitrosoruthenium (II) dichloride and its characterization. *Russ. J. Inorg. Chem.* **2008**, *53*, 1152–1159. [CrossRef]
22. Shoynkhorova, T.B.; Rogozhnikov, V.N.; Ruban, N.V.; Shilov, V.A.; Potemkin, D.I.; Simonov, P.A.; Belyaev, V.D.; Snytnikov, P.V.; Sobyanin, V.A. Composite Rh/Zr$_{0.25}$Ce$_{0.75}$O$_{2-\delta}$-η-Al$_2$O$_3$/FeCrAlloy Wire Mesh Honeycomb Module for Natural Gas, LPG and Diesel Catalytic Conversion to Syngas. *Int. J. Hydrogen Energy* **2019**, *44*, 9941–9948. [CrossRef]
23. Santos, A.C.S.F.; Damyanova, S.; Teixeira, G.N.R.; Mattos, L.V.; Noronha, F.B.; Passos, F.B.; Bueno, J.M.C. The effect of ceria content on the performance of Pt/CeO2/Al2O3 catalysts in the partial oxidation of methane. *Appl. Catal. A* **2005**, *290*, 123–132. [CrossRef]
24. Rocha, K.; Santos, J.; Meira, D.; Pizani, P.; Marques, C.; Zanchet, D.; Bueno, J. Catalytic partial oxidation and steam reforming of methane on La2O3–Al2O3 supported Pt catalysts as observed by X-ray absorption spectroscopy. *Appl. Catal. A* **2012**, *431–432*, 79–87. [CrossRef]
25. Bae, J.; Lee, S.; Kim, S.; Oh, J.; Choi, S.; Bae, M.; Kang, I.; Katikaneni, S.P. Liquid fuel processing for hydrogen production: A review. *Int. J. Hydrogen Energy* **2016**, *41*, 19990–20022. [CrossRef]
26. Rogozhnikov, V.N.; Potemkin, D.I.; Ruban, N.V.; Shilov, V.A.; Salanov, A.N.; Kulikov, A.V.; Simonov, P.A.; Gerasimov, E.Y.; Sobyanin, V.A.; Snytnikov, P.V. Post-mortem characterization of Rh/Ce0.75Zr0.25O2/Al2O3/FeCrAl wire mesh composite catalyst for diesel autothermal reforming. *Mater. Lett.* **2019**, *257*, 126715. [CrossRef]
27. Zazhigalov, S.V.; Shilov, V.A.; Rogozhnikov, V.N.; Potemkin, D.I.; Sobyanin, V.A.; Zagoruiko, A.N.; Snytnikov, P.V. Modeling of hydrogen production by diesel reforming over Rh/Ce0.75Zr0.25O2-δ-η-Al2O3/FeCrAl wire mesh honeycomb catalytic module. *Catal. Today* **2021**, *378*, 240–248. [CrossRef]
28. Zazhigalov, S.V.; Rogozhnikov, V.N.; Snytnikov, P.V.; Potemkin, D.I.; Simonov, P.A.; Shilov, V.A.; Ruban, N.V.; Kulikov, A.V.; Zagoruiko, A.N.; Sobyanin, V.A. Simulation of diesel autothermal reforming over Rh/Ce0.75Zr0.25O2-δ-η-Al2O3/FeCrAl wire mesh honeycomb catalytic module. *Chem. Eng. Process.* **2020**, *150*, 107876. [CrossRef]
29. Vita, A.; Pino, L.; Cipitì, F.; Laganà, M.; Recupero, V. Structured reactors as alternative to pellets catalyst for propane oxidative steam reforming. *Int. J. Hydrogen Energy* **2010**, *35*, 9810–9817. [CrossRef]
30. Aartun, I.; Gjervan, T.; Venvik, H.; Görke, O.; Pfeifer, P.; Fathi, M.; Holmen, A.; Schubert, K. Catalytic conversion of propane to hydrogen in microstructured reactors. *Chem. Eng. J.* **2004**, *101*, 93–99. [CrossRef]
31. Gur, T.M. Comprehensive review of methane conversion in solid oxide fuel cells: Prospects for efficient electricity generation from natural gas. *Prog. Energy Combust. Sci.* **2016**, *54*, 1–64. [CrossRef]
32. Baldinelli, A.; Barelli, L.; Bidini, G. Performance characterization and modelling of syngas-fed SOFCs (solid oxide fuel cells) varying fuel composition. *Energy* **2015**, *90*, 2070–2084. [CrossRef]
33. Yi, Y.; Rao, A.D.; Brouwer, J.; Samuelsen, G.S. Fuel flexibility study of an integrated 25 kW SOFC reformer system. *J. Power Sources* **2005**, *144*, 67–76. [CrossRef]

Article

Robust Porous TiN Layer for Improved Oxygen Evolution Reaction Performance

Gaoyang Liu [1,2,*], Faguo Hou [1,2], Xindong Wang [1,2] and Baizeng Fang [3,*]

1. State Key Laboratory of Advanced Metallurgy, University of Science and Technology Beijing, Beijing 100083, China
2. School of Metallurgical and Ecological Engineering, University of Science and Technology Beijing, Beijing 100083, China
3. Department of Chemical and Biological Engineering, University of British Columbia, 2360 East Mall, Vancouver, BC V6T 1Z3, Canada
* Correspondence: liugy@ustb.edu.cn (G.L.); baizengfang@163.com (B.F.)

Abstract: The poor reversibility and slow reaction kinetics of catalytic materials seriously hinder the industrialization process of proton exchange membrane (PEM) water electrolysis. It is necessary to develop high-performance and low-cost electrocatalysts to reduce the loss of reaction kinetics. In this study, a novel catalyst support featured with porous surface structure and good electronic conductivity was successfully prepared by surface modification via a thermal nitriding method under ammonia atmosphere. The morphology and composition characterization-confirmed that a TiN layer with granular porous structure and internal pore-like defects was established on the Ti sheet. Meanwhile, the conductivity measurements showed that the in-plane electronic conductivity of the as-developed material increased significantly to 120.8 S cm^{-1}. After IrO$_x$ was loaded on the prepared TiN-Ti support, better dispersion of the active phase IrO$_x$, lower ohmic resistance, and faster charge transfer resistance were verified, and accordingly, more accessible catalytic active sites on the catalytic interface were developed as revealed by the electrochemical characterizations. Compared with the IrO$_x$/Ti, the as-obtained IrO$_x$/TiN-Ti catalyst demonstrated remarkable electrocatalytic activity ($\eta_{10\ mA\ cm^{-2}}$ = 302 mV) and superior stability (overpotential degradation rate: 0.067 mV h^{-1}) probably due to the enhanced mass adsorption and transport, good dispersion of the supported active phase IrO$_x$, increased electronic conductivity and improved corrosion resistance provided by the TiN-Ti support.

Keywords: proton exchange membrane water electrolysis; oxygen evolution reaction; iridium oxide; titanium nitride; thermal nitriding

1. Introduction

Hydrogen is a green energy carrier which can be sustainably produced by various strategies such as photocatalysis [1–6] and electrocatalysis [7–9]. Amongst the diverse hydrogen production approaches, especially compared with the alkaline water electrolysis [10], proton exchange membrane (PEM) water electrolysis has the advantages of fast start, high current densities and energy efficiency, and low gas crossover, and is thus considered to be the most promising technology for electrocatalytic hydrogen production [11]. The electrolytic splitting of water involves two half reactions including hydrogen evolution reaction (HER) and oxygen evolution reaction (OER), and the slow kinetic process of OER on the surface of the anode limits greatly the performance improvement of PEM water electrolysis [12]. In order to compensate for the loss of electrode kinetics, PEM water electrolysis generally uses precious metals (e.g., Ir- and Ru-based materials) as catalysts, and the expensive price of catalysts is also one of the important factors restricting the development of PEM water electrolysis [13]. In order to reduce energy consumption and improve electrode performance, electrocatalysts with high catalytic activity and stable electrochemical properties are highly required.

A potential way to improve the electrocatalytic activity of OER and reduce the usage of noble metals is to load the catalytic active phase on the support [14,15]. The biggest advantage of supported catalysts is that the good dispersion of the active component can be achieved by improving the microstructure of the support material itself, thereby increasing the catalytic activity area. Meanwhile, it was reported the interactive effect between the conductive support and active phases may also accelerate the catalytic activity [16–18]. Currently, due to the corrosion caused by the high overpotential and oxygen evolution environment, ordinary carbon supports are no longer applicable. Ideal supports are required to have good corrosion resistance, good conductivity and good binding force with the active components. Increasing the potential of the OER and the oxygen-rich environment will lead to corrosion loss of the support materials; therefore, the choice of supports material can only be limited to some corrosion-resistant oxides or ceramic materials. At present, the supports under investigation mainly include SiC-Si [16] with low electron conductivity, $TinO_{2n-1}$ [17] antimony-doped tin oxide (ATO) [18] and TiC [19], etc. It was reported that with the improvement of the electron conductivity of the support, even though the loading proportion of the active component decreased from 90 wt% (SiC-Si) to 20 wt% ($TinO_{2n-1}$, ATO and TiC), and the supported catalyst still has an OER electrocatalytic activity comparable to that of the pure active component.

Other than the nanoparticle structured support, various nano- and porous structured supports have been widely researched, e.g., three-dimensional ordered macropore (3-DOM) [20], nanowires [21], etc. Compared with the nanoparticle support, the electrocatalytic activity was significantly improved due to the enhancement of the catalytic activity area and the better gas-liquid transport channel provided by the porous structure [15]. However, there are still many problems when utilizing these catalysts for electrode preparation. A crushing ultrasonic process could destroy the original microstructures. In addition, the additives of the ionomer and binder which are gas proof and electronic insulated material will lead to the poor electronic conductivity and bad mass diffusion. Recently, researchers have devoted time to developing porous matrix materials with high exposure of a large active surface area, a high electronic conductivity, and a good corrosion resistance, which are favorable for the OER. Various nanostructured supports, such as ordered porous layer [22] array [23,24], cross-linked nanowires [25], etc. have been established on carbon paper (CP), carbon cloth (CC), Ti felt, etc., and the as-obtained supports were then used to load the active phase as the integrated porous electrode [26,27]. On the one hand, the microstructure of the synthesized integrated porous electrode can be maintained and can contribute to the improved active area as well as fast mass transport. On the other hand, it avoids the usage of ionomer, which is conducive to electron transport and gas-liquid transport. It can be expected that the dispersion of the active phases, the electronic conductivity as well as the porous structure of the support have an important impact on the catalytic activity of the catalyst for the OER. Therefore, it is necessary to design catalysts from the surface composition and the microstructure regulation of the support by the optimization of the preparation method.

Recently, titanium and its alloys have been widely used to produce the bipolar plates (BPs) and the liquid and gas diffusion layer (LGDL) due to their features of excellent electronic conductivity and corrosion resistance, and they are also believed to be a potential electrocatalytic matrix material [28,29]. However, a big issue of titanium and its alloys is that a passive film with poor conductivity forms in the working environment and leads to the reduction in performance and durability [30].

The nitriding of the titanium surface has been proved to effectively reduce the contact resistance of the electrode. Especially, the nitriding under ammonia as a nitrogen source could obtain a high-quality (low oxygen, high conductivity, corrosion resistance) titanium nitride surface [31,32]. Meanwhile, the strong alkalinity and high reducibility of ammonia may have an impact on the microstructure and porosity of titanium nitride surface, which contributes to the dispersion of the active phases and thus improved active surface area [33]. In this study, the modifications of the surface composition (titanium nitride coating) as well as the surface microstructure (porous structure) of the titanium support were realized via a modified thermal nitriding method under ammonia environment. Then, the titanium nitride coated support was further loaded with the active components to fabricate an integrated porous electrode towards the OER on anode in PEM water electrolysis. From the composition and microstructure aspects, the use of the novel support not only optimizes the dispersion of the active phases, the charge as well as the mass transfer process of the OER, and finally enhances the electrocatalytic activity, but also greatly reduces the amount of noble metal used, and effectively decreases the cost.

2. Experimental Section

2.1. Materials

The commercial titanium alloy (Ti-6Al-4V) sheet was purchased from Xiamen Tmax Battery Equipments Limited, China. Nafion® perfluorinated resin solution (5 wt%) was purchased from Sigma-Aldrich (Shanghai, China). Iridium chloride acid ($H_2IrCl_6 \cdot H_2O$) and IrO_2 powders were acquired from Alfa Aesar (Shanghai, China). All the other chemicals used in the present study were purchased from Sinopharm Chemical Reagent Beijing Co., Ltd. (Beijing, China), and used as-received without further purification.

2.2. Synthesis of TiN and IrO_x/Ti

Pretreatment of the Ti sheet [34]: First, a 1 mm thick Ti sheet was cut into small pieces with a size of 10 mm × 10 mm, and then the samples were polished with 400#, 800#, 1000#, 1500# grit paper (SiC) and polishing cloth, sequentially. Next, the samples were washed by ultra-sonication in acetone, ethanol and deionized (DI) water for 30 min each time, and then dried under vacuum at 70 °C.

Nitriding of the Ti sheet: NH_3 gas was applied as the N source. Instead of using stainless steel foil, the pretreated samples were placed in a corundum crucible. Before the nitriding experiment, the crucible was placed in a drying oven and then transferred into a tube furnace filled with high-purity N_2 atmosphere for 1 h. The nitriding process was carried out in the tube furnace filled with high-purity NH_3 for 1 h at room temperature followed by heating to 600 °C at a heating rate of 5 °C min^{-1}, and nitriding for 2 h. The samples were taken out after cooling to ambient temperature. The obtained nitrided Ti sheet was marked as TiN-Ti.

Synthesis of IrO_x/TiN-Ti [25]: to load IrO_x on the nitrided Ti (TiN) sheet, 500 mg of $H_2IrCl_6 \cdot H_2O$ was firstly added to 3 mL of methanol solution, and the solution was sonicated for 1 h. Then, a piece of dried TiN sheet (1 × 1 cm^2) was soaked in the $H_2IrCl_6 \cdot H_2O$ solution for 30 min to obtain $H_2IrCl_6 \cdot H_2O$/TiN. Afterwards, $H_2IrCl_6 \cdot H_2O$/TiN-Ti was thermally annealed at 500 °C for 30 min in air. The loading density of IrO_x on the TiN-Ti could be adjusted by repeating the above-mentioned soaking-annealing processes, and the final loading of IrO_x was ca. 0.2 mg cm^{-2}. For comparison, the electrode using the un-nitrided Ti sheet as a support was prepared according to the procedures similar to those described above. The obtained IrO_x supported on the nitriding Ti sheet and the n-nitriding Ti sheet was marked as IrO_x/TiN-Ti and IrO_x/Ti, respectively.

2.3. Physiochemical Characterizations

The crystalline structures of the prepared samples were characterized by X-ray diffraction (XRD) using a Marcogroup diffractometer (MXP21 VAHF) with a Cu-Kα radiation source (λ = 1.54056 Å) to characterize catalyst crystalline structure. The morphology, particle size and composition information of samples were studied using Scanning electron microscopy (ZEISS and LEO-1530 FESEM) with EDS. The surface chemical states of the as-synthesized samples were studied by X-ray photoelectron spectroscopy (XPS) (Kratos AXIS Ultra DLD).

2.4. Electrochemical Characterizations

All the electrocatalytic tests were carried out in a three-electrode configuration in 0.5 M H_2SO_4 electrolyte solution at room temperature using an energylab XM electrochemical workstation. The as-synthesized IrO_x/TiN-Ti and IrO_x/Ti were directly used as the working electrode (WE). A 10 × 10 mm platinum plate was used as the counter electrode (CE), and Ag/AgCl (saturated) was used as the reference electrode (RE). The Ag/AgCl (saturated) reference was calibrated prior to each measurement in Ar/H_2-saturated 0.5 M H_2SO_4 solution using a clean Pt wire as the working electrode. In addition, prior to the electrochemical measurements, the electrochemical cell was purged with nitrogen to completely remove the air in the electrolyte. All electrochemical tests were performed at room temperature. All potentials in this paper were converted to reversible hydrogen electrode (RHE) reference potential according to literature [22]. For the commercial IrO_2 powder, the catalyst ink was prepared by ultrasonic treatment of the mixture of the electrocatalyst (1 mg), ethanol (5 mL) and Nafion (5 wt%, 50 µL) in an ice bath for 2 h. The working electrode (IrO_2 loading was about 0.2 mg/cm^2) was made by casting uniformly dispersed catalyst ink onto a glassy carbon electrode (area 0.283 cm^2) and being dried in air.

The details for the test protocol are as follows. First, the potentiostatic electrochemical impedance spectroscopy (PEIS) was measured at open circuit voltage (OCV) after the system was stable. Then, 50 cycles of cyclic voltammetry (CV) were recorded from 0 to 1 V (vs. RHE) at a scan rate of 50 mV/s to stabilize the OER performance of the electrocatalyst. Next, the electrochemically active area (ECSA) was obtained by CV tests at different scanning speeds (3, 5, 7, 10, 20, 50, 100, 200, 300 mV s^{-1}) in the non-Faraday potential range (OCV ± 50 mV). After that, the linear scan voltammetry (LSV) tests were carried out in a range of 1.2–1.6 V vs. RHE. Finally, the PEIS at 1.53 V vs. RHE with amplitude of 5 mV was carried out. The stability of catalysts was assessed at a constant current density of 10 mA cm^{-2} using chronopotentiometry (CP).

3. Results and Discussion

Figure 1a,b show the SEM images of the pretreated Ti sheet before and after the thermal nitriding under NH_3 atmosphere. It can be seen that the Ti sheet that has not been nitrided appears as a flat and smooth surface, and there are still scratches left by the mechanical polishing. No obvious defects and porous structures were observed. The scratches on the surface of the Ti sheet after the nitriding under NH_3 atmosphere have been completely covered by the nitride coating. In addition, the surface of the coating after nitriding is granular porous structure mainly due to the strong alkalinity and high reducibility of NH_3. Meanwhile, the high temperature thermal expansion can lead to rich internal pore-like defects between the particles, but no obvious gaps and cracks appeared. The EDS spectra shown in Figure 1c reveal that the elements present in the nitriding sample are mainly Ti, V and N. From the EDS elemental mappings of the TiN-Ti, both the Ti and N are uniformly distributed, but the specific phases in which they exist need to be further analyzed. Figure 1f,g show the cross-section SEM image and the corresponding EDS liner scans after the thermal nitriding. The results showed that there were significant pore-like defects inside the coating with a thickness of about 2.5 µm. The presence of porous morphology and defects will help to disperse the active components when used as the supporting material, which may significantly enhance the adhesion of the active phases and also increase the number of active sites that can catalyze the OER process [35].

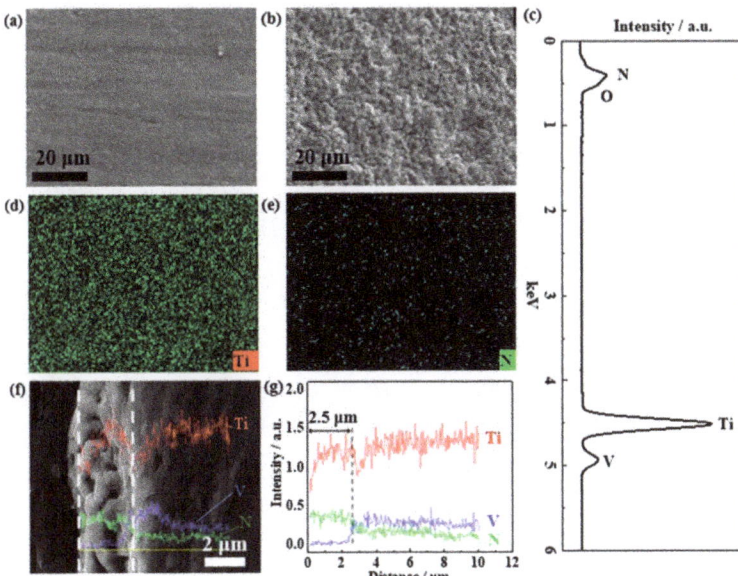

Figure 1. SEM images: (**a**) Ti sheet after the mechanical polishing, (**b**) TiN-Ti. (**c**) EDS spectra of the TiN-Ti, EDS elemental mappings of the TiN-Ti: (**d**) Ti, (**e**) N; (**f**) Cross-section SEM image and (**g**) EDS elemental liner scan of the TiN-Ti.

The surface chemical states of the Ti sheets before and after the thermal nitriding under NH_3 atmosphere were further investigated using X-ray photoelectron spectroscopy (XPS). The XPS survey spectra in Figure 2a confirm the presence of the corresponding N element in the TiN/Ti, but no N element in the Ti sheet. It indicated that the successful nitriding of the Ti surface. Figure 2b shows the XRD patterns of the Ti sheets before and after the thermal nitriding. The phase composition of the Ti sheet is mainly α-Ti (pdf#44-1294) [36], while the nitrided sample has a new phase, TiN (pdf#38-1420) [37]. In the XRD patterns, two distinct diffraction peaks were found at diffraction angles of 36.7° and 42.7°, corresponding to the (111) and (200) facets of TiN, respectively. Therefore, the SEM, EDS, XPS and XRD results confirmed that a TiN layer with rich granular porous structure and internal pore-like defects was successfully established on the Ti sheet. Meanwhile, the conductivity measurements by the four-probe method showed that the in-plane electronic conductivity increased from 2.6 S cm^{-1} of the unmodified Ti sheet to 120.8 S cm^{-1} of the Ti sheet after the thermal nitriding. The improved in-plane electronic conductivity of the TiN-Ti was further confirmed with the ICR measurement as reported by Wang et al. As it is shown in Figure 2c,d, the ICR of the TiN-Ti is significantly decreased to 3.4 mΩ cm^2, and lower than the unmodified Ti sheet (20.8 mΩ cm^2). It should be noted that even though the unmodified Ti sheet is pretreated as illustrated in the experimental part, there still could be an oxidized TiO$_x$ layer, which would result in lower in-plane electronic conductivity. The improved in-plane electronic conductivity of the TiN-Ti can be ascribed to the formation of the highly conductive TiN phase. Overall, both the promoted porous structure and enhanced electronic conductivity may contribute to more accessible active sites when the TiN-Ti is used as a support to load the active phases.

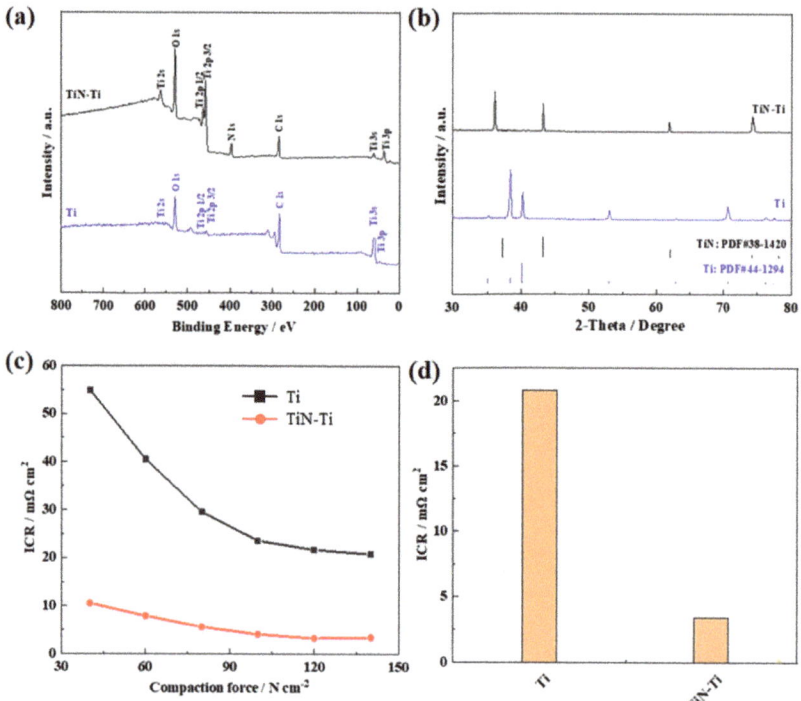

Figure 2. (a) XPS survey spectra, and (b) XRD patterns of the Ti sheet and TiN-Ti, (c) Relationship between the ICR and the compaction force, (d) ICR at 140 N cm^{-2} of the Ti and the TiN-Ti.

Figure 3a–d show the SEM images of the prepared IrO$_x$/Ti and IrO$_x$/TiN-Ti, respectively. It can be seen that catalyst layers were successfully loaded on both the Ti sheet and TiN-Ti with the soaking-annealing process, and the morphology of the IrO$_x$/Ti and IrO$_x$/TiN-Ti did not markedly alter after loading IrO$_x$ nanoparticles. For the IrO$_x$/Ti, it can be seen from Figure 3a,b that the surface remained flat but cracks and pits appeared, and it can be associated with the generated gas during the pyrolysis reaction of IrO$_x$. While for the IrO$_x$/TiN-Ti as shown in Figure 3c,d, the porous structure and rough surface can still be maintained and IrO$_x$ nanoparticles with a typical diameter of around 80–100 nm are uniformly distributed on the TiN-Ti surface. Even though the porous structure decreased after loading IrO$_x$ during the soaking-annealing process, the porous structure of the TiN-Ti could result in completely different morphology of the catalytic layers. It is believed that better dispersion of IrO$_x$ can be achieved with the TiN-Ti, and thus better ECSA and catalytic activity. The insert in Figure 3d presents the elemental mapping images and they illustrated that Ir is distributed uniformly on the TiN support with a very high density. It can be deduced that the TiN-Ti support can effectively promote the dispersion of IrO$_x$ than the un-nitrided Ti sheet.

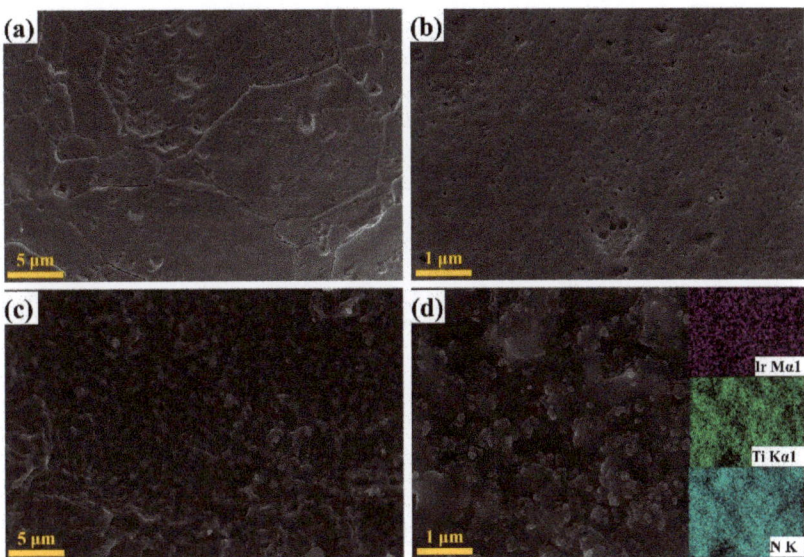

Figure 3. SEM images: (**a,b**) IrO$_x$/Ti, (**c,d**) IrO$_x$/TiN-Ti, and the inset of (**d**) is the EDS elemental mapping images of the TiN-Ti.

In order to evaluate the effect of different supports on the catalytic activity of the active phase towards the OER, the specific activity based on the same IrO$_x$ loading was used in the following discussion. The three-electrode test was carried out in 0.5 mol L^{-1} H$_2$SO$_4$ at 25 °C. The working electrode was the prepared IrO$_x$/Ti, IrO$_x$/TiN-Ti, and GC supported commercial IrO$_2$ (marked as C-IrO$_2$), and the CV scan was carried out in the double layer potential window range (0–1 V vs. RHE) and the CV test results are presented in Figure 4a. It can be seen that all electrodes showed a similar IrO$_2$ redox process, i.e., there are two pairs of redox peaks. The redox peaks located near 0.6 V correspond to Ir$_{III}$/Ir$_{IV}$ redox pairs, and redox peaks located near 0.95 V correspond to Ir$_{IV}$/Ir$_{VI}$ redox pairs [38]. The catalytic activity area can be evaluated by integrating the voltammogram in the anode part of the CV curve obtained at a scanning speed of 20 mV s^{-1}, and the calculated voltametric charges are shown in Table 1, and they were considered to be associated with the number of active sites or the catalytic active area [39]. It shows that the voltametric charges of the IrO$_x$/TiN-Ti are greater than that of the IrO$_x$/Ti, which can be attributed to the rich porous structure and internal pore-like defects possessed by the IrO$_x$/TiN-Ti, which can form a good dispersion of the IrO$_x$ nanoparticles.

Table 1. Summary of electrochemical characterization of different electrodes.

	CV			EIS		Tafel Slopes
	Q [a]	Q$_t$ [a]	Q$_o$ [a]	R$_\Omega$ [b]	R$_{ct}$ [b]	S1 [c]
IrO$_x$/TiN-Ti	340	351	303	0.493	1.331	66
IrO$_x$/Ti	112	117	96	0.562	2.781	97
C-IrO$_2$	52	53	37	0.604	3.330	107

Notes: [a] The anodic charge (Q/mC cm^{-2}), total charge (Q$_t$/mC cm^{-2}), outer charge (Q$_o$/mC cm^{-2}) of all prepared electrodes calculated from the cyclic voltammograms. [b] Ohmic resistance (R$_\Omega$/ohm cm^2), charge-transfer resistance (R$_{ct}$/ohm cm^2). [c] Tafel slopes (mV dec^{-1}).

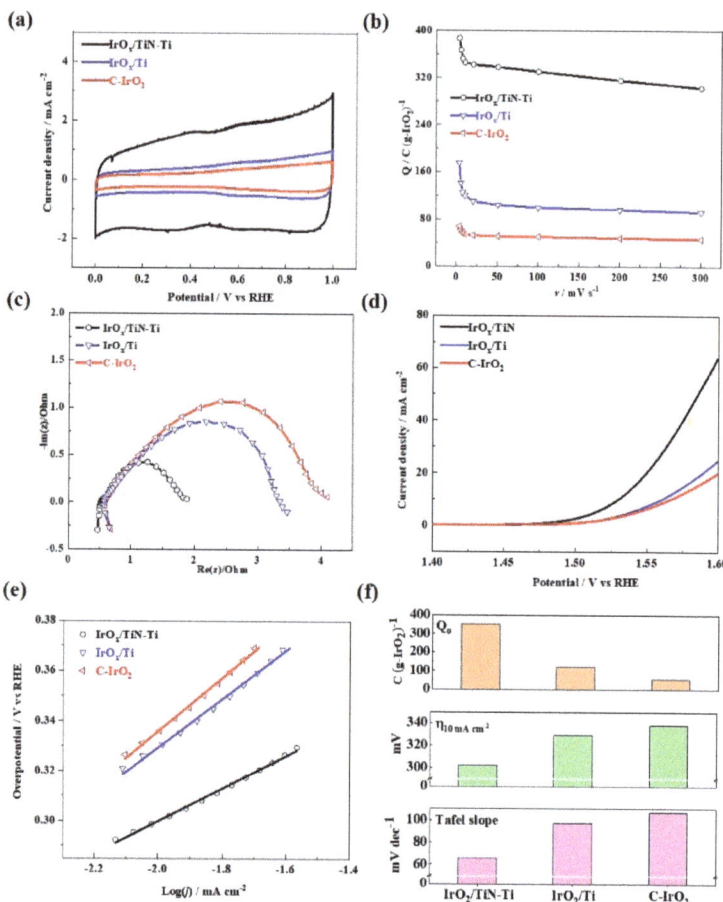

Figure 4. (a) Representative CVs recorded with a scan rate of 20 mV s^{-1}, (b) The plots of the charge as a function of the scan rate, (c) Nyquist diagram measured at 1.53 V vs. RHE during oxygen evolution, the inset is the equivalent circuit, (d) iR-corrected steady-state polarization curves, (e) Tafel plots and (f) Comparison of Q_o, $\eta_{10\ mA\ cm^{-2}}$, Tafel slopes of the IrO$_x$/TiN-Ti, IrO$_x$/Ti and C-IrO$_2$. All the electrochemical measurements were tested in 0.5 mol L^{-1} H$_2$SO$_4$ at 25 °C.

Figure 4b shows the voltametric charges for the C-IrO$_2$ and the supported catalysts obtained at different potential scan rates. It could be seen that the charges show a significant decrease trend as the sweep speed increases, which is due to the hysteresis of charge transfer or the relatively slow transfer speed [38]. At high sweep speeds, charge transfer can only occur at the active interface ("outer" surface) in direct contact with the electrolyte, corresponding to the outer surface charge (Q_o); At low sweep speed, proton exchange energy occurs at the active interface of the entire electrode (the "internal" interface and the "outside" surface of the proton conduction) slowly, corresponding to the total electrode charge (Q_t) [18,21]. Table 1 and Figure 4f list the calculated Q_o and Q_t of the synthetic electrodes according to literature. Compared with the IrO$_x$/Ti, both the Q_o and Q_t have been significantly improved, indicating that the unique porous structure and internal pore-like defects in the TiN-Ti promote the dispersion of the active components, as well as the accessibility of the catalytic sites. In addition, according to the literature [40–45], the presence of a porous structure with abundant macropores (or mesopores with a large pore size) in the catalyst support (i.e., TiN-Ti) facilitates fast mass transport, resulting in improved electrocatalytic performance.

Figure 4c shows the EIS behavior of the different electrodes, and the simulation data results obtained using the equivalent circuit in the insert are listed in Table 1. As shown in Table 1, compared with the IrO$_x$/Ti and the C-Ti, the R_Ω of the IrO$_x$/TiN-Ti is significantly decreased, which is mainly due to the fact that the conductivity of the TiN-Ti is lower than that of the unmodified Ti sheet. It has been reported that the Ti surfaces can be easily oxidized and the passive film with poor conductivity could be formed in the anodic working environment [32,37]. For the different electrodes, it can be seen that compared with the C-IrO$_2$, the R_{ct} decreased for both the supported electrodes IrO$_x$/Ti and IrO$_x$/TiN-Ti, indicating the charge transfer has been promoted due to the interactive effect between the support and active phases [17,18]. Meanwhile, the lowest R_{ct} was obtained with the IrO$_x$/TiN-Ti. It can be deduced that both the optimized microstructure and the enhanced electronic conductivity of the porous TiN layer are conducive to the charge transfer process at the heterogeneous catalytic interface.

Steady state polarization curves of the three prepared electrodes were recorded in the potential region for the OER (1.4–1.6 V vs. RHE) as shown in Figure 4d. At a current density of 10 mA cm^{-2}, the overpotentials of the IrO$_x$/TiN-Ti, IrO$_x$/Ti, and C-IrO$_2$ were 302 mV, 329 mV, and 338 mV, respectively (Figure 4f). Compared with the recently reported electrocatalysts towards the OER, the IrO$_x$/TiN-Ti in this study shows remarkable catalytic activity [46–48]. The Tafel slope reflects the reaction mechanism of the OER process, and it is generally used to evaluate reaction kinetics of electro-catalytic materials. Figure 4e,f show the Tafel curves for the different electrodes. Generally, the IrO$_x$/TiN-Ti exhibited better kinetic parameters than the IrO$_x$/Ti and C-IrO$_2$. The IrO$_x$/TiN-Ti shows the smallest Tafel slopes (ca. 66 mV dec^{-1}). It suggests the reaction mechanism due to the adsorption contribution of intermediates. Overall, the loading of IrO$_x$ nanoparticles on the TiN-Ti could accelerate the reaction kinetics, and thus remarkably enhance the catalytic activity.

The stability is a key challenge for electrocatalytic materials to meet the demanding targets of practical applications in PEM water electrolyzers [49]. The electrochemical stability of the prepared electrodes was evaluated by chronopotentiometric (CP) method at a constant current density of 10 mA cm^{-2} in 0.5 M H$_2$SO$_4$ electrolyte. The results are shown in Figure 5a.

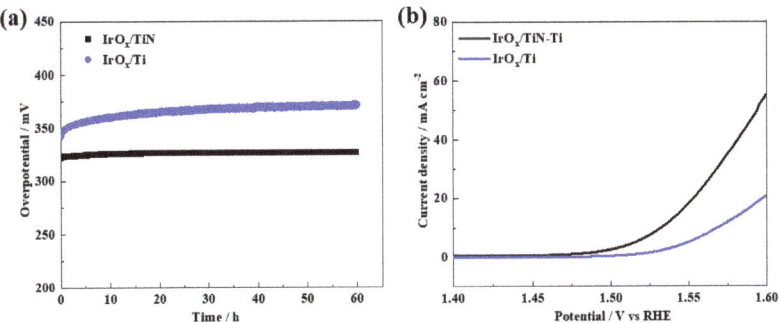

Figure 5. (**a**) Chronopotentiometric curves of the IrO$_x$/TiN-Ti and IrO$_x$/Ti recorded at a constant current density of 10 mA cm^{-2}. (**b**) iR-corrected steady-state polarization curves of the IrO$_x$/TiN-Ti and IrO$_x$/Ti after 60 h of OER.

During the first 10 h of stability testing, the overpotential required to achieve current densities of 10 mA cm^{-2} increases significantly for both the IrO$_x$/Ti and IrO$_x$/TiN-Ti mainly due to the bubble effect, which reduced the catalyst utilization. After this stable period, the IrO$_x$/TiN-Ti shows negligible degradation and good stability until 60 h, and the overpotential required to achieve current density of 10 mA cm^{-2} slightly increases from 322 mV to 326 mV. The whole process of the overpotential degradation rate was only 0.067 mV h^{-1}, which is much smaller than that of the IrO$_x$/Ti (from 342 mV to 371 mV, and the overpotential degradation rate was 0.483 mV h^{-1}). It indicated that the IrO$_x$/TiN-Ti can continue to operate stably without a sharp increase. Figure 5b presents the steady-state polarization curves of the IrO$_x$/TiN-Ti and IrO$_x$/Ti after 60 h of the OER. Compared with Figure 4d, there is less attenuation of catalytic activity before and after the 60 h stability test for the IrO$_x$/TiN-Ti than IrO$_x$/Ti. Overall, the IrO$_x$/TiN-Ti has been proved to show remarkable electrocatalytic activity and superior stability. The application of support materials with rich porous microstructure and the enhanced surface conductivity could have important consequences for PEM water electrolysis technology and production of green hydrogen.

4. Conclusions

In this study, a TiN layer with rich granular porous structure and internal pore-like defects was successfully established on the Ti sheet via the thermal nitriding method under NH$_3$ atmosphere. The strong alkalinity and high reducibility of NH$_3$ not only result in porous microstructured surfaces, but also lead to a high-quality TiN layer with sufficient electronic conductivity. All these features were characterized and verified by SEM, EDS, XPS and XRD analyses. Furthermore, the novel TiN-Ti was used as a support for loading IrO$_x$ active phases towards the OER. The electrochemical tests revealed that significant enhancement of the OER activity was obtained for the IrO$_x$/TiN-Ti compared with the un-nitrided TiN supported catalysts. The remarkable catalytic activity is ascribed to the more accessible active sites created as well as the faster charge transfer on the catalytic interfaces, which are mainly due to the good dispersion, and high electronic conductivity provided by the porous structured TiN layer. The IrO$_x$/TiN-Ti also showed good durability under a current density of 10 mA cm^{-2} in a period of 60 h due to the high corrosion resistance of the TiN.

Author Contributions: Conceptualization, G.L. and F.H.; methodology, F.H.; software, G.L. and F.H.; validation, G.L. and F.H.; formal analysis, G.L. and F.H.; investigation, G.L. and F.H.; resources, G.L.; data curation, G.L. writing—original draft preparation, G.L. and B.F.; writing—review and editing, G.L., B.F. and X.W.; visualization, G.L. and B.F.; supervision, G.L.; project administration, G.L.; funding acquisition, G.L. All authors have read and agreed to the published version of the manuscript.

Funding: This study was financially supported by Fundamental Research Funds for the Central Universities (No. FRF-TP-20-010A1), and the Communication Program for Young Scientist in USTB (No. QNXM20220015).

Data Availability Statement: Not applicable.

Conflicts of Interest: The authors declare no conflict of interest.

References

1. Liao, G.; Gong, Y.; Zhang, L.; Gao, H.; Yang, G.; Fang, B. Semiconductor polymeric graphitic carbon nitride photocatalysts: The "holy grail" for the photocatalytic hydrogen evolution reaction under visible light. *Energy Environ. Sci.* **2019**, *12*, 2080–2147. [CrossRef]
2. Liao, G.; Li, C.; Li, X.; Fang, B. Emerging polymeric carbon nitride Z-scheme systems for photocatalysis. *Cell Rep. Phys. Sci.* **2021**, *2*, 100355. [CrossRef]
3. Shandilya, P.; Sambyal, S.; Sharma, R.; Mandyal, P.; Fang, B. Properties, optimized morphologies and advanced strategies for enhanced photocatalytic applications of WO$_3$ based photocatalysts. *J. Hazard. Mater.* **2022**, *428*, 128218. [CrossRef] [PubMed]
4. Liao, G.; Li, C.; Liu, S.; Fang, B.; Yang, H. Z-scheme systems: From fundamental principles to characterization, synthesis, and photocatalytic fuel-conversion applications. *Phys. Rep.* **2022**, *983*, 1–41. [CrossRef]
5. Liao, G.; Li, C.; Liu, S.; Fang, B.; Yang, H. Emerging frontiers of Z-scheme photocatalytic systems. *Trends Chem.* **2022**, *4*, 111–127. [CrossRef]
6. Liao, G.; Tao, X.; Fang, B. An innovative synthesis strategy for highly efficient and defects-switchable hydrogenated TiO$_2$ photocatalysts. *Matter* **2022**, *5*, 377–379. [CrossRef]

7. Liu, G.; Hou, F.; Peng, S.; Wang, X.; Fang, B. Synthesis, physical properties and electrocatalytic performance of nickel phosphides for hydrogen evolution reaction of water electrolysis. *Nanomaterials* **2022**, *12*, 2935. [CrossRef]
8. Lu, L.; Zheng, H.; Li, Y.; Zhou, Y.; Fang, B. Ligand-free synthesis of noble metal nanocatalysts for electrocatalysis. *Chem. Eng. J.* **2023**, *451*, 138668. [CrossRef]
9. Lu, L.; Zou, S.; Fang, B. The critical impacts of ligands on heterogeneous nanocatalysis: A review. *ACS Catal.* **2021**, *11*, 6020–6258. [CrossRef]
10. Sahoo, P.; Tan, J.-B.; Zhang, Z.-M.; Singh, S.K.; Lu, T.-B. Engineering the Surface Structure of Binary/Ternary Ferrite Nanoparticles as High-Performance Electrocatalysts for the Oxygen Evolution Reaction. *ChemCatChem* **2018**, *10*, 1075. [CrossRef]
11. Carmo, M.; Fritz, D.; Mergel, J.; Stolten, D. A comprehensive review on PEM water electrolysis. *Int. J. Hydrogen Energy* **2013**, *38*, 4901–4934. [CrossRef]
12. Li, X.; Hao, X.; Abudula, A.; Guan, G. Nanostructured catalysts for electrochemical water splitting: Current state and prospects. *J. Mater. Chem. A* **2016**, *4*, 11973–12000. [CrossRef]
13. Song, J.; Wei, C.; Huang, Z.; Liu, C.; Zeng, L.; Wang, X.; Xu, Z. A review on fundamentals for designing oxygen evolution electrocatalysts. *Chem. Soc. Rev.* **2020**, *49*, 2196–2214. [CrossRef] [PubMed]
14. Liao, G.; Fang, J.; Li, S.; Li, Q.; Xu, Z.; Fang, B. Ag-based nanocomposites: Synthesis and applications in catalysis. *Nanoscale* **2019**, *11*, 7062–7096. [CrossRef]
15. Hao, C.; Lv, H.; Mi, C.; Song, Y.; Ma, J. Investigation of mesoporous niobium-doped TiO_2 as an oxygen evolution catalyst support in an SPE water electrolyzer. *ACS Sustain. Chem. Eng.* **2016**, *4*, 746–756. [CrossRef]
16. Polonský, J.; Mazúr, P.; Paidar, M.; Bouzek, K. Investigation of β-SiC as an anode catalyst support for PEM water electrolysis. *J. Solid State Electrochem.* **2014**, *18*, 2325–2332. [CrossRef]
17. Krishnan, P.; Advani, S.G.; Prasad, A.K. Magneli phase Ti_nO_{2n-1} as corrosion-resistant PEM fuel cell catalyst support. *J. Solid State Electrochem.* **2012**, *16*, 2515–2521. [CrossRef]
18. Xu, J.; Aili, D.; Li, Q.; Pan, C.; Christensen, E.; Jensen, J.O.; Zhang, W.; Liu, G.; Wang, X.; Bjerrum, N.J.; et al. Antimony doped tin oxide modified carbon nanotubes as catalyst supports for methanol oxidation and oxygen reduction reactions. *J. Mater. Chem. A* **2013**, *1*, 9737–9745. [CrossRef]
19. Ma, L.; Sui, S.; Zhai, Y. Preparation and characterization of Ir/TiC catalyst for oxygen evolution. *J. Power Sources* **2008**, *177*, 470–477. [CrossRef]
20. Xu, J.; Aili, D.; Li, Q.; Christensen, E.; Jensen, J.O.; Zhang, W.; Hansen, M.K.; Liu, G.; Wang, X.; Bjerrum, N.J. Oxygen evolution catalysts on supports with a 3-D ordered array structure and intrinsic proton conductivity for proton exchange membrane steam electrolysis. *Energy Environ. Sci.* **2014**, *7*, 820–830. [CrossRef]
21. Liu, G.; Xu, J.; Wang, Y.; Wang, X. An oxygen evolution catalyst on an antimony doped tin oxide nanowire structured support for proton exchange membrane liquid water electrolysis. *J. Mater. Chem. A* **2015**, *3*, 20791–20800. [CrossRef]
22. Liu, G.; Yang, Z.; Wang, X.; Fang, B. Ordered porous TiO_2@C layer as an electrocatalyst support for improved stability in PEMFCs. *Nanomaterials* **2021**, *11*, 3462. [CrossRef] [PubMed]
23. Lei, C.; Zheng, Q.; Cheng, F.; Hou, Y.; Yang, B.; Li, Z.; Wen, Z.; Lei, L.; Chai, G.; Feng, X. High-performance metal-free nanosheets array electrocatalyst for oxygen evolution reaction in acid. *Adv. Funct. Mater.* **2020**, *30*, 2003000. [CrossRef]
24. Lu, Z.; Shi, Y.; Yan, C.; Guo, C.; Wang, Z. Investigation on IrO_2 supported on hydrogenated TiO_2 nanotube array as OER electro-catalyst for water electrolysis. *Int. J. Hydrogen Energy* **2017**, *42*, 3572–3578. [CrossRef]
25. Yu, Z.; Xu, J.; Li, Y.; Wei, B.; Zhang, N.; Li, Y.; Bondarchuk, O.; Miao, H.; Araujo, A. Ultrafine oxygen-defective iridium oxide nanoclusters for efficient and durable water oxidation at high current densities in acidic media. *J. Mater. Chem. A* **2020**, *8*, 24743–24751. [CrossRef]
26. Li, R.; Xu, J.; Lu, C.; Huang, Z.; Wu, Q.; Ba, J.; Tang, T.; Meng, D.; Luo, W. Amorphous NiFe phosphides supported on nanoarray-structured nitrogen-doped carbon paper for high-performance overall water splitting. *Electrochim. Acta* **2020**, *357*, 136873. [CrossRef]
27. Artyushkova, K.; Workman, M.J.; Matanovic, I.; Dzara, M.J.; Ngo, C.; Pylypenko, S.; Serov, A.; Atanassov, P. Role of surface chemistry on catalyst/ionomer interactions for transition metal–nitrogen–carbon electrocatalysts. *ACS Appl. Energy Mater.* **2018**, *1*, 68–77. [CrossRef]
28. Teuku, H.; Alshami, I.; Goh, J.; Masdar, M.S.; Loh, K.S. Review on bipolar plates for low-temperature polymer electrolyte membrane water electrolyzer. *Int. J. Energy Res.* **2021**, *45*, 20583–20600. [CrossRef]
29. Wakayama, H.; Yamazaki, K. NiTiP-coated Ti as low-cost bipolar plates for water electrolysis with polymer electrolyte membranes. *Electrocatalysis* **2022**, *13*, 479–485. [CrossRef]
30. Wakayama, H.; Yamazaki, K. Low-cost bipolar plates of Ti_4O_7-coated Ti for water electrolysis with polymer electrolyte membranes. *ACS Omega* **2021**, *6*, 4161–4166. [CrossRef]
31. Avasarala, B.; Haldar, P. Electrochemical oxidation behavior of titanium nitride based electrocatalysts under PEM fuel cell conditions. *Electrochim. Acta* **2010**, *55*, 9024–9034. [CrossRef]
32. Toops, T.J.; Brady, M.; Zhang, F.; Meyer, H.M.; Ayers, K.; Roemer, A.; Dalton, L. Evaluation of nitrided titanium separator plates for proton exchange membrane electrolyzer cells. *J. Power Sources* **2014**, *272*, 954–960. [CrossRef]
33. Laidani, N.; Perriere, J.; Lincot, D.; Gicquel, A.; Amouroux, J. Nitriding of bulk titanium and thin titanium films in a NH_3 low pressure plasma. *Appl. Surf. Sci.* **1989**, *36*, 520–529. [CrossRef]

34. Bi, J.; Yang, J.; Liu, X.; Wang, D.; Yang, Z.; Liu, G.; Wang, X. Development and evaluation of nitride coated titanium bipolar plates for PEM fuel cells. *Int. J. Hydrogen Energy* **2021**, *46*, 1144–1154. [CrossRef]
35. Touni, A.; Grammenos, O.-A.; Banti, A.; Karfaridis, D.; Prochaska, C.; Lambropoulou, D.; Pavlidou, E.; Sotiropoulos, S. Iridium oxide-nickel-coated titanium anodes for the oxygen evolution reaction. *Electrochim. Acta* **2021**, *390*, 138866. [CrossRef]
36. Gao, X.; Tong, W.; Ouyang, X.; Wang, X. Facile fabrication of a superhydrophobic titanium surface with mechanical durability by chemical etching. *RSC Adv.* **2015**, *5*, 84666–84672. [CrossRef]
37. Zhao, C.; Zhu, Y.; Yuan, Z.; Li, J. Structure and tribocorrosion behavior of Ti/TiN multilayer coatings in simulated body fluid by arc ion plating. *Surf. Coat. Technol.* **2020**, *403*, 126399. [CrossRef]
38. Liu, G.; Xu, J.; Jiang, J.; Peng, B.; Wang, X. Nanosphere-structured composites consisting of Cs-substituted phosphotungstates and antimony doped tin oxides as catalyst supports for proton exchange membrane liquid water electrolysis. *Int. J. Hydrogen Energy* **2014**, *39*, 1914–1923. [CrossRef]
39. Xu, J.; Liu, G.; Li, J.; Wang, X. The electrocatalytic properties of an IrO_2/SnO_2 catalyst using SnO_2 as a support and an assisting reagent for the oxygen evolution reaction. *Electrochim. Acta* **2012**, *59*, 105–112. [CrossRef]
40. Nouri-Khorasani, A.; Bonakdarpour, A.; Fang, B.; Wilkinson, D. Rational design of multimodal porous carbon for the interfacial microporous layer of fuel cell oxygen electrodes. *ACS Appl. Mater. Interfaces* **2022**, *44*, 9084–9096. [CrossRef]
41. Fang, B.; Daniel, L.; Bonakdarpour, A.; Govindarajan, R.; Sharman, J.; Wilkinson, D. Dense Pt nanowire electrocatalysts for improved fuel cell performance using a graphitic carbon nitride-decorated hierarchical nanocarbon support. *Small* **2021**, *17*, 2102288. [CrossRef] [PubMed]
42. Lu, L.; Wang, B.; Wu, D.; Zou, S.; Fang, B. Engineering porous Pd-Cu nanocrystals with tailored three-dimensional catalytic facets for highly efficient formic acid oxidation. *Nanoscale* **2021**, *13*, 3709–3722. [CrossRef] [PubMed]
43. Yu, S.; Song, S.; Li, R.; Fang, B. The lightest solid meets the lightest gas: An overview of carbon aerogels and their composites for hydrogen related applications. *Nanoscale* **2020**, *12*, 19536–19556. [CrossRef]
44. Zhang, Y.; Wang, X.; Luo, F.; Tan, Y.; Zeng, L.; Fang, B.; Liu, A. Rock salt type $NiCo_2O_3$ supported on ordered mesoporous carbon as a highly efficient electrocatalyst for oxygen evolution reaction. *Appl. Catal. B: Environ.* **2019**, *256*, 117852. [CrossRef]
45. Fang, B.; Kim, J.; Kim, M.; Yu, J. Hierarchical nanostructured carbons with meso-macroporosity: Design, characterization and applications. *Acc. Chem. Res.* **2013**, *46*, 1397–1406. [CrossRef] [PubMed]
46. Tariq, M.; Wu, Y.; Ma, C.; Ali, M.; Zaman, W.Q.; Abbas, Z.; Ayub, K.S.; Zhou, J.; Wang, G.; Cao, L.; et al. Boosted up Stability and Activity of Oxygen Vacancy Enriched RuO_2/MoO_3 Mixed Oxide Composite for Oxygen Evolution Reaction. *Int. J. Hydrog. Energy* **2020**, *45*, 17287–17298. [CrossRef]
47. Tariq, M.; Zaman, W.Q.; Wu, Y.; Nabi, A.; Abbas, Z.; Iqbal, W.; Sun, W.; Hao, Z.; Zhou, Z.H.; Cao, L.; et al. Facile Synthesis of $IrO2$ Nanoparticles Decorated @ WO3 as Mixed Oxide Composite for Outperformed Oxygen Evolution Reaction. *Int. J. Hydrog. Energy* **2019**, *44*, 31082–31093. [CrossRef]
48. Genova-Koleva, R.V.; Alcaide, F.; Álvarez, G.; Cabot, P.L.; Grande, H.J.; Martínez-Huerta, M.V.; Miguel, O. Supporting $IrO2$ and IrRuOx Nanoparticles on $TiO2$ and Nb-Doped $TiO2$ Nanotubes as Electrocatalysts for the Oxygen Evolution Reaction. *J. Energy Chem.* **2019**, *34*, 227–239. [CrossRef]
49. Cherevko, S.; Geiger, S.; Kasian, O.; Kulyk, N.; Grote, J.-P.; Savan, A.; Shrestha, B.R.; Merzlikin, S.; Breitbach, B.; Ludwig, A.; et al. Oxygen and hydrogen evolution reactions on Ru, RuO_2, Ir, and IrO_2 thin film electrodes in acidic and alkaline electrolytes: A comparative study on activity and stability. *Catal. Today* **2016**, *262*, 170–180. [CrossRef]

Article

The Study of Thermal Stability of Mn-Zr-Ce, Mn-Ce and Mn-Zr Oxide Catalysts for CO Oxidation

T. N. Afonasenko [1], D. V. Glyzdova [1], V. L. Yurpalov [1], V. P. Konovalova [2], V. A. Rogov [2], E. Yu. Gerasimov [2] and O. A. Bulavchenko [2,*]

[1] Center of New Chemical Technologies BIC, Boreskov Institute of Catalysis, Neftezavodskaya st., 54, Omsk 644040, Russia
[2] Boreskov Institute of Catalysis SB RAS, Lavrentiev Ave. 5, Novosibirsk 630090, Russia
* Correspondence: obulavchenko@catalysis.ru

Abstract: MnO_x-CeO_2, MnO_x-ZrO_2, MnO_x-ZrO_2-CeO_2 oxides with the Mn/(Zr + Ce + Mn) molar ratio of 0.3 were synthesized by coprecipitation method followed by calcination in the temperature range of 400–800 °C and characterized by XRD, N_2 adsorption, TPR, TEM, and EPR. The catalytic activity was tested in the CO oxidation reaction. It was found that MnO_x-CeO_2, MnO_x-ZrO_2-CeO_2, MnO_x-ZrO_2 catalysts, calcined at 400–500 °C, 650–700 °C and 500–650 °C, respectively, show the highest catalytic activity in the reaction of CO oxidation. According to XRD and TEM results, thermal stability of catalysts is determined by the temperature of decomposition of the solid solution $Mn_x(Ce,Zr)_{1-x}O_2$. The TPR-H_2 and EPR methods showed that the high activity in CO oxidation correlates with the content of easily reduced fine MnO_x particles in the samples and the presence of paramagnetic defects in the form of oxygen vacancies. The maximum activity for each series of catalysts is associated with the start of solid solution decomposition. Formation of active phase shifts to the high-temperature region with the addition of zirconium to the MnO_x-CeO_2 catalyst.

Keywords: Mn-Ce; Mn-Zr; Mn-Zr-Ce; thermal stability; CO oxidation

1. Introduction

Catalytic CO oxidation is one of the most effective techniques for removing this pollutant, which is formed during the incomplete combustion of hydrocarbon fuels [1]. Materials based on noble metals are often used as a catalyst for CO oxidation, but the price of such samples and restricted availability of the noble metals makes their use economically unreasonable, helping to prevent their wider application. Thus, development of more economical, cheaper and affordable catalysts is of great importance [1,2]. An example of such non-noble systems can be manganese-containing compounds in the form of pristine or mixed oxides, which are widely studied as effective catalysts for the CO oxidation [3,4], as well as systems for soot oxidation [5], combustion of volatile organic compounds [6–10] and NO_x removal [11–13]. The catalytic properties of Mn-containing systems are due to the ability of manganese to exist in the form of various stoichiometric (MnO_2, Mn_5O_8, Mn_2O_3, Mn_3O_4, MnO and their polymorphs) and nonstoichiometric oxides, which are characterized by a variable oxidation state of manganese between +4 and +2 [14].

The calcination temperature is an important parameter that determines the properties of manganese catalysts. In the case of manganese oxides, it has a significant effect on the surface area of the samples, the states of manganese ions, and ethe nature and content of reactive oxygen species, including adsorbed oxygen, oxygen vacancies, and mobile oxygen. These factors determine the catalytic activity of MnO_x [15]. The effects of high-temperature treatment on mixed manganese-containing oxides can lead to phase transformations associated with the formation and subsequent decomposition of joint phases, which should also affect the catalytic activity. Thus, a significant increase in activity was found for MnO_x-Al_2O_3 catalysts calcined at 950–1000 °C in the reaction of CO and hydrocarbons (butane, benzene, and cumene) oxidation [16]. It was established earlier [17] that the $Mn_{3-x}Al_xO_4$ (x~1.5) mixed oxide is formed during high-temperature treatment. Subsequent cooling in air leads to the addition of oxygen to the $Mn_{3-x}Al_xO_4$ phase and its partial decomposition with the formation of highly dispersed β-Mn_3O_4 particles, which increases the amount of weakly bound oxygen, causing an increase in catalytic activity.

MnO_x-CeO_2 are the most catalytically active manganese-containing oxides [18]. The incorporation of manganese cations into the ceria lattice with the formation of solid solution causes a decrease in crystallinity, creating more lattice defects and oxygen vacancies, which significantly improves the ability of CeO_2 to accumulate oxygen, and also increases the mobility of oxygen on its surface [19]. However, MnO_x-CeO_2 has low thermal stability. Calcination above 500 °C leads to its deactivation, due to the decomposition of solid solution and a sharp decrease in the surface area [7,20,21]. One of the ways to improve the thermal stability of MnO_x-CeO_2 is by the addition of zirconium, which inhibits the agglomeration of oxide particles, forming a joint phase with CeO_2 [22–24]. Zirconium also has a similar effect on manganese oxide [25]. According to Zeng et al. [26], the significant activity improvement in propane oxidation after zirconium addition to MnO_x (after calcination at 500 °C) was attributed to the formation of solid solution, the superior redox ability, a higher oxygen mobility and a more abundant oxygen vacancy. Previously, we found [27] that $Mn_{0.12}Zr_{0.88}O_x$ samples calcined at 650–700 °C exhibit the highest catalytic activity in CO oxidation compared to catalysts calcined at other temperatures, which is explained by the partial decomposition of solid solution and formation of highly dispersed MnO_x particles on its surface.

Based on the properties of MnO_x-CeO_2 and MnO_x-ZrO_2, it can be assumed that the MnO_x-ZrO_2-CeO_2 oxide should combine high activity and thermal stability. Long et al. [28] showed that MnO_x-ZrO_2-CeO_2 oxides, compared to MnO_x-ZrO_2 and MnO_x-CeO_2 samples calcined at the same temperature of 500 °C, have the highest catalytic activity in the oxidation of chlorobenzene due to the high content of Mn4+ and surface oxygen, and higher mobility of oxygen in the solid solution. A similar comparison of supported catalysts calcined at 500 °C in the process of NO removal [29] showed the advantage of MnO_x/CeO_2 and MnO_x/CeO_2-ZrO_2 compared to MnO_x/ZrO_2; according to the authors, this is associated with a higher dispersion of manganese oxide and a higher content of weakly bound oxygen in MnO_x/CeO_2 and MnO_x/CeO_2-ZrO_2 catalysts. These comparative studies demonstrate the high activity of the MnO_x-ZrO_2-CeO_2 system; however, the calcination temperature of such catalysts usually does not exceed 550 °C [30,31]. Moreover, information about the thermal stability of MnO_x-ZrO_2-CeO_2 is limited to the studies of its properties at two or three calcination temperatures (500–550 °C, 800–900 °C). Such studies are carried out to estimate the resistance of the sample to possible overheating [24] or to simulate the "aging" of the catalyst [22]. To the best of our knowledge, no such studies of changes in the structural and catalytic properties of MnO_x-ZrO_2-CeO_2 under the influence of calcination temperature in comparison with the properties of MnO_x-CeO_2 and MnO_x-ZrO_2 oxides have been reported. These kinds of studies are important in developing a fundamental understanding of the thermal stability of the Mn-containing catalyst.

In current work we study the effect of a calcination temperature in the range of 400–800 °C on the catalytic activity and structural properties of MnO_x-ZrO_2-CeO_2, MnO_x-CeO_2, and MnO_x-ZrO_2 catalysts for CO oxidation. Comparison of three series of catalysts would help to determine the structure—activity relationship, including the role of the cationic composition in formation of solid solution, the temperature of decomposition, and the change in the catalytically active states.

2. Materials and Methods

2.1. Catalytic Synthesis

The samples were prepared by the precipitation method. A joint solution of $ZrO(NO_3)_2$, $Ce(NO_3)_3$, and $Mn(NO_3)_2$ salts was obtained, and the NH_4OH solution was gradually added into the mixture with continual stirring until the pH reached 10. The precipitation process was carried out at 80 °C. Stirring of the suspension was continued for 1 hour, then H_2O_2 was added dropwise to ensure the completeness of precipitation. The amount of H_2O_2 corresponded to the molar ratio of H_2O_2:(Mn + Zr + Ce). The suspension was finally kept without stirring for 2 hours. The obtained precipitate was filtered off and washed with distilled water on the filter to pH = 6–7. The samples were dried at 120 °C (2 h) and calcined in a muffle furnace at 400–800 °C for 4 h. The molar ratio of Mn:Zr:Ce in the prepared samples was 0.3:0.35:0.35.

Mn-Zr and Mn-Ce binary samples with the molar ratio Mn:Zr(Ce) 0.3:0.7 were prepared by a similar way.

The samples were designated as Mn-Zr-Ce-T, Mn-Ce-T and Mn-Zr-T, where T is the calcination temperature (in °C).

2.2. Catalytic Characterization

The specific surface area of the catalysts (S_{BET}) was determined by the Brunauer –Emmett–Teller method using nitrogen adsorption isotherms measured at liquid nitrogen temperature. The studies were carried out using an ASAP 2400 automated system (Micromeritics Instrument Corp., Norcross, GA, USA).

X-ray diffraction analysis (XRD) of Mn-Zr, Mn-Ce and Mn-Zr-Ce catalysts was performed on a Bruker D8 Advance diffractometer using Cu-Kα radiation (λ = 1.5418 Å) in the 2θ angle range of 15–90° at scan step 0.05°. The accumulation time was 4 s.

Transmission electron microscope (TEM) analysis was performed by ThemisZ electron microscope (Thermo Fisher Scientific, Waltham, MA, USA) equipped with a corrector of spherical aberrations, with the accelerating voltage of 200 kV. Images were recorded using Ceta 16 CCD sensor (Thermo Fisher Scientific, Waltham, MA, USA). Elemental maps were obtained using energy dispersive spectrometer SuperX (Thermo Fisher Scientific, Waltham, MA, USA).

Thermally programmed reduction (TPR-H_2) was carried out in a quartz reactor using a flow unit equipped with a thermal conductivity detector. A gas mixture (10 vol.% H_2 in Ar) was fed to the reactor at a rate of 40 mL/min. The heating rate was 10 °C/min from room temperature to 900 °C.

Electron paramagnetic resonance (EPR) spectroscopy was performed at 25 °C on X-band (~9.7 GHz) EMXplus spectrometer (Bruker, Karlsruhe, Germany) with an ER 4105 DR resonator at a microwave power of 2.0 mW, a modulation frequency of 100 kHz and an amplitude of modulation of 1.0 G. The spectrometer was controlled by a personal computer and the Bruker WinEPR Acquisition program. The results were processed using the Bruker WinEPR Processing software. The EPR spectra were simulated by means of the Bruker WinEPR SimFonia (v.1.2). The spectra were recorded for ~20 mg of each catalyst placed in a quartz tube with an outer diameter of 5 mm.

Catalytic tests of the samples in the CO oxidation reaction were carried out in a glass flow reactor (170 mm × Ø 10 mm). The initial gas mixture included 1 vol.% of CO and 99 vol.% of air. The flow rate of the gas mixture through the reactor during the catalytic tests was changed in the range of 253–487 mL/min to vary the contact time. The reaction temperature was 150 °C. The CO conversion for each sample was determined at 3 different contact times (τ). A catalyst fraction of 0.4–0.8 mm was used. The catalyst weight varied from 0.2 to 1.5 g. To avoid overheating during the exothermic reaction of CO oxidation, the catalyst was mixed with quartz granules of the same fractional composition. The temperature in the catalyst bed was controlled and regulated using a chromel-alumel thermocouple connected to a heat controller.

The reaction mixture was analyzed by gas chromatography. The mixture was separated on a packed column filled with CaA zeolite (3 m). The unreacted amount of CO was determined using a thermal conductivity detector. The reaction rate (activity) R

($cm^3(CO)/(g*s)$) was calculated from the CO conversion determined at different flow rates of the reaction mixture and taking into account the weight of the catalyst:

$$R = [C_o - C] \times V/(60 \times m_{cat}),$$

where $C = C_o \times (1 - ((P_o - P/P_o)))$, m_{cat} is the catalyst weight (g), V is the flow rate of the reaction mixture (mL/min), C_o is the initial CO concentration in the gas mixture (vol.%), C is the CO concentration in the outlet flow (vol.%), P_o is the peak area on the chromatogram corresponding to the CO concentration in the initial reaction mixture, and P is the peak area on the chromatogram corresponding to the CO concentration in the outlet flow.

3. Results

3.1. Catalytic Properties

Figure 1 shows data on the catalytic activity of Mn-Ce, Mn-Zr and Mn-Zr-Ce samples calcined at 400–800 °C in the CO oxidation reaction. It can be noted that the form of the observed dependence is individual for each series of samples.

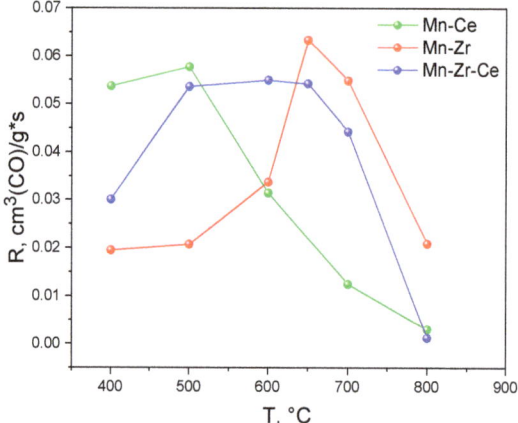

Figure 1. Catalytic activity of Mn-Ce, Mn-Zr and Mn-Zr-Ce samples, calcined at different temperatures, in CO oxidation at 150 °C.

In the case of Mn-Ce system, the highest value of catalytic activity is typical for catalysts calcined at 400–500 °C and is 5.4–5.8×10^{-2} $cm^3(CO)/g*s$. A further increase in the calcination temperature causes a sharp decrease in catalytic activity: the activity decreases to 3.2×10^{-2}, 1.3×10^{-2} and 0.3×10^{-2} $cm^3(CO)/g*s$ for Mn-Ce samples calcined at 600, 700 and 800 °C, respectively.

The effect of the calcination temperature of Mn-Zr catalysts on their activity in the CO oxidation reaction differs from that observed for the Mn-Ce system. The catalytic activity of Mn-Zr catalysts calcined at 400–500 °C remains approximately at a constant level of 2.0–2.1×10^{-2} $cm^3(CO)/g*s$. Raising the calcination temperature to 600 °C leads to an increase in activity up to 3.4×10^{-2} $cm^3(CO)/g*s$. A further increase in the calcination temperature to 650 °C causes an almost twofold increase in the catalytic activity, so that the parameter reaches the value of 6.3×10^{-2} $cm^3(CO)/g*s$. At the same time, the activity of the catalysts calcined at 700 °C slightly decreases to 5.5×10^{-2} $cm^3(CO)/g*s$. A higher temperature treatment at 800 °C causes a sharp loss of activity to a value of 2.1×10^{-2} $cm^3(CO)/g*s$, which coincides with the activity of the samples treated at 400 and 500 °C. Thus, the dependence of the catalytic activity of Mn-Zr catalysts on the calcination temperature has the form of an extremum; the highest activity is achieved for samples calcined at 650–700 °C.

In the case of Mn-Zr-Ce catalysts, an increase in the calcination temperature from 400 to 500 °C leads to an increase in activity from 3.0 to 5.4×10^{-2} cm^3(CO)/g*s. A further increase in the treatment temperature up to 650 °C has no significant effect on the catalytic activity, while the Mn-Zr-Ce catalysts treated at 700 °C lose their activity, so that R does not exceed 4.4×10^{-2} cm^3(CO)/g*s. Calcination at 800 °C leads to a further sharp drop in activity to 0.1×10^{-2} cm^3(CO)/g*s. Thus, the temperature range required for the formation of the most catalytically active states in Mn-Zr-Ce samples is 500–650 °C.

Comparison of the kinetic data obtained for Mn-Ce, Mn-Zr and Mn-Zr-Ce oxides indicates that each of the studied systems is characterized by the calcination temperature range, where its samples demonstrate the highest catalytic activity relative to the samples of the other two systems. Such an optimal calcination temperature range for Mn-Ce catalysts is 400–500 °C. Mn-Zr catalysts are preferably calcined at 650–700 °C, while Mn-Zr-Ce triple oxide catalysts are in an intermediate position between Mn-Zr and Mn-Ce systems and exhibit high catalytic activity only in the calcination temperature range of 500–650 °C.

In order to explain the observed changes in the catalytic activity of Mn-Ce, Mn-Zr and Mn-Zr-Ce oxides depending on the calcination temperature, the changes in the structural properties of these systems were considered and compared.

3.2. Structural and Microstructural Properties

Figure 2a shows X-ray diffraction patterns of Mn-Ce catalysts. Peaks with maxima at 2θ = 28.6, 33.1, 47.5, 56.5, 59.2, 69.5, 76.9, 79.10, and 88.6°, corresponding to 111, 002, 022, 113, 222, 004, 133, 024, and 224 reflections of CeO$_2$ fluorite structure (space group Fm-3m, PDF № 431002), are observed in the XRD patterns of all samples. An increase in the calcination temperature leads to a narrowing of the diffraction peaks and their shift towards smaller angles. As the temperature increases to 800 °C, additional peaks appear at 2θ = 23.2, 38.5, 55.2 and 65.9°, corresponding to 112, 004, 444, and 226 reflections of Mn$_2$O$_3$ (space group Ia-3, PDF № 41-1442), and 2θ = 36.1, corresponding to the most intense 121 reflection of Mn$_3$O$_4$ oxide (space group I4$_1$/amd, PDF № 41-1442).

Diffraction patterns of Mn-Zr-Ce catalysts are shown in Figure 2b. It can be seen that catalysts calcined at 400–700 °C have similar X-ray patterns: broad peaks with maxima at 2θ = 28.9, 33.5, 48.2, 57.2, 60.0, 70.5, 78.0, 80.4, and 89.3°, corresponding to 111, 002, 022, 113, 222, 004, 133, 024, and 224 reflections of CeO$_2$ fluorite structure, are observed. A low-intensity Mn$_3$O$_4$ reflection appears at 2θ = 36.3°, and its intensity increases with an increase in the calcination temperature. Significant changes occur in the diffraction patterns of Mn-Zr-Ce-800: the reflections of Mn$_3$O$_4$ appear and the fluorite reflections split. The latter may be related to decomposition of the initial oxide.

Figure 2c contains X-ray diffraction patterns of the Mn-Zr catalysts. The calcination of the samples at 400–500 °C leads to the appearance of a wide halo at 2θ = 25–40°, which indicates the formation of an X-ray amorphous state. The amorphous phase partially remains and induces an appearance of a background in the diffraction pattern of the sample calcined at 600 °C. The diffraction pattern of this sample also contains peaks at 2θ = 30.5, 35.4, 50.9, 60.6, 63.6, 74.9, 83.0, and 85.7°, which correspond to the ZrO2 oxide. Since the broadening of diffraction peaks, from XRD data it is not possible to correctly distinguish the cubic and tetragonal modification of zirconia. Diffraction signals of Mn$_2$O$_3$ oxide also appear at 2θ = 32.9, 38.2, 55.1, and 65.7°, corresponding to 222, 004, 444 and 226 reflections. These phases are present in the XRD patterns of Mn-Zr-650 and Mn-Zr-700 samples. High-temperature treatment at 800 °C causes the appearance of additional peaks at 2θ = 30.2, 34.6, 35.8, 50.2, 50.7, 59.3, 60.2, 62.9, 72.9, and 74.6°, corresponding to 011, 002, 110, 112, 020, 013, 121, 022, 004, and 220 reflections of the tetragonal modification of t-ZrO$_2$ oxide (space group P4$_2$/nmc, PDF № 54-1089). The monoclinic modification of m-ZrO$_2$ also forms, as evidenced by the presence of a peak at 2θ = 28.2°, corresponding to the 111 reflection (space group P 2$_1$/c, PDF № 37-1484).

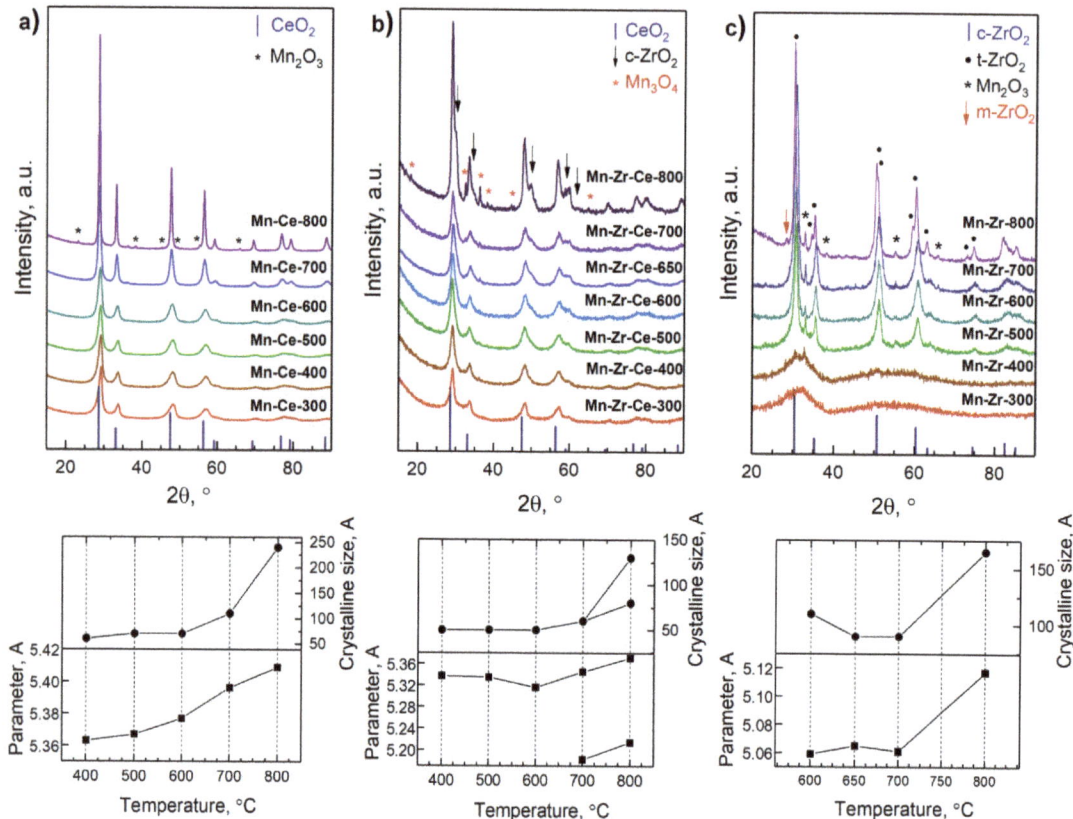

Figure 2. XRD data for (**a**) Mn-Ce, (**b**) Mn-Zr-Ce and (**c**) Mn-Zr catalysts activated at different temperatures.

The structural characteristics of the catalysts calculated by the Rietveld method are given in Table 1. For Mn-Zr-600 and Mn-Zr-700, the observed lattice parameter of ZrO_2 catalysts is 5.059–5.061 Å, which is much less than the literature parameter for pristine zirconium oxide (5.110 Å). The difference in the lattice parameters may indicate the formation of the $Mn_yZr_{1-y}O_{2-\delta}$ solid solution, since the ionic radii of Zr^{4+} and Mn^{3+} are 0.84 Å and 0.66 Å, respectively [32]. The lattice parameter increases with an increase in the calcination temperature and becomes close to the value characteristic of pristine ZrO_2 at 800 °C. Simultaneously, an increase in the content of the Mn_2O_3 phase to 9 wt.% is observed. These three factors indicate the decomposition of the $Mn_yZr_{1-y}O_{2-\delta}$ solid solution, accompanied by the diffusion of Mn ions from the volume of oxide and the Mn_2O_3 phase formation: a decrease in the manganese content in the mixed oxide, a decrease in symmetry of zirconia, and the extrication of m-ZrO_2.

Table 1. Structural and microstructural characteristics of Mn-Ce, Mn-Zr and Mn-Zr-Ce samples.

Sample	Phases, wt%	Lattice Constant of Ce(Mn,Zr)O$_2$, Å	CSR, Å	S$_{BET}$, m$^2 \times$ g^{-1}
Mn-Zr-Ce-400	Ce(Mn,Zr)O$_2$ Mn$_3$O$_4$ (traces)	5.337(1) -	50 130	146
Mn-Zr-Ce-500	Ce(Mn,Zr)O$_2$ Mn$_3$O$_4$ (traces)	5.335(1) -	50 130	110
Mn-Zr-Ce-600	Ce(Mn,Zr)O$_2$ Mn$_3$O$_4$ (traces)	5.316(1) -	50 140	88
Mn-Zr-Ce-650	57% Ce(Mn,Zr)O$_2$ 6% Mn$_3$O$_4$ 37% c-Zr(Ce,Mn)O$_2$	5.330(1) 5.169(1)	70 260 60	74
Mn-Zr-Ce-700	55% Ce(Mn,Zr)O$_2$ 8% Mn$_3$O$_4$ 37% Ce$_{x2}$(Mn$_{y2}$,Zr$_{z2}$)O$_2$	5.345(1) - 5.182(3)	80 270 70	63
Mn-Zr-Ce-800	44% Ce(Mn,Zr)O$_2$ 13% Mn$_3$O$_4$ 43% Ce$_{x2}$(Mn$_{y2}$,Zr$_{z2}$)O$_2$	5.371(1) - 5.214(1)	130 340 80	12
Mn-Zr-400	Amorphous	-	-	304
Mn-Zr-500	Amorphous Mn$_2$O$_3$	-	-	306
Mn-Zr-600	Amorphous Zr(Mn)O$_2$ Mn$_2$O$_3$	5.059(1) [1]	- 110 375	176
Mn-Zr-650	98% Zr(Mn)O$_2$ 2% Mn$_2$O$_3$	5.065(1) [1]	90 410	111
Mn-Zr-700	98% Zr(Mn)O$_2$ 2% Mn$_2$O$_3$	5.061(1) [1]	90 340	66
Mn-Zr-800	77% t-Zr(Mn)O$_2$ 14% m-ZrO$_2$ 9% Mn$_2$O$_3$	5.117(2) [1]	165 195 210	24
Mn-Ce-400	Ce(Mn)O$_2$	5.360(1)	60	73
Mn-Ce-500	Ce(Mn)O$_2$	5.362(1)	70	57
Mn-Ce-600	Ce(Mn)O$_2$	5.377(1)	70	41
Mn-Ce-700	Ce(Mn)O$_2$	5.397(1)	110	20
Mn-Ce-800	91% Ce(Mn)O$_2$ 2% Mn$_3$O$_4$ 7% Mn$_2$O$_3$	5.409(1)	240 - 330	9

[1] lattice parameter was calculated in cubic approximation or corrected to cubic lattice.

Similar trends have been established for the Mn-Ce and Mn-Zr-Ce series. In the case of Mn-Ce-400, the lattice parameter of the oxide is 5.360(1) Å, which is noticeably lower than that of pristine cerium oxide CeO$_2$ (PDF № 431002, a = 5.411 Å) and indicates the formation of Mn$_y$Ce$_{1-y}$O$_{2-\delta}$ solid solution. An increase in the synthesis temperature causes an increase in the oxide lattice parameter to 5.406 Å. The gradual increase in the lattice constant to the value of pristine CeO$_2$ phase indicates the diffusion of manganese cations from the solid solution structure to its surface. At 800 °C, Mn cations form crystalline manganese oxides Mn$_2$O$_3$ and Mn$_3$O$_4$, which are detected by XRD. In the case of the Mn-Zr-Ce series, the lattice parameters vary from 5.182(1) to 5.371(1) Å, indicating the formation of a Mn$_y$Zr$_x$Ce$_{1-y}$O$_{2-\delta}$ solid solution. An increase in the calcination temperature from 400 to 600 °C leads to a decrease in the lattice constant from 5.337(1) to 5.316(1) Å and a simultaneous increase in the content of manganese oxide. The observed effect is associated with the decomposition of solid solution by diffusion of Mn cations from the initial oxide and the appearance of a "new" $Mn_{y2}Zr_{x2}Ce_{1-y2}O_{2-\delta2}$ solid solution. Catalysts treated at 700–800 °C contain phases with lattice parameters of 5.345(1)–5.371(1) and 5.182(3)–5.214(1) Å, which tend to the values for pristine CeO_2 and ZrO_2 oxides. In this case, two mixed oxides are formed; the first one is based on ceria and the second one is based on zirconia. In addition, some heterogeneity in composition cannot be excluded for catalysts synthesized at lower temperatures. The average sizes of coherent scattering regions (CSR) grow with an increase in the calcination temperature from 50–110 Å at 400–600 °C to 130–240 Å at 800 °C. The introduction of Zr reduces the average CSR size of the solid solution formed at 700–800 °C as compared to the Mn-Ce catalysts.

Thus, it can be concluded that all three series are characterized by similar changes in the phase composition, depending on the synthesis temperature: an increase in temperature leads to decomposition of $Mn_yZr_xCe_{1-y}O_{2-\delta}$ solid solutions with the formation of Mn_2O_3 and Mn_3O_4 crystalline oxides and cerium/zirconium oxides. Based on the lattice parameters' change, it can be assumed that there is a temperature range, one lower than the temperature of crystalline manganese oxides formation, at which the decomposition of the initial solid solutions has already begun and manganese cations are on the oxide surface in the form of amorphous MnO_x particles (Table 1, Figure 2). It should be noted that an increase in the calcination temperature leads to a sharper change in the lattice parameters of the solid solutions formed in the Mn-Ce-Zr and Mn-Zr samples (at 700 °C) compared to the solid solution existing in the Mn-Ce system, which is characterized by a linear change in the lattice constant.

As expected, the specific surface area of the catalysts decreases with increasing the calcination temperature for all series (Table 1). The S_{BET} of X-ray amorphous Mn-Zr samples obtained at 400–500 °C is 304–306 $m^2 \times g^{-1}$. A further increase in the calcination temperature to 600 °C causes phase crystallization and a decrease in specific surface area to 176 $m^2 \times g^{-1}$. The decomposition of a solid solution at 800 °C leads to a decrease in S_{BET} down to 24 $m^2 \times g^{-1}$. The dependence of the specific surface area on the synthesis temperature of Mn-Ce catalysts is close to linear: the S_{BET} value is 72 $m^2 \times g^{-1}$ for the sample calcined at 400 °C and decreases to 9 $m^2 \times g^{-1}$ for the catalyst calcined at 800 °C. The specific surface area of the Mn-Zr-Ce triple oxide system calcined at 400 °C is 146 $m^2 \times g^{-1}$. Further, it decreases linearly with an increase in the calcination temperature within the temperature range of a solid solution existence up to 700 °C (63 $m^2 \times g^{-1}$). Raising the temperature to 800 °C leads to a sharp drop in S_{BET} to 12 $m^2 \times g^{-1}$. Accordingly, the observed changes in the specific surface area correlate with the phase transformations occurring in mixed oxides during an increase in the calcination temperature, and the thermal stability of each of the studied oxides is determined by the decomposition temperature of the corresponding solid solution.

The most active catalysts in each series (Mn-Ce-500, Mn-Zr-Ce-600, Mn-Zr-700) were studied by high-resolution transmission electron microscopy coupled with EDS mapping (Figure 3a–c). EDS mapping shows that for all cases there are areas enriched by Mn cations and a mixed Mn-Ce-Zr region. The latter is associated with formation of solid solution. In the case of Mn-Ce-500 and Mn-Zr-700 catalysts, the content of Mn (x) in mixed oxide $Mn_x(Ce,Zr)_{1-x}O_2$ is 0.25–0.35. For Mn-Zr-Ce-600, several Mn-Zr-Ce regions can be distinguished in the region of the solid solution enriched in Ce and Zr cations. These results correlate well with XRD analysis (Table 1). In addition to areas containing three cations, one can distinguish regions enriched in manganese. Figure 3d–f illustrates the formation of various manganese oxides: Mn_3O_4, Mn_2O_3 and Mn_5O_8 for Mn-Ce-500, Mn-Zr-700, and Mn-Zr-Ce-600 catalysts, respectively. For Mn-Zr-Ce-600 catalyst, Mn_5O_8 was not detected by XRD (Table 1); the Mn oxide is probably in an XRD amorphous state. The binary catalysts are characterized by the presence of manganese oxides in the form of nanorods with particle sizes of 10–30 nm in diameter and 100–200 nm in length. For a ternary Mn-Zr-Ce-600 catalyst, manganese oxide has a rounded particle shape 20–30 nm in size.

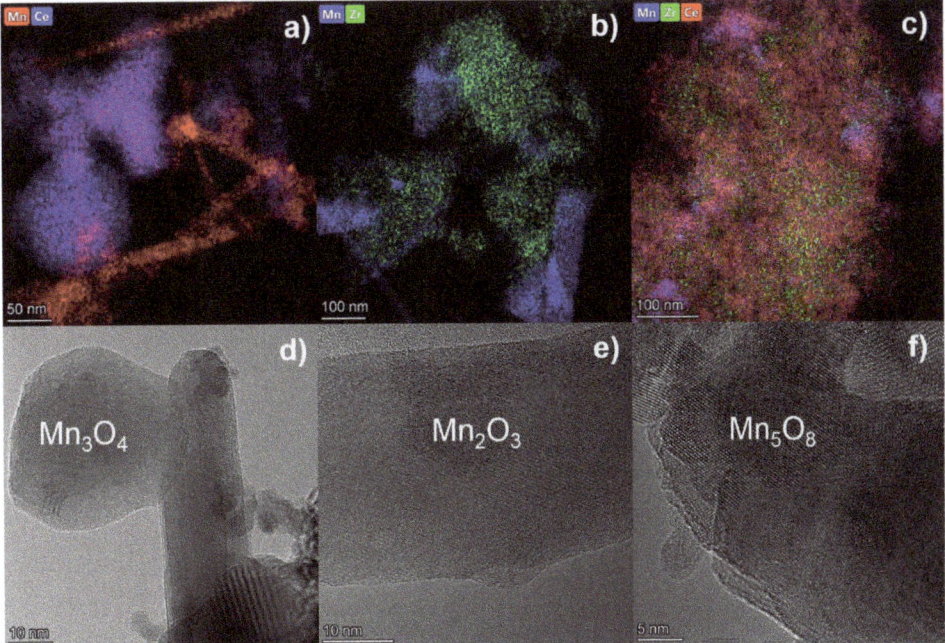

Figure 3. (a–c) TEM image; (d–f) EDS mapping pattern for (a,d) Mn-Ce-500, (d,e) Mn-Zr-700 and (c,f) Mn-Zr-Ce-600 catalysts.

3.3. Temperature-Programmed Reduction

Mn-Ce, Mn-Zr, and Mn-Zr-Ce catalysts synthesized at 500, 600 and 700 °C were studied by TPR-H$_2$ in order to evaluate the differences in their redox properties. The choice of these calcination temperatures is due to the fact that each of the temperatures is optimal for achieving the highest catalytic activity for one of the three studied systems. The TPR-H$_2$ profiles are shown in Figure 4.

The TPR curves of Mn-Ce catalysts show an intensive signal at high temperatures of ~600–900 °C with a maximum at 818 °C, which is explained by the reduction of Ce^{4+} ions in ceria [33–37]. In the range from 100 to 500 °C, the TPR profiles contain a broad signal of partially unseparated peaks with the main maxima at ~260–270 °C and 350–360 °C. These signals are associated with the consumption of hydrogen due to the sequential partial reduction of manganese cations present in the CeO$_2$-based solid solution. The appearance of a signal at ~260–270 °C corresponds to the reduction of Mn^{4+}/Mn^{3+} ions to Mn^{3+}/Mn^{2+}, while the peak at ~350–360 °C is due to the Mn^{3+}/Mn^{2+} → Mn^{2+} transition [34]. In the case of Mn-Ce catalysts calcined at 500 and 600 °C, the first signal with a maximum at 260–270 °C has a shoulder at low temperatures (~220 °C). The consumption of hydrogen in this region can be explained by the reduction of finely dispersed MnO$_x$ particles, which are probably formed on the surface of the solid solution during its decomposition. It should be noted that this signal does not appear on the TPR curve of the sample calcined at 700 °C, which may be due to the formation of Mn$_3$O$_4$ crystalline phase. The TPR profiles of Mn-Ce samples calcined at 600 and 700 °C also contain an additional low-intensity signal at ~500–520 °C, which can be explained by the reduction of Ce^{4+} ions located on the ceria surface or in the subsurface layer of the samples [33,34].

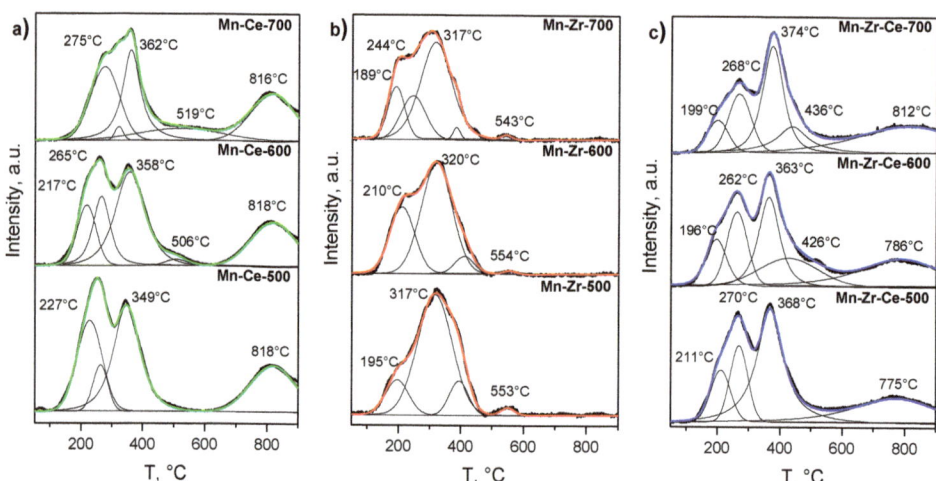

Figure 4. TPR-H$_2$ curves for (**a**) Mn-Ce, (**b**) Mn-Zr and (**c**) Mn-Zr-Ce catalysts, calcined at 500, 600 and 700 °C. Diferent colours corresponds to different series.

The TPR-H$_2$ curves of Mn-Zr series are characterized by the presence of a complex, wide, intense signal in the reduction temperature range from ~100 to 500 °C, one associated with the occurrence of parallel processes of manganese reduction. Manganese can exist both in the composition of a solid solution based on ZrO$_2$ and in nonstoichiometric highly-dispersive MnO$_x$ particles [38,39], which can be formed during the decomposition of a solid solution at T ≥ 500 °C [40]. Similar to the Mn-Ce catalysts, sequential reduction processes of manganese ions Mn^{4+}/Mn^{3+} → Mn^{3+}/Mn^{2+} → Mn^{2+} also occur in the system [39]. It is most likely that the intense consumption of hydrogen at ~320 °C is a consequence of the Mn^{3+}/Mn^{2+} phase transformation. It should be noted that the TPR-H$_2$ curves of all Mn-Zr samples contain a low-intensity signal at ~550 °C, which can be associated with the partial reduction of zirconium oxide [40,41].

The reduction curves of Mn-Zr-Ce triple systems are shown in Figure 4c. A characteristic feature of the TPR profiles of Mn-Zr-Ce catalysts is the appearance of a broad low-intensity signal of the partial reduction of Ce^{4+} in the temperature range of 600–900 °C (T$_{max}$~770–810 °C) [33,34]. The TPR profiles of the Mn-Zr-Ce systems, similarly to the TPR curves recorded for the Mn-Ce and Mn-Zr binary systems, also contain a wide signal of partially unseparated peaks in the temperature range of 100–500 °C, which are associated with the sequential phase transformations of manganese oxides: MnO$_2$/Mn$_2$O$_3$ → Mn$_3$O$_4$ (the maximum at 260–270 °C) and Mn$_3$O$_4$ → MnO (the maximum at 360–370 °C) [42,43], as well as a gradual change in the oxidation state of manganese cations in the Mn$_y$Zr$_x$Ce$_{1-y}$O$_{2-\delta}$ structure of solid solution. All Mn-Zr-Ce samples are also characterized by hydrogen consumption at low temperatures of 150–300 °C, which is expressed in the appearance of a shoulder on the TPR curves. We assume that this feature is explained by the reduction of highly dispersed MnO$_x$ particles, which are formed during the decomposition of the Mn$_y$Zr$_x$Ce$_{1-y}$O$_{2-\delta}$ solid solution and exist in a highly dispersed state on its surface. An increase in the calcination temperature of Mn-Zr-Ce catalysts to 600 °C leads to the appearance of an additional signal of hydrogen absorption with a maximum at ~430 °C. This signal may appear due to the peaks overlap for the reduction of manganese and Ce^{4+} ions located in the surface layer of the catalyst [44].

By fitting the TPR profiles with individual components, it was possible to calculate the fraction of the low-temperature signal at T~200 °C for each series of the samples (Table 2). In the case of Mn-Ce-500, the content of weakly bound oxygen is ~27% toward the total hydrogen uptake. Its amount decreases to ~18% with an increase in the calcination temperature to 600 °C, and a further increase in the calcination temperature causes the complete disappearance of the low-temperature signal. In the case of Mn-Zr catalysts, an increase in the synthesis temperature from 500 to 600 °C leads to an increase in the amount of weakly bound oxygen in the system from ~17 to ~29%, which can occur due to an increase in the amount of fine MnO_x particles. The total content of low-temperature signals observed in the TPR profile of Mn-Zr-700 is ~37%. An analysis of the TPR curves of the Mn-Zr-Ce triple oxide catalysts showed that the content of weakly bound oxygen in the samples calcined at 500 and 600 °C is ~12%, which corresponds to the states of solid solutions before their decomposition and may indicate the formation of dispersed amorphous MnO_x particles. A further increase in the calcination temperature to 700 °C leads to a decrease in the amount of the low-temperature signal to ~4%.

Table 2. TPR data for Mn-Ce, Mn-Zr and Mn-Zr-Ce catalysts, calcined at 500, 600 and 700 °C.

Catalyst	Total Hydrogen Uptake, mmol $(H_2) \times g^{-1}$	Content of Weakly Bound Oxygen, %
Mn-Ce-500	2.00×10^{-3}	27
Mn-Ce-600	1.76×10^{-3}	18
Mn-Ce-700	1.63×10^{-3}	0
Mn-Zr-500	1.38×10^{-3}	17
Mn-Zr-600	1.17×10^{-3}	29
Mn-Zr-700	1.37×10^{-3}	37
Mn-Zr-Ce-500	2.19×10^{-3}	12
Mn-Zr-Ce-600	2.17×10^{-3}	12
Mn-Zr-Ce-700	2.09×10^{-3}	4

3.4. Electron Paramagnetic Resonance Spectroscopy

EPR spectra obtained for Mn-Ce, Mn-Zr and Mn-Zr-Ce samples (Figure 5) are similar for each series, but differ depending on the chemical composition of the catalyst. Mn-Ce and Mn-Zr-Ce samples have signals of sextet in EPR spectra, characteristic for Mn^{2+} cations [22,33], while in Mn-Zr samples there are no signals with a width less than 500 G, including Zr^{3+} [45] from the monoclinic phase of zirconium oxide.

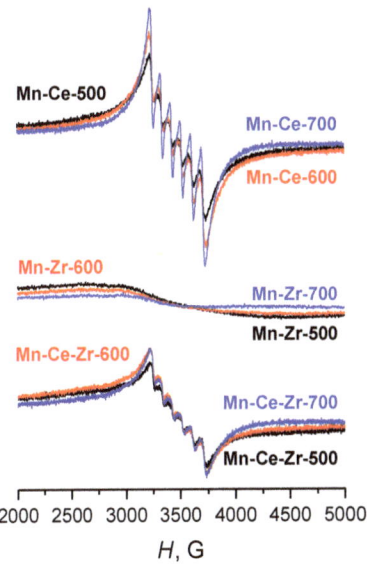

Figure 5. EPR spectra of Mn-Ce, Mn-Zr and Mn-Zr-Ce catalysts, calcined at 500 °C, 600 °C and 700 °C.

A more thorough analysis and simulation of the EPR spectra made it possible to establish detailed differences in the spectra for the samples of the same composition activated at different temperatures. In particular, it was able to identify the components of the spectra associated with the signals of Mn^{2+}, Mn^{4+} ions and paramagnetic defects (oxygen vacancies), as well as to evaluate their contribution to the overall spectrum. Mn^{3+} cations are EPR-silent under the conditions of the current experiment. It is necessary to use high-frequency EPR spectrometers for their registration [46].

The experimental and simulated EPR spectra of Mn-Ce catalysts are shown in Figure 6. The spectra are represented as a superposition of three types of signals with the following calculated parameters: $g = 2.08$–2.10, $\Delta H_{pp} = 600$ G (paramagnetic defects/oxygen vacancies [45,47]); $g = 1.998$, $\Delta H_{pp} = 2000$ G (Mn^{4+} cations [48]), as well as two Mn^{2+} cation signals [22,33] with $g = 2.00$, $A = 92$ G and different line widths of $\Delta H_{pp} = 240$ G and $\Delta H_{pp} = 80$ G (Mn-Ce-500), $\Delta H_{pp} = 65$ G (Mn-Ce-600), $\Delta H_{pp} = 55$ G. The narrowing of Mn^{2+} EPR-line in the spectra is usually associated with an increase in the interaction between Mn and Ce cations, namely, with a more active incorporation of manganese ions into the structure of cerium oxide [29]. The narrowing of Mn^{2+} signal lines with an increase in the calcination temperature for the Mn-Ce series can also be explained by a decrease in the dipole-dipole interaction of the nearest Mn^{2+} cations as a result of a decrease in their concentration in the two-component phase due to the diffusion of paramagnetic cations from the structure of the solid solution [49] and/or their partial transition to the EPR-inactive Mn^{3+} form. This phenomenon is also confirmed by some decrease in the total concentration of paramagnetic species in the sample upon calcination above 600 °C (Table 3). Another reason for the decrease in the linewidth of Mn^{2+} ions can be related to the decrease in the sizes of manganese(II)-containing clusters [49,50]. In the case of Mn-Ce catalysts, an increase in the calcination temperature leads to the increase of Mn^{2+} signal contribution to the total spectrum with a simultaneous decrease in the fraction of Mn^{4+} signal, which represents manganese (IV) ions in a solid solution and/or in the dispersed phase of MnO_x [48]. The decrease in the intensity of the manganese (IV) signal can presumably be explained by the agglomeration of MnO_x particles formed as a result of the solid solution decomposition and/or partial sintering of more dispersed particles of manganese oxides. All that leads to a strong broadening of Mn^{4+} EPR line up to the complete disappearance of the recorded signal of such particles. Similarly, the proportion of paramagnetic defects (oxygen vacancies, V_O) signal in the series decreases from 17 to 4% (Table 3). These defects can act as precursors for active oxidation sites, since they are able to adsorb oxygen with the formation of its active forms [47]. Typically, such sites are characterized by lower values of the g-factor and linewidth [47], however, at high concentrations of paramagnetic defects or other sites, an EPR-signal can be broadened with a shift of the peak to a weaker field, since the spin system begins to exhibit ferromagnetic properties [45]. It should be noted that the decrease in the fraction of V_O signal with an increase in the calcination temperature of the Mn-Ce catalysts is in good agreement with the decrease in the amount of weakly bound oxygen in these samples, which was determined by TPR-H_2.

The Mn-Zr samples demonstrate a significant difference compared to the Mn-Ce systems: in particular, there are no signals of Mn^{2+} ions in the spectra (Figure 7). In this case, signals of Mn^{4+} and paramagnetic defects/oxygen vacancies are observed. The proportion of paramagnetic oxygen vacancies in V_O/Mn^{4+} ratio increases with an increase in the calcination temperature of the samples (Table 3), which also correlates well with the patterns established by TPR-H_2 of Mn-Zr catalysts.

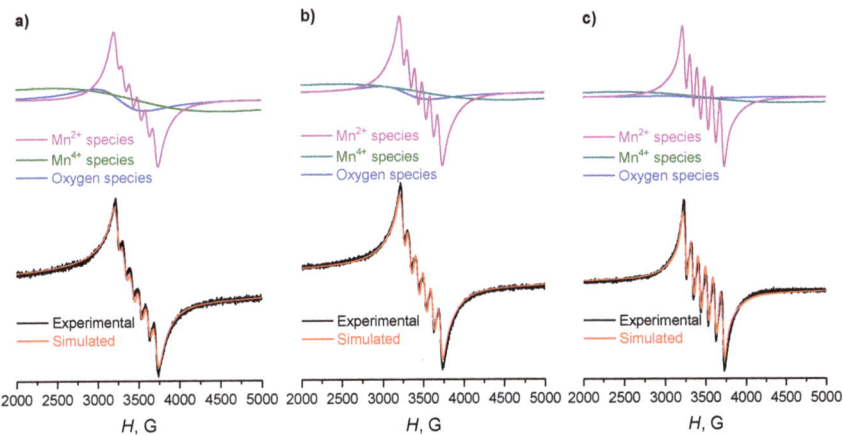

Figure 6. Experimental and simulated EPR spectra of Mn-Ce catalysts, calcined at (**a**) 500 °C, (**b**) 600 °C and (**c**) 700 °C. Different components of the spectra are presented above each simulated curve.

Table 3. The components of simulated EPR spectra and their contribution to overall spectrum.

Catalyst	The Contribution of the Species to Overall EPR Spectrum, %			Intensity (Double Integral), % rel.
	V_O [1]	Mn^{4+} [2]	Mn^{2+} [3]	
Mn-Ce-500	17	53	30	100
Mn-Ce-600	12	45	43	105
Mn-Ce-700	4	46	50	75
Mn-Zr-500	8	92	-	80
Mn-Zr-600	10	90	-	60
Mn-Zr-700	20	80	-	30
Mn-Ce-Zr-500	13	71	16	110
Mn-Ce-Zr-600	13	68	19	110
Mn-Ce-Zr-700	12	60	28	75

[1] V_O—oxygen vacancies; g = 2.08-2.10, ΔH_{pp} = 600 G. [2] g = 1.998, ΔH_{pp} = 2000 G. [3] g = 2.00, A = 92 G, ΔH_{pp} = 240 G and ΔH_{pp} = 80 G (Mn-Ce-500), ΔH_{pp} = 65 G (Mn-Ce-600), ΔH_{pp} = 55 G (Mn-Ce-700).

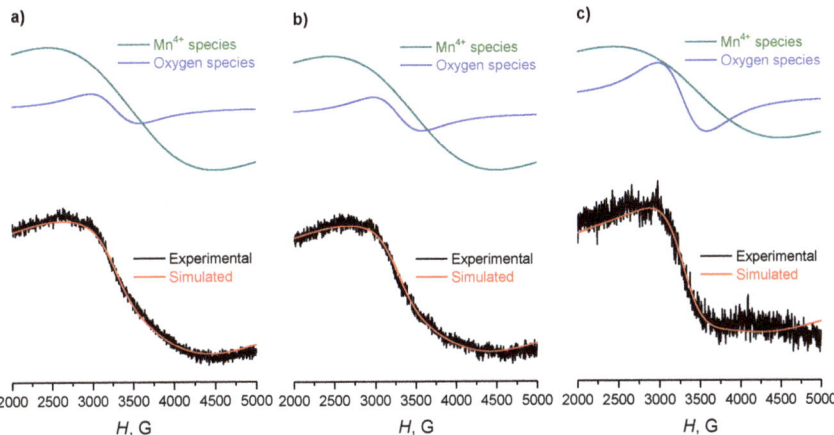

Figure 7. Experimental and simulated EPR spectra of Mn-Zr catalysts, calcined at (**a**) 500 °C, (**b**) 600 °C and (**c**) 700 °C. Different components of the spectra are presented above each simulated curve.

The EPR spectra of the Mn-Zr-Ce triple oxide system (Figure 8) are similar to the spectra of Mn-Ce binary catalysts. However, the ratio of narrow signals of divalent and broad signals of tetravalent manganese ions differs (Table 3). The proportion of EPR-detected Mn^{4+} in these samples compared to other signals is significantly higher compared to the double Mn-Ce system. In general, for a triple oxide catalyst, no significant differences are observed depending on the calcination temperature up to 600 °C. At calcination temperature of 700 °C the contribution of manganese (II) signal slightly increases with a simultaneous decrease in the contribution of signals from paramagnetic defects and Mn^{4+} cations.

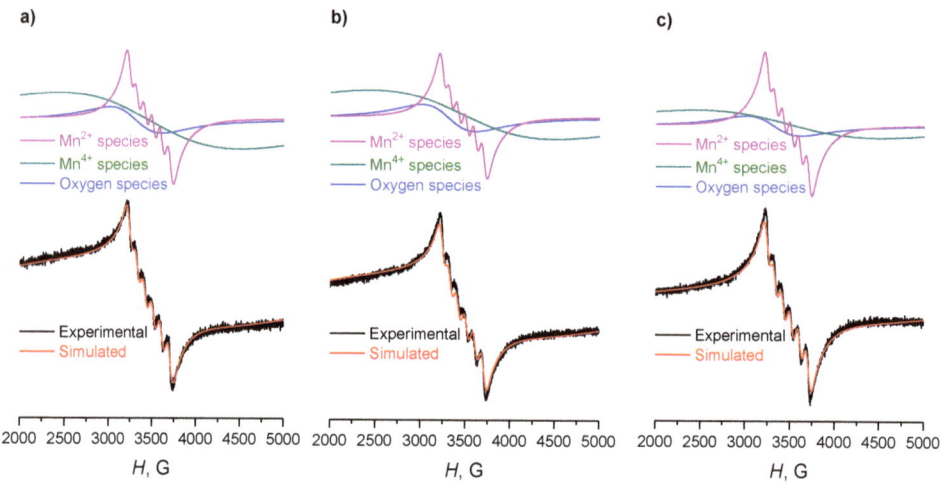

Figure 8. Experimental and simulated EPR spectra of Mn-Ce-Zr catalysts, calcined at (**a**) 500 °C, (**b**) 600 °C and (**c**) 700 °C. Different components of the spectra are presented above each simulated curve.

4. Discussion

A comparative study of the calcination temperature effect on the catalytic activity of MnO_x-CeO_2, MnO_x-ZrO_2 and MnO_x-ZrO_2-CeO_2 oxides in CO oxidation reaction showed that the activities of the MnO_x-CeO_2 and MnO_x-ZrO_2-CeO_2 catalysts calcined at 500 °C are close and significantly exceed the activity of the MnO_x-ZrO_2 sample, which is in accordance with the literature data [29]. At the same time, an increase in the synthesis temperature of the MnO_x-CeO_2 catalyst from 400 to 500 °C affects the lattice parameter of the CeO_2-based solid solution, in that it indicates the diffusion of manganese from its structure. The revealed changes in structural properties contribute to the increase in catalytic activity. The deactivation of MnO_x-CeO_2, observed with a further increase in the calcination temperature, is caused by the intensification of solid solution decomposition, which leads to a significant decrease in the specific surface area and the amount of weakly bound oxygen. On the contrary, the X-ray amorphous state with a high specific surface area in the MnO_x-ZrO_2 samples prepared at 400–500 °C has a low catalytic activity. Crystallization of the solid solution phase based on ZrO_2 and its subsequent decomposition at 650–700 °C cause a sharp increase in activity. Based on the TPR-H_2 data, it can be argued that such behavior of MnO_x-ZrO_2 corresponds to the appearance of fine MnO_x particles on the solid solution surface, that increases the content of easily reduced weakly bound oxygen. The decrease in the activity of MnO_x-ZrO_2 after calcination at 800 °C is due to the completion of the solid solution decomposition and transformation of the cubic modification of ZrO_2 into the monoclinic one.

The thermal stability of the MnO_x-ZrO_2-CeO_2 system is in intermediate position between the thermal stability of the binary MnO_x-CeO_2 and MnO_x-ZrO_2 oxides. The high activity of MnO_x-ZrO_2-CeO_2 observed after calcination at 500 °C is retained up to 650 °C in contrast to MnO_x-CeO_2. As in the case of binary oxides, the $Mn_yZr_xCe_{1-y}O_{2-\delta}$ solid solution gradually decomposes with the diffusion of manganese cations from its structure and the formation of fine MnO_x particles. The shift of the catalytic activity profile towards higher synthesis temperatures is caused by the presence of zirconium, since zirconium ions stabilize the fluorite structure by replacing cerium ions and prevent particle agglomeration, as was mentioned above [22,23]. Indeed, the average CSR size is 50 Å in the temperature range from 400 to 600 °C, while it has a value of 70 Å for MnO_x-CeO_2.

An analysis of the structural properties of MnO_x-CeO_2, MnO_x-ZrO_2 and MnO_x-ZrO_2-CeO_2 catalysts shows that all three series are characterized by similar changes in the phase composition, depending on the synthesis temperature: an increase in temperature leads to the decomposition of $Mn_yZr_xCe_{1-y}O_{2-\delta}$ solid solutions with the formation of Mn_2O_3 or Mn_3O_4 crystalline oxides and cerium/zirconium oxides. Changes in the lattice parameter of mixed oxide (Figure 2) indicate release of Mn ions from the volume of solid solutions and formations of highly dispersed MnO_x. The maximum activity for each series is associated with the beginning of solid-solution decomposition. The formation of active phase shifts to the high-temperature region with the addition of Zr to the MnOx-CeO2 catalyst due to growth of the temperature of solid solution decomposition. On the other hand, an introduction of Ce into catalyst leads to the formation of crystalline mixed oxide based on the fluorite structure at lower temperature than in the case of MnO_x-ZrO_2 system. The latter effect enhances activity of the MnO_x-ZrO_2-CeO_2 catalysts prepared at 400–500 °C.

The correlation of the catalytic activity of the MnO_x-CeO_2, MnO_x-ZrO_2 and MnO_x-ZrO_2-CeO_2 catalysts with the content of weakly bound oxygen and paramagnetic oxygen vacancies in the sample, as well as the ambiguous interrelation between the catalytic activity and the specific surface area, allow us to claim that the dependences of the activity on the calcination temperature are primarily determined by phase transformations occurring in the studied systems. It can be concluded that the highest value of catalytic activity for each of the considered Mn-containing systems is achieved for the samples containing manganese both in the form of ions in the structure of the corresponding substitutional solid solution and in the form of highly dispersed MnO_x particles, formed as a result of manganese segregation on the surface of solid solution during its decomposition. This combination of catalytically active states of manganese in the samples of all studied oxides causes the presence of the highest content of easily reduced reactive oxygen, which was established by TPR-H_2. In accordance with the EPR data, changes in the catalytic activity of the studied oxides correlate with the content of paramagnetic defects, presumably in the form of oxygen vacancies, which can be located both in the composition of the solid solution and the3 MnO_x clusters. In the case of a solid solution, the formation of vacancies can occur due to the diffusion of manganese cations from its structure.

A comparison of MnO_x-CeO_2, MnO_x-ZrO_2 and MnO_x-ZrO_2-CeO_2 showed that MnO_x-ZrO_2-CeO_2 triple oxide system combines the properties of binary oxides, where the presence of zirconium makes it possible to increase the thermal stability of MnO_x-CeO_2 from 500 °C up to 650 °C without significant loss of catalytic activity.

5. Conclusions

The influence of the calcination temperature of MnO_x-CeO_2, MnO_x-ZrO_2 and MnO_x-ZrO_2-CeO_2 catalysts on their structural properties and catalytic activity in the CO oxidation reaction has been studied and compared. The highest catalytic activity of MnO_x-CeO_2 is observed for the samples activated at 400–500 °C. In the case of MnO_x-ZrO_2, the optimal calcination temperature range is 650–700 °C, and for the Mn-Zr-Ce triple oxide system this interval corresponds to 500–650 °C. It has been established by XRD that the thermal stability of the studied oxides is limited by the onset temperature of decomposition of the corresponding solid solution, where manganese ions diffuse from its structure to the surface in the form of fine MnO_x particles. Comparison of the oxides calcined at 500, 600, and 700 °C by TPR-H_2 and EPR methods showed that the high activity of each of the three studied systems at a specific calcination temperature correlates with the content of easily reduced fine MnO_x particles and the presence of paramagnetic oxygen vacancies. The MnO_x-ZrO_2-CeO_2 catalyst showed thermal stability up to 650 °C without significant loss of catalytic activity due to the presence of zirconium in its composition.

Author Contributions: Conceptualization, O.A.B.; formal analysis, T.N.A., D.V.G., V.L.Y. and V.A.R.; investigation, O.A.B., T.N.A., D.V.G., V.L.Y., V.A.R., V.P.K. and E.Y.G.; writing—original draft preparation, T.N.A. and D.V.G.; writing—review and editing, O.A.B.; visualization, O.A.B. All authors have read and agreed to the published version of the manuscript.

Funding: This work was supported by the Russian Science Foundation, grant 21-73-10218.

Institutional Review Board Statement: Not applicable.

Informed Consent Statement: Not applicable.

Acknowledgments: The X-ray structural studies were carried out using facilities of the shared research center "National Center of Investigation of Catalysts" at Boreskov Institute of Catalysis. The EPR experiments were performed using equipment of the Omsk Regional Center of Collective Usage SB RAS.

Conflicts of Interest: The authors declare no conflict of interest.

References

1. Soliman, N.K. Factors affecting CO oxidation reaction over nanosized materials: A review. *J. Mater. Sci. Technol.* **2019**, *8*, 2395–2407. [CrossRef]
2. Kim, H.J.; Jang, M.G.; Shin, D.; Han, J.W. Design of Ceria Catalysts for Low-Temperature CO Oxidation. *ChemCatChem* **2020**, *12*, 11–26. [CrossRef]
3. Liu, X.; Lu, J.; Qian, K.; Huang, W.; Luo, M. A comparative study of formaldehyde and carbon monoxide complete oxidation on MnO_x-CeO_2 catalysts. *J. Rare Earths* **2009**, *27*, 418–424. [CrossRef]
4. Mobini, S.; Meshkani, F.; Rezaei, M. Supported Mn catalysts and the role of different supports in the catalytic oxidation of carbon monoxide. *Chem. Eng. Sci.* **2019**, *197*, 37–51. [CrossRef]
5. Lin, X.; Li, S.; He, H.; Wu, Z.; Wu, J.; Chen, L.; Ye, D.; Fu, M. Evolution of oxygen vacancies in MnO_x-CeO_2 mixed oxides for soot oxidation. *Appl. Catal. B Environ.* **2018**, *223*, 91–102. [CrossRef]
6. Yang, X.; Yu, X.; Jing, M.; Song, W.; Liu, J.; Ge, M. Defective $Mn_xZr_{1-x}O_2$ Solid Solution for the Catalytic Oxidation of Toluene: Insights into the Oxygen Vacancy Contribution. *ACS Appl. Mater. Interfaces* **2019**, *11*, 730–739. [CrossRef] [PubMed]
7. Tang, X.; Li, Y.; Huang, X.; Xu, Y.; Zhu, H.; Wang, J.; Shen, W. MnO_x-CeO_2 mixed oxide catalysts for complete oxidation of formaldehyde: Effect of preparation method and calcination temperature. *Appl. Catal. B Environ.* **2006**, *62*, 265–273. [CrossRef]
8. Lin, F.; Wang, Z.; Zhang, Z.; Xiang, L.; Yuan, D.; Yan, B.; Wang, Z.; Chen, G. Comparative investigation on chlorobenzene oxidation by oxygen and ozone over a MnO_x/Al_2O_3 catalyst in the presence of SO_2. *Environ. Sci. Technol.* **2021**, *55*, 3341–3351. [CrossRef]
9. Yin, R.; Sun, P.; Cheng, L.; Liu, T.; Zhou, B.; Dong, X. A Three-dimensional melamine sponge modified with MnOx mixed graphitic carbon nitride for photothermal catalysis of formaldehyde. *Molecules* **2022**, *27*, 5216. [CrossRef]
10. Kulchakovskaya, E.V.; Dotsenko, S.S.; Liotta, L.F.; La Parola, V.; Galanov, S.I.; Sidorova, O.I.; Vodyankina, O.V. Synergistic effect in Ag/Fe–MnO_2 catalysts for ethanol oxidation. *Catalysts* **2022**, *12*, 872. [CrossRef]
11. Qi, G.; Yang, R.T.; Chang, R. MnO_x-CeO_2 mixed oxides prepared by co-precipitation for selective catalytic reduction of NO with NH_3 at low temperatures. *Appl. Catal. B Environ.* **2004**, *51*, 93–106. [CrossRef]
12. Zhao, B.; Ran, R.; Wu, X.; Weng, D.; Wu, X.; Huang, C. Comparative study of Mn/TiO_2 and Mn/ZrO_2 catalysts for NO oxidation. *Catal. Commun.* **2014**, *56*, 36–40. [CrossRef]
13. Li, S.; Zheng, Z.; Zhao, Z.; Wang, Y.; Yao, Y.; Liu, Y.; Zhang, J.; Zhang, Z. CeO_2 nanoparticle-lLoaded MnO_2 nanoflowers for selective catalytic reduction of NO_x with NH_3 at low temperatures. *Molecules* **2022**, *27*, 4863. [CrossRef]
14. Frey, K.; Iablokov, V.; Sáfrán, G.; Osán, J.; Sajó, I.; Szukiewicz, R.; Chenakin, S.; Kruse, N. Nanostructured MnO_x as highly active catalyst for CO oxidation. *J. Catal.* **2012**, *287*, 30–36. [CrossRef]
15. Tang, W.; Wu, X.; Li, D.; Wang, Z.; Liu, G.; Liu, H.; Chen, Y. Oxalate route for promoting activity of manganese oxide catalysts in total VOCs' oxidation: Effect of calcination temperature and preparation method. *J. Mater. Chem. A* **2014**, *2*, 2544–2554. [CrossRef]
16. Tsyrul'nikov, P.G.; Sal'nikov, V.A.; Drozdov, V.A.; Stuken, S.A.; Bubnov, A.V.; Grigorov, E.I.; Kalinkin, A.V.; Zaikovskii, V.I. Investigation of thermal activation of aluminum-manganese total oxidation catalysts. *Kinet. Catal.* **1991**, *32*, 387–394.
17. Bulavchenko, O.A.; Afonasenko, T.N.; Tsyrul'nikov, P.G.; Tsybulya, S.V. Effect of heat treatment conditions on the structure and catalytic properties of MnO_x/Al_2O_3 in the reaction of CO oxidation. *Appl. Catal. A Gen.* **2013**, *459*, 73–80. [CrossRef]
18. Silva, A.M.T.; Marquesy, R.R.N.; Quinta-Ferreira, R.M. Catalytic Wet Oxidation of Acrylic Acid: Studies with Manganese-based Oxides. *Inter. J. Chem. Reactor Eng.* **2003**, *1*, 1–8. [CrossRef]

19. Bulavchenko, O.A.; Afonasenko, T.N.; Osipov, A.R.; Pochtar', A.A.; Saraev, A.A.; Vinokurov, Z.S.; Gerasimov, E.Y.; Tsybulya, S.V. The formation of Mn-Ce oxide catalysts for co oxidation by oxalate route: The role of manganese content. *Nanomaterials* **2021**, *11*, 988. [CrossRef]
20. Zhang, H.; Wang, J.; Cao, Y.; Wang, Y.; Gong, M.; Chen, Y. Effect of Y on improving the thermal stability of MnO_x-CeO_2 catalysts for diesel soot oxidation. *Chin. J. Catal.* **2015**, *36*, 1333–1341. [CrossRef]
21. Wu, X.; Liu, S.; Weng, D.; Lin, F.; Ran, R. MnO_x-CeO_2-Al_2O_3 mixed oxides for soot oxidation: Activity and thermal stability. *J. Hazard. Mater.* **2011**, *187*, 283–290. [CrossRef] [PubMed]
22. Rao, T.; Shen, M.; Jia, L.; Hao, J.; Wang, J. Oxidation of ethanol over Mn-Ce-O and Mn-Ce-Zr-O complex compounds synthesized by sol-gel method. *Catal. Commun.* **2007**, *8*, 1743–1747. [CrossRef]
23. Jo, S.H.; Shin, B.; Shin, M.C.; Van Tyne, C.J.; Lee, H. Dispersion and valence state of $MnO_2/Ce_{(1-x)}Zr_xO_2$-TiO_2 for low temperature NH_3-SCR. *Catal. Commun.* **2014**, *57*, 134–137. [CrossRef]
24. Liberman, E.Y.; Kleusov, B.S.; Naumkin, A.V.; Zagaynov, I.V.; Konkova, T.V.; Simakina, E.A.; Izotova, A.O. Thermal stability and catalytic activity of the MnO_x–CeO_2 and the MnO_x–ZrO_2–CeO_2 highly dispersed materials in the carbon monoxide oxidation reaction. *Inorg. Mater. Appl. Res.* **2021**, *12*, 468–476. [CrossRef]
25. Zhao, G.; Li, J.; Zhu, W.; Ma, X.; Guo, Y.; Liu, Z.; Yang, Y. Mn_3O_4 doped with highly dispersed Zr species: A new non-noble metal oxide with enhanced activity for three-way catalysis. *New J. Chem.* **2016**, *40*, 10108–10115. [CrossRef]
26. Zeng, K.; Li, X.; Wang, C.; Wang, Z.; Guo, P.; Yu, J.; Zhang, C.; Zhao, X.S. Three-dimensionally macroporous $MnZrO_x$ catalysts for propane combustion: Synergistic structure and doping effects on physicochemical and catalytic properties. *J. Colloid Interface Sci.* **2020**, *572*, 281–296. [CrossRef]
27. Afonasenko, T.N.; Bulavchenko, O.A.; Gulyaeva, T.I.; Tsybulya, S.V.; Tsyrul'nikov, P.G. Effect of the calcination temperature and composition of the MnO_x–ZrO_2 system on its structure and catalytic properties in a reaction of carbon monoxide oxidation. *Kinet. Catal.* **2018**, *59*, 104–111. [CrossRef]
28. Long, G.; Chen, M.; Li, Y.; Ding, J.; Sun, R.; Zhou, Y.; Huang, X.; Han, G.; Zhao, W. One-pot synthesis of monolithic Mn-Ce-Zr ternary mixed oxides catalyst for the catalytic combustion of chlorobenzene. *Chem. Eng. J.* **2019**, *360*, 964–973. [CrossRef]
29. Shen, B.; Zhang, X.; Ma, H.; Yao, Y.; Liu, T. A comparative study of Mn/CeO_2, Mn/ZrO_2 and Mn/Ce-ZrO_2 for low temperature selective catalytic reduction of NO with NH_3 in the presence of SO_2 and H_2O. *J. Environ. Sci.* **2013**, *25*, 791–800. [CrossRef]
30. Cao, F.; Xiang, J.; Su, S.; Wang, P.; Sun, L.; Hu, S.; Lei, S. The activity and characterization of MnO_x-CeO_2-ZrO_2/γ-Al_2O_3 catalysts for low temperature selective catalytic reduction of NO with NH_3. *Chem. Eng. J.* **2014**, *243*, 347–354. [CrossRef]
31. Hou, Z.; Feng, J.; Lin, T.; Zhang, H.; Zhou, X.; Chen, Y. The performance of manganese-based catalysts with $Ce_{0.65}Zr_{0.35}O_2$ as support for catalytic oxidation of toluene. *Appl. Surf. Sci.* **2018**, *434*, 82–90. [CrossRef]
32. Shannon, R.D. Revised Effective Ionic Radii and Systematic Studies of Interatomic Distances in Halides and Chalcogenides. *Acta Cryst.* **1976**, *A32*, 751–767. [CrossRef]
33. Kaplin, I.Y.; Lokteva, E.S.; Golubina, E.V.; Shishova, V.V.; Maslakov, K.I.; Fionov, A.V.; Isaikina, O.Y.; Lunin, V.V. Efficiency of manganese modified CTAB-templated ceria-zirconia catalysts in total CO oxidation. *Appl. Surf. Sci.* **2019**, *485*, 432–440. [CrossRef]
34. Wang, Z.; Shen, G.; Li, J.; Liu, H.; Wang, Q.; Chen, Y. Catalytic removal of benzene over CeO_2-MnO_x composite oxides prepared by hydrothermal method. *Appl. Catal. B Environ.* **2013**, *138–139*, 253–259. [CrossRef]
35. Aneggi, E.; Boaro, M.; De Leitenburg, C.; Dolcetti, G.; Trovarelli, A. Insights into the redox properties of ceria-based oxides and their implications in catalysis. *J. Alloys Compd.* **2006**, *408–412*, 1096–1102. [CrossRef]
36. Venkataswamy, P.; Rao, K.N.; Jampaiah, D.; Reddy, B.M. Nanostructured manganese doped ceria solid solutions for CO oxidation at lower temperatures. *Appl. Catal. B Environ.* **2015**, *162*, 122–132. [CrossRef]
37. Zhu, H.; Qin, Z.; Shan, W.; Shen, W.; Wang, J. Pd/CeO_2-TiO_2 catalyst for CO oxidation at low temperature: A TPR study with H_2 and CO as reducing agents. *J. Catal.* **2004**, *225*, 267–277. [CrossRef]
38. Gutiérrez-Ortiz, J.I.; de Rivas, B.; López-Fonseca, R.; Martín, S.; González-Velasco, J.R. Structure of Mn-Zr mixed oxides catalysts and their catalytic performance in the gas-phase oxidation of chlorocarbons. *Chemosphere* **2007**, *68*, 1004–1012. [CrossRef]
39. Huang, X.; Li, L.; Liu, R.; Li, H.; Lan, L.; Zhou, W. Optimized synthesis routes of MnO_x-ZrO_2 hybrid catalysts for improved toluene combustion. *Catalysts* **2021**, *11*, 1037. [CrossRef]
40. Bulavchenko, O.A.; Vinokurov, Z.S.; Afonasenko, T.N.; Tsyrul'nikov, P.G.; Tsybulya, S.V.; Saraev, A.A.; Kaichev, V.V. Reduction of mixed Mn-Zr oxides: In situ XPS and XRD studies. *Dalt. Trans.* **2015**, *44*, 15499–15507. [CrossRef]
41. Li, W.B.; Chu, W.B.; Zhuang, M.; Hua, J. Catalytic oxidation of toluene on Mn-containing mixed oxides prepared in reverse microemulsions. *Catal. Today* **2004**, *93–95*, 205–209. [CrossRef]
42. Sun, W.; Li, X.; Mu, J.; Fan, S.; Yin, Z.; Wang, X.; Qin, M.; Tadé, M.; Liu, S. Improvement of catalytic activity over Mn-modified $CeZrO_x$ catalysts for the selective catalytic reduction of NO with NH_3. *J. Colloid Interface Sci.* **2018**, *531*, 91–97. [CrossRef] [PubMed]
43. Azalim, S.; Franco, M.; Brahmi, R.; Giraudon, J.M.; Lamonier, J.F. Removal of oxygenated volatile organic compounds by catalytic oxidation over Zr-Ce-Mn catalysts. *J. Hazard. Mater.* **2011**, *188*, 422–427. [CrossRef] [PubMed]
44. Zhong, L.; Fang, Q.; Li, X.; Li, Q.; Zhang, C.; Chen, G. Influence of preparation methods on the physicochemical properties and catalytic performance of Mn-Ce catalysts for lean methane combustion. *Appl. Catal. A Gen.* **2019**, *579*, 151–158. [CrossRef]
45. del Silva-Calpa, L.R.; Zonetti, P.C.; Rodrigues, C.P.; Alves, O.C.; Appel, L.G.; de Avillez, R.R. The $Zn_xZr_{1-x}O_{2-y}$ solid solution on m-ZrO_2: Creating O vacancies and improving the m-ZrO_2 redox properties. *J. Mol. Catal. A Chem.* **2016**, *425*, 166–173. [CrossRef]

46. Azamat, D.V.; Badalyan, A.G.; Dejneka, A.; Trepakov, V.A.; Jastrabik, L.; Frait, Z. High-frequency electron paramagnetic resonance investigation of Mn^{3+} centers in $SrTiO_3$. *J. Phys. Chem. Solids.* **2012**, *73*, 822–826. [CrossRef]
47. Liu, Y.; Xia, C.; Wang, Q.; Zhang, L.; Huang, A.; Ke, M.; Song, Z. Direct dehydrogenation of isobutane to isobutene over Zn-doped ZrO2 metal oxide heterogeneous catalysts. *Catal. Sci. Technol.* **2018**, *8*, 4916–4924. [CrossRef]
48. Zhecheva, E.; Stoyanova, R.; Gorova, M. Microstructure of $Li_{1+x}Mn_{2-x}O_4$ cathode materials monitored by EPR of Mn^{4+}. In *Materials for Lithium-Ion Batteries*; Springer: Dordrecht, The Netherlands, 2000; pp. 485–489. [CrossRef]
49. AitMellal, O.; Oufni, L.; Messous, M.Y.; Matei, E.; Rostas, A.M.; Galca, A.C.; Secu, M. Temperature-induced phase transition and tunable luminescence properties of Ce^{3+}-Mn^{2+}-Zr^{4+} tri-doped $LaPO_4$ phosphor. *Opt. Mater.* **2022**, *129*, 112567. [CrossRef]
50. Saab, E.; Aouad, S.; Abi-Aad, E.; Zhilinskaya, E.; Aboukaïs, A. Carbon black oxidation in the presence of Al_2O_3, CeO_2, and Mn oxide catalysts: An EPR study. *Catal. Today* **2007**, *119*, 286–290. [CrossRef]

Article

Nonstoichiometry Defects in Double Oxides of the A_2BO_4-Type

Aleksandr S. Gorkusha [1,2], Sergey V. Tsybulya [1,2,*], Svetlana V. Cherepanova [1], Evgeny Y. Gerasimov [1] and Svetlana N. Pavlova [1]

[1] Boreskov Institute of Catalysis SB RAS, 630090 Novosibirsk, Russia
[2] Department of Physics, Novosibirsk State University, 630090 Novosibirsk, Russia
* Correspondence: tsybulya@catalysis.ru

Abstract: Double oxides with the structure of the Ruddlesden–Popper (R-P) layered perovskite $A_{n+1}B_nO_{3n+1}$ attract attention as materials for various electrochemical devices, selective oxygen-permeable ceramic membranes, and catalytic oxidative reactions. In particular, Sr_2TiO_4 layered perovskite is considered a promising catalyst in the oxidative coupling of methane. Our high-resolution transmission electron microscopy (HRTEM) studies of Sr_2TiO_4 samples synthesized using various methods have shown that their structure often contains planar defects disturbing the periodicity of layer alternation. This is due to the crystal-chemical features of the R-P layered perovskite-like oxides whose structure is formed by n consecutive layers of perovskite $(ABO_3)_n$ in alternating with layers of rock-salt type (AO) in various ways along the c crystallographic direction. Planar defects can arise due to a periodicity violation of the layers alternation that also leads to a violation of the synthesized phase stoichiometry. In the present work, a crystallochemical analysis of the possible structure of planar defects is carried out, structures containing defects are modeled, and the effect of such defects on the X-ray diffraction patterns of oxides of the A_2BO_4 type using Sr_2TiO_4 is established as an example. For the calculations, we used the method of constructing probabilistic models of one-dimensionally disordered structures. For the first time, the features of diffraction were established, and an approach was demonstrated for determining the concentration of layer alternation defects applicable to layered perovskite-like oxides of the A_2BO_4 type of any chemical composition. A relation has been established between the concentration of planar defects and the real chemical composition (nonstoichiometry) of the Sr_2TiO_4 phase. The presence of defects leads to the Ti enrichment of particle volume and, consequently, to the enrichment of the surface with Sr. The latter, in turn, according to the data of a number of authors, can serve as an explanation for the catalytic activity of Sr_2TiO_4 in the oxidative coupling of methane.

Keywords: Ruddlesden–Popper phases; X-ray powder diffraction; HRTEM; defects; 1D simulation; nonstoichiometry; oxidative coupling of methane

Citation: Gorkusha, A.S.; Tsybulya, S.V.; Cherepanova, S.V.; Gerasimov, E.Y.; Pavlova, S.N. Nonstoichiometry Defects in Double Oxides of the A_2BO_4-Type. *Materials* **2022**, *15*, 7642. https://doi.org/10.3390/ma15217642

Academic Editor: Daniela Kovacheva

Received: 3 October 2022
Accepted: 27 October 2022
Published: 31 October 2022

Publisher's Note: MDPI stays neutral with regard to jurisdictional claims in published maps and institutional affiliations.

Copyright: © 2022 by the authors. Licensee MDPI, Basel, Switzerland. This article is an open access article distributed under the terms and conditions of the Creative Commons Attribution (CC BY) license (https://creativecommons.org/licenses/by/4.0/).

1. Introduction

Complex oxides of the general formula $A_{n+1}B_nO_{3n+1}$ with the layered Ruddlesden–Popper (R-P) structure attract considerable attention of researchers as promising materials for electrochemical and magnetic devices, selective oxygen-conducting membranes and also as catalysts for various processes [1–4]. In particular, oxides of the R-P series (with the compositions A = Ca, Ba, Sr, B = Ti, Sn) are considered promising catalysts for the oxidative coupling of methane (OCM) [5–10], a one-stage method for producing ethane and ethylene (C2), for which an active search for effective catalysts is currently underway [11–13]. It is believed that the features of the R-P structure, which consists of alternating blocks with structures of the ABO_3 perovskite-type and the AO rock salt type [14], determine the high activity and selectivity of catalysts in OCM [9,10]. Earlier in [15], we showed that in single-phase or containing a minimum amount of impurity phases, Sr_2TiO_4 samples could be obtained using a synthesis procedure that includes the stages of mechanochemical activation of precursors and high-temperature calcination.

However, the structural features of layered perovskites allow the presence of various types of defects, and the real structure of the resulting phases can be the cause of their specific physical and/or chemical (including catalytic) properties.

Nonstoichiometry defects in the structure of R-P oxides have been previously the subject of a large number of experimental and theoretical studies [16–21].

It has been repeatedly shown by HRTEM that extended (planar) defects can occur in these structures, violating the periodicity in the alternation of layers [19,20].

In particular, our preliminary studies [15,22] also showed that the structure of some Sr_2TiO_4 samples obtained using the mechanochemical activation of precursors contains planar defects, leading to nonstoichiometry of the formed phases both in terms of the ratio of cations and oxygen content. In principle, quantitative estimates of the content of planar defects can be obtained from X-ray powder diffraction data if a relationship is established between the presence of planar defects of one type or another and changes in the diffraction patterns [23]. At present, there is no such analysis for structures of the A_2BO_4 type.

In this paper, we consider the influence of planar defects associated with the violation of the order in the alternation of layers in structures of the A_2BO_4 type (using Sr_2TiO_4 as an example) on their diffraction patterns and propose a method for estimating the concentration of defects from X-ray powder diffraction data.

2. Materials and Methods

2.1. Sample Synthesis

The samples under study were selected from a series of samples synthesized in [15].

As the starting materials, $SrCO_3$, TiO_2 (rutile), and $TiO(OH)_2$ were used to prepare the Sr_2TiO_4 samples. To provide a target stoichiometry of samples, corresponding amounts of starting compounds were taken on the basis of their thermal analysis. The stoichiometric mixtures of the starting chemicals were mixed and then activated in two modes.

In the first mode (sample No. 1), an APF-5 low-energy mill was used. The starting compounds are $SrCO_3$ and $TiO(OH)_2$. Activation was carried out with zirconium balls 5 mm in diameter in water-cooled iron drums with a volume of 25 mL. The ratio of the mass of the balls to the mass of the mixture was 10, the drum rotation speed was 800 rpm, and activation was carried out within 10 min.

In the second mode (sample No. 2) of activation, a high-voltage planetary mill AGO-2 was used. The starting compounds are $SrCO_3$ and TiO_2. Activation was carried out with zirconium balls 5 mm in diameter in water-cooled iron drums with a volume of 150 mL. The ratio of the mass of the balls to the mass of the mixture was 20, the drum rotation speed was 1200 rpm, and activation was carried out for 20 min.

Before each synthesis, a preliminary treatment of the drums and balls with the corresponding mixture was performed to cover the surface of the drums and balls with a layer of the initial mixture. This minimizes the contamination of the samples with Fe and Zr due to their rubbing during mechanochemical activation.

Activated mixtures were pressed in tablets and annealed at 1100 °C for 4 h.

The choice of samples with different backgrounds was conscious. On the one hand, the chosen samples were almost single-phase; on the other hand, they had differences in the real structure, as shown below.

2.2. Sample Characterization

2.2.1. XRD

Diffraction experiments were performed using synchrotron radiation (λ = 1.5369 Å) in the Bragg–Brentano geometry at the Siberian Center for Synchrotron and Terahertz Radiation [24]. The refinement of the averaged crystal structure, including the lattice parameters and diffraction patterns, was analyzed using the Rietveld method using the GSAS-II program [25].

Simulation of diffraction patterns in the presence of layer alternation defects was performed according to the method described in detail in the monograph [26]. The diffraction patterns were calculated based on a statistical model of a one-dimensional (1D) disordered crystal. The calculations use the fact that the scattering region from a two-dimensional (2D) periodic layer is localized along the rods passing through the nodes of the two-dimensional periodic reciprocal lattice, which are determined by integers h and k. The same localization of the scattering region is also preserved for a one-dimensional disordered crystal. The model of a 1D-disordered crystal is presented as a statistical sequence of two-dimensional periodic layers of various types. As a rule for generating a statistical sequence of layers, a Markov chain of the nth order is used. The zero-order S = 0 means a completely random sequence of layers. A Markov chain with an order greater than zero is used to set the short-range order in the alternation of layers or in the methods of their superposition methods. In our calculations, we used the first-order Markov chain (S = 1), which makes it possible to create both strictly ordered structures and structures with layer alternation defects.

2.2.2. HRTEM

The structure and microstructure of the samples were studied by HRTEM using a ThemisZ electron microscope (Thermo Fisher Scientific, Waltham, MA, USA) with an accelerating voltage of 200 kV and a limiting resolution of 0.07 nm. Images were recorded using a Ceta 16 CCD array (Thermo Fisher Scientific, USA). High-angle annular dark-field imaging (HAADF STEM image) was performed using a standard ThemisZ detector. The instrument is equipped with a SuperX (Thermo Fisher Scientific, USA) energy-dispersive characteristic X-ray spectrometer (EDX) with a semiconductor Si detector with an energy resolution of 128 eV.

For electron microscope studies, sample particles were deposited on perforated carbon substrates fixed on copper grids using a UZD-1UCH2 ultrasonic disperser, which made it possible to achieve a uniform distribution of particles over the substrate surface. The sample is placed in an alcohol drop which is deposited on an ultrasonic disperser. After an increase in the ultrasonic frequency, the mixture of alcohol and sample in the form of vapor falls onto a standard copper grid.

Fast Fourier Transform Images (FFT—images) were made in Digital Micrograph software (Gatan, Pleasanton, CA, USA) from a selected area in the HRTEM images.

3. Results and Discussion

3.1. Interpretation of Experimental Data

3.1.1. X-ray Diffraction Analysis

According to the preliminary X-ray phase analysis (Figure 1), samples 1 and 2 were practically single-phase layered Sr_2TiO_4 perovskites. Small peaks of impurity phases are marked with asterisks: a peak at about 25.1° corresponds to an interplanar distance of 3.53 Å and is probably associated with the strontium carbonate phase [PDF 05-0418]; a peak with position 30.8° was not identified, unfortunately. However, a more detailed comparison of the X-ray diffraction patterns (Figure 1) showed that they are not identical—the positions of some reflections of the two samples differ (see the bar diagram in Figure 1). In addition, the peaks in the X-ray diffraction pattern of sample 2 are noticeably broader. The ratios of the peak heights for the samples also differ somewhat, which is especially noticeable for the pair of the strongest reflections, 103 and 110. The crystal structure models were refined by the Rietveld method (Figure S1a,b, Table S1). But for sample 2, some experimental peaks are shifted relative to the positions calculated from the refined average values of the lattice parameters. This effect is especially noticeable for peak 004 (Figure 2). The observed differences in the diffraction patterns of the two samples can be associated with the features of the real structure of the Sr_2TiO_4 synthesized. Therefore, we carried out studies using HRTEM.

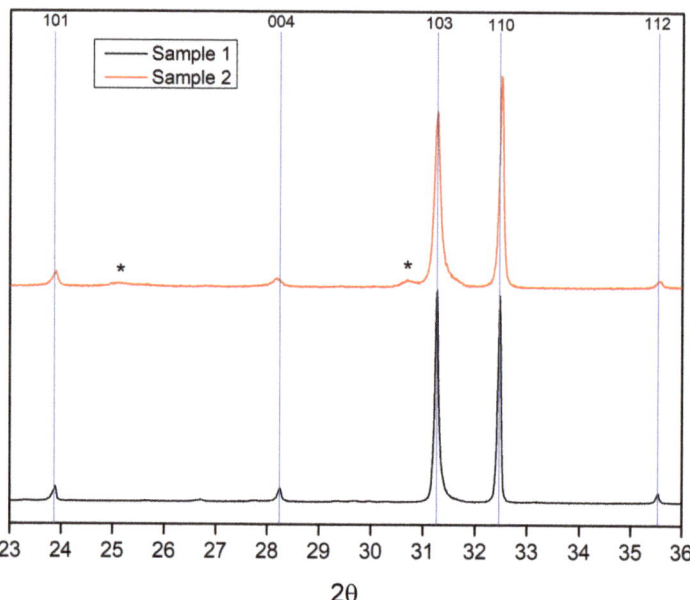

Figure 1. Comparison of the X-ray diffraction patterns of samples 1 and 2. The lines show the positions of the Sr_2TiO_4 reflections for sample 1. Asterisks indicate the weak peaks of impurity phases.

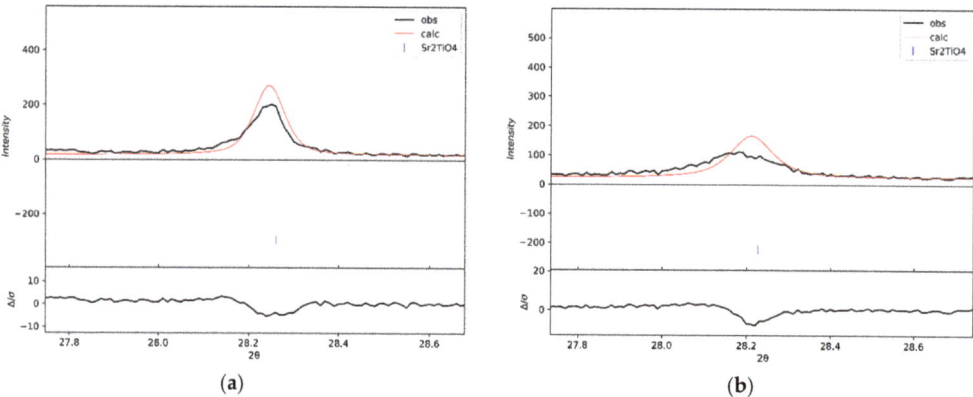

Figure 2. Positions of the experimental and calculated 004 XRD peaks based on the results of the refinement of the sample structures by the Rietveld method. (**a**) Sample 1; (**b**) Sample 2.

3.1.2. Electron Microscopy

HRTEM data have shown that both samples exhibit characteristic planar defects disrupting the structure periodicity and layers alternation in the [001] direction (Figures 3 and 4), but they are more numerous in sample 2. Similar results for systems with a perovskite structure were observed in [27,28].

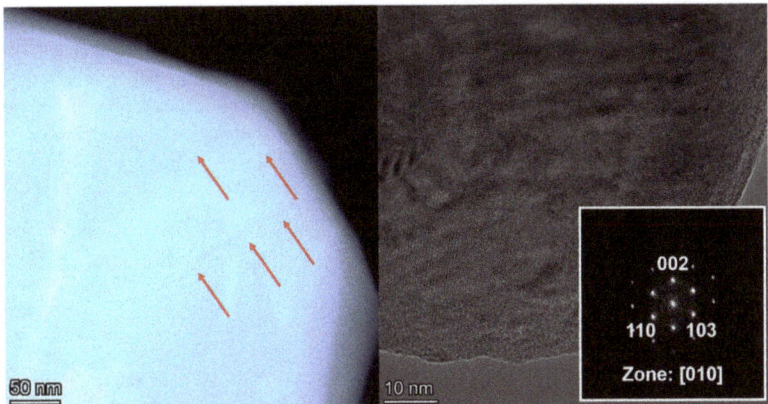

Figure 3. HAADF-STEM and HRTEM images of sample 1. Red arrows show planar defects in [001] direction, white square inset shows the corresponding FFT image.

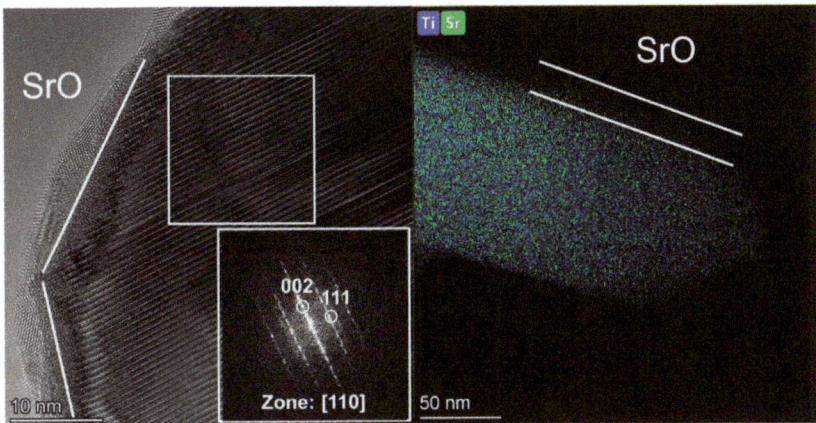

Figure 4. HRTEM image and EDX mapping of sample 2. The white square inset shows the corresponding FFT image with planar defects in [001] directions.

The HRTEM data also showed that the surface of the particles is covered with an oxide shell different from the main phase. Measurements of the interplanar distances of the coating showed that the oxide shell corresponds to SrO. It can be formed as a result of the incomplete decomposition of hydrocarbonate phases or $SrCO_3$ (traces of which are detected by XRD, Figure 1) due to the action of an electron beam on the sample under study in the microscope chamber. A detailed study of the reasons for the formation of such heterogeneous systems is presented in [29].

Partial segregation of divalent cations (Sr, Ca, etc.) on the surface due to the mechanisms proposed in [29] causes the formation of vacancy structures in the perovskite matrix that could lead to the formation of planar defects, both in the ABO_3 [30] and in the R-P structures [20]. This effect can arise due to the structural features of perovskite-like oxides as a result of the synthesis of complex oxides and their subsequent processing.

What can be the reason for the existence of such defects in Sr_2TiO_4? Planar defects associated with a violation of the order of alternation of layers can arise due to the peculiarities of the Ruddlesden–Popper structural series.

3.1.3. Structural Features of the Ruddlesden–Popper Series

The R-P phases are the layered perovskite-like oxides with the general formula $A_{n+1}B_nO_{3n+1}$. Their structure consists of $nABO_3$ perovskite layers situated between two AO rock-salt layers along the **c** crystallographic direction. The number of perovskite polyhedral units determines the phase peculiarities. The structures of two representatives of this series for our systems, namely Sr_2TiO_4 (n = 1) and $Sr_4Ti_3O_{10}$ (n = 3), are shown in Figure 5.

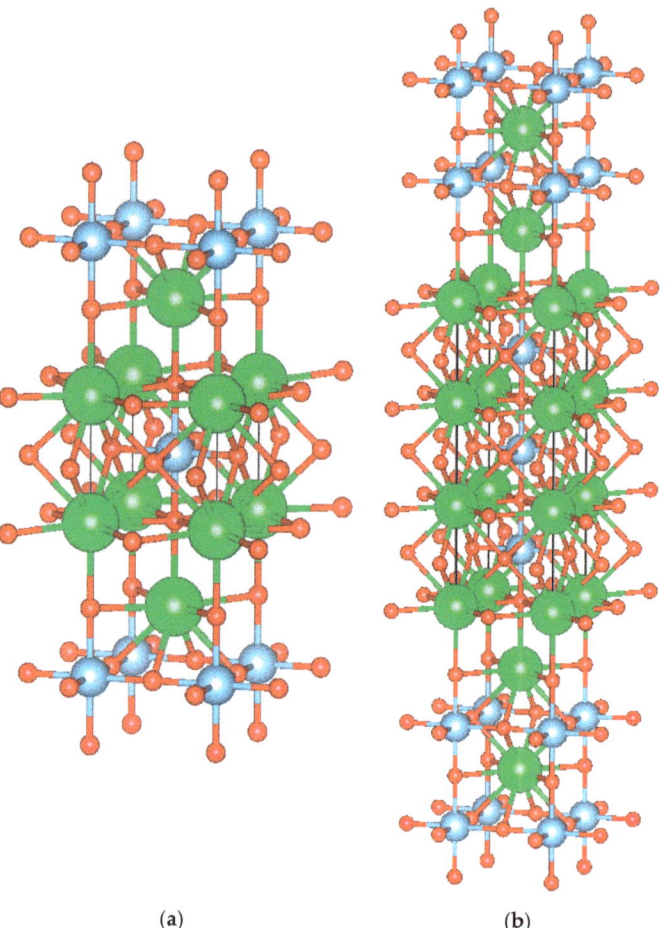

(a) (b)

Figure 5. Crystal structure of (**a**) Sr_2TiO_4; (**b**) $Sr_4Ti_3O_{10}$. Where: large (green) balls—strontium; medium (blue) balls—titanium; small (red) balls—oxygen [31,32].

Sr-Ti-O phases of this R-P series differ in the length of perovskite-like fragments. The presence of such identical perovskite-like layers suggests that layer alternation defects can quite easily occur in these structures. That is, fragments with one or two (or more) perovskite-like layers joining each other can appear. This may explain the observed violation of periodicity in electron microscopy images.

The formation of periodicity violation defects raises the question of their effect on X-ray powder diffraction patterns and the possibility of calculating their concentrations from diffraction data.

3.2. Simulation of XRD Patterns of Sr_2TiO_4 Containing Defects of Layer Alternation

Violation of the regular order in the alternation of layers can lead to various diffraction effects that affect the shape of the X-ray diffraction pattern profile. These are, for example, the broadening and/or shift of diffraction peaks with certain Miller indices [33] or the appearance of diffuse scattering maxima [23], among others. A technique that allows us to analyze the type of defects and estimate the concentration of planar defects is based on the simulation of diffraction patterns for one-dimensionally disordered structures [34].

3.2.1. Real Structure Model of Sr_2TiO_4

Modeling of the diffraction patterns was carried out using the program [28], which implements the method for constructing a full profile of X-ray diffraction patterns for highly dispersed and partially disordered objects. For the calculations, layers with the $SrTiO_3$ perovskite structure (layer A) and with the SrO structure (layer B) were chosen as "building blocks."

The ideal structure of Sr_2TiO_4 was set as a model with strict alternation of layers A and B (the fractions of layers $W_A = W_B = 0.5$, conditional probability of the existence of layer B after layer A $P_{AB} = 1$, as well as the probability of the existence of layer A after layer B $P_{BA} = 1$, which means that the probabilities $P_{AA} = 0$ and $P_{BB} = 0$).

An imperfect structure implies the appearance of layer A after layer A with some probability. Therefore, the fraction of layers A should be more than half. Thus, the structure with planar defects was specified as a particle of the same size but with the following conditions:

- the fraction of layer A $W_A = 0.5 + \delta$, and layer B, in turn—$W_B = 0.5 - \delta$; here δ is varied parameter;
- the probability of occurrence of layer B after layer B $P_{BB} = 0$ (but the probability of occurrence of layer A after layer A $P_{AA} \neq 0$ and depends, obviously, on the value of δ).

With the introduction of such conditions, with a certain probability, particles with an increased, in comparison with the ideal structure, length (thickness) of the perovskite-like layer can appear.

Obviously, the δ parameter is proportional to the defect concentration, which in this case can be defined as the fraction of SrO layers absent in the structure Sr_2TiO_4 (or the fraction of $SrTiO_3$ layers additionally intercalated into the structure).

Specific parameters of the model used for calculations (lattice parameters, layer thicknesses, atomic coordinates, particle sizes) are given in Table S3.

3.2.2. Results of Simulation of Diffraction Patterns

The calculated diffraction patterns (several characteristic reflections taken as an example) obtained during the simulation are shown in Figure 6.

According to the simulation data (Table 1), it can be seen that in the presence of defects, reflections from the different families of planes behave differently. The positions of some peaks shift while others remain in their place, determined only by the average parameters of the lattice. Peaks with zero indices l, such as 110, 200, and 220, do not change their positions, while reflections with a nonzero index l shift, some of which are very noticeable (for a given value of δ). In particular, the 004 peak shifts to the smaller angles and broadens (Figure 6), as observed in the experimental X-ray diffraction pattern of sample 2 (Figure 2b).

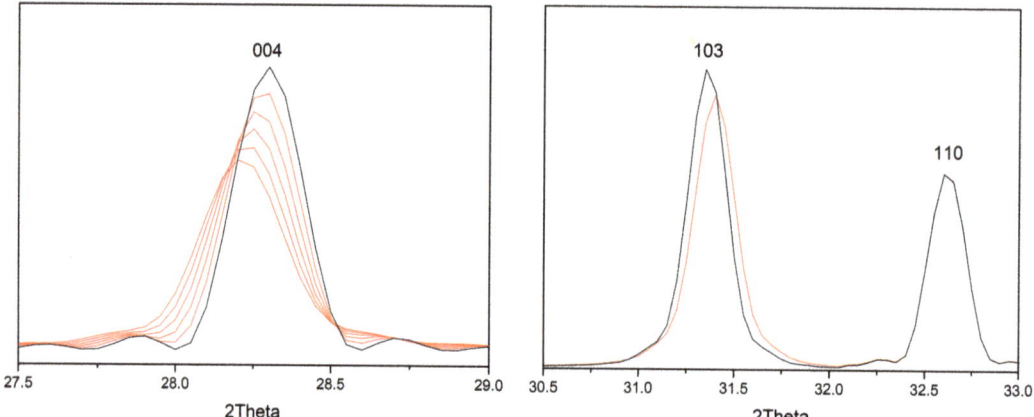

Figure 6. Some X-ray diffraction peaks calculated for Sr_2TiO_4 at different values of δ. XRD patterns for perfect (δ = 0) and imperfect structures are shown by black and red colors, respectively (δ value from 0 to 0.01 for 004 peaks, and δ = 0.01 for 103 and 110 peaks).

Table 1. Position of diffraction maxima.

2θ (°) without Defects, δ = 0	2θ (°) with Defects, δ = 0.01	h	k	l	Δ(2θ) (°)
23.90	23.88	1	0	1	−0.02
28.20	28.11	0	0	4	−0.09
31.25	31.29	1	0	3	+0.04
32.51	32.51	1	1	0	0.00
35.55	35.57	1	1	2	+0.02
42.96	42.87	0	0	6	−0.09
43.5	43.56	1	1	4	+0.06
46.64	46.64	2	0	0	0.00
53.06	53.07	2	1	1	+0.01
54.89	54.81	1	1	6	−0.08
55.35	55.39	2	0	4	+0.04
56.02	56.07	1	0	7	+0.05
57.24	57.21	2	1	3	−0.03
65.26	65.19	2	0	6	−0.07
68.08	68.08	2	2	0	0.00

The simulation results show that our test samples can have different concentrations of defects. Accordingly, the diffraction patterns were simulated using the program [33] for samples under study by varying the parameter δ, lattice parameters, and atomic coordinates. The results of the model optimization are shown below.

3.2.3. Compliance of the Calculated Diffraction Pattern with Experimental Data

Figure 7 demonstrates characteristic fragments of both samples' experimental and calculated X-ray diffraction patterns. For sample 1, the XRD pattern calculated based on the model of the perfect crystal structure of Sr_2TiO_4 (δ = 0) corresponds well with the experimental data (Figure 7a). For sample 2, the best agreement with the experiment is achieved at δ = 0.003 (Figure 7b). The calculated position of the 004 peaks for the defect structure model coincides with the experimental one and the positions of other peaks. Deviations of the maxima of the theoretical peaks from the experimental ones are within 0.01–0.02° (Table 2), which corresponds to the experimental error in determining the positions of the peaks (see the position of diffraction peaks for sample 1 in Table S2).

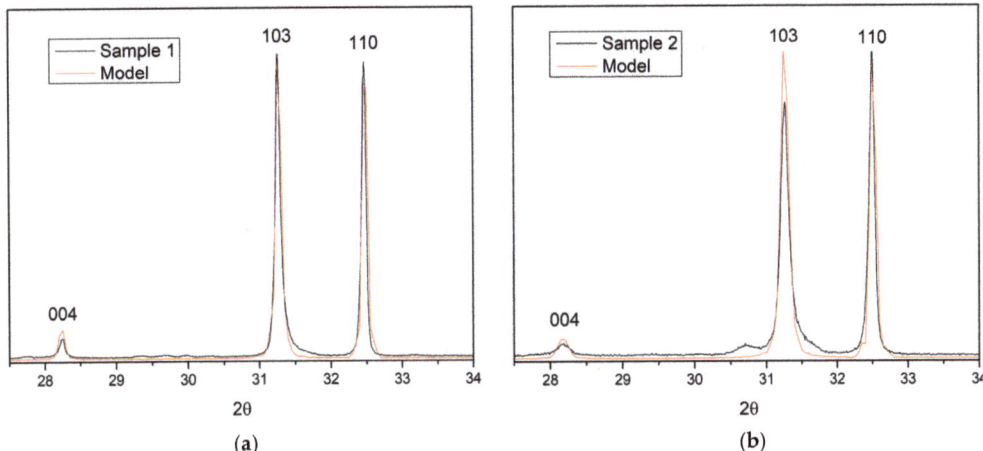

Figure 7. Comparison of experimental and simulated δ = 0.003 x-ray diffraction patterns for samples 1 (a) and 2 (b) (with δ = 0 and δ = 0.003, respectively).

Table 2. The positions of diffraction peaks for sample 2 calculated using the Rietveld method (average lattice parameters) and by modeling diffraction patterns for one-dimensionally disordered structures (at δ = 0.003).

h	k	l	Experiment	2θ (°) Rietveld Model	Defect Structure Model	Δ(2θ) (°) (Experiment-Rietveld Model)	Δ(2θ) (°) (Experiment-Defect Model)
1	0	1	23.87	23.91	23.89	−0.04	−0.02
0	0	4	28.17	28.23	28.17	−0.06	0.00
1	0	3	31.27	31.27	31.27	0.00	0.00
1	1	0	32.49	32.51	32.51	−0.02	−0.02
1	1	2	35.55	35.56	35.55	−0.01	0.00
0	0	6	42.89	42.91	42.90	−0.02	−0.01
1	1	4	43.55	43.58	43.55	−0.03	0.00
2	0	0	46.63	46.64	46.64	−0.01	−0.01
2	1	1	53.06	53.07	53.06	−0.01	0.00
1	1	6	54.85	54.85	54.84	0.00	0.01
2	0	4	55.39	55.41	55.38	−0.02	0.01
1	0	7	56.05	56.12	56.07	−0.07	−0.02
2	1	3	57.24	57.23	57.22	0.01	0.02
2	0	6	65.22	65.22	65.22	0.00	0.00
2	2	0	68.08	68.09	68.08	−0.01	0.00

When optimizing the model, we first paid attention to the positions of the diffraction peaks since the preliminary calculations showed that the shifts of some peaks were a characteristic indicator of the presence of defects in the violation of the alternation of layers. Perhaps in the future, it is necessary to make a complete analysis of the influence of these defects on the intensity of maxima. Qualitatively, it is seen that when these defects are introduced, the ratios of intensities of some reflections change, and they become closer to the experimentally observed ones. Still, a detailed analysis of this effect remains to be done. This may be important if we assume that the structure also contains point defects (vacancies). The number of point defects can be judged from the ratio of the intensities of the diffraction peaks. But planar defects also affect intensities. Only taking into account the contribution of planar defects, the number of which can be independently estimated, as our work shows, from the shift of peaks, we should proceed to the analysis of point defects if we are talking about their determination from diffraction data. Let us return once again to the nature of the detected defects. Since we are talking about a decrease in the relative number of layers of the SrO type in the defective crystal structure, this leads to a violation of the stoichiometry of the resulting phase. As for the calculation of the chemical composition of the phase containing defects, it is based on the general formula normalized to the Ti content (=1):

$$[(0.5 - \delta)SrO + (0.5 + \delta)SrTiO_3]/(0.5 + \delta) = SrO - [2\delta/(0.5 + \delta)]SrO + SrTiO_3.$$

It is immediately clear here that the difference from the stoichiometry of Sr_2TiO_4 is $-[2\delta/(0.5 + \delta)]SrO$. For $\delta = 0.003$, the difference from stoichiometry in Sr and O is -0.012.

Strontium ions that are not included in the crystal structure segregate on the surface, modifying their composition and properties. It is important to note that a number of authors [7–10] associate the catalytic activity of Sr_2TiO_4 oxides in the oxidative coupling of methane precisely with the segregation of Sr on the particle surface.

Since peak positions are of such great importance in terms of determining the concentration of defects, it is obvious that it is desirable to use precision experimental data obtained using synchrotron sources. At a low defect concentration (about 1% as in our case), the deviations of the position of the peak do not exceed $0.1°$, which, of course, requires special attention to measurement accuracy.

4. Conclusions

In this work, a crystal-chemical analysis of the possible structure of planar defects has been performed, defect structures were simulated, and the effects of such defects on the X-ray diffraction patterns of oxides of the A_2BO_4 type were established using Sr_2TiO_4 as an example. For the calculations, we used the method of constructing probabilistic models of one-dimensionally disordered structures. The most general qualitative results in this work, related to A_2BO_4 (R-P) oxides of any chemical composition, consisted of determining the diffraction signs of the presence of this type of defects, namely, the shift of peaks with specific Miller indices (see Table 1) with respect to their positions determined using the average lattice parameters. Of course, to obtain quantitative estimates, simulations must be carried out for each specific compound, taking into account specific lattice parameters and specific compositions.

A relation has been established between the concentration of planar defects and the real chemical composition (nonstoichiometry) of the Sr_2TiO_4 phase. The presence of defects leads to the Ti enrichment of particle volume and, consequently, to the enrichment of the surface with Sr. The latter can explain the catalytic activity of Sr_2TiO_4 in the oxidative coupling of methane.

Supplementary Materials: The supporting information can be downloaded at: https://www.mdpi.com/article/10.3390/ma15217642/s1.

Author Contributions: A.S.G. was responsible for processing and interpreting the XRD data, performed simulations of diffraction patterns, and contributed to the writing; S.V.T. carried out crystal chemical analysis, contributed to the writing; S.V.C. was responsible for the methodology for conducting simulations of diffraction patterns, contributed to the writing; E.Y.G. responsible for conducting HRTEM research and interpretation, contributed to the writing; S.N.P. was responsible for the synthesis of samples, contributed to the writing. All authors have read and agreed to the published version of the manuscript.

Funding: The work was carried out with the support of the Ministry of Science and Higher Education of the Russian Federation, state assignment: AAAA-A21-121011390053-4.

Institutional Review Board Statement: Not applicable.

Informed Consent Statement: Not applicable.

Data Availability Statement: Data available upon request.

Acknowledgments: The authors are grateful to A.N. Shmakov for his help in the synchrotron experiment.

Conflicts of Interest: The authors declare no conflict of interest.

References

1. Yatoo, M.A.; Skinner, S.J. Ruddlesden-Popper phase materials for solid oxide fuel cell cathodes: A short review. *Mater. Today Proc.* **2022**, *56*, 3747–3754. [CrossRef]
2. Lee, K.H.; Kim, S.W. Ruddlesden-Popper phases as thermoelectric oxides: Nb-doped $SrO(SrTiO_3)n$ (n = 1,2). *J. Appl. Phys.* **2006**, *100*, 063717. [CrossRef]
3. Sorkh-Kaman-Zadeh, A.; Dashtbozorg, A. Facile chemical synthesis of nanosize structure of Sr2TiO4 for degradation of toxic dyes from aqueous solution. *J. Mol. Liq.* **2016**, *223*, 921–926. [CrossRef]
4. Kwak, B.S.; Do, J.Y.; Park, N.K.; Kang, M. Surface modification of layered perovskite Sr_2TiO_4 for improved CO_2 photoreduction with H_2O to CH_4. *Sci. Rep.* **2017**, *7*, 16370. [CrossRef] [PubMed]
5. Xu, J.; Xi, R.; Xiao, Q.; Xu, X.; Liu, L.; Li, S.; Gong, Y.; Zhang, Z.; Fang, X.; Wang, X. Design of strontium stannate perovskites with different fine structures for the oxidative coupling of methane (OCM): Interpreting the functions of surface oxygen anions, basic sites and the structure–reactivity relationship. *J. Catal.* **2022**, *408*, 465–477. [CrossRef]
6. Lim, S.; Choi, J.-W.; Suh, D.J.; Lee, U.; Song, K.H.; Ha, J.-M. Low-temperature oxidative coupling of methane using alkaline earth metal oxide-supported perovskites. *Catal. Today* **2020**, *352*, 127–133. [CrossRef]
7. Li, X.-H.; Fujimoto, K. Low Temperature Oxidative Coupling of Methane by Perovskite Oxide. *Chem. Lett.* **1994**, 1581–1584. [CrossRef]
8. Yang, W.-M.; Yan, Q.-J.; Fu, X.-C. Oxidative coupling of methane over Sr-Ti, Sr-Sn perovskites and corresponding layered perovskites. *React. Kinet. Mech. Catal. Lett.* **1995**, *54*, 21–27. [CrossRef]
9. Ivanov, D.; Isupova, L.; Gerasimov, E.Y.; Dovlitova, L.; Glazneva, T.; Prosvirin, I. Oxidative methane coupling over Mg, Al, Ca, Ba, Pb-promoted $SrTiO_3$ and Sr_2TiO_4: Influence of surface composition and microstructure. *Appl. Catal. A Gen.* **2014**, *485*, 10–19. [CrossRef]
10. Ivanova, Y.A.; Sutormina, E.F.; Rudina, N.A.; Nartova, A.V.; Isupova, L.A. Effect of preparation route on Sr_2TiO_4 catalyst for the oxidative coupling of methane. *Catal. Commun.* **2018**, *117*, 43–48. [CrossRef]
11. Kondratenko E v Schlüter, M.; Baerns, M.; Linke, D.; Holena, M. Developing catalytic materials for the oxidative coupling of methane through statistical analysis of literature data. *Catal. Sci. Technol.* **2015**, *5*, 1668–1677. [CrossRef]
12. Alexiadis, V.; Chaar, M.; van Veen, A.; Muhler, M.; Thybaut, J.; Marin, G. Quantitative screening of an extended oxidative coupling of methane catalyst library. *Appl. Catal. B* **2016**, *199*, 252–259. [CrossRef]
13. Gambo, Y.; Jalil, A.A.; Triwahyono, S.; Abdulrasheed, A.A. Recent advances and future prospect in catalysts for oxidative coupling of methane to ethylene: A review. *J. Ind. Eng. Chem.* **2018**, *59*, 218–229. [CrossRef]
14. Ruddlesden, S.N.; Popper, P. New compounds of the K_2NiF_4 type. *Acta Cryst.* **1957**, *10*, 538. [CrossRef]
15. Pavlova, S.; Ivanova, Y.; Tsybulya, S.; Chesalov, Y.; Nartova, A.; Suprun, E.; Isupova, L. Sr2TiO4 Prepared Using Mechanochemical Activation: Influence of the Initial Compounds' Nature on Formation, Structural and Catalytic Properties in Oxidative Coupling of Methane. *Catalysts* **2022**, *12*, 929. [CrossRef]
16. Tilley, R.J.D. An electron microscope study of perovskite-related oxides in the Sr-Ti-O system. *J. Solid State Chem.* **1977**, *21*, 293–301. [CrossRef]
17. Čeh, M.; Kolar, D. Solubility of CaO in $CaTiO_3$. *J. Mater. Sci.* **1994**, *29*, 6295–6300. [CrossRef]
18. Fujimoto, M.; Tanaka, J.; Shirasaki, S. Planar Faults and Grain Boundary Precipitation in Non-Stoichiometric (Sr, Ca)TiO$_3$ Ceramics. *Jpn. J. Appl. Phys.* **1988**, *27*, 1162–1166. [CrossRef]
19. Mccoy, M.A.; Grimes, R.W.; Lee, W.E. Phase stability and interfacial structures in the SrO–$SrTiO_3$ system. *Philos. Mag. A* **1997**, *75*, 833–846. [CrossRef]
20. Šturm, S.; Rečnik, A.; Scheu, C.; Čeh, M. Formation of Ruddlesden–Popper faults and polytype phases in SrO-doped $SrTiO_3$. *J. Mater. Res.* **2000**, *15*, 2131–2139. [CrossRef]
21. Wood, N.D.; Teter, D.M.; Tse, J.S.; Jackson, R.A.; Cooke, D.J.; Gillie, L.J.; Parker, S.C.; Molinari, M. An atomistic modelling investigation of the defect chemistry of $SrTiO_3$ and its Ruddlesden-Popper phases, $Sr_{n+1}Ti_nO_{3n+1}$ (n = 1–3). *J. Solid State Chem.* **2021**, *303*, 122523. [CrossRef]
22. Gorkusha, A.S. Use of the synthesis method for the formation of layers of perovskite-like oxide Sr_2TiO_4 and its activity in the reaction of oxidative coupling of methane. In Proceedings of the International Scientific Student Conference, NSU, Novosibirsk, Russia, 10–20 April 2022.
23. Tsybulya, S.V.; Cherepanova, S.V.; Hasin, A.A.; Zaykovski, V.I.; Parmon, V.N. Struktura geterogennih kogerentnih sostoyaniy v visokodispersnih chasticah metallicheskogo kobalta. *Dokl. Akad. Nauk* **1999**, *366*, 216–220.
24. Siberian Center for Synchrotron and Terahertz Radiation. Available online: https://ssrc.biouml.org/#! (accessed on 20 October 2020).
25. Toby, B.H.; von Dreele, R.B. GSAS-II: The genesis of a modern open-source all purpose crystallography software package. *J. Appl. Cryst.* **2013**, *46*, 544–549. [CrossRef]
26. Drits, V.A.; Tchoubar, C. *X-ray Diffraction by Disordered Lamellar Structures: Theory and Applications to Microdivided Silicates and Carbons*; Springer: Berlin/Heidelberg, Germany, 1990. [CrossRef]
27. Gerasimov EYu Rogov, V.A.; Prosvirin, I.P.; Isupova, L.A.; Tsybulya, S.V. Microstructural Changes in $La_{0.5}Ca_{0.5}Mn_{0.5}Fe_{0.5}O_3$ Solid Solutions under the Influence of Catalytic Reaction of Methane Combustion. *Catalysts* **2019**, *9*, 563. [CrossRef]
28. Bokhimi; Portilla, M. Oxygen and the Formation of New Ordered Perovskite-Based Structures in the Bi-Sr-O System. *J. Solid State Chem.* **1993**, *105*, 371–377. [CrossRef]

29. Koo, B.; Kim, K.; Kim, J.K.; Kwon, H.; Han, J.W.; Jung, W. Sr Segregation in Perovskite Oxides: Why It Happens and How It Exists. *Joule* **2018**, *2*, 1476–1499. [CrossRef]
30. Gerasimov, E.Y.; Isupova, L.A.; Tsybulya, S.V. Microstructural features of the $La_{1-x}Ca_xFeO_{3-\delta}$ solid solutions prepared via Pechini route. *Mater. Res. Bull.* **2015**, *70*, 291–295. [CrossRef]
31. Kawamura, K.; Yashima, M.; Fujii, K.; Omoto, K.; Hibino, K.; Yamada, S.; Hester, J.R.; Avdeev, M.; Miao, P.; Torii, S.; et al. Structural origin of the anisotropic and isotropic thermal expansion of K_2NiF_4-type $LaSrAlO_4$ and Sr_2TiO_4. *Inorg. Chem.* **2015**, *54*, 3896–3904. [CrossRef]
32. Ruddlesden, S.N.; Popper, P. The compound Sr3Ti2O7 and its structure. *Acta Cryst.* **1958**, *11*, 54–55. [CrossRef]
33. Cherepanova, S.V.; Tsybulya, S.V. Simulation of X-ray Powder Diffraction Patterns for One-Dimensionally Disordered Crystals. *Mater. Sci. Forum* **2004**, *443–444*, 87–90. [CrossRef]
34. Kakinoki, J.; Komura, Y. Intensity of X-ray Diffrection by an One-Dimensionally Disordered Crystal. *J. Phys. Soc. Jpn.* **1952**, *7*, 30–35. [CrossRef]

Article

High Throughput Preparation of Ag-Zn Alloy Thin Films for the Electrocatalytic Reduction of CO_2 to CO

Jiameng Sun [1], Bin Yu [1], Xuejiao Yan [2], Jianfeng Wang [1], Fuquan Tan [1], Wanfeng Yang [1], Guanhua Cheng [1,*] and Zhonghua Zhang [1,*]

[1] Key Laboratory for Liquid-Solid Structural Evolution and Processing of Materials (Ministry of Education), School of Materials Science and Engineering, Shandong University, Jinan 250061, China
[2] Taian Institute of Supervision & Inspection on Product Quality, Taian 271000, China
* Correspondence: guanhua.cheng@sdu.edu.cn (G.C.); zh_zhang@sdu.edu.cn (Z.Z.)

Abstract: Ag-Zn alloys are identified as highly active and selective electrocatalysts for CO_2 reduction reaction (CO_2RR), while how the phase composition of the alloy affects the catalytic performances has not been systematically studied yet. In this study, we fabricated a series of Ag-Zn alloy catalysts by magnetron co-sputtering and further explored their activity and selectivity towards CO_2 electroreduction in an aqueous $KHCO_3$ electrolyte. The different Ag-Zn alloys involve one or more phases of Ag, AgZn, Ag_5Zn_8, $AgZn_3$, and Zn. For all the catalysts, CO is the main product, likely due to the weak CO binding energy on the catalyst surface. The Ag_5Zn_8 and $AgZn_3$ catalysts show a higher CO selectivity than that of pure Zn due to the synergistic effect of Ag and Zn, while the pure Ag catalyst exhibits the highest CO selectivity. Zn alloying improves the catalytic activity and reaction kinetics of CO_2RR, and the $AgZn_3$ catalyst shows the highest apparent electrocatalytic activity. This work found that the activity and selectivity of CO_2RR are highly dependent on the element concentrations and phase compositions, which is inspiring to explore Ag-Zn alloy catalysts with promising CO_2RR properties.

Keywords: high throughput; electrocatalysis; CO_2 reduction; magnetron sputtering; silver-zinc alloy

Citation: Sun, J.; Yu, B.; Yan, X.; Wang, J.; Tan, F.; Yang, W.; Cheng, G.; Zhang, Z. High Throughput Preparation of Ag-Zn Alloy Thin Films for the Electrocatalytic Reduction of CO_2 to CO. *Materials* 2022, *15*, 6892. https://doi.org/10.3390/ma15196892

Academic Editor: Ilya V. Mishakov

Received: 3 September 2022
Accepted: 30 September 2022
Published: 4 October 2022

Publisher's Note: MDPI stays neutral with regard to jurisdictional claims in published maps and institutional affiliations.

Copyright: © 2022 by the authors. Licensee MDPI, Basel, Switzerland. This article is an open access article distributed under the terms and conditions of the Creative Commons Attribution (CC BY) license (https:// creativecommons.org/licenses/by/ 4.0/).

1. Introduction

With the worldwide consumption of fossil fuels, the increasing emission of carbon dioxide (CO_2) in the atmosphere is becoming a more and more serious environmental threat [1–3]. Artificial conversion of CO_2 into fuels and chemical feedstocks is essential to reduce the concentration of CO_2 and mitigate the greenhouse effect. The available methods include electrochemical, photoelectrochemical, thermochemical, and biological conversions, of which electrochemical reduction is regarded as a clean and effective method due to its mild reaction conditions and ability to use renewable electricity [4–6].

Electrochemical CO_2 reduction reaction (CO_2RR) can yield a variety of products through two-, four-, six-, and eight-electron pathways, such as carbon monoxide (CO), methane (CH_4), formic acid (HCOOH), methanol (CH_3OH) and C2/C3 compounds [3,7–9]. CO is the simplest and one of the most common products of CO_2RR due to the sluggish kinetics of C–C coupling, which can be further electrocatalytically reduced to high-value chemicals or used as a component of syngas to produce hydrocarbon fuels (e.g., aldehydes, methanol, and phosgene) by the Fischer-Tropsch process [10,11]. Therefore, it is important to study efficient catalysts to selectively reduce CO_2 to CO ($CO_2 + 2H^+ + 2e^- \leftrightarrow CO + H_2O$) [12]. However, the first step of CO_2RR requires a larger overpotential compared to hydrogen evolution reaction (HER: $2H^+ + 2e^- \leftrightarrow H_2$), which will inevitably occur as a main competitive reaction in the aqueous CO_2 reduction system [13,14]. The challenges of CO_2RR are the high overpotential, kinetically sluggish multi-electron transfer process, and the selectivity against the HER. Developing promising CO_2RR catalysts with high activity and selectivity will be the key point to efficiently utilizing CO_2.

Various metals (e.g., Au, Ag, Zn) have been identified as potential electrocatalysts to overcome the sluggish kinetics of CO_2 activation and improve the selectivity of CO [12,15,16]. The Sabatier principle shows that adsorbed CO (CO*) is an important reaction intermediate for the production of hydrocarbons and oxygen-containing compounds, and optimal binding energy of CO* (E_B[CO]) can reduce the yield of other products and lead to higher selectivity of CO [17]. Au and Ag show a high selectivity of CO in CO_2RR, due to the low HER activity and weak E_B[CO] [18]. Ag is a more cost-effective catalyst for selective reduction of CO_2 to CO compared to Au, which is more than 70 times more expensive than Ag [19]. Zn processes the ability to form CO in CO_2RR, which is an earth-abundant metal and much cheaper than Ag. Although Zn does not show satisfying activity and selectivity, it is one of the most promising non-previous metals for CO_2RR in the CO-forming class [20]. Jaramillo and Nørskov et al. reported that the binding energy of adsorbed COOH (COOH*, E_B[COOH]) is a key descriptor for CO generation efficiency in the CO_2RR [21]. In the volcano plot of E_B[COOH], Ag and Zn appear on the weak- and strong-binding legs of the plot respectively [12], therefore, Ag-Zn alloys would have a more appropriate E_B[COOH] to improve the activity of CO_2RR than either single metal (Ag or Zn) [12]. Taking advantage of the synergistic effect, Ag-Zn alloys may be able to increase the CO selectivity and activity of CO_2RR while reducing the catalyst cost.

High-throughput catalyst screening technique is a promising scientific method, which has been developed for the discovery and optimization of catalysts. Using this method can speed up the catalyst discovery process in exploring the complex catalyst composition space, such as alloying elements and surface molecules [22,23]. Applying the concept of high throughput in material preparation, samples with different element concentrations can be prepared directly at one time, significantly simplifying the sample preparation process, and providing convenience for the subsequent exploration of electrochemical properties.

In this work, to identify the relation between the CO_2RR activity and the element concentrations as well as phase compositions of Ag-Zn alloys, we applied the concept of high throughput in material preparation and synthesized a series of Ag-Zn bimetallic alloys by composition gradient sputtering. The selectivity and activity of the CO_2RR on catalysts with different element concentrations and phase compositions were compared for the efficient electrochemical production of CO.

2. Experimental

2.1. Catalysts Preparation

The Ag-Zn alloy catalysts were prepared by magnetron sputtering using carbon fiber paper (CFP) as the substrate. CFP was ultrasonically cleaned in ethanol for 5 min and dried at 60 °C in air. Subsequently, Ag and Zn were directly deposited onto the CFP using direct-current magnetron sputtering equipment (SKY Technology Development Co., Ltd., Shenyang, China) with a working distance of 6 cm. High-purity Ag and Zn (99.99 wt.%, Beijing Dream Material Technology Co., Ltd., Beijing, China) were used as the sputtering targets. The normal inclination angle between the target and CFP substrate was 20°, and the sample holder stood still to obtain samples with concentration gradients [24,25]. The sputtering was operated at room temperature for 1 h, with a power of 25 W for the Ag target and 35 W for the Zn target, and an Ar pressure of 1 Pa at a flow rate of 30 cm^3 min^{-1}. Similarly, pure Ag and Zn were also sputtered on CFP, respectively. The sputtering power was 25 W for Ag and 35 W for Zn at a rotating speed of 5 revolutions per minute for the substrate. Finally, we obtained uniform Ag and Zn films with a mass loading of around 0.7 mg cm^{-2}, and Ag-Zn alloy films with different concentrations and a mass loading of about 1.4 mg cm^{-2}.

2.2. Material Characterizations

X-ray diffraction (XRD) patterns were recorded by an XD-3 diffractometer equipped with Cu Kα radiation (Beijing Purkinje General Instrument Co., Ltd., Beijing, China). The microstructures and chemical compositions of the samples were characterized by a scanning electron microscope (SEM, COXEM EM-30, Daejeon, Korea) equipped with an energy dispersive X-ray (EDX) analyzer.

2.3. Electrochemical Measurements

Electrochemical measurements were conducted on a CHI 760E potentiostat in an H-type cell with an ion-exchange membrane separator (Nafion 117 membrane, FuelCell Store, TX, America) inserted between the cathodic and anodic chambers. 0.5 M $KHCO_3$ solution was used as the electrolyte in both the cathode and anode compartments. The sputtering samples, a Pt foil, and a double-junction saturated calomel electrode (SCE, saturated KCl, TIANJINAIDA R0232, Tianjin, China) were used as the working, counter, and reference electrodes, respectively. The potentials reported in this work were converted to be against the reversible hydrogen electrode (RHE) according to the following equation:

$$E_{RHE} = E_{SCE} + 0.241 + 0.059 \times PH \quad (1)$$

A chronoamperometric electrolysis test was carried out at the working electrode with a geometric area of 1 cm^2. Before the test, the fresh electrolyte solution was bubbled with high-purity CO_2 for 15 min, and the test was under a continuous CO_2 (99.9%) gas flow of 20 mL min^{-1}. The gas-phase products from the gastight cathodic compartment were directly introduced into a gas chromatograph (GC, Lunan Instrument GC-7820, Jinan, China) which used high-purity Ar as the carrier gas for all compartments and analyzed using a thermal conductivity detector (TCD) and two flame ionization detectors (FID1 and FID2). The amount of CO gas was estimated from the FID1 data and H_2 gas from the TCD data. An external standard method was adopted to quantify the products. The gas-phase products were injected into GC every 12 min during the CO_2RR in the chronoamperometry measurement mode.

3. Results and Discussion

3.1. Preparation and Characterization of the Ag-Zn Alloys as well as Pure Ag, Zn Films

As illustrated in Figure 1a, the Ag-Zn alloy film was sputtered on CFP by a co-sputtering method. The macroscopic picture of the sputtering film in Figure 1b, shows a color gradient, indicating a corresponding concentration gradient in different regions. The sample appears black gray and bright silver on the side near the Zn target and the Ag target, respectively, and shows a color transition between the two regions. The concentration gradient distribution of the obtained Ag-Zn alloy film is shown in Figure 1c,d based on EDX analysis, and several respective results (e.g., $Ag_{10}Zn_{90}$, $Ag_{20}Zn_{80}$, $Ag_{30}Zn_{70}$, $Ag_{35}Zn_{65}$, $Ag_{40}Zn_{60}$, $Ag_{45}Zn_{55}$, $Ag_{55}Zn_{45}$, $Ag_{70}Zn_{30}$, and $Ag_{85}Zn_{15}$) are shown in Figure S1. In addition, pure Ag and Zn films with a uniform composition were prepared by sputtering with the single respective metal target. The macroscopic view of the pure Zn and Ag films shows that the film color is consistent with the respective metal target (Figure S2).

The morphologies and structures of the Ag-Zn alloy films were investigated by SEM. It obviously shows that the surface of pure Zn film is grainy (Figure 2a), while it is smooth and without obvious features for the pure Ag film (Figure 2b). As shown in Figure S3, with the Ag concentration increasing, the roughness of the film surface decreases and becomes smooth gradually. When the atomic percentage of Ag reaches 20%, the film becomes relatively smooth, and a further increase in Ag concentration shows little effect on the morphology. The $Ag_{45}Zn_{55}$ alloy shows similar surface morphology as that of pure Ag film as shown in Figure 2c. The EDX mappings (Figure 2d–f) show both Ag and Zn coat well on the fiber surface of the CFP.

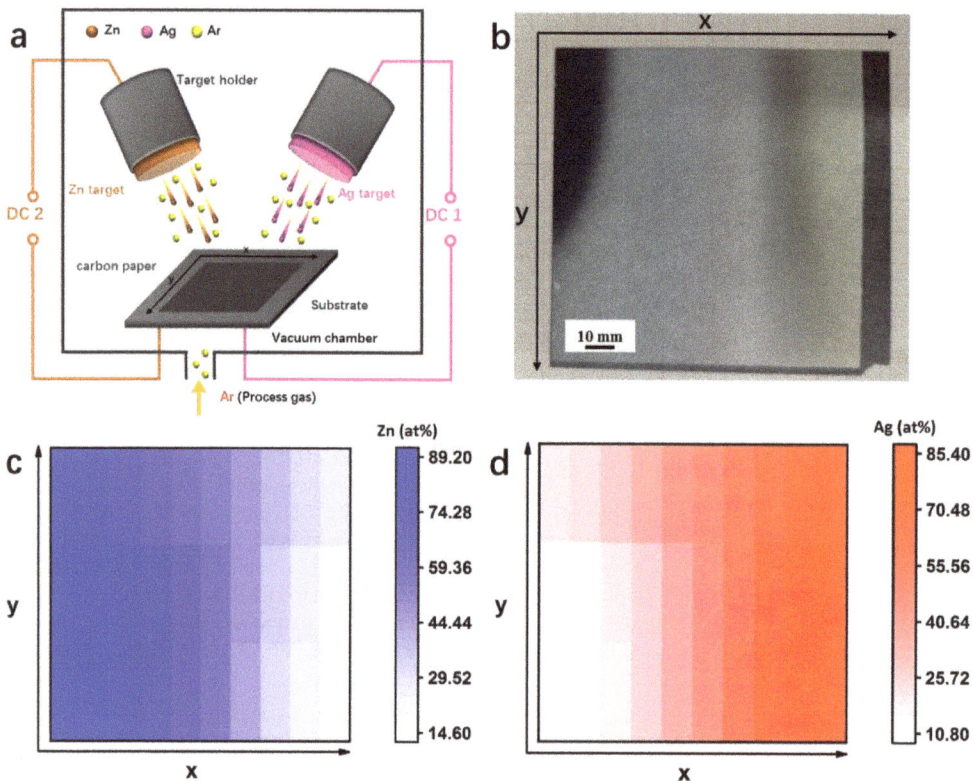

Figure 1. (a) Schematical illustration showing the preparation process of the Ag-Zn film catalysts by co-sputtering. (b) Macroscopic view of the Ag-Zn film after co-sputtering. (c,d) Concentration gradient distribution of (c) Zn and (d) Ag elements.

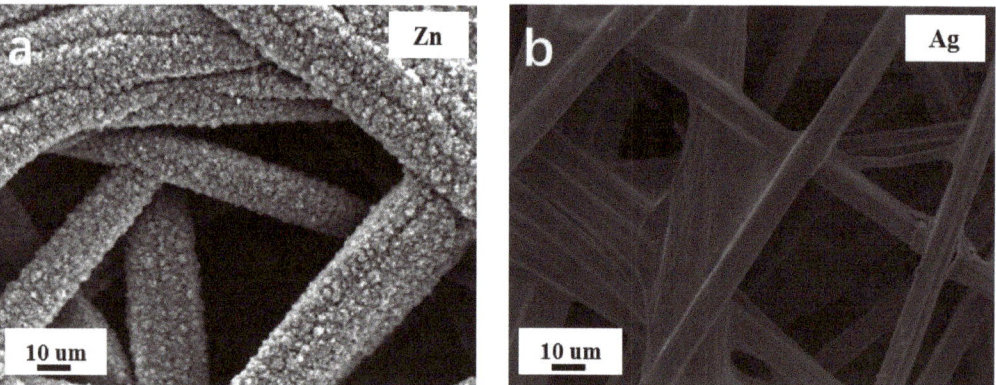

Figure 2. *Cont.*

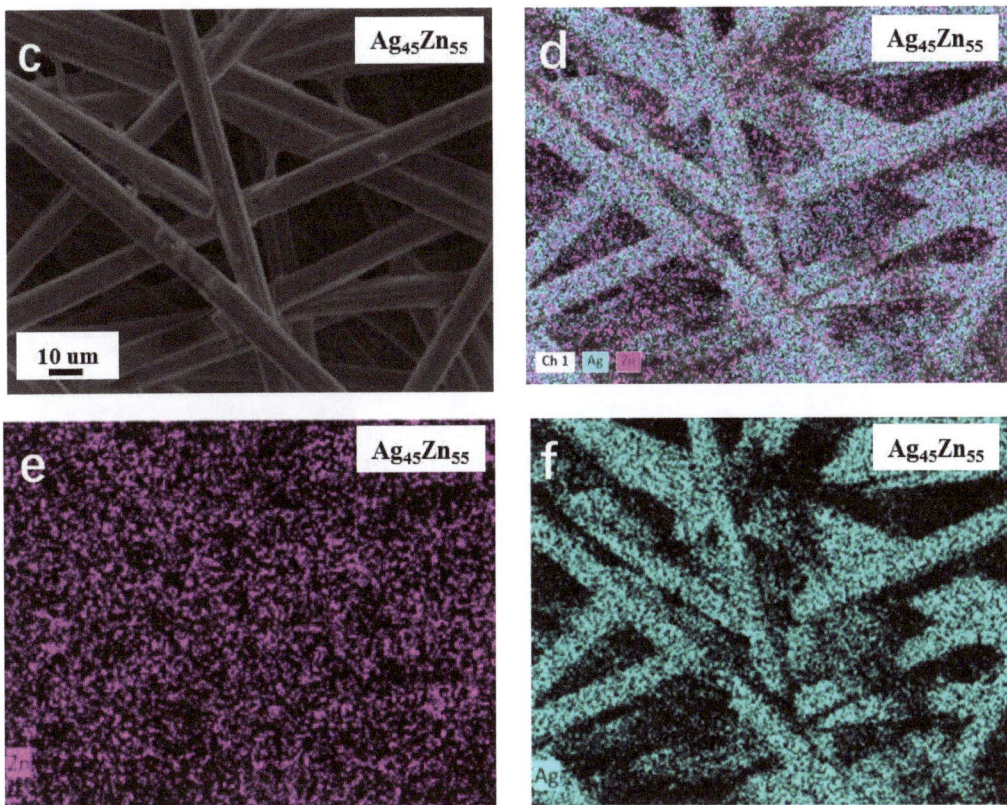

Figure 2. (**a–c**) SEM images of the (**a**) Zn, (**b**) Ag, and (**c**) Ag$_{45}$Zn$_{55}$ films. (**d–f**) EDX mapping images of the Ag$_{45}$Zn$_{55}$ film.

The phase constitutions of the selected samples with different concentrations were characterized by XRD in Figure 3a. The characteristic peaks at 18.0°, 26.4°, and 54.5° are ascribed to CFP (Figure S4). The diffraction peaks of pure Ag film at 38.1°, 44.3°, 64.4°, 77.4°, and 81.5° are consistent with the standard Ag phase (PDF #65-2871). For the Ag$_{55}$Zn$_{45}$, Ag$_{70}$Zn$_{30}$, and Ag$_{85}$Zn$_{15}$ catalysts, the diffraction peaks of Ag are observed to shift towards higher 2θ angles, indicating the formation of Ag(Zn) solid solution alloy associated with the lattice distortion/contraction (Figure 3b) [26]. As shown in Figure 3c,d, other catalysts with a smaller atomic ratio of Ag to Zn consist of the AgZn phase (Ag$_{45}$Zn$_{55}$, Ag$_{40}$Zn$_{60}$, and Ag$_{35}$Zn$_{65}$), Ag$_5$Zn$_8$ phase (Ag$_{30}$Zn$_{70}$ and Ag$_{20}$Zn$_{80}$), AgZn$_3$ phase (Ag$_{10}$Zn$_{90}$) and Zn phase (pure Zn catalyst) according to their diffraction peak positions. Therefore, a series of Ag-Zn alloy films with different compositions and phases were successfully fabricated through the high throughput co-sputtering method.

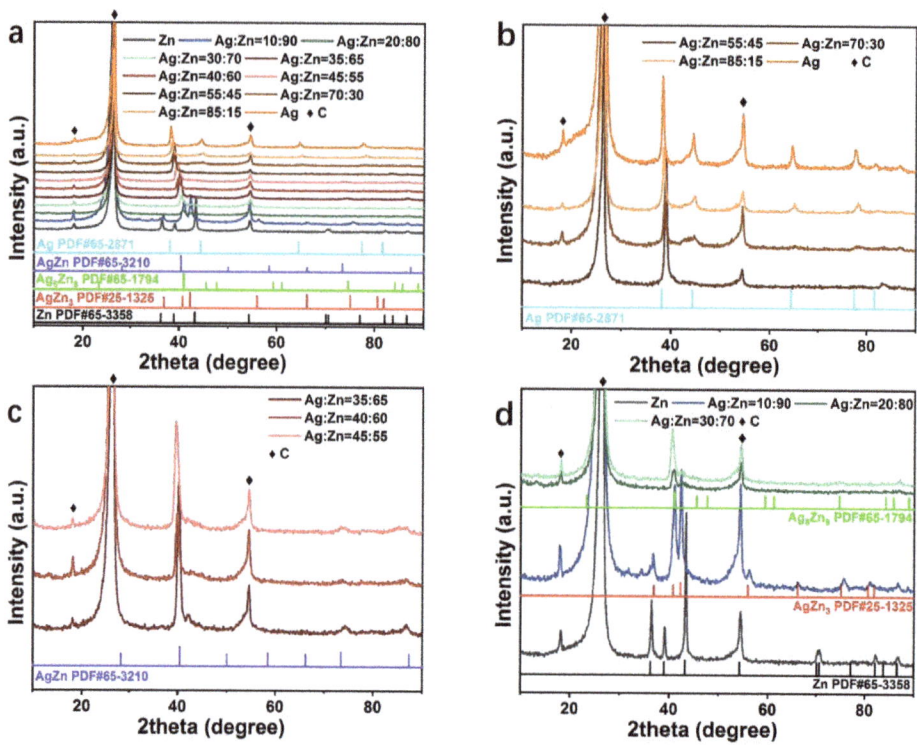

Figure 3. (a) XRD patterns of the pure Zn, pure Ag, and Ag-Zn alloy catalysts with different concentrations. (b–d) XRD patterns of the (b) Ag, $Ag_{55}Zn_{45}$, $Ag_{70}Zn_{30}$, $Ag_{85}Zn_{15}$, (c) $Ag_{35}Zn_{65}$, $Ag_{40}Zn_{60}$, $Ag_{45}Zn_{55}$, and (d) $Ag_{10}Zn_{90}$, $Ag_{20}Zn_{80}$, $Ag_{30}Zn_{70}$, and Zn samples.

3.2. Electrochemical CO_2RR Performance over Ag-Zn Alloys as well as Pure Ag, Zn Films in the H-Cell

The electrocatalytic performance of CO_2RR was explored over the Ag-Zn alloys in CO_2-saturated 0.5 M $KHCO_3$ by using a gas-tight H-cell. For comparison, the electrochemical CO_2RR performance of the pure Ag and Zn films was also evaluated under the same reaction conditions. Potentiostatic experiments were conducted at 7 different potentials in the range of −0.6 to −1.2 V vs. RHE. The faradaic efficiency of x (FE_x) production and its partial current density (j_x) was calculated as follows [12,27]:

$$FE_x\ (\%) = \frac{n_x(\text{mol}) \cdot N \cdot F\ (\text{C mol}^{-1})}{Q\ (\text{C})} \times 100 = \frac{0.1315 \cdot V\ \left(\frac{\text{mL}}{\text{min}}\right) \cdot R_x}{1000 \cdot i_{\text{total}}\ (\text{mA})} \times 100 \qquad (2)$$

$$j_x\ \left(\text{mA cm}^{-2}\right) = \frac{i_x(\text{mA})}{A\ (\text{cm}^2)} = \frac{i_{\text{total}}\ (\text{mA}) \cdot FE_x}{A\ (\text{cm}^2)} \qquad (3)$$

where n_x is the amount of product x (mol), N is the number of electrons transferred for product x formation (N = 2 for product CO and H_2), F is the Faraday constant (96,485 C mol^{-1}), Q is the charge passed to produce x (C), V is the actual flow rate during the test, R_x is the concentration of product x from the cathode compartment (obtained from GC data), i_x is the current value of product x (mA), and A is the geometric surface area (cm^2). In general, the reduction mechanism of CO_2 to CO is as follows [28]:

$$CO_2(g) + * + H^+(aq) + e^- \leftrightarrow COOH^* \qquad (4)$$

$$COOH^* + H^+(aq) + e^- \leftrightarrow CO^* + H_2O(l) \qquad (5)$$

$$CO^* \leftrightarrow CO(g) + * \qquad (6)$$

where * denotes an active site. A promising catalyst for selectively reducing CO_2 to CO needs to appropriately immobilize $COOH^*$, therefore, to facilitate the formation of CO^* in the next elementary step (Equation (5)), and to bind CO weakly so that to release the active site easily (Equation (6)). That is, a relatively stronger binding of $COOH^*$ than CO^* is necessary for efficient CO_2RR [17,18].

The faradaic efficiency of CO (FE_{CO}), H_2 (FE_{H_2}) as well as CO and H_2 ($FE_{CO\&H_2}$) of CO_2RR was plotted as a function of applied potentials in Figure 4. CO and H_2 are the major products of the Ag-Zn alloy, indicating that the Ag-Zn alloy is a CO-forming catalyst, akin to Ag and Zn [29]. For all Ag-Zn film electrodes, H_2 is the major product at less negative potentials due to the high overpotential required for CO_2RR [18,30]. Whereas under the moderate potentials, CO is the major product. There is enough electrochemical driving force to reduce CO_2 to CO and sufficient CO_2 supply at the electrode interface so that the reaction rate of CO_2 to CO exceeds that of HER generated from water [31]. At more negative potentials, the mass-transport limitations of CO_2 may suppress CO yielding [18,28].

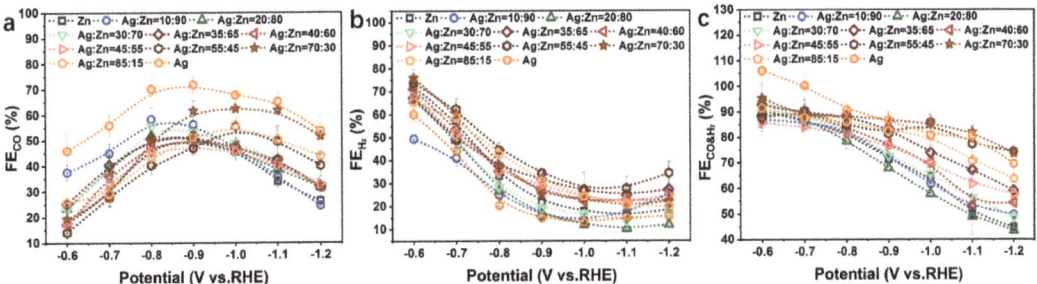

Figure 4. (a) FE_{CO}, (b) FE_{H_2} and (c) $FE_{CO\&H_2}$ for the pure Zn, pure Ag, and Ag-Zn alloy catalysts with different element concentrations.

As shown in Figure 4a, different phase compositions and concentrations of the Ag-Zn alloy catalysts exhibit different trends of FE_{CO} at $-0.6 \sim -1.2$ V vs. RHE. When the Ag content is high and Ag/Ag(Zn) solid solution phase is displayed, the maximum FE_{CO} appears in the range of $-0.9 \sim -1.0$ V vs. RHE, compared to the range of $-0.8 \sim -0.9$ V vs. RHE for other AgZn and Zn phases. For the Ag(Zn) phase catalysts, pure Ag catalyst shows the highest FE_{CO} over the entire potential range investigated (Figure 4a), and the addition of Zn to Ag decreases CO yielding (Figure S5a), showing that Ag is more effective at reducing CO_2 to CO than the Ag-Zn alloys or Zn. This trend can be explained by the changing of $E_B[CO]$ [17,28]. The volcano-type relation of the partial current density of CO as a function of $E_B[CO]$ shows that the catalyst surface with a weak $E_B[CO]$ can be limited by the activation of CO_2 due to the instability of the $COOH^*$ intermediate, whereas surfaces with a strong $E_B[CO]$ are limited by highly stabilized CO^* [17]. Both Ag and Zn have weak $E_B[CO]$ and belong to the category of surfaces limited by the activation of CO_2. Ag is closer to the top of the volcano plot compared to Zn, and better for CO_2 activation, which explains why Ag is more effective for CO formation [31,32]. When the alloy films consist of the AgZn phase, FE_{CO} decreases with the increase of Ag concentration at the intermediate potentials, while has little change at low and high potentials (Figure S5b). Pure Zn catalyst shows the lowest current efficiency for CO production, and the small amount addition of Ag (10 at.%) significantly improves the FE_{CO} by ~100% at $-0.6 \sim -0.7$ V vs. RHE (Figure S5c). At higher potentials, the $AgZn_3$ phase catalyst shows the closest FE_{CO} to pure Ag catalyst, but with the potential decreasing, it is consistent with the pure Zn. The CO yielding of the Ag_5Zn_8 catalyst is higher than that of pure Zn and increases with the increase of Ag concentration at lower potentials. This indicates that the $AgZn_3$ phase catalyst is favorable for CO production at higher potentials, while the Ag_5Zn_8 phase catalyst is favorable at lower potentials. The scenario of increased CO selectivity of the Ag-Zn alloys compared to Zn at intermediate potentials may originate from the electron transfer from Ag to Zn [33], due to the lower work function of Ag (4.26 eV) than Zn (4.33 eV) [34–36]. The transferred electrons are found to fill the bonding states of localized d-orbitals and exhibit a conspicuous upshift toward the Fermi level, which enhances the adsorption of the intermediate COOH*, resulting in a higher catalytic activity for CO production compared to pure Zn catalysts [35,37].

Figure 4b shows the changes of FE_{H_2} with potentials, which are basically the same for all Ag-Zn film electrodes. FE_{H_2} decreases rapidly with the potential decreasing, and tends to flatten or even slightly rebound at lower potentials. This indicates that water decomposition is more favorable at less negative potentials, while more active sites are occupied by CO_2 and related intermediates at more negative potentials. Pure Ag, Ag_5Zn_8 ($Ag_{30}Zn_{70}$ and $Ag_{20}Zn_{80}$ catalysts), $AgZn_3$ ($Ag_{10}Zn_{90}$ catalyst), and pure Zn phase catalysts exhibit lower FE_{H_2}, whereas that of Ag(Zn) solid solution and AgZn phase catalysts is higher (Figure S6), indicating that the former can inhibit HER better.

It can be seen from Figure 4c that the combined faradaic efficiency of CO and H_2 can reach over 80% at the less negative potential range, and almost no other products are produced. This is due to the weak E_B[CO] at lower overpotential, leading to the rapid desorption of CO from the surface once generated, and therefore, it is difficult to carry out further reduction [31]. In addition, other complex products require multiple proton-electron transfer steps and a larger driving force (more negative potentials) [18]. The $FE_{CO\&H_2}$ for Ag/Ag(Zn) solid solution alloy catalysts decreases by 20–30% at a more negative potential (Figure S7a), that of AgZn phase catalysts decreases by about 30% (Figure S7b), and that of Ag_5Zn_8, $AgZn_3$ as well as pure Zn phase catalysts decreases by about 40% (Figure S7c). This is ascribed to the formation of other liquid products, demonstrating an unfavorable competition between the further reduction of CO* and desorption of CO [18,31]. An explanation is that a higher overpotential provides a higher driving force and improves the electron transfer, which enhances CO* electroreduction, competing with the non-faradaic step of CO desorption [31].

Figure 5a presents the overall electrode activity of the Ag-Zn alloys as well as pure Zn and Ag as a function of the applied potentials. For all electrodes, a well-defined exponential increase in the overall activity is observed with decreasing potentials. With the addition of Zn in Ag, the total current densities (j_{total}) increase for the Ag/Ag(Zn) solid solution phase catalysts at all potentials (Figure S8a). The catalysts with other phases exhibit a similar level of j_{total}, which is higher than that of the pure Ag catalyst (Figure S8b,c). Notably, the $AgZn_3$ phase catalyst ($Ag_{10}Zn_{90}$) exhibits an especially high j_{total}. This indicates that Ag-Zn alloying will improve the activity of the catalysts, and the $AgZn_3$ phase catalyst has the highest activity.

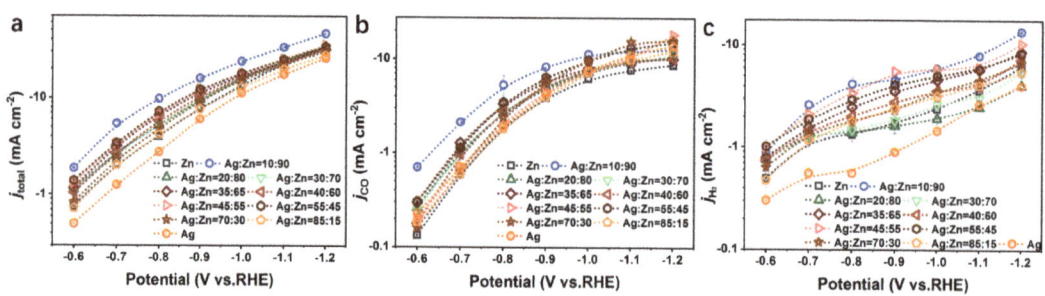

Figure 5. (a) j_{total}, (b) j_{CO}, and (c) j_{H_2} for the pure Zn, pure Ag, and Ag-Zn alloy catalysts with different element concentrations.

The partial current density towards specific CO_2RR products is another important figure of merit which is useful in comparing CO_2RR catalysts. Because the partial current density is directly proportional to the turnover frequency (TOF) of a certain product, the presentation of the data is useful in obtaining insights into the kinetics and mechanisms of the reduction reactions [18]. The partial current densities of CO (j_{CO}) and H_2 (j_{H_2}) are presented in Figure 5b,c in the form of a semi-log Tafel plot, respectively. The observed trend of j_{CO} is similar for all catalysts (Figures 5b and S9). In the less negative potential region, the TOF of CO formation increases with the decrease of potentials, whereas in the region of lower potentials, this growth of TOF gradually slows down and reaches a plateau, which is assigned to a CO_2-mass-transport limitation in aqueous solution due to its low solubility at atmospheric pressure [31,38], consistent with FE_{CO} in Figure 4a. To obtain further insights into these catalysts, their Tafel plots were investigated and presented in Figure S10. A small Tafel slope is beneficial in practical applications because it will lead to a much faster increase in the increment of CO_2 reduction rate with increasing overpotential [13]. For the catalysts with the Ag/Ag(Zn) solid solution phase, j_{CO} increases as a function of overpotential with a slope of 233, 221, 188, and 187 mV dec^{-1} for the Ag, $Ag_{85}Zn_{15}$, $Ag_{70}Zn_{30}$, and $Ag_{55}Zn_{45}$ catalysts, respectively in the lower overpotential region (Figure S10a). It is observed that the Tafel slope decreases gradually with a higher Zn content, and when the Ag-Zn alloy behaves as the AgZn phase, the $Ag_{40}Zn_{60}$ catalyst shows the lowest Tafel slope which is only 165 mV dec^{-1} (Figure S10b). While for the Ag_5Zn_8 and $AgZn_3$ phase catalysts, the Tafel slope is even larger than that of the pure Zn (Figure S10c). These results indicate that the alloying of Ag with Zn can improve the reaction kinetics of CO_2RR, and the enhanced degree depends on the phase composition and element concentration.

As shown in Figure 5c, j_{H_2} of the catalysts with different phase compositions forms transient platforms in different potential regions. The platform of pure Ag and AgZn phase catalysts appears between $-0.7 \sim -0.8$ V and $-0.9 \sim -1.1$ V vs. RHE, respectively (Figure S11a,b), and that of Ag_5Zn_8, $AgZn_3$ and Zn phase catalysts appears between $-0.7 \sim -0.9$ V vs. RHE (Figure S11c). This reflects the observed shift in selectivity from hydrogen to CO in this platform region, as discussed earlier [18]. In the more negative potential region, an increase in the TOF of H_2 is observed, which explains the slight rebound of H_2 and the production of other products in this potential region.

4. Conclusions

In summary, a series of Ag-Zn thin alloy film catalysts with different concentrations were prepared by a simple co-sputtering method, and the effect of different element concentrations and phase compositions on the activity and selectivity of CO_2RR was analyzed. It can be seen from the results that different phase compositions have affected the trend of FE_{CO} and the selectivity from hydrogen to CO with varying potentials. Pure Ag catalyst exhibits the best CO_2 selectivity, and the alloying with Zn does not significantly increase the activity of CO_2RR for the Ag/Ag(Zn) phase catalysts. The Ag_5Zn_8 and $AgZn_3$ phase catalysts show higher CO selectivity than pure Zn due to the synergistic effect of the Ag-Zn alloy. The alloying of Ag with Zn could improve the activity of the catalysts and the reaction kinetics of CO_2RR, and the $AgZn_3$ phase catalyst has the highest activity. The present work could provide guidelines for the design of Ag-Zn alloy catalysts for efficient CO_2RR.

Supplementary Materials: The following are available online at https://www.mdpi.com/article/10.3390/ma15196892/s1, Figure S1: Typical EDX spectra of (a) $Ag_{10}Zn_{90}$, (b) $Ag_{20}Zn_{80}$, (c) $Ag_{30}Zn_{70}$, (d) $Ag_{35}Zn_{65}$, (e) $Ag_{40}Zn_{60}$, (f) $Ag_{45}Zn_{55}$, (g) $Ag_{55}Zn_{45}$, (h) $Ag_{70}Zn_{30}$ and (i) $Ag_{85}Zn_{15}$ films, Figure S2: Macroscopic view of the (a) Zn film and (b) Ag film, Figure S3: SEM images of (a) $Ag_{10}Zn_{90}$, (b) $Ag_{20}Zn_{80}$, (c) $Ag_{30}Zn_{70}$, (d) $Ag_{35}Zn_{65}$, (e) $Ag_{40}Zn_{60}$, (f) $Ag_{45}Zn_{55}$, (g) $Ag_{55}Zn_{45}$, (h) $Ag_{70}Zn_{30}$ and (i) $Ag_{85}Zn_{15}$ films, Figure S4: XRD pattern of the CFP, Figure S5: FE_{CO} for the (a) Ag, Ag(Zn), (b) AgZn, (c) Ag_5Zn_8, $AgZn_3$, and Zn phase catalysts, Figure S6: FE_{H2} for the (a) Ag, Ag(Zn), (b) AgZn, (c) Ag_5Zn_8, $AgZn_3$, and Zn phase catalysts, Figure S7: $FE_{CO\&H2}$ for the (a) Ag, Ag(Zn),

(b) AgZn, (c) Ag$_5$Zn$_8$, AgZn$_3$, and Zn phase catalysts, Figure S8: j_{total} for the (a) Ag, Ag(Zn), (b) AgZn, (c) Ag$_5$Zn$_8$, AgZn$_3$, and Zn phase catalysts, Figure S9: j_{CO} for the (a) Ag, Ag(Zn), (b) AgZn, (c) Ag$_5$Zn$_8$, AgZn$_3$, and Zn phase catalysts, Figure S10: Tafel plots of the (a) Ag, Ag(Zn), (b) AgZn, (c) Ag$_5$Zn$_8$, AgZn$_3$ and Zn phase catalysts, Figure S11: j_{H2} for the (a) Ag, Ag(Zn), (b) AgZn, (c) Ag$_5$Zn$_8$, AgZn$_3$, and Zn phase catalysts.

Author Contributions: Conceptualization, J.S.; methodology, J.S., B.Y., X.Y., F.T., J.W. and W.Y.; formal analysis, J.S., B.Y., X.Y., F.T., J.W. and W.Y.; investigation, J.S., B.Y., X.Y., F.T., J.W. and W.Y.; writing—original draft preparation, J.S.; writing—review and editing, J.S., G.C. and Z.Z.; supervision, Z.Z., G.C. and W.Y.; project administration, Z.Z.; funding acquisition, Z.Z. and G.C. All authors have read and agreed to the published version of the manuscript.

Funding: The authors gratefully acknowledge the financial support from the Natural Science Foundation of Shandong Province (ZR2021QE229), the National Natural Science Foundation of China (51871133), China Postdoctoral Science foundation (2022M710077), Taishan Scholar Foundation of Shandong Province, and the Key Research and Development Program of Shandong Province (2021ZLGX01).

Institutional Review Board Statement: Not applicable.

Informed Consent Statement: Not applicable.

Data Availability Statement: All data are available within the paper and its Supplementary Information files or from the corresponding authors upon request.

Conflicts of Interest: The authors declare no conflict of interest.

References

1. Turner, J.A. A Realizable Renewable Energy Future. *Science* **1999**, *285*, 687–689. [CrossRef] [PubMed]
2. Spinner, N.S.; Vega, J.A.; Mustain, W.E. Recent Progress in the Electrochemical Conversion and Utilization of CO_2. *Catal. Sci. Technol.* **2012**, *2*, 19–28. [CrossRef]
3. Peter, S.C. Reduction of CO_2 to Chemicals and Fuels: A Solution to Global Warming and Energy Crisis. *ACS Energy Lett.* **2018**, *3*, 1557–1561. [CrossRef]
4. Tu, W.G.; Zhou, Y.; Zou, Z.G. Photocatalytic Conversion of CO_2 into Renewable Hydrocarbon Fuels: State-of-the-Art Accomplishment, Challenges, and Prospects. *Adv. Mater.* **2014**, *26*, 4607–4626. [PubMed]
5. Lahijani, P.; Zainal, Z.A.; Mohammadi, M.; Mohamed, A.R. Conversion of the Greenhouse Gas CO_2 to the Fuel Gas CO via the Boudouard Reaction: A Review. *Renew. Sustain. Energy Rev.* **2015**, *41*, 615–632.
6. Gonzales, J.N.; Matson, M.M.; Atsumi, S. Nonphotosynthetic Biological CO_2 Reduction. *Biochemistry* **2019**, *58*, 1470–1477. [CrossRef] [PubMed]
7. Hirunsit, P.; Soodsawang, W.; Limtrakul, J. CO_2 Electrochemical Reduction to Methane and Methanol on Copper-Based Alloys: Theoretical insight. *J. Phys. Chem. C* **2015**, *119*, 8238–8249. [CrossRef]
8. Lee, S.M.; Lee, H.; Kim, J.; Ahn, S.H.; Chang, S.T. All-Water-Based Solution Processed Ag Nanofilms for Highly Efficient Electrocatalytic Reduction of CO_2 to CO. *Appl. Catal. B Environ.* **2019**, *259*, 118045. [CrossRef]
9. Ramdin, M.; Morrison, A.R.T.; de Groen, M.; van Haperen, R.; de Kler, R.; van den Broeke, L.J.P.; Trusler, J.P.M.; de Jong, W.; Vlugt, T.J.H. High Pressure Electrochemical Reduction of CO_2 to Formic Acid/Formate: A Comparison Between Bipolar Membranes and Cation Exchange Membranes. *Ind. Eng. Chem. Res.* **2019**, *58*, 1834–1847. [CrossRef]
10. Centi, G.; Quadrelli, E.A.; Perathoner, S. Catalysis for CO_2 Conversion: A Key Technology for Rapid Introduction of Renewable Energy in the Value Chain of Chemical Industries. *Energy Environ. Sci.* **2013**, *6*, 1711–1731. [CrossRef]
11. Gao, Y.; Li, F.; Zhou, P.; Wang, Z.; Zheng, Z.; Wang, P.; Liu, Y.; Dai, Y.; Whangbo, M.-H.; Huang, B. Enhanced Selectivity and Activity for Electrocatalytic Reduction of CO_2 to CO on an Anodized Zn/Carbon/Ag Electrode. *J. Mater. Chem. A* **2019**, *7*, 16685–16689. [CrossRef]
12. Jo, A.; Kim, S.; Park, H.; Park, H.-Y.; Jang, J.H.; Park, H.S. Enhanced Electrochemical Conversion of CO_2 to CO at Bimetallic Ag-Zn Catalysts Formed on Polypyrrole-Coated Electrode. *J. Catal.* **2021**, *393*, 92–99. [CrossRef]
13. Liu, K.; Wang, J.; Shi, M.; Yan, J.; Jiang, Q. Simultaneous Achieving of High Faradaic Efficiency and CO Partial Current Density for CO_2 Reduction via Robust, Noble-Metal-Free Zn Nanosheets with Favorable Adsorption Energy. *Adv. Energy Mater.* **2019**, *9*, 1900276. [CrossRef]
14. Lu, Q.; Rosen, J.; Jiao, F. Nanostructured Metallic Electrocatalysts for Carbon Dioxide Reduction. *Chemcatchem* **2015**, *7*, 38–47. [CrossRef]
15. Hori, Y.; Wakebe, H.; Tsukamoto, T.; Koga, O. Electrocatalytic Process of CO Selectivity in Electrochemical Reduction of CO_2 at Metal-Electrodes in Aqueous-Media. *Electrochim. Acta.* **1994**, *39*, 1833–1839. [CrossRef]

16. Jones, J.P.; Prakash, G.K.S.; Olah, G.A. Electrochemical CO_2 Reduction: Recent Advances and Current Trends. *Isr. J. Chem.* **2014**, *54*, 1451–1466. [CrossRef]
17. Peterson, A.A.; Nørskov, J.K. Activity descriptors for CO_2 Electroreduction to Methane on Transition-Metal Catalysts. *J. Phys. Chem. Lett.* **2012**, *3*, 251–258. [CrossRef]
18. Hatsukade, T.; Kuhl, K.P.; Cave, E.R.; Abram, D.N.; Jaramillo, T.F. Insights into the Electrocatalytic Reduction of CO_2 on Metallic Silver Surfaces. *Phys. Chem. Chem. Phys.* **2014**, *16*, 13814–13819. [CrossRef]
19. Lee, C.Y.; Zhao, Y.; Wang, C.Y.; Mitchell, D.R.G.; Wallace, G.G. Rapid Formation of Self-Organised Ag Nanosheets with High Efficiency and Selectivity in CO_2 Electroreduction to CO. *Sustain. Energy Fuels* **2017**, *1*, 1023–1027. [CrossRef]
20. Zhong, H.X.; Ghorbani-Asl, M.; Ly, K.H.; Zhang, J.C.; Ge, J.; Wang, M.C.; Liao, Z.Q.; Makarov, D.; Zschech, E.; Brunner, E.; et al. Synergistic Electroreduction of Carbon Dioxide to Carbon Monoxide on Bimetallic Layered Conjugated Metal-Organic Frameworks. *Nat. Commun.* **2020**, *11*, 1409. [CrossRef]
21. Feaster, J.T.; Shi, C.; Cave, E.R.; Hatsukade, T.; Abram, D.N.; Kuhl, K.P.; Hahn, C.; Nørskov, J.K.; Jaramillo, T.F. Understanding Selectivity for the Electrochemical Reduction of Carbon Dioxide to Formic Acid and Carbon Monoxide on Metal Electrodes. *ACS Catal.* **2017**, *7*, 4822–4827. [CrossRef]
22. Hitt, J.L.; Li, Y.C.; Tao, S.; Yan, Z.; Gao, Y.; Billinge, S.J.L.; Mallouk, T.E. A High Throughput Optical Method for Studying Compositional Effects in Electrocatalysts for CO_2 Reduction. *Nat. Commun.* **2021**, *12*, 1114. [CrossRef]
23. Gerken, J.B.; Shaner, S.E.; Masse, R.C.; Porubsky, N.J.; Stahl, S.S. A Survey of Diverse Earth Abundant Oxygen Evolution Electrocatalysts Showing Enhanced Activity from Ni-Fe Oxides Containing a Third Metal. *Energy Environ. Sci.* **2014**, *7*, 2376–2382. [CrossRef]
24. Sun, J.M.; Yu, B.; Tan, F.Q.; Yang, W.F.; Cheng, G.H.; Zhang, Z.H. High Throughput Preparation of Ni-Mo alloy Thin Films as Efficient Bifunctional Electrocatalysts for Water Splitting. *Int. J. Hydrogen Energy* **2022**, *47*, 15764–15774. [CrossRef]
25. Gao, H.; Yan, X.J.; Niu, J.Z.; Zhang, Y.; Song, M.J.; Shi, Y.J.; Ma, W.S.; Qin, J.Y.; Zhang, Z.H. Scalable Structural Refining via Altering Working Pressure and In-situ Electrochemically-Driven Cu-Sb Alloying of Magnetron Sputtered Sb Anode in Sodium Ion Batteries. *Chem. Eng. J.* **2020**, *388*, 124299. [CrossRef]
26. Liu, N.; Yin, K.; Si, C.; Kou, T.; Zhang, Y.; Ma, W.; Zhang, Z. Hierarchically Porous Nickel-Iridium-Ruthenium-Aluminum Alloys with Tunable Compositions and Electrocatalytic Activities towards the Oxygen/Hydrogen Evolution Reaction in Acid Electrolyte. *J. Mater. Chem. A* **2020**, *8*, 6245–6255. [CrossRef]
27. Lamaison, S.; Wakerley, D.; Kracke, F.; Moore, T.; Zhou, L.; Lee, D.U.; Wang, L.; Hubert, M.A.; Aviles Acosta, J.E.; Gregoire, J.M.; et al. Designing a Zn-Ag Catalyst Matrix and Electrolyzer System for CO_2 Conversion to CO and Beyond. *Adv. Mater.* **2022**, *34*, 2103963. [CrossRef]
28. Hansen, H.A.; Varley, J.B.; Peterson, A.A.; Norskov, J.K. Understanding Trends in the Electrocatalytic Activity of Metals and Enzymes for CO_2 Reduction to CO. *J. Phys. Chem. Lett.* **2013**, *4*, 388–392. [CrossRef]
29. Hori, Y. CO_2 Reduction Using Electrochemical Approach. In *Solar to Chemical Energy Conversion*, 1st ed.; Springer: New York, NY, USA, 2016; pp. 191–211.
30. Kuhl, K.P.; Hatsukade, T.; Cave, E.R.; Abram, D.N.; Kibsgaard, J.; Jaramillo, T.F. Electrocatalytic Conversion of Carbon Dioxide to Methane and Methanol on Transition Metal Surfaces. *J. Am. Chem. Soc.* **2014**, *136*, 14107–14113. [CrossRef]
31. Hatsukade, T.; Kuhl, K.P.; Cave, E.R.; Abram, D.N.; Feaster, J.T.; Jongerius, A.L.; Hahn, C.; Jaramillo, T.F. Carbon Dioxide Electroreduction Using a Silver-Zinc Alloy. *Energy Technol.* **2017**, *5*, 955–961. [CrossRef]
32. Giamello, E.G.; Fubini, B. Heat of Adsorption of Carbon Monoxide on Zinc Oxide Pretreated by Various Methods. *J. Chem. Soc. Faraday Trans. I* **1983**, *79*, 1995–2003. [CrossRef]
33. Park, S.A.; Lim, H.; Kim, Y.T. Enhanced Oxygen Reduction Reactionactivity Due to Electronic Effects Between Ag and Mn_3O_4 in Alkaline Media. *ACS Catal.* **2015**, *5*, 3995–4002. [CrossRef]
34. Tang, W.; Huang, D.L.; Wu, L.L.; Zhao, C.Z.; Xu, L.L.; Gao, H.; Zhang, X.T.; Wang, W.B. Surface Plasmon Enhanced Ultraviolet emission and Observation of Random Lasing from Self-Assembly Zn/ZnO Composite Nanowires. *Crystengcomm* **2011**, *13*, 2336–2339. [CrossRef]
35. Zhao, Z.; Lu, G. Computational Screening of Near-Surface Alloys for CO_2 Electroreduction. *ACS Catal.* **2018**, *8*, 3885–3894. [CrossRef]
36. Guo, W.; Shim, K.; Kim, Y.-T. Ag Layer Deposited on Zn by Physical Vapor Deposition with Enhanced CO Selectivity for Electrochemical CO_2 Reduction. *Appl. Surf. Sci.* **2020**, *526*, 146651. [CrossRef]
37. Zhang, Z.; Wen, G.; Luo, D.; Ren, B.; Zhu, Y.; Gao, R.; Dou, H.; Sun, G.; Feng, M.; Bai, Z.; et al. "Two Ships in a Bottle" Design for Zn-Ag-O Catalyst Enabling Selective and Long-Lasting CO_2 Electroreduction. *J. Am. Chem. Soc.* **2021**, *143*, 6855–6864. [CrossRef]
38. Lamaison, S.; Wakerley, D.; Blanchard, J.; Montero, D.; Rousse, G.; Mercier, D.; Marcus, P.; Taverna, D.; Giaume, D.; Mougel, V.; et al. High-Current-Density CO_2-to-CO Electroreduction on Ag-Alloyed Zn Dendrites at Elevated Pressure. *Joule* **2020**, *4*, 395–406. [CrossRef]

Article

A New Insight into the Mechanisms Underlying the Discoloration, Sorption, and Photodegradation of Methylene Blue Solutions with and without BNO$_x$ Nanocatalysts

Andrei T. Matveev [1,*], Liubov A. Varlamova [1], Anton S. Konopatsky [1], Denis V. Leybo [1], Ilia N. Volkov [1], Pavel B. Sorokin [1], Xiaosheng Fang [2] and Dmitry V. Shtansky [1,*]

[1] Research Laboratory Inorganic Nanomaterials, National University of Science and Technology (MISIS), Leninskiy Prospect 4, 119049 Moscow, Russia
[2] Department of Materials Science, Fudan University, Shanghai 200433, China
* Correspondence: matveev.at@misis.ru (A.T.M.); shtansky@shs.misis.ru (D.V.S.)

Abstract: Methylene blue (MB) is widely used as a test material in photodynamic therapy and photocatalysis. These applications require an accurate determination of the MB concentration as well as the factors affecting the temporal evolution of the MB concentration. Optical absorbance is the most common method used to estimate MB concentration. This paper presents a detailed study of the dependence of the optical absorbance of aqueous methylene blue (MB) solutions in a concentration range of 0.5 to 10 mg·L^{-1}. The nonlinear behavior of optical absorbance as a function of MB concentration is described for the first time. A sharp change in optical absorption is observed in the range of MB concentrations from 3.33 to 4.00 mg·L^{-1}. Based on the analysis of the absorption spectra, it is concluded that this is due to the formation of MB dimers and trimers in the specific concentration range. For the first time, a strong, thermally induced discoloration effect of the MB solution under the influence of visible and sunlight was revealed: the simultaneous illumination and heating of MB solutions from 20 to 80 °C leads to a twofold decrease in the MB concentration in the solution. Exposure to sunlight for 120 min at a temperature of 80 °C led to the discoloration of the MB solution by more than 80%. The thermally induced discoloration of MB solutions should be considered in photocatalytic experiments when tested solutions are not thermally stabilized and heated due to irradiation. We discuss whether MB is a suitable test material for photocatalytic experiments and consider this using the example of a new photocatalytic material—boron oxynitride (BNO$_x$) nanoparticles—with 4.2 and 6.5 at.% of oxygen. It is shown that discoloration is a complex process and includes the following mechanisms: thermally induced MB photodegradation, MB absorption on BNO$_x$ NPs, self-sensitizing MB photooxidation, and photocatalytic MB degradation. Careful consideration of all these processes makes it possible to determine the photocatalytic contribution to the discoloration process when using MB as a test material. The photocatalytic activity of BNO$_x$ NPs containing 4.2 and 6.5 at.% of oxygen, estimated at ~440 μmol·g^{-1}·h^{-1}. The obtained results are discussed based on the results of DFT calculations considering the effect of MB sorption on its self-sensitizing photooxidation activity. A DFT analysis of the MB sorption capacity with BNO$_x$ NPs shows that surface oxygen defects prevent the sorption of MB molecules due to their planar orientation over the BNO$_x$ surface. To enhance the sorption capacity, surface oxygen defects should be eliminated.

Keywords: methylene blue; discoloration; sorption; photodegradation; BNO$_x$ nanocatalyst; DFT calculations

Citation: Matveev, A.T.; Varlamova, L.A.; Konopatsky, A.S.; Leybo, D.V.; Volkov, I.N.; Sorokin, P.B.; Fang, X.; Shtansky, D.V. A New Insight into the Mechanisms Underlying the Discoloration, Sorption, and Photodegradation of Methylene Blue Solutions with and without BNO$_x$ Nanocatalysts. *Materials* 2022, 15, 8169. https://doi.org/10.3390/ma15228169

Academic Editor: Ilya V. Mishakov

Received: 17 October 2022
Accepted: 15 November 2022
Published: 17 November 2022

Publisher's Note: MDPI stays neutral with regard to jurisdictional claims in published maps and institutional affiliations.

Copyright: © 2022 by the authors. Licensee MDPI, Basel, Switzerland. This article is an open access article distributed under the terms and conditions of the Creative Commons Attribution (CC BY) license (https://creativecommons.org/licenses/by/4.0/).

1. Introduction

Methylene blue (MB) is a widely used phenothiazinium dye that finds applications as a photosensitizer, as well as a redox and optical redox indicator in analytical chemistry and in trace analyses of anionic surfactants [1]. MB is also used for anticancer treatments in photodynamic therapy [2–6]. The widespread industrial use of dyes leads to their inevitable release into the environment. According to available estimates, about 10–15% of more than 0.7 million tons of dyes produced annually worldwide are released into the environment [7]. At the same time, MB is known to be a toxic and carcinogenic pollutant, which requires precise control over its concentration, as well as efficient removal or degradation to less toxic substances. Adsorption and photodegradation are effective ways to purify water from toxic pollutants. Various materials, such as activated carbon and coal [8,9], as well as natural and renewable biomaterials [10–12], are used as adsorbents for MB removal. Various adsorbents for organic and inorganic substances have been tested [13–15]. Recently, it was shown that hexagonal boron nitride (h-BN) is also a good adsorbent for organics [16–18].

The discoloration of dye solutions is one of the main methods of studying adsorption and photodegradation, and MB dye is a widely used test material for these reactions. For MB photodegradation, a wide variety of nanocatalytic assemblies, mainly consisting of binary and ternary metal oxides, have been studied [19,20]. A number of publications note that MB is not a suitable test material for photocatalytic experiments, since it becomes colorless when irradiated with visible light [21,22]. On the one hand, the self-photodiscoloration of MB contributes to its degradation and reduces the severity of the problem of its accumulation in the environment. On the other hand, the self-photodegradation of MB should be taken into account to avoid errors in the assessment of the photocatalytic activity of the studied photocatalyst. This requires a deep understanding of the photo decolorization of MB solutions.

It has recently been shown that boron oxynitride (BNO_x) nanoparticles (NPs) are good substrates for photocatalysis under UV irradiation [20]. BNO_x is a relatively cheap, chemically inert, and environmentally friendly material. Here, we studied its photocatalytic activity using MB as a test material. BNO_x NPs containing 4.2 and 6.5 at.% of oxygen were studied as photocatalysts for MB degradation under UV illumination. The main objectives of the study were (i) a detailed study of the optical absorbance of aqueous MB solutions at various concentrations; (ii) a study of the self-discoloration effect of MB solutions (including heating) under visible and artificial sunlight illumination; (iii) an analysis of the applicability of MB as a test material for photocatalytic experiments; (iv) an investigation of the photocatalytic activity of BNO_x NPs in the photodegradation of MB solutions under UV illumination; (v) a study of the effect of oxygen on the photocatalytic and sorption capacity of BNO_x NPs; (vi) an analysis of the stability of BNO_x NPs and the possibility of their reuse in photocatalytic experiments; (vii) and to answer the question of whether MB is a suitable test material for photocatalytic experiments.

It has been shown that the discoloration of MB solutions is a complex process involving the following mechanisms: thermally induced MB photodegradation, MB absorption on BNO_x NPs, self-sensitizing MB photooxidation, and photocatalytic MB degradation. Taking into account all of these mechanisms, the photocatalytic activity of BNO_x NPs containing 4.2 and 6.5 at.% of oxygen is estimated to be as high as ~440 $\mu mol \cdot g^{-1} \cdot h^{-1}$. The obtained results are discussed based on DFT calculations, taking into account the effect of MB sorption on its self-sensitizing photooxidation activity.

2. Materials and Methods
2.1. Materials

MB in the form of a hydrochloride salt (with three water molecules) was acquired from Rushim (Moscow, Russia). MB has a molecular weight of 319.85 $g \cdot mol^{-1}$. MB is a cationic thiazine dye with the molecular formula $C_{16}H_{18}N_3ClS$. It is highly water-soluble and forms a stable solution with water at room temperature. MB has an amino autochrome unit and has a maximum of optical absorption absorbance at of 663 nm [1]. BNO_x NPs with various oxygen concentrations were synthesized with the low-temperature ammonolysis of boric acid, as described elsewhere [23]. Boric acid was treated with gaseous ammonia at room temperature to produce an ammonium borate hydrate (ABH) phase. Heating the ABH phase in ammonia led to successive dehydration and, starting from a temperature of

550 °C, the formation of h-BNO$_x$ NPs. The size of the resulting BNO$_x$ NPs, as well as the oxygen content, depended on the maximum heating temperature. BNO$_x$ NPs synthesized at 650 °C were designated sample BNO$_1$. The h-BNO$_x$ nanopowder, further annealed in ammonia at 1100 °C for 1 h, was designated sample BNO$_2$.

2.2. Materials Characterization

The sample phase composition was determined with a SmartLab diffractometer (Rigaku, Tokyo, Japan) using Cu-Kα radiation and a graphite monochromator. X-ray diffraction (XRD) patterns were recorded in symmetrical mode and analyzed using the PDXL software, (Version 2.8.4.0) (Rigaku, Tokyo, Japan). Fourier-transform infrared (FTIR) spectra were recorded based on powder samples using a Vertex 70v vacuum spectrometer (Bruker, Billerica, MA, USA) in the range of 400–4000 cm^{-1} with a partial internal reflection device. The chemical composition was analyzed with an X-ray photoelectron spectroscopy (XPS, 18725 Lake Drive East, Chanhassen, MN, USA) using a Versa Probe III (PHI) instrument equipped with a monochromatic Al Kα X-ray source ($h\nu$ = 1486.6 eV). Atomic concentrations were determined from survey spectra using the relative sensitivity factors of the elements. The integral intensities of the XPS B1s, N1s, O1s, and C1s peaks were used for analysis. The specific surface area was determined with the Brunauer–Emmett–Teller (BET) nitrogen adsorption method using a NOVA 1200e instrument (Quantachrome Instruments, Boynton Beach, FL, USA).

2.3. Spectrophotometric Measurements

The ultraviolet–visible (UV-vis) absorption spectra were recorded on an UVmini-1240 spectrophotometer (Shimadzu, Tokyo, Japan) using a 1 cm quartz cuvette. Fluorescence spectra were recorded on a Cary Eclipse fluorescence spectrophotometer (Agilent Technologies, Santa Clara, CA, USA).

Diffuse reflectance spectra in the ultraviolet and visible regions (DRS UV-vis) were recorded on a V-750 spectrophotometer (Jasco, JASCO Corporation, Tokyo, Japan) in the wavelength of 200–800 nm with a resolution of 1 nm. Barium sulfate was used as a standard.

2.4. Discoloration and Photocatalytic Measurements

The discoloration and photodegradation of MB aqueous solutions were studied under UV, solar, and visible light. A 50 W low-pressure mercury lamp with a main emission line at a wavelength of 254 nm was used as a UV source. An Osram Ultra-Vitalux lamp (Munich, Germany) was employed as a source of simulated sunlight. For visible light illumination, a filter was used that cuts off the UV part of the spectrum with wavelengths shorter than 420 nm.

For photocatalytic measurements under visible and sunlight illumination, 10 mg of BNO$_1$ NPs were ultrasonically dispersed in 25 mL of distilled water, and then 25 mL of an MB aqueous solution was added. The MB concentration in the resulting solution was 10 mg·L^{-1}. The experiments were carried out in Duran glasses. The same protocol was used for photocatalytic measurements under UV irradiation, with the only difference being that quartz glasses and 5 mg of BNO$_1$ or BNO$_2$ nanopowders were used for each solution. When the solution was illuminated, its temperature increased. A water-cooled cell was used to stabilize the solution temperature at 20 or 80 °C. The solution temperature was controlled by a thermocouple. The loss of water due to illumination-induced evaporation was compensated after each illumination stage before taking an aliquot for analysis. The catalyst mass-specific activity was calculated as the number of moles of degraded MB divided by the weight of the catalyst and the degradation time. To evaluate the stability and reusability of the BNO$_x$ nanocatalysts under UV illumination, four successive photocatalytic cycles were performed accordingly to the above protocol. After each cycle, the solution was centrifuged at 9000 rpm for 15 min, the supernatant was taken with a syringe, and fresh MB solution was added.

The following chemicals were used as scavengers: isopropyl alcohol (IPA) as an ·OH scavenger, dimethyl sulfoxide as an electron (e$^-$) scavenger, disodium ethylenediaminetetraacetate (Na-EDTA) to remove positive charge carriers (h$^+$), and benzoquinone to neutralize superoxide anion ·O$^{\cdot}_2$. The scavengers tests were performed for 20 min of UV illumination of MB solutions (50 mL with an MB concentration of 10 mg·L^{-1}) containing 5 mg of BNO$_1$ NPs.

2.5. Computational Methods

Density functional theory (DFT) calculations were performed using the VASP package with the PBE functional and a plane wave cutoff of 400 eV [24–26]. Spin-polarized calculations were used. During relaxation, the atomic positions and lattice parameters were optimized.

3. Results and Discussion

3.1. Spectrophotometry of MB Solutions

The concentration of dyes in a solution is most often determined by light absorption using the spectroscopic method. The MB molecule is planar and exists as a cation in aqueous solutions. At a certain concentration, MB molecules tend to form sandwich-like dimers, trimers, or higher oligomers [27], also called H-type aggregates. Quite recently, it was suggested that the MB molecule exists in two mesomere forms, which differ in their electric charge distribution [28]. The MB monomer, dimer, trimer, n-oligomers, and mesomeres have different molar attenuation coefficients; therefore, it can be expected that a change in their solution concentration can lead to a nonlinear dependence of solution optical absorption on MB concentration. Figure 1a shows the concentration dependence of the normalized optical absorption of MB aqueous solutions in an MB concentration range of 0.5 to 10 mg·L^{-1} (from 1.56 × 10^{-6} to 3.13 × 10^{-5} mol·L^{-1}). Here and below, C$_0$ is the absorbance of the initial MB solution, and C is the absorbance at a given time of illumination.

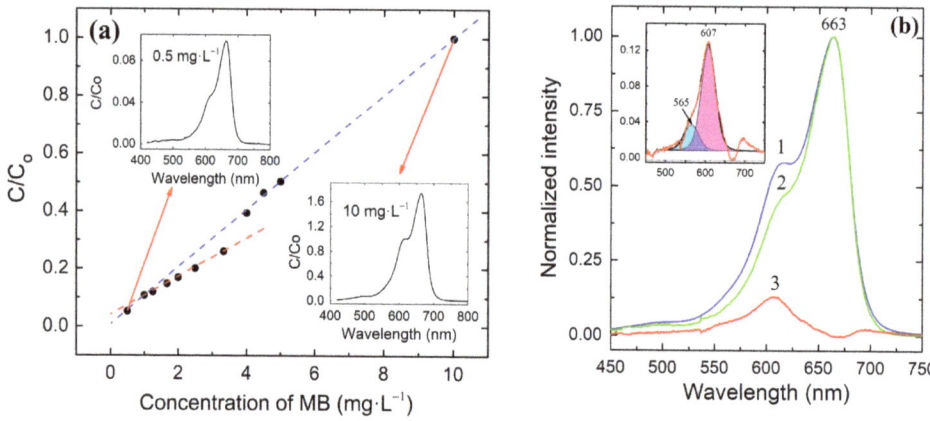

Figure 1. Concentration dependence of the normalized absorbance of MB aqueous solutions in a concentration range of 0.5 to 10 mg·L^{-1} (from 1.56 × 10^{-6} to 3.13 × 10^{-5} mol·L^{-1}) at 25 °C (**a**). The insets in (**a**) show absorbance spectra at the lowest and highest measured concentrations. Normalized absorbance of MB solution (**b**): 10 mg·L^{-1} (curve 1) and 0.5 mg·L^{-1} (curve 2). Curve 3 represents the difference spectrum obtained by subtracting curve 2 from curve 1. Inset in (**b**) shows the deconvolution of difference spectrum 3.

In the entire range of studied concentrations, the dependence is not linear. At concentrations above 1.0 mg·L^{-1}, the absorbance deviates from the initial trend and follows a line with a lower slope up to 3.33 mg·L^{-1}. Between 3.33 and 4.00 mg·L^{-1}, the C_t/C_o value sharply increases (dotted area in Figure 1a), and then, the optical absorbance follows a line with approximately the same slope as it does at low concentrations ranging from 0.5 to 1.0 mg·L^{-1}.

To identify the cause of optical absorption deviating from linearity in the range of 3.33–4.00 mg·L^{-1}, a normalized absorption spectrum for the 0.5 mg·L^{-1} solution (curve 1 in Figure 1b) was subtracted from the normalized absorption spectrum of the 10 mg·L^{-1} solution (curve 2). The spectrum obtained after subtraction (curve 3) was fitted using two components at 607 nm and 565 nm. These peaks almost coincide with those reported for the dimer [29] and trimer [29,30], respectively. Thus, we can conclude that the observed deviation of the concentration curve from the linear Beer–Lambert law is due to the formation of dimeric and trimeric molecular associates. The fractions of the monomer, dimer, and trimer in the 10 mg·L^{-1} MB solution, estimated from the peak areas, are 91.1%, 7.1%, and 1.8%, respectively. To the best of our knowledge, this is the first mention of the nonlinearity of the optical absorption of MB solutions in a low concentration range. An additional analysis of the data presented in [29] also shows a change in the relative content of the monomers, dimers, and trimers, but the authors did not pay attention to this fact. The trimer fraction increases with the increasing MB concentration in two steps: first, a small step above approximately 0.5×10^{-5} mol·L^{-1} (determined from Figure 1 in [29]), and then, a second strong step in a range of 1.2×10^{-5} to 4.5×10^{-5} mol·L^{-1}, which correlates well with the step observed in Figure 1. The fraction of the monomer accordingly decreases stepwise with the increasing MB concentration.

It should be noted that the available data on MB agglomeration and polymerization are rather contradictory. The presence of at least three absorbing species (monomers, dimers, and trimers) has been observed in a concentration range of 6.0×10^{-7} mol·L^{-1} to 6.0×10^{-2} mol·L^{-1}, and it was suggested that trimerization occurs simultaneously with dimerization due to the reaction of dimers with monomers [30]. Heger et al. observed a very small fraction of dimers, but also a steadily increasing fraction of trimers above 1.0×10^{-5} mol·L^{-1} [29]. It has recently been shown that the fraction of dimers increases from 1.0×10^{-6} mol·L^{-1}, reaches a maximum at approximately 1.0×10^{-4} mol·L^{-1}, and then decreases, while the fraction of tetramers and oligomers constantly increases above 1.0×10^{-6} mol·L^{-1}, but trimers were not observed [27]. Therefore, it would be very speculative to propose a detailed model explaining the nonlinear behavior of absorbance as a function of MB concentration. For this, additional studies of MB agglomeration and polymerization are required. In addition, accurate measurements of the molar attenuation coefficients of each n-dimensional MB type are required.

It should also be noted that we did not observe an absorption peak with a maximum at 600 nm, observed elsewhere and attributed to the tetramer [27].

It is important to note that the error in estimating the MB concentration from the optical absorption tests reaches 1 mg·L^{-1} without taking into account the change in the slope of the concentration curve. Given the toxicity of MB, such an error may be important in some applications, such as biomedicine.

To further explain the observed feature of the absorption/concentration curve, we obtained and analyzed the luminescence spectra of the three solutions (with MB concentrations of 1, 3.33, and 5.0 mg·L^{-1} near the curve inflection) at three excitation wavelengths: 250, 320, and 365 nm (Figure 2). At an excitation wavelength of 250 nm and an MB concentration of 1 mg·L^{-1}, the luminescence peak is observed at 683 nm. This is a characteristic MB emission [31]. With an increase in the MB concentration, the peak intensity increases, and its position shifts by 4–9 nm: 683 nm at 1 mg·L^{-1}, 687 nm at 3.33 mg·L^{-1}, and 692 nm at 5 mg·L^{-1}. This behavior is observed for all studied excitation wavelengths. We consider these results as additional evidence of oligomer formation since it is known that the fluorescence lines of dimers shift toward longer wavelengths relative to the fluorescence line of the monomeric form [32]. In addition to a peak at approximately 690 nm, MB exhibits a strong and broad fluorescence zone in a range of 450–650 nm when excited at a wavelength of

365 nm [33]. Fluorescence at 690 nm was associated with electron transitions from dimethyl amino groups to the central aromatic ring, i.e., along the longer molecule axis, while fluorescence at 550 nm was assigned to a transition along the shorter molecule axis, when electrons from sulfur move to nitrogen [33]. In the range of 450–650 nm, we observed only a very weak luminescence zone (inset in Figure 2c), which indicates the almost complete absence of an electron transition between sulfur and nitrogen (in contrast to [28,33]). This can be explained by the lower MB concentration (by one order of magnitude) used in our studies. It is important to note that the emission of *leuco*-MB at 460 nm upon excitation at a wavelength of 320 nm was not observed [31]. This eliminates the possibility that the abrupt increase in absorbance is due to the oxidation of colorless *leuco*-MB to color MB. The narrow peaks observed at 543 nm (excitation at 250 nm) and at ~420 nm (excitation at 365 nm) are insensitive to MB concentration and have a full width at a half maximum (FWHM) value of approximately 15 nm, which is typical for the Raman scattering of light in water (Figure 2a,c) [34].

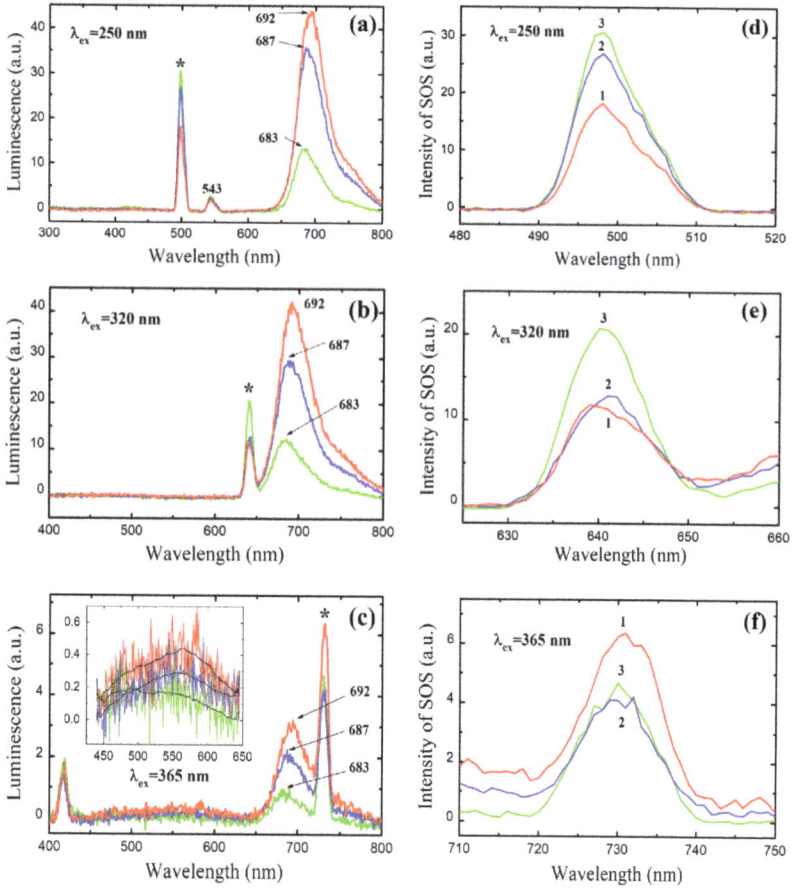

Figure 2. Luminescence spectra (**a–c**) of MB aqueous solutions with MB concentrations of 1 mg·L^{-1} (green line), 3.33 mg·L^{-1} (blue line), and 5 mg·L^{-1} (red line) under excitation wavelengths of 250 (**a**), 320 (**b**), and 365 (**c**) nm. Asterisks mark second-order scattering. Panels (**d–f**) show enlarged peaks of second-order scattering; numbers 1, 2, and 3 correspond to MB concentrations of 5, 3.33, and 1 mg·L^{-1}.

Additional narrow and strong peaks are observed at 500 nm, 640 nm, and 730 nm when excited at wavelengths of 250, 320, and 365 nm, respectively. These peaks are marked with asterisks in Figure 2a–c and are also shown at enlarged scales in Figure 2d–f. The intensity of these peaks depends on the MB concentration and the excitation wavelength. We assume that these maxima are due to the second-order scattering (SOS) of light, as they are observed at wavelengths twice that of the excitation wavelength. Indeed, at an excitation wavelength of 250 nm (Figure 2d), all peaks are observed at 498 nm; i.e., they are shifted by 2 nm toward a shorter wavelength. In the past, SOS was commonly observed in spectroscopic measurements and was considered a kind of interference phenomenon until it was shown that the SOS intensity of the aqueous solution of the ion-association complex of the Se (IV)–I$^-$–rhodamine B system is sensitive to trace amounts of Se [35]. Since then, SOS peaks have been successfully used to study the structure and concentration of various colloids, including macromolecules, nanoparticles, quantum dots, and, especially, organic polymers such as proteins [36–40].

In contrast to luminescence (Figure 2a), at 250 nm excitation, the intensity of the SOS peak at 500 nm demonstrates an inverse dependence on the MB concentration (Figure 2d). A similar effect was observed at a strong dilution of humate solutions and was explained by an increase in the number of scattering centers upon dilution [40]. In this regard, the observed reverse order of SOS intensities suggests an association of MB molecules at a low MB concentration. With an increase in the MB concentration, as discussed above, oligomers are formed, and associates are destroyed due to the steric effect. More reliable conclusions require additional SOS spectroscopic studies of MB solutions at various concentrations.

3.2. Discoloration of MB Solutions under Visible Light and Sunlight

An MB aqueous solution with a concentration of 10 mg·L^{-1} was exposed to visible light and sunlight at temperatures of 20 and 80 °C for 120 min. The solution absorption spectra are shown in Figure 3. When illuminated, the intensities of the main absorbance peak located at 664 nm and the shoulder at approximately 600 nm decrease. The position of the main peak does not change at 20 °C but slightly shifts toward a shorter wavelength at 80 °C, demonstrating a weak hypsochromic effect, more pronounced under visible light. These changes in the absorbance spectra suggest the degradation of chromophore moieties in the MB molecule.

The time dependences of the normalized optical absorbance of the MB solutions are depicted in Figure 4. For comparison, the normalized absorbance values of the MB solution heated in the dark to 80 °C are shown with black symbols. Heating the MB solution in the dark did not affect its color. Illuminating the MB solution at 20 °C for 120 min led to a decrease in absorption by 37% (V20) and 53% (S20). When the solution temperature was raised to 80°C, the absorption decreased by 69% (V80) and 83% (S80). Thus, the MB solution rapidly decolorized under visible light and even more rapidly under sunlight (containing some UV component), with decolorization greatly accelerated with the increasing temperature. The decolorization rate (observed from the curve slopes) also increased with heating, and the absorption curves do not tend toward any asymptotic limit. To the best of our knowledge, this is the first observation of strong thermal photodegradation in an MB solution. MB thermal degradation with an efficiency of more than 80% has been reported, but only in the presence of a catalyst [41]. Thus, when studying the photocatalytic degradation of MB solutions, it is necessary to take into account the self-decomposition of MB under visible light and sunlight in order to avoid an incorrect assessment of the material photocatalytic activity if the temperature of the solutions is unstable and increases with prolonged illumination.

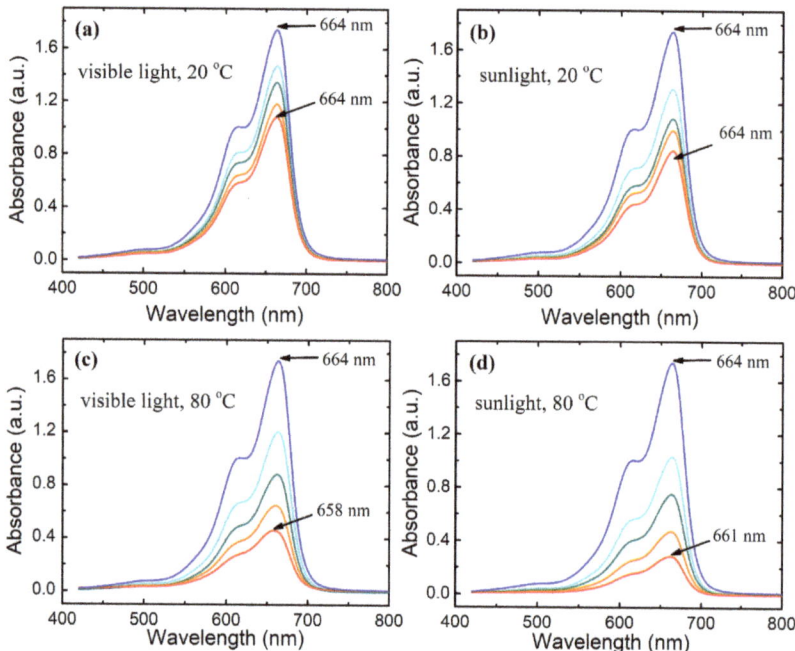

Figure 3. Absorption spectra of MB aqueous solution with a concentration of 10 mg·L^{-1} under visible light and sunlight illumination at temperatures of 20 and 80 °C for 120 min. Lighting time from upper blue curve to lower red curve: 0, 30, 60, 90, and 120 min. (**a**) illumination by visible light at 20 °C; (**b**) illumination by sunlight at 20 °C; (**c**) illumination by visible light at 80 °C; (**d**) illumination by sunlight at 80 °C.

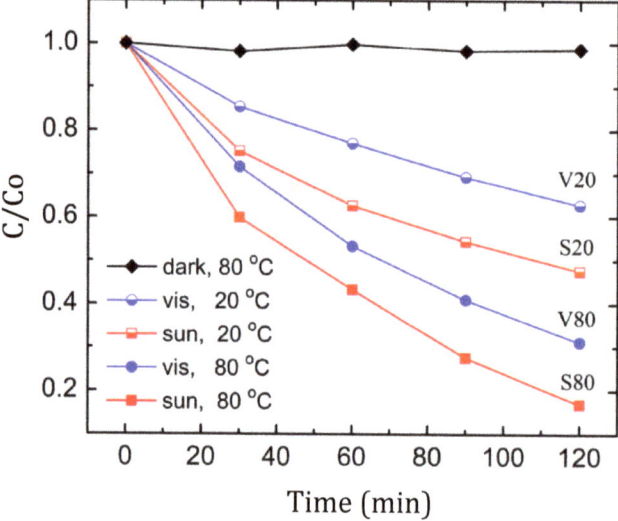

Figure 4. Temporal changes in the normalized absorbance of MB solution (10 mg·L^{-1}) in the dark at 80 °C (black symbols) and under illumination with visible (V) light and sunlight (S) at 20 and 80 °C.

It is generally accepted that the photodecomposition of organic compounds occurs as a result of their interaction with active species formed during the light-activated process. Most often, such species are hydroxyl (·OH), superoxide (·O_2^-), and peroxide (HO_2·) radicals, as well as holes h^+ [42]. The positions of the LUMO and HOMO of MB were estimated as −0.88 and 1.55 eV, respectively [43]. The MB bandgap is 2.43 eV, and visible light photons can transfer energy to electrons and facilitate their transition from HOMO to LUMO orbitals. Thus, MB can be used as a photosensitizer in various applications, including phototherapy [2–6,44] and water photo-splitting [45]. In the visible-light-driven self-decomposition process, MB apparently acts as a self-photosensitizer. The formation of MB oligomers and mesomeres, as well as the temperature change in the dielectric constant of water [28], provokes a change in the electron charge distribution in the MB monomer and affects this photodecomposition process.

To understand the MB photodegradation process, the solution fluorescence spectra were obtained at excitation wavelengths of 320 and 365 nm after solution illumination with visible light and sunlight for 120 min (Figure 5a,b). At both excitation wavelengths, the intensity of the main MB fluorescence peak at 696 nm decreases in the following order: V20→V80→S20→S80. This sequence differs from the normalized absorbance shown in Figure 4: V20→S20→V80→S80. This indicates that after illumination new species appear in the solutions, which contribute differently to absorbance and luminescence. When excited by light with a wavelength of 320 nm, the maximum emission bands are observed in the range of 450–460 nm (Figure 5a). Fluorescence at ~450 nm for *leuco*-MB and ~460 nm for *leuco*-Thionine (Th) has been reported upon excitation at a wavelength of 320 nm. At both excitation wavelengths (320 and 365 nm), the luminescence peak observed at 696 nm for an unilluminated MB solution gradually shifts toward a shorter wavelength up to approximately 680 nm. This is due to the demethylation of the MB molecule [22]. According to the results of FTIR spectroscopy measurements, the decrease in the intensity of the 690 nm peak may be associated with several decomposition stages occurring in different parts of the MB molecular [46]. Sequential demethylation results in the formation of structurally related by-products such as Asure B, Azure A, Azure C, and thionine [47–49], which causes charge redistribution and a shift in electron density to the nitrogen atom in the central aromatic ring. This electron transition causes luminescence at ~550 nm upon 365 nm excitation [33], which can be seen in Figure 5b. The peaks at 565 and 570 nm are also associated with changes in the aromatic rings of the MB molecule [22,33]. Thus, the emission maxima observed at 560–580 nm upon excitation at a wavelength of 365 nm (Figure 5b) indicate a complex stepwise degradation of the MB molecule.

The SOS peaks marked with asterisks for the respective excitation wavelengths (Figure 5a,b) are shown at a larger scale in Figure 5c,d. The intensity of the SOS peaks at 640 nm after excitation at a wavelength of 320 (Figure 5c) decreases in the same order as the intensity of the luminescence peaks (Figure 5a). Note that the intensity of the SOS peaks is higher than the SOS peak of the unilluminated MB solution. This indicates an increase in the number of scattering centers after irradiating the MB solution since the intensity of the SOS peak correlates with the concentration of the scattering centers, but the intensity of the luminescence of these centers is significantly lower than the luminescence of the MB molecule. At an excitation wavelength of 365 nm, an intense SOS peak at 730 nm is observed only in a solution illuminated with sunlight at 80 °C (S80) for 120 min. The strong dependence of light scattering on the excitation wavelength is confirmed by the fact that the SOS spectra have a pronounced maximum at a certain wavelength [39].

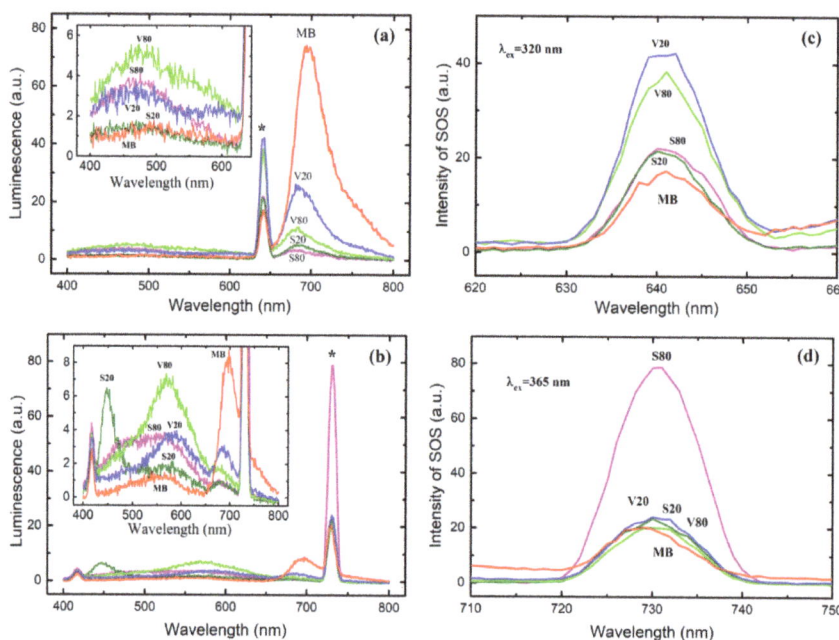

Figure 5. Luminescence spectra upon excitation at wavelengths of 320 nm (**a**) and 365 nm (**b**) of MB aqueous solutions (10 mg·L^{-1}) after illumination with visible (V) light and sunlight (S) at 20 °C and 80 °C for 120 min. Asterisks mark second-order scattering. The insets show magnified parts of the spectra. The enlarged second-order scattering peaks are (**c**) and (**d**).

3.3. BNO$_x$ Photocatalyst

BNO$_x$ NPs with differing oxygen content were studied as photocatalysts. NPs designated BNO$_1$ were synthesized using the low-temperature ammonolysis of boric acid at a sintering temperature of 650 °C, as described elsewhere [23]. Sample BNO$_2$ was prepared by annealing a portion of the BNO$_1$ powder in ammonia at 1100 °C for 1 h. Figure 6 represents TEM images of the BNO$_1$ (a) and BNO$_2$ (b) samples.

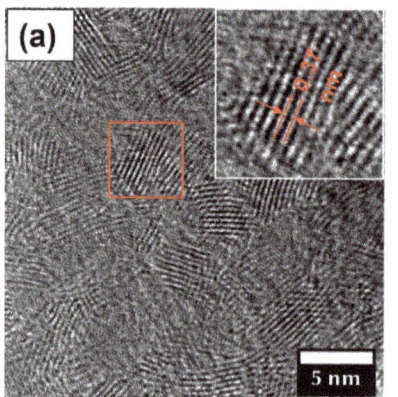
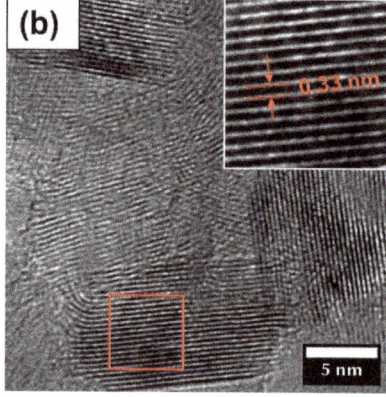

Figure 6. TEM images of the BNO$_1$ (**a**) and BNO$_2$ (**b**) samples. Insets show enlarged areas marked with red rectangles (red lines and arrows mark atomic planes).

The TEM analysis shows that the BNO$_1$ and BNO$_2$ samples are composed of nanocrystals with an average size of approximately 5 and 10 nm, respectively. Insets show enlarged areas marked with red rectangles. An interlayer spacing was determined to be 0.37 and 0.33 nm for the BNO$_1$ and BNO$_2$ samples, respectively.

Figure 7 represents XRD patterns: the FTIR and XPS spectra of these samples are denoted as 1 and 2 for BNO$_1$ and BNO$_2$, respectively.

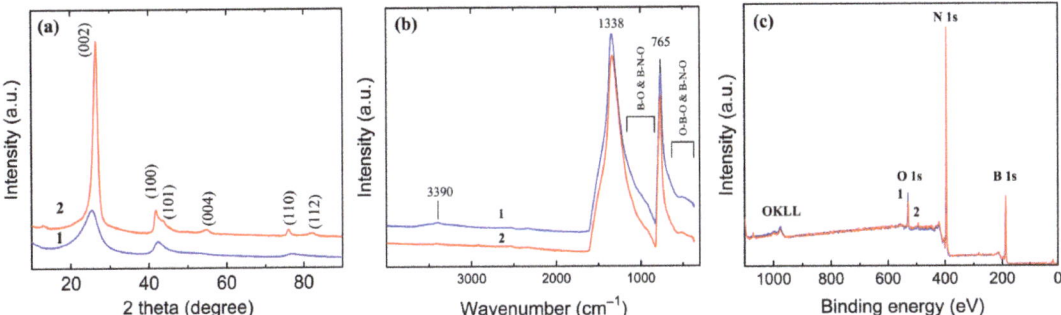

Figure 7. XRD patterns (Cu-Kα radiation) (**a**), FTIR (**b**), and XPS (**c**) spectra of the BNO$_1$ (1) and BNO$_2$ (2) nanopowders.

The XRD analysis shows that sample BNO$_2$ is nanocrystalline h-BN (Figure 7a, curve 2). The interlayer spacing along the c-axis is 0.33 nm, and the average crystal size, determined by the Debye–Scherrer equation, is approximately 10 nm. The XRD pattern of sample BNO$_1$ shows broadened, low-intensity peaks typical of the turbostratic h-BN structure [23,50]. This sample has an interlayer distance along the c-axis of 0.35 nm and an average crystal size of 6 nm. The XRD data are in good agreement with the results of the TEM analysis. The FTIR spectrum of sample BNO$_2$ (Figure 7b) demonstrates an intense peak at 1338 cm^{-1} due to in-plane B-N stretching vibrations and a narrow peak at 765 cm^{-1} due to out-of-plane B-N-B bending vibrations. The absence of other peaks indicates an almost pure BN phase. The FTIR spectrum of sample BNO$_1$ (Figure 7b) demonstrates a small peak at 3390 cm^{-1}, attributed to asymmetric O-H stretching vibrations or an overtone of B-O trigonal vibrations, and shoulders in the range of 1200–840 cm^{-1} and 670–400 cm^{-1} assigned to B-O stretching vibrations and B-O-B and B-N-O bending vibrations, respectively [23]. The XPS analysis (Table 1) shows that samples BNO$_1$ and BNO$_2$ consist of boron, nitrogen, and oxygen in the following amounts: 6.5 (1) and 4.2 at. % (2). The nitrogen content of sample BNO$_1$ is lower than the boron content, which indicates that oxygen mostly substitutes nitrogen rather than boron. Indeed, in oxidized BN, oxygen atoms substitute nitrogen atoms and form B-O bonds [51]. A more detailed description of the initial structure and its transformations during heat treatment can be found elsewhere [23].

Table 1. Elemental composition of BNO$_x$ samples synthesized at 650 (BNO$_1$) and 1100 °C (BNO$_2$). The rest is carbon from adsorbed carbonaceous species.

Sample	Content, at. %		
	B	N	O
BNO$_1$	48.0	44.8	6.5
BNO$_2$	47.7	46.6	4.2

To study the effect of MB sorption on MB degradation, two measurements were carried out: with sorption in the dark for an hour followed by sunlight illumination (curve 1 in Figure 8a) and with sunlight without sorption in the dark (curve 2 in Figure 8a). For comparison, the discoloration curve of the MB solution without a catalyst is also shown (curve 3 in Figure 8a).

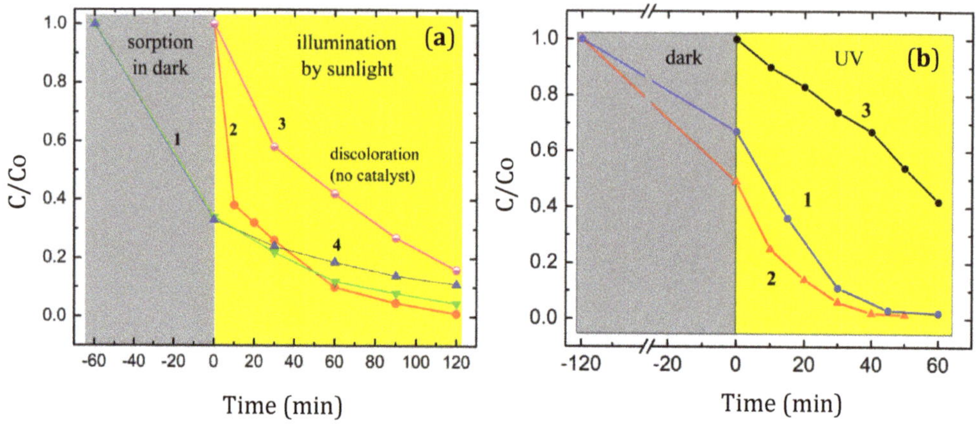

Figure 8. Time-dependences of the normalized absorbance of MB solutions (0.5 mg MB in 50 mL H_2O) containing 10 mg of BNO_1 under sunlight illumination at 80 °C after 1 h of sorption in the dark (curve 1) and without sorption in the dark (curve 2). For comparison, the discoloration of the MB solution without a catalyst is shown (curve 3). Curve 4 shows the photocatalytic degradation of the solution under visible light illumination (**a**). Time-dependences of the normalized absorbance of MB solutions (0.5 mg MB in 50 mL H_2O) containing 5 mg of BNO_1 (curve 1) and 5 mg of BNO_2 (curve 2) under UV illumination at 80 °C after 2 h of sorption in the dark (**b**). Curve 3 shows the discoloration of the MB solution under UV irradiation without a catalyst.

The results obtained show that the discoloration rate without the sorption stage in the dark (curve 2 in Figure 8a) of the BNO_1-containing MB solution is higher than the discoloration rate after sorption in the dark. This implies that the sorption of MB molecules on the BNO_1 surface deactivated some of the active centers involved in the photodegradation process. Within one hour of illumination, the discoloration of the MB solution reached 90%. A comparison with curve 3 in Figure 8a (without a catalyst) clearly shows that the discoloration occurs not only due to the MB photocatalytic degradation, but also due to the discoloration of the MB solution itself, and ignoring this fact introduces a significant error to the assessment of the photocatalytic activity of the catalyst. The specific catalyst mass activity was calculated taking into account the discoloration of the MB solution under sunlight according to the following equation:

$$(C_3(t) - C_2(t)) \times m_{MB}/m_{cat}/\Delta t \qquad (1)$$

where C_2 and C_3 are the normalized absorbance values of the MB solution with a photocatalyst (curve 2) and without a photocatalyst (curve 3) at time t, Δt is the solution irradiation time, and m_{MB} and m_{cat} are the masses of the MB and catalyst in solution. The specific photocatalytic mass activity of the BNO_1 nanopowder during the photodegradation of an MB solution (10 mg·L^{-1}) under sunlight illumination for an hour was calculated to be 15 mg·g^{-1}·h^{-1} (50 µmol·g^{-1}·h^{-1}). Note that, without taking into account the MB solution discoloration, the catalyst activity would be three times higher.

Figure 8b shows the time-dependent discoloration of MB solutions containing BNO_1 and BNO_2 NPs under UV illumination after MB sorption in the dark (curves 1 and 2). An estimate of the band gap values in samples BNO_1 and BNO_2 based on the diffuse reflectance spectra (Figure 9) yielded 4.7 and 5.2 eV, respectively. The energy of the generated UV photons using a low-pressure mercury lamp generating UV light at a wavelength of 254 nm was 4.88 eV. This value exceeds the band gap of sample BNO_1 and is large enough to excite electrons from the valence band to the conduction band. However, in the case of sample BNO_2, with a band gap of 5.2 eV, the energy of the UV photons is insufficient for the direct generation of photoelectrons. The DFT simulation (see below) shows that, when an MB molecule is adsorbed on the BNO_x surface, the nitrogen atom of the central MB ring forms a strong chemical bond with the boron atom closest to the oxygen defect. This causes a redistribution of the electron density near the oxygen defect and leads to the formation of an interband state. Thus, it can be assumed that MB sorption on the surface of BNO_x NPs leads to the formation of additional levels near the conduction band, which provide photoexcitation with lower energy photons.

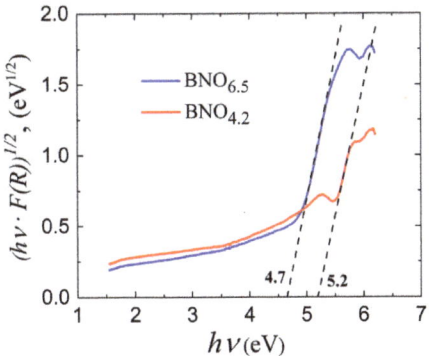

Figure 9. Tauc plots of the BNO_1 and BNO_2 nanopowders containing 6.5 and 4.2 at.% of oxygen.

Curve 3 in Figure 8b shows the discoloration of the MB solution (without a catalyst) under UV irradiation. As with exposure to sunlight, there was a strong discoloration (about 60%) within an hour. Approximately the same discoloration rate was observed under sunlight for an hour, but in these experiments, the catalyst was taken twice as much (10 and 5 mg for sunlight and UV experiments, respectively). Therefore, the discoloration rate under UV illumination is twice as high as under sunlight. Discoloration is a complex process and includes the following mechanisms: thermally induced MB photodegradation, MB absorption on BNO_x NPs, self-sensitizing MB photooxidation, and photocatalytic MB degradation. Careful consideration of all these processes makes it possible to determine the photocatalytic contribution to the discoloration process, and in this case, MB can be used as a test material.

Taking all MB discoloration mechanisms into account, and using Equation (1), the specific photocatalytic mass activity of the BNO_1 and BNO_2 nanopowders during half an hour of illumination was calculated to be ~140 mg·g^{-1}·h^{-1} (440 µmol·g^{-1}·h^{-1}). Photocatalytic activity in the MB degradation of various catalytic systems is shown in Table 1.

Pure BN is an indirect semiconductor with a bandgap of about 6 eV, which is much higher than the photon energy of visible light. Doping with oxygen reduces the band gap, and with a high oxygen content, it can be only 2.1 eV [52]. To estimate the band gaps of the BNO_1 and BNO_2 nanopowders, diffuse reflectance spectra (DRS) were collected. Figure 9 shows a Tauc plot,

$$((F(R_\infty) \times h\nu)^{1/\gamma} = B(h\nu - E_g),$$

converted from DRS using the Kubelka–Munk function ($F(R_\infty)$),

$$F(R_\infty) = \frac{(1-R_\infty)^2}{2R_\infty},$$

where $R_\infty = R_{sample}/R_{standart}$ is the reflectance of an infinitely thick specimen, h is Planck's constant, ν is the photon frequency, E_g is the band gap energy, and B is a constant. For the indirect semiconductors, $\gamma = 2$ [53].

From the Tauc plot, the band gap energies of samples BNO$_1$ and BNO$_2$ were determined to be 5.2 and 4.7 eV, respectively. Since these values exceed the energy of visible light, it was assumed that the observed photocatalytic activity of BNO$_1$ is associated with the presence of a UV component in the sunlight spectrum. To evaluate this effect, the photocatalytic degradation of the MB solution in the presence of the BNO$_1$ nanocatalyst was measured under visible light illumination (curve 4 in Figure 8a). It can be seen that the degradation rate slightly decreased. Note that a noticeable photodegradation of an MB aqueous solution was observed under irradiation with a laser beam with a wavelength of 670 nm [54]. It has been suggested that the photobleaching of an MB aqueous solution is a photodynamic process [54–57], and MB is a powerful photosensitizer that generates reactive oxygen species (ROS), including singlet oxygen $^1O_2$ and superoxide anion $\cdot O^{\cdot}_2$ [58,59]. The ROS generated during MB photosensitization can attack the material itself and lead to photochemical reactions on its surface (the so-called self-sensitized photooxidation). This explains the discoloration of the MB solution under visible or solar light illumination.

To assess the stability and reusability of the BNO$_x$ nanocatalysts, four successive cycles of degradation of MB solutions under UV illumination were carried out. The obtained results are shown in Figure 10a. After four cycles, the degradation ability of the catalysts remained at the 98% level, which indicates their high stability.

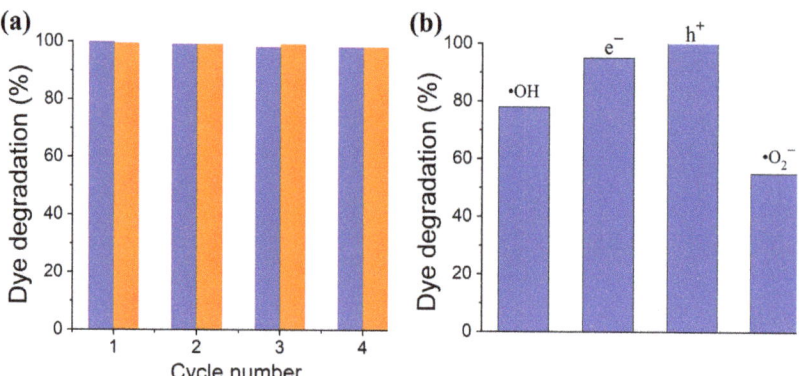

Figure 10. Photocatalytic activity and stability of BNO$_1$ and BNO$_2$ nanoparticles during the degradation of MB under UV irradiation (45 min) in four successive cycles (blue and orange colors correspond to 6.5 and 4.2 at.% of oxygen) (**a**); results of scavenger testing of the BNO$_1$ sample (**b**).

In the process of the photodegradation of organic dyes in the presence of a wide-gap photocatalyst, the following main reaction stages are usually considered [60–62]:

$$\text{Catalyst} + h\nu \rightarrow e^-_{CB} + h^+_{VB} \tag{2}$$

$$(O_2)_{ads} + e^-_{CB} \rightarrow \cdot O^{\cdot}_2 \tag{3}$$

$$H_2O + h^+_{VB} \rightarrow H^+ + \cdot OH \tag{4}$$

$$\cdot O^{\cdot}_2 + H^+ \rightarrow \cdot OOH \tag{5}$$

$$\cdot OOH + \cdot OOH \rightarrow H_2O_2 + O_2 \tag{6}$$

$$H_2O_2 \rightarrow 2 \cdot OH \tag{7}$$

$$\text{organic dye} + \cdot OH \rightarrow CO_2 + H_2O \tag{8}$$

The band gap of BN is large enough for the photolytic formation of superoxide radicals ($\cdot O_2^-$) from adsorbed oxygen. According to reaction (2), BN absorbs UV light and generates electrons in the conduction band and holes in the valence band as charge carriers. The electrons interact with adsorbed oxygen to form superoxide radicals (3). The holes interact with the H$_2$O molecule adsorbed on the BN surface to form a hydrogen ion and hydroxyl radicals (4). The superoxide radical (2) interacts with a hydrogen ion (4) to form hydroperoxy radicals (5) and, hence, generates hydrogen peroxide and molecular oxygen (6). Hydrogen peroxide is then decomposed into hydroxyl radicals by UV irradiation (7). Hydroxyl radicals are strong oxidants and decompose organic dyes (8).

To determine the primary reaction in the MB photodecomposition process in the presence of the BNO$_1$ catalyst, scavenger tests were performed (Figure 10b). The addition of a hole scavenger did not affect the photodegradation process, while the electron scavenger only slightly reduced the degradation efficiency. In contrast, scavengers of ·OH and, especially, ·O$^-{}_2$ species significantly reduced the degradation kinetics. Accordingly, we assume that these radicals make the main contribution to the dye degradation process. It was mentioned above that the superoxide anion ·O$^-{}_2$ is also generated by MB. Thus, in an MB-BNO$_x$ system, the superoxide anion is generated via both processes: MB photosensitization and BNO$_x$-catalyzed photolysis. This explains its great contribution to MB photodegradation.

Experiments on the photocatalytic UV degradation of MB solutions (Figure 8b) show that the sorption capacity of BNO$_x$ NPs depends on the oxygen content, and as it increases from 4.2 to 6.5 at.% (curve 1), the sorption capacity decreases. This is exactly the opposite of what one would expect since the specific surface area of BNO$_1$ is almost 1.4 times higher than that of BNO$_2$ (122.56 and 89.89 m^2·g^{-1}). The XPS analysis (Table 1) showed that the nitrogen content in the BNO$_2$ sample increased relative to boron, while the oxygen content decreased. Obviously, high-temperature annealing in ammonia led to the substitution of nitrogen for part of the oxygen, and the substitution mainly affected the oxygen atoms located on the surface. This means that the lower sorption capacity of the BNO$_1$ sample is associated with higher oxygen content on the surface. In this regard, it should be noted that surface oxygen defects do not change activity, since the photocatalytic activity of both materials is almost the same (Figure 8b). During photocatalysis, a photocatalyst is also exposed to the active particles formed, which usually leads to its oxidation and degradation. The high stability of the BNO$_2$ photocatalyst is expected until it is oxidized to an oxygen content comparable to sample BNO$_1$.

As can be seen from the comparison catalysts in Table 2, BNO$_x$ NPs are an efficient photocatalyst for MB degradation under UV irradiation.

Table 2. Photodegradation of MB over various photocatalysts under UV irradiation.

Photocatalyst	Light Source	%MB Degraded@Time	Reference
CuO/Bi$_2$O$_3$ Nanocomposite	UV-C irradiation	88.32%@120 min	[60]
5% PTh/ZnO	250 W high-pressure mercury lamp	95%@180 min	[61]
ZnO-NR/ACF Nanocomposites	UV irradiation	99%@120 min	[62]
γ-Fe$_3$/Fe$_3$O$_4$/SiO$_2$ (Ar modified)	UV irradiation	87.5%@120 min	[63]
70% CeO$_2$/g-C$_3$N$_4$ Z-scheme Heterojunction	UV irradiation	90.1%@180 min	[64]
BNO$_x$ nanoparticles	50 W low-pressure mercury lamp	100%@45 min	This work

3.4. Computational Analysis of MB Sorption on BNO$_x$

As noted above, oxygen defects on the BNO$_x$ surface prevent the sorption of MB molecules. At first glance, this is surprising since MB exists in the solution as a cation, and one would expect increased sorption due to negatively charged oxygen substituents. To elucidate the sorption process of MB molecules on BNO$_x$, we calculated the sorption energy depending on the orientation of the MB molecule using DFT. A layer of oxidized BN (6.5 at.% of O) was used as a model system. During the simulation, various possibilities for the location of the MB molecule on the BN surface were considered. The sorption process of the MB molecule on the oxidized BN differs from the process on pure h-BN. In the case of a defect-free BN surface, a flat MB molecule stands on an edge at an angle of about 45 degrees to the plane. This orientation makes it possible to create a denser packing and, as a result, increases the sorption capacity of BN with respect to BNO_x.

In the case of BNO_x, the MB molecule is oriented parallel to the surface so that its aromatic system is above the BN rings (Figure 11a). This orientation is most likely due to the mutual coordination of conjugated π-systems over each other. In addition, the result of our simulation showed that the nitrogen atom of the central MB ring is bound to the boron atom nearest to the oxygen defect and forms a chemical bond with a bonding energy of 2.7 eV and a bond length of 1.55 Å (Figure 11b), which is 0.1 Å less than the B-N bond in BN. Thus, the MB molecule strongly binds to the surface of the oxidized BN, as evidenced by both the distance between the atoms and the binding energy of the molecule to the surface. A strong bond is possible due to the redistribution of the charge on the MB molecule and the redistribution of the electron density near the oxygen defect in the BN. Despite the strong chemical binding, the location of the MB molecule is such that it occupies a large surface area of the BNO_x, which, accordingly, reduces its sorption capacity.

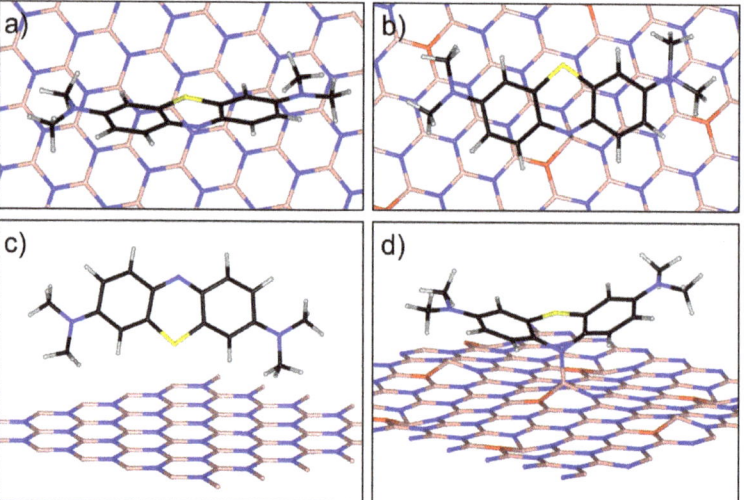

Figure 11. Methylene blue molecule adsorbed on the surface of pure (**a**,**c**) and oxidized BN (**b**,**d**), top (**a**,**b**) and side (**c**,**d**) view, respectively. Oxygen defects are marked in red and nitrogen in blue.

4. Conclusions

The optical absorbance of methylene blue (MB) aqueous solutions in a concentration range of 0.5 to 10 mg·L^{-1} and photolytic effects leading to discoloration of MB solutions with and without boron oxynitride (BNO_x) nanoparticles (NPs) were studied under various types of illumination (visible light, sunlight, and UV light). It was shown for the first time that in an MB concentration range of 3.33 to 4.00 mg·L^{-1}, there is a violation of the linear dependence of optical absorption on the MB concentration, which is due to the formation of dimeric and trimeric molecular associates. This must be taken into account in order to correctly assess the MB concentration. The fractions of the monomer, dimer, and trimer in the MB solution with a concentration of 10 mg·L^{-1}, estimated from the absorption peak areas, are approximately 91.1%, 7.1%, and 1.8%, respectively.

The MB solutions discolorized when they were illuminated in a wide spectral range, from visible light to the UV-B range (254 nm). This process is thermally dependent, and the discoloration rate in visible light and sunlight nearly doubles as the temperature rises from 20 to 80 °C. MB discoloration may be due to its self-sensitized photooxidation, in which MB, when illuminated, generates reactive oxygen species that oxidize MB molecules. Although thermally induced MB discoloration has been demonstrated only at 20 and 80 °C, it is clear that it occurs at any temperature in this range, but with less efficiency. This effect can be easily exploited in practice, either by using an excess of industrial heat or with focused sunlight.

A DFT analysis of MB sorption capacity on BNO_x NPs shows that surface oxygen defects prevent the sorption of MB molecules. This is due to the planar orientation of the MB molecule above the BNO_x surface. The calculations also show that the MB molecule is chemically bound to the BNO_x surface by the boron atom nearest to the oxygen defect. A strong electrostatic interaction changes the electronic configuration of the MB molecule and increases its self-sensitizing activity. This explains the enhanced photodegradation of MB in visible light in the presence of BNO_x nanoparticles.

The discoloration process of MB involves the following mechanisms: thermally induced MB photodegradation, MB absorption on BNO_x nanoparticles, self-sensitizing MB photooxidation, and photocatalytic MB degradation. Accounting for all these processes makes it possible to reveal the contribution of the photocatalyst to the discoloration process, and in this case, MB can be used as a test material.

Taking into account all these mechanisms of MB discoloration, the photocatalytic activity of BNO_x NPs containing 4.2 and 6.5 at.% of oxygen was studied under UV irradiation of MB aqueous solutions. The specific mass activity of both types of NPs is approximately 140 mg·g^{-1}·h^{-1} (440 μmol·g^{-1}·h^{-1}). The high photocatalytic activity of BNO_x NPs in a wide range of oxygen substitutions, combined with their high stability, makes them promising metal-free photocatalysts for water treatment.

Author Contributions: Methodology: A.T.M. and L.A.V.; experiments: A.T.M., A.S.K., D.V.L. and I.N.V.; experimental results analysis: A.T.M.; theoretical results analysis: P.B.S.; writing—original draft preparation: A.T.M. and L.A.V.; writing—review and editing: X.F. and D.V.S. All authors have read and agreed to the published version of the manuscript.

Funding: This research was funded by the Russian Science Foundation (No. 21-49-00039). X.F. thanks the financial support from the National Natural Science Foundation of China (No. 12061131009).

Institutional Review Board Statement: Not applicable.

Informed Consent Statement: Not applicable.

Data Availability Statement: There are no data to report.

Acknowledgments: The authors would like to acknowledge the Russian Science Foundation and the National Natural Science Foundation of China.

Conflicts of Interest: The authors declare no conflict of interest.

References

1. Khan, I.; Saeed, K.; Zekker, I.; Zhang, B.; Hendi, A.H.; Ahmad, A.; Ahmad, S.; Zada, N.; Ahmad, H.; Shah, L.A.; et al. Review on Methylene Blue: Its Properties, Uses, Toxicity and Photodegradation. *Water* **2022**, *14*, 242. [CrossRef]
2. Tardivo, J.P.; Del Giglio, A.; de Oliveira, C.S.; Gabrielli, D.S.; Junqueira, H.C.; Tada, D.B.; Severino, D.; de Fátima Turchiello, R.; Baptista, M.S. Methylene Blue in Photodynamic Therapy: From Basic Mechanisms to Clinical Applications. *Photodiagnosis Photodyn. Ther.* **2005**, *2*, 175–191. [CrossRef]
3. Dos Santos, A.F.; Terra, L.F.; Wailemann, R.A.M.; Oliveira, T.C.; Gomes, V.d.M.; Mineiro, M.F.; Meotti, F.C.; Bruni-Cardoso, A.; Baptista, M.S.; Labriola, L. Methylene Blue Photodynamic Therapy Induces Selective and Massive Cell Death in Human Breast Cancer Cells. *BMC Cancer* **2017**, *17*, 194. [CrossRef] [PubMed]
4. Francisco, C.M.L.; Gonçalves, J.M.L.A.; Brum, B.S.; Santos, T.P.C.; Lino-dos-Santos-Franco, A.; Silva, D.F.T.; Pavani, C. The Photodynamic Efficiency of Phenothiazinium Dyes Is Aggregation Dependent. *New J. Chem.* **2017**, *41*, 14438–14443. [CrossRef]

5. Coronel, A.; Catalán-Toledo, J.; Fernández-Jaramillo, H.; Godoy-Martínez, P.; Flores, M.E.; Moreno-Villoslada, I. Photodynamic Action of Methylene Blue Subjected to Aromatic-Aromatic Interactions with Poly(Sodium 4-Styrenesulfonate) in Solution and Supported in Solid, Highly Porous Alginate Sponges. *Dyes Pigments* **2017**, *147*, 455–464. [CrossRef]
6. Cwalinski, T.; Polom, W.; Marano, L.; Roviello, G.; D'Angelo, A.; Cwalina, N.; Matuszewski, M.; Roviello, F.; Jaskiewicz, J.; Polom, K. Methylene Blue—Current Knowledge, Fluorescent Properties, and Its Future Use. *J. Clin. Med.* **2020**, *9*, 3538. [CrossRef]
7. Bayomie, O.S.; Kandeel, H.; Shoeib, T.; Yang, H.; Youssef, N.; El-Sayed, M.M.H. Novel Approach for Effective Removal of Methylene Blue Dye from Water Using Fava Bean Peel Waste. *Sci. Rep.* **2020**, *10*, 7824. [CrossRef]
8. Rafatullah, M.; Sulaiman, O.; Hashim, R.; Ahmad, A. Adsorption of Methylene Blue on Low-Cost Adsorbents: A Review. *J. Hazard. Mater.* **2010**, *177*, 70–80. [CrossRef]
9. El Messaoudi, N.; El Khomri, M.; Fernine, Y.; Bouich, A.; Lacherai, A.; Jada, A.; Sher, F.; Lima, E.C. Hydrothermally Engineered Eriobotrya Japonica Leaves/MgO Nanocomposites with Potential Applications in Wastewater Treatment. *Groundw. Sustain. Dev.* **2022**, *16*, 100728. [CrossRef]
10. Elmorsi, R.R.; El-Wakeel, S.T.; Shehab El-Dein, W.A.; Lotfy, H.R.; Rashwan, W.E.; Nagah, M.; Shaaban, S.A.; Sayed Ahmed, S.A.; El-Sherif, I.Y.; Abou-El-Sherbini, K.S. Adsorption of Methylene Blue and Pb2+ by Using Acid-Activated Posidonia Oceanica Waste. *Sci. Rep.* **2019**, *9*, 3356. [CrossRef]
11. Loutfi, M.; Mariouch, R.; Mariouch, I.; Belfaquir, M.; ElYoubi, M.S. Adsorption of Methylene Blue Dye from Aqueous Solutions onto Natural Clay: Equilibrium and Kinetic Studies. *Mater. Today Proc.* **2022**, *in press*. [CrossRef]
12. Aragaw, T.A.; Alene, A.N. A Comparative Study of Acidic, Basic, and Reactive Dyes Adsorption from Aqueous Solution onto Kaolin Adsorbent: Effect of Operating Parameters, Isotherms, Kinetics, and Thermodynamics. *Emerg. Contam.* **2022**, *8*, 59–74. [CrossRef]
13. Foroutan, R.; Peighambardoust, S.J.; Latifi, P.; Ahmadi, A.; Alizadeh, M.; Ramavandi, B. Carbon Nanotubes/β-Cyclodextrin/MnFe$_2$O$_4$ as a Magnetic Nanocomposite Powder for Tetracycline Antibiotic Decontamination from Different Aqueous Environments. *J. Environ. Chem. Eng.* **2021**, *9*, 106344. [CrossRef]
14. Peighambardoust, S.J.; Foroutan, R.; Peighambardoust, S.H.; Khatooni, H.; Ramavandi, B. Decoration of Citrus Limon Wood Carbon with Fe$_3$O$_4$ to Enhanced Cd2+ Removal: A Reclaimable and Magnetic Nanocomposite. *Chemosphere* **2021**, *282*, 131088. [CrossRef]
15. Foroutan, R.; Zareipour, R.; Mohammadi, R. Fast Adsorption of Chromium (VI) Ions from Synthetic Sewage Using Bentonite and Bentonite/Bio-Coal Composite: A Comparative Study. *Mater. Res. Express* **2018**, *6*, 025508. [CrossRef]
16. Li, J.; Xiao, X.; Xu, X.; Lin, J.; Huang, Y.; Xue, Y.; Jin, P.; Zou, J.; Tang, C. Activated Boron Nitride as an Effective Adsorbent for Metal Ions and Organic Pollutants. *Sci. Rep.* **2013**, *3*, 3208. [CrossRef]
17. Marchesini, S.; Wang, X.; Petit, C. Porous Boron Nitride Materials: Influence of Structure, Chemistry and Stability on the Adsorption of Organics. *Front. Chem.* **2019**, *7*, 160. [CrossRef]
18. Xiong, J.; Di, J.; Zhu, W.; Li, H. Hexagonal Boron Nitride Adsorbent: Synthesis, Performance Tailoring and Applications. *J. Energy Chem.* **2020**, *40*, 99–111. [CrossRef]
19. Din, M.I.; Khalid, R.; Najeeb, J.; Hussain, Z. Fundamentals and Photocatalysis of Methylene Blue Dye Using Various Nanocatalytic Assemblies- a Critical Review. *J. Clean. Prod.* **2021**, *298*, 126567. [CrossRef]
20. Matveev, A.T.; Konopatsky, A.S.; Leybo, D.V.; Volkov, I.N.; Kovalskii, A.M.; Varlamova, L.A.; Sorokin, P.B.; Fang, X.; Kulinich, S.A.; Shtansky, D.V. Amorphous MoSxOy/h-BNxOy Nanohybrids: Synthesis and Dye Photodegradation. *Nanomaterials* **2021**, *11*, 3232. [CrossRef]
21. Yan, X.; Ohno, T.; Nishijima, K.; Abe, R.; Ohtani, B. Is Methylene Blue an Appropriate Substrate for a Photocatalytic Activity Test? A Study with Visible-Light Responsive Titania. *Chem. Phys. Lett.* **2006**, *429*, 606–610. [CrossRef]
22. Sáenz-Trevizo, A.; Pizá-Ruiz, P.; Chávez-Flores, D.; Ogaz-Parada, J.; Amézaga-Madrid, P.; Vega-Ríos, A.; Miki-Yoshida, M. On the Discoloration of Methylene Blue by Visible Light. *J. Fluoresc.* **2019**, *29*, 15–25. [CrossRef] [PubMed]
23. Matveev, A.T.; Permyakova, E.S.; Kovalskii, A.M.; Leibo, D.; Shchetinin, I.V.; Maslakov, K.I.; Golberg, D.V.; Shtansky, D.V.; Konopatsky, A.S. New Insights into Synthesis of Nanocrystalline Hexagonal BN. *Ceram. Int.* **2020**, *46*, 19866–19872. [CrossRef]
24. Kresse, G.; Furthmüller, J. Efficient Iterative Schemes for Ab Initio Total-Energy Calculations Using a Plane-Wave Basis Set. *Phys. Rev. B* **1996**, *54*, 11169–11186. [CrossRef] [PubMed]
25. Shimojo, F.; Hoshino, K.; Zempo, Y. Ab Initio Molecular-Dynamics Simulation Method for Complex Liquids. *Comput. Phys. Commun.* **2001**, *142*, 364–367. [CrossRef]
26. Kresse, G.; Furthmüller, J. Efficiency of Ab-Initio Total Energy Calculations for Metals and Semiconductors Using a Plane-Wave Basis Set. *Comput. Mater. Sci.* **1996**, *6*, 15–50. [CrossRef]
27. Fernández-Pérez, A.; Marbán, G. Visible Light Spectroscopic Analysis of Methylene Blue in Water; What Comes after Dimer? *ACS Omega* **2020**, *5*, 29801–29815. [CrossRef]
28. Fernández-Pérez, A.; Valdés-Solís, T.; Marbán, G. Visible Light Spectroscopic Analysis of Methylene Blue in Water; the Resonance Virtual Equilibrium Hypothesis. *Dyes Pigments* **2019**, *161*, 448–456. [CrossRef]
29. Heger, D.; Jirkovský, J.; Klán, P. Aggregation of Methylene Blue in Frozen Aqueous Solutions Studied by Absorption Spectroscopy. *J. Phys. Chem. A* **2005**, *109*, 6702–6709. [CrossRef]
30. Braswell, E. Evidence for Trimerization in Aqueous Solutions of Methylene Blue. *J. Phys. Chem.* **1968**, *72*, 2477–2483. [CrossRef]
31. Lee, S.-K.; Mills, A. Luminescence of *Leuco*-Thiazine Dyes. *J. Fluoresc.* **2003**, *13*, 375–377. [CrossRef]

32. Yuzhakov, V.I. Association of Dye Molecules and Its Spectroscopic Manifestation. *Russ. Chem. Rev.* **1979**, *48*, 1076–1091. [CrossRef]
33. Tsuchiya, N.; Kuwabara, K.; Hidaka, A.; Oda, K.; Katayama, K. Reaction Kinetics of Dye Decomposition Processes Monitored inside a Photocatalytic Microreactor. *Phys. Chem. Chem. Phys.* **2012**, *14*, 4734. [CrossRef]
34. Lobyshev, V.I.; Shikhlinskaya, R.E.; Ryzhikov, B.D. Experimental Evidence for Intrinsic Luminescence of Water. *J. Mol. Liq.* **1999**, *82*, 73–81. [CrossRef]
35. Shaopu, L.; Zhongfang, L.; Ming, L. Analytical Application of Double Scattering Spectra of Double Scattering Spectra of Ion-Association Complex 1: Selenium (IV)-Iˆ-Rhodamine B System. *Acta Chim. Sin.* **1995**, *53*, 1178.
36. Luo, H.Q.; Liu, S.P.; Li, N.B.; Liu, Z.F. Resonance Rayleigh Scattering, Frequency Doubling Scattering and Second-Order Scattering Spectra of the Heparin–Crystal Violet System and Their Analytical Application. *Anal. Chim. Acta* **2002**, *468*, 275–286. [CrossRef]
37. Long, X.; Zhang, H.; Bi, S. Frequency Doubling Scattering and Second-Order Scattering Spectra of Phosphato-Molybdate Heteropoly Acid–Protein System and Their Analytical Application. *Spectrochim. Acta. A. Mol. Biomol. Spectrosc.* **2004**, *60*, 1631–1636. [CrossRef]
38. Liu, J.F.; Li, N.B.; Luo, H.Q. Resonance Rayleigh Scattering, Second-Order Scattering and Frequency Doubling Scattering Spectra for Studying the Interaction of Erythrosine with and Its Analytical Application. *Spectrochim. Acta. A. Mol. Biomol. Spectrosc.* **2011**, *79*, 631–637. [CrossRef]
39. Zhou, J.-F.; Li, N.-B.; Luo, H.-Q. Analytical Application of Resonance Rayleigh Scattering, Frequency Doubling Scattering, and Second-Order Scattering Spectra for the Sodium Alginate-CTAB System. *Anal. Lett.* **2011**, *44*, 637–647. [CrossRef]
40. Morozova, M.A.; Tumasov, V.N.; Kazimova, I.V.; Maksimova, T.V.; Uspenskaya, E.V.; Syroeshkin, A.V. Second-Order Scattering Quenching in Fluorescence Spectra of Natural Humates as a Tracer of Formation Stable Supramolecular System for the Delivery of Poorly Soluble Antiviral Drugs on the Example of Mangiferin and Favipiravir. *Pharmaceutics* **2022**, *14*, 767. [CrossRef]
41. Luo, X.; Zhang, S.; Lin, X. New Insights on Degradation of Methylene Blue Using Thermocatalytic Reactions Catalyzed by Low-Temperature Excitation. *J. Hazard. Mater.* **2013**, *260*, 112–121. [CrossRef] [PubMed]
42. Herrmann, J.-M. Fundamentals and Misconceptions in Photocatalysis. *J. Photochem. Photobiol. Chem.* **2010**, *216*, 85–93. [CrossRef]
43. Lee, Y.Y.; Moon, J.H.; Choi, Y.S.; Park, G.O.; Jin, M.; Jin, L.Y.; Li, D.; Lee, J.Y.; Son, S.U.; Kim, J.M. Visible-Light Driven Photocatalytic Degradation of Organic Dyes over Ordered Mesoporous $Cd_xZn_{1-x}S$ Materials. *J. Phys. Chem. C* **2017**, *121*, 5137–5144. [CrossRef]
44. Junqueira, H.C.; Severino, D.; Dias, L.G.; Gugliotti, M.S.; Baptista, M.S. Modulation of Methylene Blue Photochemical Properties Based on Adsorption at Aqueous Micelle Interfaces. *Phys. Chem. Chem. Phys.* **2002**, *4*, 2320–2328. [CrossRef]
45. Barakat, N.A.M.; Tolba, G.M.K.; Khalil, K.A. Methylene Blue Dye as Photosensitizer for Scavenger-Less Water Photo Splitting: New Insight in Green Hydrogen Technology. *Polymers* **2022**, *14*, 523. [CrossRef]
46. Yu, Z.; Chuang, S.S.C. Probing Methylene Blue Photocatalytic Degradation by Adsorbed Ethanol with In Situ IR. *J. Phys. Chem. C* **2007**, *111*, 13813–13820. [CrossRef]
47. Rauf, M.A.; Meetani, M.A.; Khaleel, A.; Ahmed, A. Photocatalytic Degradation of Methylene Blue Using a Mixed Catalyst and Product Analysis by LC/MS. *Chem. Eng. J.* **2010**, *157*, 373–378. [CrossRef]
48. Mills, A.; Hazafy, D.; Parkinson, J.; Tuttle, T.; Hutchings, M.G. Effect of Alkali on Methylene Blue (C.I. Basic Blue 9) and Other Thiazine Dyes. *Dyes Pigments* **2011**, *88*, 149–155. [CrossRef]
49. Mondal, S.; Reyes, M.E.D.A.; Pal, U. Plasmon Induced Enhanced Photocatalytic Activity of Gold Loaded Hydroxyapatite Nanoparticles for Methylene Blue Degradation under Visible Light. *RSC Adv.* **2017**, *7*, 8633–8645. [CrossRef]
50. Shtansky, D.V.; Tsuda, O.; Ikuhara, Y.; Yoshida, T. Crystallography and Structural Evolution of Cubic Boron Nitride Films during Bias Sputter Deposition. *Acta Mater.* **2000**, *48*, 3745–3759. [CrossRef]
51. Makarova, A.A.; Fernandez, L.; Usachov, D.Y.; Fedorov, A.; Bokai, K.A.; Smirnov, D.A.; Laubschat, C.; Vyalikh, D.V.; Schiller, F.; Ortega, J.E. Oxygen Intercalation and Oxidation of Atomically Thin H-BN Grown on a Curved Ni Crystal. *J. Phys. Chem. C* **2019**, *123*, 593–602. [CrossRef]
52. Weng, Q.; Kvashnin, D.G.; Wang, X.; Cretu, O.; Yang, Y.; Zhou, M.; Zhang, C.; Tang, D.-M.; Sorokin, P.B.; Bando, Y.; et al. Tuning of the Optical, Electronic, and Magnetic Properties of Boron Nitride Nanosheets with Oxygen Doping and Functionalization. *Adv. Mater.* **2017**, *29*, 1700695. [CrossRef]
53. Makuła, P.; Pacia, M.; Macyk, W. How To Correctly Determine the Band Gap Energy of Modified Semiconductor Photocatalysts Based on UV–Vis Spectra. *J. Phys. Chem. Lett.* **2018**, *9*, 6814–6817. [CrossRef]
54. Zhang, L.Z.; Tang, G.-Q. The Binding Properties of Photosensitizer Methylene Blue to Herring Sperm DNA: A Spectroscopic Study. *J. Photochem. Photobiol. B* **2004**, *74*, 119–125. [CrossRef]
55. Fisher, A.M.R.; Murphree, A.L.; Gomer, C.J. Clinical and Preclinical Photodynamic Therapy. *Lasers Surg. Med.* **1995**, *17*, 2–31. [CrossRef]
56. Wainwright, M. The Emerging Chemistry of Blood Product Disinfection. *Chem. Soc. Rev.* **2002**, *31*, 128–136. [CrossRef]
57. Wainwright, M. Phenothiazinium Photosensitisers: Choices in Synthesis and Application. *Dyes Pigments* **2003**, *57*, 245–257. [CrossRef]
58. Kearns, D.R. Physical and Chemical Properties of Singlet Molecular Oxygen. *Chem. Rev.* **1971**, *71*, 395–427. [CrossRef]
59. Wasserman, H.H.; Scheffer, J.R.; Cooper, J.L. Singlet Oxygen Reactions with 9,10-Diphenylanthracene Peroxide. *J. Am. Chem. Soc.* **1972**, *94*, 4991–4996. [CrossRef]
60. Poorsajadi, F.; Sayadi, M.H.; Hajiani, M.; Rezaei, M.R. Synthesis of CuO/Bi_2O_3 Nanocomposite for Efficient and Recycling Photodegradation of Methylene Blue Dye. *Int. J. Environ. Anal. Chem.* **2020**, 1–14. [CrossRef]

61. Faisal, M.; Harraz, F.A.; Jalalah, M.; Alsaiari, M.; Al-Sayari, S.A.; Al-Assiri, M.S. Polythiophene Doped ZnO Nanostructures Synthesized by Modified Sol-Gel and Oxidative Polymerization for Efficient Photodegradation of Methylene Blue and Gemifloxacin Antibiotic. *Mater. Today Commun.* **2020**, *24*, 101048. [CrossRef]
62. Albiss, B.; Abu-Dalo, M. Photocatalytic Degradation of Methylene Blue Using Zinc Oxide Nanorods Grown on Activated Carbon Fibers. *Sustainability* **2021**, *13*, 4729. [CrossRef]
63. Sanad, M.M.S.; Farahat, M.M.; El-Hout, S.I.; El-Sheikh, S.M. Preparation and Characterization of Magnetic Photocatalyst from the Banded Iron Formation for Effective Photodegradation of Methylene Blue under UV and Visible Illumination. *J. Environ. Chem. Eng.* **2021**, *9*, 105127. [CrossRef]
64. Wei, X.; Wang, X.; Pu, Y.; Liu, A.; Chen, C.; Zou, W.; Zheng, Y.; Huang, J.; Zhang, Y.; Yang, Y.; et al. Facile Ball-Milling Synthesis of CeO_2/g-C_3N_4 Z-Scheme Heterojunction for Synergistic Adsorption and Photodegradation of Methylene Blue: Characteristics, Kinetics, Models, and Mechanisms. *Chem. Eng. J.* **2021**, *420*, 127719. [CrossRef]

Review

Removal of Pharmaceuticals and Personal Care Products (PPCPs) by Free Radicals in Advanced Oxidation Processes

Jiao Jiao [1], Yihua Li [1,*], Qi Song [1], Liujin Wang [2], Tianlie Luo [2], Changfei Gao [3], Lifen Liu [4] and Shengtao Yang [1,*]

[1] Key Laboratory of Pollution Control Chemistry and Environmental Functional Materials for Qinghai-Tibet Plateau of the National Ethnic Affairs Commission, School of Chemistry and Environment, Southwest Minzu University, Chengdu 610041, China

[2] State of Environmental Protection Key Laboratory of Synergetic Control and Joint Remediation for Soil & Water Pollution, College of Ecology and Environment, Chengdu University of Technology, Chengdu 610059, China

[3] School of Environmental and Material Engineering, Yantai University, Yantai 264005, China

[4] Key Laboratory of Industrial Ecology and Environmental Engineering, Ministry of Education, School of Ocean Science and Technology, Dalian University of Technology, Panjin 124221, China

* Correspondence: yihuali@swun.edu.cn (Y.L.); yangst@pku.edu.cn (S.Y.)

Abstract: As emerging pollutants, pharmaceutical and personal care products (PPCPs) have received extensive attention due to their high detection frequency (with concentrations ranging from ng/L to μg/L) and potential risk to aqueous environments and human health. Advanced oxidation processes (AOPs) are effective techniques for the removal of PPCPs from water environments. In AOPs, different types of free radicals (HO·, $SO_4^{·-}$, $O_2^{·-}$, etc.) are generated to decompose PPCPs into non-toxic and small-molecule compounds, finally leading to the decomposition of PPCPs. This review systematically summarizes the features of various AOPs and the removal of PPCPs by different free radicals. The operation conditions and comprehensive performance of different types of free radicals are summarized, and the reaction mechanisms are further revealed. This review will provide a quick understanding of AOPs for later researchers.

Keywords: pharmaceuticals and personal care products; advanced oxidation processes; free radicals; water treatment

1. Introduction

Pharmaceutical and personal care products (PPCPs) are attracting increasing concern [1–3] due to the fact that they have been extensively detected in aqueous environments, solids and sediments [4–10]. PPCPs are defined as widespread chemicals including pharmaceuticals (such as hormones, antibiotics, antidepressants, non-steroidal anti-inflammatory drugs, and lipid regulators) and personal care products (such as preservatives, disinfectants, fragrances, and sunscreens) [2,11]. PPCPs are widely used in high quantities throughout the world, and are known to be released into aquatic environments from multiple discharges, including domestic wastewater, pharmaceutical wastewater [11], daily washing, swimming, excreting after human ingestion [12], livestock, aquaculture and households (excretion and littering) [13]. Meanwhile, in terms of household medicine, the inappropriate disposal of pharmaceutical products could adversely infect the environment and increase the risk of accidental poisoning [14]. It was revealed that domestic sewage was the primary source of PPCP emissions in the surface water of China [15,16]. These pollutants, along with their transformed intermediate products, have been prevalent in most environmental matrices [17].

To evaluate the per-capita emission rates of some PPCPs, in Korea, Subedi et al. [18] found that the per-capita emission rates of triclocarban and acetaminophen (ACE) were 158 and 59 μg/capita/day, respectively. In the long run, trace concentrations of 1 ng/L~100 μg/L of PPCPs in aqueous environments pose potential risks to animals and human health [19].

The detection results of partial PPCPs in various aquatic environments from different countries were summarized in Table 1. PPCPs have been widely detected in surface water, groundwater, and even drinking water, with concentrations ranging from ng/L to mg/L. Additionally, the concentrations of PPCPs in sediment and soil are at the level of mg/kg [19–22]. Recently, Chaves et al. [23] revealed the frequencies of detection of PPCPs in surface water with 104 articles (Figure 1a). Among them, carbamazepine (CBZ), diclofenac (DCF), sulfamethoxazole (SMX), caffeine (CAF), ACE and ibuprofen (IBP) have the highest detection frequency. The maximum concentrations (ng/L) of the most frequent PPCPs in each analyzed continent are shown in Figure 1b. Although the concentrations of PPCPs are not enough to cause acute toxicity to humans or animals, most of the PPCPs will gradually accumulate in the aqueous organisms for their refractory properties, causing potential threats to human health [24–26]. Meanwhile, the ubiquitous presence of PPCPs would cause trepidation in maintaining the homeostasis of the ecological environment [27]. Thus, it is of great importance to remove PPCPs from the environment. In recent decades, the number of publications on PPCPs had been increased annually (Figure 2).

Table 1. The detection of partial PPCPs in various aquatic environments from different countries.

Location	Aquatic Environment	Chemical	Category	Concentration	Ref.
Poland	Groundwater	N,Ndiethyl-meta-toluamide(DEET)	Mosquito and insect repellants	17.28 µg/L	[28]
		17β-oestradiol	Hormones	48 ng/L	
	Surface water	CAF	Stimulants	29.9955 µg/L	
		Bisphenol A (BPA)	Hormones	3.113 µg/L	
	Drinking water	Azithromycin (AZM)	Antibiotics	193 ng/L	
		Paracetamol	Non-steroidal anti-inflammatory drugs	173 ng/L	
Brazil		IBP	Non-steroidal anti-inflammatory drugs	224 ng/L	[29]
		CAF	Stimulants	159 ng/L	
	Surface water	Avobenzone (ABZ)	Sunscreen agents	340 ng/L	
		Glibenclamide (GBC)	Hypoglycemic drugs	50–120 ng/L	
	Drinking water	Nimesulide (NI)	Non-steroidal anti-inflammatory drugs	181 ng/L	
		Methylparaben	Preservatives	234 ng/L	
		ABZ	Sunscreen agents	290 ng/L	
China	Surface water	SMX	Antibiotics	<LOQ–2.92 ng/L	[30]
		4-n-nonylphenol	Hormones	9.90–457.40 ng/L	
		Salicylic acid (SA)	Analgesics	2.92–34.12 ng/L	
	Drinking water	Sulfamethoxypyridazine	Antibacterial	107.14 ng/L	[31]
		Lincomycin	Antibiotics	1.00–29.32 ng/L	
Vietnam	Surface water	SMX	Antibiotics	<LOQ–2.18 ng/L	[32]
		CBZ	Antiepileptics	<LOQ–57.4 ng/L	
		LCM	Antibiotics	<LOQ–378 ng/L	
		SMZ	Antibiotics	3.65–2778 ng/L	

Table 1. *Cont.*

Location	Aquatic Environment	Chemical	Category	Concentration	Ref.
India	Groundwater	Ketoprofen (KPF)	Non-steroidal anti-inflammatory drugs	<LOQ–23.4 ng/L	[33]
		IBP	Non-steroidal anti-inflammatory drugs	<LOQ–49.4 ng/L	
		CAF	Stimulant	15.2–262 ng/L	
Pakistan	Groundwater	Tigecycline	Antibiotics	21.3 ng/L	[34]
		Ciprofloxacin (CIP)	Antibiotics	18.2 ng/L	
United States	Drinking water	CBZ	Antiepileptic	51 ng/L	[35]
		DCF	Non-steroidal anti-inflammatory drugs	1.2 ng/L	
		SMX	Antibiotics	110 ng/L	
		Naproxen	Non-steroidal anti-inflammatory drugs	32 ng/L	

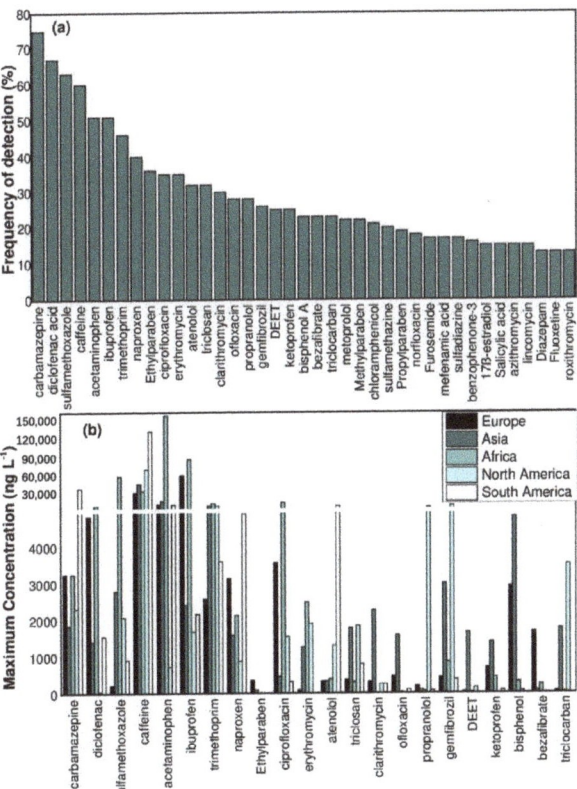

Figure 1. Detection results of PPCPs in surface water. (**a**) Detection frequencies of 104 articles; (**b**) the maximum concentrations (ng/L) of the most detected PPCPs in each analyzed continent. Reproduced with permission from [23].

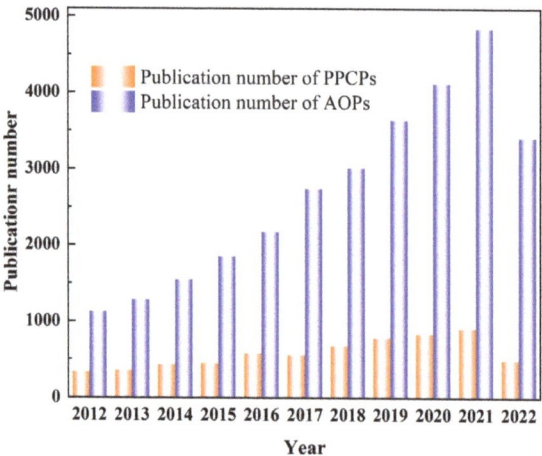

Figure 2. Number of publications on PPCPs and AOPs from 2012 to 2022 (topic keywords "pharmaceutical and personal care products" and "advanced oxidation processes" searched from web of science), data updated by 28 September 2022.

Wastewater treatment technologies, such as sedimentation [36], adsorption, biodegradation, sand filtration and membrane separation [37], have shown excellent removal performance in terms of conventional contaminants. Nevertheless, the removal performances of PPCPs are not satisfactory yet due to the high biotoxicity and pseudo-persistence properties of PPCPs. Most of the PPCPs were discharged to surface water after treated by municipal sewage treatment plants [38,39]. To remove PPCPs from aqueous environments, efficient approaches need to be explored. Advanced oxidation processes (AOPs), as deep treatment technologies, are acknowledged as among the most promising technologies in terms of the removal of PPCPs [40–43]. In recent decades, the number of publications focused on AOPs has grown exponentially (Figure 2). For instance, Somathilake et al. [44] used the UV-assisted ozone oxidation process for the removal of CBZ in treated domestic wastewater. The complete removal efficiency was achieved in the beginning as 0.5 min, which yielded an excellent mineralization extent of CBZ. In the process of oxidation, the degradation efficiencies of PPCPs are realized mostly by the strong oxidization of free radicals, which are in situ generated during the reaction processes under appropriate conditions such as high temperature, high pressure, microelectricity, ultrasound, light irradiation and catalysts [45–47]. According to different mechanisms, AOPs could be classified into photochemical oxidation [48,49], catalytic wet oxidation [50,51], ultrasonic oxidation [52,53], ozone oxidation [54,55], electrochemical oxidation [56,57], Fenton oxidation [58,59], persulfate-based oxidation [60,61], radiation oxidation [62,63], photoelectrocatalysis [64,65], etc. Due to their high reaction rates, inducible chain reactions and final products, AOPs could effectively remove PPCPs and decompose them into micromolecule compounds from complex environments (Figure 3) [66]. In AOPs, different free radicals are generated in different processes, such as hydroxyl radical (HO·), sulfate radical ($SO_4·^-$), superoxide radical ($O_2·^-$), and chlorine radical (Cl·).

Herein, the removal performance of PPCPs of different free radicals generated in various AOPs were analyzed and summarized systematically by in-depth analysis of the reaction mechanisms. Current status, future directions, perspectives and challenges of AOPs were further discussed.

Figure 3. The illustration of the removal of PPCPs by free radicals in AOPs.

2. Various AOPs

AOPs is the general term for the photochemical oxidation process, the electrochemical oxidation process, the wet air oxidation process, the ultrasonic oxidation process, the gamma ray/electron beam radiation process, the Fenton oxidation process, the ozone oxidation process, the persulfate-based oxidation process, etc. AOPs are employed either independently or in combination with other chemical processes for the removal of PPCPs. Prior to revealing the degradation mechanism of PPCPs, it is needed to analyze and summarize the generation of possible free radicals in different AOPs. Figure 4 summarized the possible free radicals generated in different AOPs.

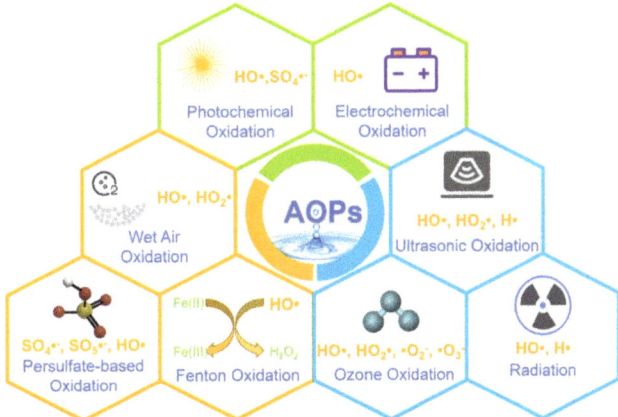

Figure 4. Possible free radicals generated in different AOPs.

2.1. Photochemical Oxidation

Photochemical oxidation processes can be achieved by two approaches—the photocatalytic oxidation method and the photo-excited oxidation method. The former uses a semiconductor (such as TiO_2 [67,68] and WO_3 [69,70]) as a photocatalyst. When the semiconductor makes contact with water, strongly oxidizing free radicals (i.e., HO·) are generated on its surface (as shown in Equations (1) and (2) [71,72]), which react with PPCPs and degrade them in water. As among the most promising photocatalyst materials, nano-TiO_2 is highly utilized to remove PPCPs from water, and nano-TiO_2 has been studied by numerous studies [73–75].

$$TiO_2 + h\nu \rightarrow e_{CB}^- + h_{VB}^+ \qquad (1)$$

$$h_{VB}^+ + OH^- \rightarrow HO\cdot \qquad (2)$$

The latter method aims at enhancing the oxidation potential of oxidants under ultraviolet (UV) irradiation. In the process, free radicals with strong oxidizing properties such as $O_2\cdot^-$ and $HO\cdot$ are generated by different oxidants [76]—during which, when H_2O_2 and peroxysulfate are employed as oxidants separately, $HO\cdot$ and $SO_4\cdot^-$ are the main free radicals, respectively (Equations (3) and (4)) [77,78]. Otherwise, when peroxymonosulfate (PMS) is employed as an oxidant (Equation (4)), both $HO\cdot$ and $SO_4\cdot^-$ are the main free radicals (Equation (5)) [78].

$$H_2O_2 + h\nu \rightarrow 2\ HO\cdot \tag{3}$$

$$S_2O_8^{2-} + h\nu \rightarrow 2\ SO_4\cdot^- \tag{4}$$

$$HSO_5^- + h\nu \rightarrow HO\cdot + SO_4\cdot^- \tag{5}$$

New insights were provided by these innovative methods for the potential application of photochemical oxidation in water pollution remediation. For instance, improving the structure and properties of catalysts would have an impact on the generation of free radicals, and further on the removal of PPCPs. Fu et al. [79] investigated a three-dimensional core-shell composite material for PPCPs degradation, which exhibited a >90% removal efficiency of CBZ. Wang et al. [80] synthesized $Sr@TiO_2/UiO-66-NH_2$ heterostructures as a photocatalyst for the degradation of ACE. In the report, strontium titanate was used as a precursor to obtain the heterostructures, resulting in the anchoring and dispersing of $Sr@TiO_2$ on the surface of $UiO-66-NH_2$. The process exhibited an excellent removal rate towards ACE (over 90%). This method provides a new platform for the application of MOF materials in photocatalytic oxidation processes.

With mild reaction conditions and high oxidation capacity, the photochemical oxidation process is acknowledged as an eco-friendly method. Nevertheless, it has several limitations: (1) most of the catalysts used in photochemical oxidation are nanoparticles that are difficult to recover; (2) the electron–hole pairs generated by light are easily recombined and inactivated; (3) the UV radiation has a narrow absorption range and a low utilization rate of light energy. These limitations are urgently to be improved in later research.

2.2. Electrochemical Oxidation

The electrochemical oxidation method is among the most acclaimed methods to remove PPCPs from aqueous environments. In the process, free radicals (i.e., $HO\cdot$) are generated by the decomposition of H_2O and simultaneously the oxidation of hydroxyl ions via direct or indirect electrochemical oxidation. The reactions were expressed in Equations (6) and (7) [78,81].

$$H_2O \rightarrow HO\cdot + H^+ + e^- \tag{6}$$

$$OH^- \rightarrow HO\cdot + e^- \tag{7}$$

Recently, Guo et al. [82] fabricated single-atom copper (Cu) and nitrogen (N) atom-codoped graphene (Cu@NG) and used it as electrocatalytic anode. The efficient degradation of ACE was achieved with a current of 15 mA within 90 min. Although this method was beneficial to improve the electrocatalytic performance (100% degradation efficiency), there was a lack of toxicity analysis of intermediate products. Xia et al. [83] prepared a self-made $Ti/SnO_2-Sb_2O_3/\alpha,\beta-Co-PbO_2$ electrode to degrade norfloxacin (NOR). After electrolysis for 60 min, the removal efficiency of NOR, chemical oxygen demand (COD) and total organic carbon (TOC) were 85.29%, 43.65% and 41.89%, respectively. However, as the results revealed, in the processes of decomposing PPCPs, the toxic intermediates which perhaps harm the environment might be produced because of the low TOC removal efficiency. Therefore, later studies are expected to investigate how to achieve high efficiency of TOC.

Due to the simple assembly, easy operation and convenient control, the electrochemical oxidation device provides a facile approach to large-scale application. Nonetheless, satisfactory decomposition of PPCPs may need high energy consumption and high equipment cost, and there is a shortage of electrochemical oxidation that needs to be further solved.

Furthermore, the preparation of inexpensive and efficient electrode materials for practical engineering applications is also needed.

2.3. Wet Air Oxidation

Wet air oxidation (WAO) uses O_2 as an oxidant at high temperature (473–593 K) and high pressure (20–200 bar) to generate free radicals (i.e., HO·) [84]. A radical chain reaction is involved in the process of WAO [85]. In general, alkyl radical (R·) and hydroperoxide radicals (HO_2·) are generated by the reaction of PPCPs and O_2, which is known as a chain induced reaction. The reaction was expressed in Equation (8) [78]. HO_2· can be converted to HO· via the chain transfer reaction, as shown in Equations (9) and (10).

$$RH + O_2 \rightarrow R\cdot + HO_2\cdot \quad (8)$$

$$2\, HO_2\cdot \rightarrow H_2O_2 + O_2 \quad (9)$$

$$HO_2\cdot + H_2O_2 \rightarrow HO\cdot + H_2O + O_2 \quad (10)$$

Zhu et al. [86] employed the WAO method to treat antibiotic wastewater, with the optimum TOC removal reached being 87.3%. Likewise, Boucher et al. [87] used the WAO method as a pretreatment technology in the removal of pharmaceuticals from hospital effluents. Significant removal rates (>90%) were achieved for all pharmaceuticals under 300 °C within 60 min. This experiment indicated the potential application of WAO to remove pharmaceuticals from hospital wastewater, which could effectively prevent the release of pharmaceuticals into surface water.

The WAO method has the advantages of eco-friendliness and splendid degradation performance; however, the operation conditions are harsh, which leads to high cost in the practical application. Thus, in further studies, inexpensive WAO technologies should be explored to realize the large-scale application for the removal of PPCPs in sewage treatment plants.

2.4. Ultrasonic Oxidation

The ultrasonic oxidation method is a process that applies acoustic waves with frequencies ranging from 15 kHz to 1 MHz at high temperature and high pressure to remove refractory PPCPs by oxidants (i.e., HO·) [52]. During the ultrasonic oxidation of pure water, the reaction chains progressed, as expressed in Equations (11)–(14) [88].

$$H_2O +))) \rightarrow HO\cdot + H\cdot \quad (11)$$

$$HO\cdot + H\cdot \rightarrow H_2O \quad (12)$$

$$HO\cdot + HO\cdot \rightarrow H_2O_2 \quad (13)$$

$$H\cdot + H\cdot \rightarrow H_2 \quad (14)$$

In the above process, additional free radicals are also generated in the gas phase when the solution is saturated with O_2. The reactions were expressed in Equations (15)–(18) [89].

$$O_2 +))) \rightarrow O + O \quad (15)$$

$$O + H_2O \rightarrow HO\cdot + HO\cdot \quad (16)$$

$$O_2 + H\cdot \rightarrow HO_2\cdot \quad (17)$$

$$HO_2\cdot + HO_2\cdot \rightarrow H_2O_2 + O_2 \quad (18)$$

In recent research, Sierra et al. [90] presented a low-frequency ultrasonic oxidation to remove cephalexin (CPX) and doxycycline (DOX) from water, and the pharmaceuticals were completely degraded at a frequency of 40 kHz. The system had a significant effect on antibiotics elimination, which could diminish the potential risk to ecosystems. Camargo-

Perea et al. [91] utilized ultrasonic oxidation to degrade seven types of pharmaceuticals with different chemical structures in distilled water environments. The highest removal of DCF was obtained at 24.4 W (a complete degradation efficiency was achieved within 30 min). To extend practical applications, experiments on various aqueous mixtures need to be explored.

The ultrasonic oxidation method has typical advantages such as no addition of chemicals and selective degradation depending on the nature of various PPCPs. Meanwhile, the ultrasonic oxidation method is simple to operate and convenient to use, and can degrade toxic PPCPs into small molecules with less toxicity or no toxicity. In addition, the ultrasonic oxidation method could be utilized as an assistive technology in combination with other AOPs to remove PPCPs from water.

However, ultrasonic degradation of wastewater is still at the laboratory stage—the degradation mechanism, reaction kinetics, reactor design and amplification of the ultrasonic degradation process have not been sufficiently studied. Moreover, it takes a lot of energy to produce ultrasonic waves. The above shortages make ultrasonic degradation difficult to realize practically in environmental remediation.

2.5. Gamma Ray/Electron Beam Radiation

Gamma ray/electron beam radiation can activate H_2O molecules, and cause the ionization of plenty of H_2O molecules simultaneously in a few seconds to generate free radicals (HO·, H·) [92,93]. The reactions were expressed in Equation (19).

$$8\,H_2O \rightarrow 4\,HO\cdot + 2\,e_{aq}^- + 2\,H\cdot + H_2 + H_2O_2 + 2\,H_3O^+ \tag{19}$$

Chen et al. [92] used the electron beam method for the degradation of benzothiazole (BTH) in aqueous solution. Experiments showed that the method had an effective removal rate (up to 90%) towards BTH when the electron beam reaches 5 kGy. Toxicity calculations exhibited that the toxicity of most of the intermediates had been significantly reduced after radiation. As shown in Figure 5, most of the produced intermediates were non-toxic during the degradation of BTH, although there are still a few intermediates with higher toxicity than BTH. Trojanowicz et al. [93] utilized gamma ray radiation for the removal of endocrine disruptor BPA from wastewater. The degradation rate of BPA reached more than 90% in 5.5 min. These novel approaches provide new platforms for the removal of PPCPs from water.

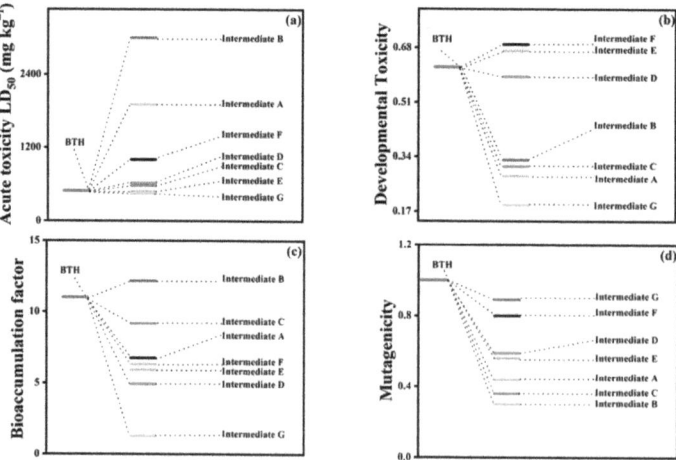

Figure 5. The results of the toxicity calculation of BTH and degradation intermediates in the electron beam irradiation system: (**a**) oral rat LD50; (**b**) developmental toxicity; (**c**) bioaccumulation factor; and (**d**) mutagenicity. Reproduced with permission from [92].

Compared with other AOPs, the advantage of radiation processes is that it can simultaneously generate HO·, e_{aq}^- and H· with high efficiency, which will help to achieve high degradation rates towards target contaminants. However, the defects of gamma ray radiation are also obvious, such as the requirement of a long exposure time, the regular replacement of radionuclides, and the persistent potential risk of radiation contamination.

2.6. Fenton Oxidation

The mixture of Fe^{2+} and H_2O_2 is defined as Fenton's reagent [94]. The core process of Fenton oxidation is the reaction of Fe^{2+} and H_2O_2—during which, H_2O_2 is activated by Fe^{2+} to generate free radical HO· and $HO_2·$, which are performed as the primary product and the secondary product, respectively (Equations (20)–(22)) [78,94].

$$Fe^{2+} + H_2O_2 \rightarrow Fe^{3+} + HO· + HO^- \quad (20)$$

$$H_2O_2 + HO· \rightarrow H_2O + HO_2· \quad (21)$$

$$2\ HO· \rightarrow H_2O_2 \quad (22)$$

In recent research, the removal of CBZ and CAF from tap water by the Fenton oxidation process was studied by Sönmez et al. [95]. The results showed that the removal efficiencies of CBZ and CAF were calculated as 99.77% and 99.66%, respectively. The optimum performance was exhibited when the concentrations of H_2O_2 and Fe^{2+} were either 0.6 mg H_2O_2/L corresponding to 8 mg Fe^{2+}/L or 7.5 mg H_2O_2/L corresponding to 6 mg Fe^{2+}/L. The reaction conditions of Fenton oxidation are mild without high temperature or high pressure, and the device is also easy to operate independently or combined with other treatment technologies. However, there are several disadvantages of Fenton oxidation, such as the limit of the acidic condition and the production of a large amount of iron-containing sludge which is difficult to remove [96]. To overcome these disadvantages, Fe-based porous catalysts are used to replace the role of the dissolved ion Fe^{2+}, to get access to recycle, based on which the process are called as Fenton-like oxidation process, including photo-Fenton oxidation and electro-Fenton oxidation [97]. In the above Fenton-like oxidation processes, the removal of PPCPs in aqueous solution can be achieved within an expanded applicable pH range, and the degradation performance is also superior to that of the classical Fenton oxidation process.

2.7. The Ozone Oxidation Process

The ozone oxidation process is a widely used AOP in the removal of PPCPs. Ozone has high oxidizing ability with a redox potential of 2.07 eV [98]. In the process of ozone oxidation, HO·, $O_2·^-$, $O_3·^-$ and $HO_2·$ are generated by chain reactions, as shown in Equations (23)–(27) [99], which can decompose recalcitrant PPCPs efficiently.

$$O_3 + OH^- \rightarrow HO_2^- + O_2 \quad (23)$$

$$3\ HO_2^- + 2\ O_3 \rightarrow 3\ O_2·^- + 3\ HO_2· \quad (24)$$

$$HO_2· \rightarrow O_2·^- + H^+ \quad (25)$$

$$O_2·^- + O_3 \rightarrow O_3·^- + O_2 \quad (26)$$

$$O_3·^- + H_2O \rightarrow HO· + O_2 + OH^- \quad (27)$$

Yang et al. [100] employed a catalytic ozone oxidation/membrane filtration process to degrade SMX in aqueous environments. The catalyst was prepared by the method of impregnation combined with in situ precipitation. The degradation efficiency was up to 81.3% after the treatment of the oxidation–filtration combined process, which provides new ideas for the combination of AOPs and other technologies. Paucar et al. [101] investigated the degradation performance of the ozone oxidation process towards PPCPs in the secondary effluent of a sewage treatment plant. The results exhibited that the ozone oxidation process is absolutely capable of decomposing a wide range of PPCPs in 10–15 min. These attempts provide novel methods for future studies [102–104].

The ozone oxidation process is a promising technology to decompose PPCPs from water environments, which has no secondary pollution. However, to date, there is no consensus on the mechanisms of the ozone oxidation process, and this needs further investigation. Meanwhile, the low utilization efficiency of ozone is also a significant issue for further water treatment applications.

2.8. Persulfate-Based Oxidation

Persulfate, including PDS (peroxydisulfate, $S_2O_8^{2-}$) and PMS (HSO_5^-), can be activated to produce free radical $SO_4 \cdot^-$, which has the characteristic of strong oxidation. Commonly, the persulfate can be activated by methods including thermal activation [105,106], mechanochemical activation [107,108], carbonaceous materials activation [109,110], alkali activation [111,112], electrochemical activation [113,114], UV activation [115,116], and transition metal activation [117,118]. The possible activation mechanisms were shown in Figure 6. Generally, in addition to $SO_4 \cdot^-$, free radicals including $SO_5 \cdot^-$ and HO· can also be generated during various processes of persulfate activation [119,120].

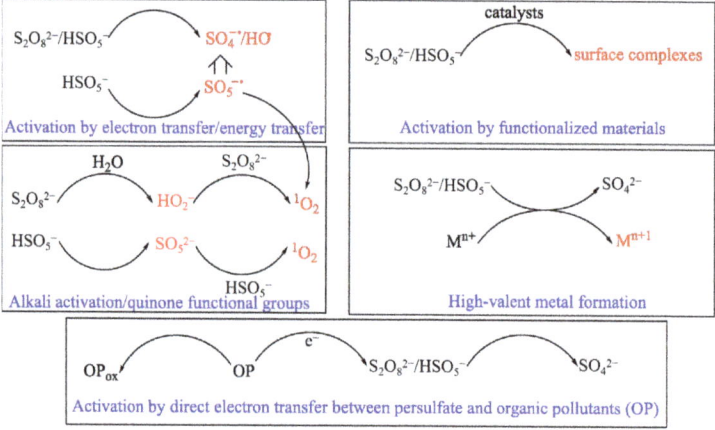

Figure 6. Possible activation mechanisms of persulfate. Reproduced with permission from [78].

In a recent study [121], the heat-activated PMS process was introduced to remove ACE from water environments, during which sodium tetraborate was used as a catalyst. The results indicated that the degradation efficiencies of ACE were significantly high (nearly 100%) in multiple mediums including ultrapure water, lake water and groundwater within 15 minutes of reaction time. Weng et al. [122] utilized Fe^0 to activate the persulfate oxidation system via hydrodynamic cavitation. The removal rate of tetracycline (TC) was up to 97.80%. These studies have shed light on the potential implementation of the persulfate oxidation process on the removal of PPCPs.

Despite the fact that the persulfate-based oxidation process has been widely reported, it is still at the experimental stage. The industrial application needs to be studied thoroughly. Furthermore, later studies on persulfate-based oxidation processes should focus on evaluating their comprehensive performance, i.e., activation rates of persulfate, energy cost, toxicity and yield of by-products.

To evaluate the electrical energy per order (E_{EO}) values of various AOPs, significant differences among AOP efficiency were observed by Miklos et al. [123]. As shown in Figure 7, based on reported E_{EO} values, AOPs were classified into (1) AOPs with median E_{EO} values of <1 kWh/m^3, (2) processes with median E_{EO} values in the range of 1–100 kWh/m^3 and (3) median E_{EO} values of >100 kWh/m^3 (i.e., UV-based photocatalysis, ultrasound, and microwave-based AOPs), which are considered as not (yet) energy efficient AOPs. The research provides an excellent figure of merit to directly compare and evaluate AOPs based on energy efficiency.

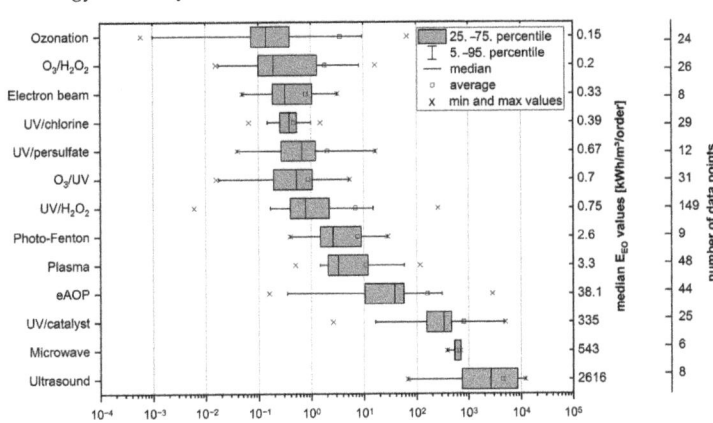

Figure 7. Overview of published EEO values of different AOPs sorted according to median values. For O$_3$- and UV-based AOP data, only substances resistant to direct ozonation/photolysis are shown. Median values and number of data points are reported on the second and third y-axis, respectively. Reproduced with permission from [123].

Overall, to efficiently remove PPCPs from aqueous environments, there has been a lot of effort to develop different types of AOPs or their combined technologies. Table 2 summarized the dominant free radicals, experimental conditions, and removal efficiencies of different processes on the removal of several PPCPs with the highest detection frequencies. However, among all the developed AOPs, there is still no process which can simultaneously achieve the goals of high efficiency, low cost and simple operation on the removal of PPCPs. Thus, further studies should focus on combining different types of free radicals and explore more advanced processes to make AOPs more considerable for industrial application.

Table 2. Degradation compilation of frequently detected PPCPs via various AOPs.

PPCP	AOPs	Dominant Radicals	Concentration of PPCPs	Reaction Time (min)	pH	Removal Rate (%)	Ref.
CBZ	Photochemical oxidation	HO·, O$_2$·$^-$	8.75 mg/L	45	6	99.77	[124]
	Persulfate-based + photochemical oxidation	SO$_4$·$^-$, HO·, O$_2$·$^-$	5 ppm	30	7	97.1	[125]
	UV/chlorine oxidation	HO·, Cl·	40 μM	60	7	83.9	[126]
	Electrochemical oxidation	HO·, SO$_4$·$^-$	1 μM	5	2	100	[127]
	Photo Fenton oxidation	HO·	285 ng/L	30	6–7	94	[128]
	Photo-assisted ozone oxidation	HO·	5 mg/L	2	7 ± 0.2	>99.99	[44]
	Sono-photocatalytic oxidation	O$_2$·$^-$	10 ppm	240	7	93.37	[129]
	Electron beam radiation	HO·, H·	75 mg/L	/	6.3	99.9	[130]
DCF	Photochemical oxidation	HO·	20 mg/L	270	5	76	[131]
	Sonoelectrochemical oxidation	HO·	50 μg/L	5	4	96.8	[132]
	Fenton-like oxidation	HO·	20 mg/L	60	5	86.62	[133]
	Ozone oxidation	HO·	29.6 mg/L	60	7	73.3	[134]
	Persulfate-based oxidation	SO$_4$·$^-$	20 mg/L	15	5	90	[135]

Table 2. Cont.

PPCP	AOPs	Dominant Radicals	Concentration of PPCPs	Reaction Time (min)	pH	Removal Rate (%)	Ref.
SMX	Photochemical oxidation	HO·, $O_2·^-$	10 mg/L	60	7.5	84	[136]
	Electrochemical oxidation	HO·	74.45 mg/L	51.49	4.78	100	[137]
	Ultrasonic oxidation	HO·	10 μM	60	7	97	[138]
	Photo-Fenton oxidation	HO·	25 mg/L	15	3	98.06	[139]
	Persulfate-based oxidation	$SO_4·^-$	0.04 mM	120	3.4	100	[140]
	Gamma ray radiation	HO·	39.48 μM	/	6.7	88.6	[141]

3. Different Free Radicals in the Removal of PPCPs

3.1. Hydroxyl Radical (HO·)

Free radical HO· has a redox potential of 2.80 V and extremely strong oxidizing potential [142]. In AOPs, the HO· is mainly generated via hydrogen abstraction and hydroxylation [143]. The HO· is a strong oxidant which can react with the unsaturated carbon-carbon bond, the carbon-nitrogen bond, the carbon-sulfur bond and other chemical bonds to decompose PPCPs, with non-selective chemical oxidation [144]. The Fenton oxidation process is a typical AOP, which generates HO· by chain reactions between H_2O_2 and Fe^{2+} under acidic conditions. Nevertheless, the Fenton oxidation process has the disadvantages of using excessive amounts of Fe^{2+} and H_2O_2, which resulted in low utilization of H_2O_2. Compared to the conventional Fenton oxidation process, the required H_2O_2 could be in situ generated by the electro-Fenton process, which saves cost and improves decomposition efficiency. Currently, Cui et al. [145] designed an electro-Fenton (EF) oxidation device to degrade carbamazepine (CBZ) in water. In the report, FeS_2/carbon felt was used as the cathode and Ti/IrO_2-RuO_2 was used as the anode of the EF device. The reaction between H_2O_2 and Fe^{2+} was accelerated by the cathode, which helped to produce more HO· to remove CBZ. The possible degradation pathways of CBZ were shown in Figure 8. Under the attack of HO·, CBZ molecules were finally mineralized and decomposed into harmless and small molecules with a degradation rate of 99.99%.

Du et al. [146] investigated Fe/Fe_3C@PC hybrid materials with core-shell structures as catalysts for the degradation of sulfadimethoxine (SMT) in the non-homogeneous Fenton process. The MIL-101(Fe) precursor was prepared by the solvothermal method. Then, the activated precursor was put in a tube furnace under a flowing argon atmosphere and heated to 800 °C for 6 h at a heating rate of 5 °C min^{-1} to produce Fe/Fe_3C@PC. The removal of SMT was up to 96% at pH = 4 with a microcurrent of 25 mA. Furthermore, the degradation rate decreased with the increase in pH, due to decomposition into H_2O and O_2 when pH $\geq$ 4.5, which hindered the generation of HO· [147]. The oxidation potential of HO· decreased with the increase in pH (E^0 = +2.8 V at pH = 0 and E^0 = +1.98 V at pH = 14) [148]. As shown in Figure 9a, the dominant free radical was HO·. Density functional theory (DFT) calculations indicated the presence of internal microelectrolysis (IME) in Fe/Fe_3C@PC hybrid materials has greatly promoted the activation of H_2O_2 to generate HO·. During the electrolysis process, SMT molecules were attacked by HO· and ultimately decomposed into innocuous CO_2 and H_2O.

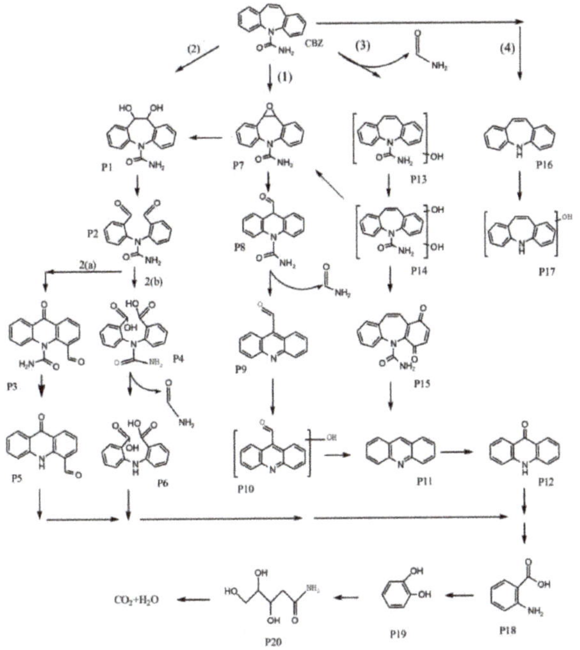

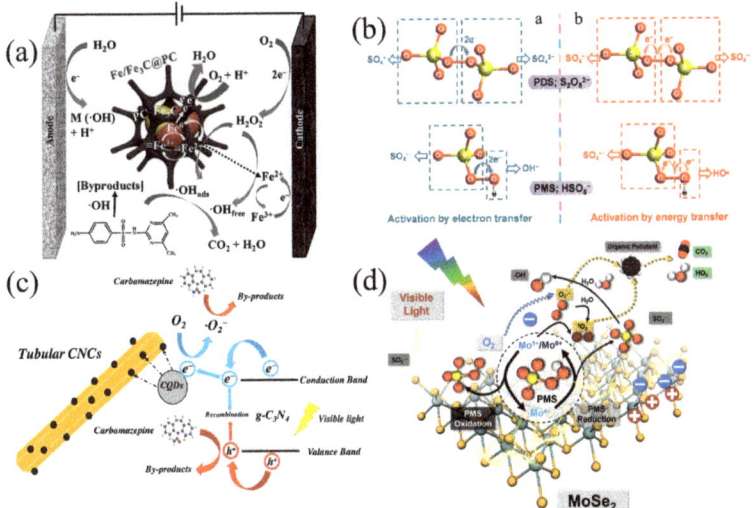

Figure 8. Degradation pathway of CBZ in the EF process with FeS$_2$/CF as a cathode. Reproduced with permission from [145].

Figure 9. Illustrations of mechanisms of different AOPs. (**a**) Proposed scheme of the mechanism of the hetero-EF process catalyzed by Fe/Fe$_3$C@PC. Reproduced with permission from [146]. (**b**) Activation of PDS and PMS through **a** electron transfer and **b** energy transfer reactions. Reproduced with permission from [149]. (**c**) Schematic illustration of the photocatalytic mechanism using carbon quantum dots modified tubular graphitic carbon nitride as material. Reproduced with permission from [150]. (**d**) Possible mechanisms of the visible light-driven MoS$_2$/PMS system. Reproduced with permission from [151].

In summary, HO·-based AOPs can effectively remove PPCPs from water environments. Nevertheless, there are still some difficulties in HO· radicals-based AOPs. On the one hand, HO·-based AOPs have several limitations. For instance, to the best of our knowledge, HO·-based AOPs have the disadvantages of a large amount of reagents, no selectivity towards target substances, and narrow pH conditions (pH 3~4). Therefore, further optimization and exploration of degradation conditions are expected. On the other hand, the identification of HO· is significantly imprecise. Recently, Chen et al. [152] found that in the UV-based AOPs, the widely used scavenger alcohol will accidentally generate H_2O_2 in the process of eliminating HO·. These generated H_2O_2 will be photodissociated into HO·, thus affecting the accuracy of HO· quantification. Therefore, researchers should select suitable scavenger to eliminate HO· like N-butyl alcohol, and more accurate identification technologies should be proposed and promoted. Lastly, the HO·-based AOPs are still at the experimental research stage, when it comes to industrial application, existing problems such as high operation cost need to be solved. In the later research, how to realize practical application and improve the selectivity oxidation of target PPCPs are the current challenges of Fenton-like technologies. Meanwhile, efficient and stable catalysts should be developed to increase utilization efficiency and reduce energy cost.

3.2. Sulfate Radical ($SO_4 \cdot^-$)

Studies on sulfate radical ($SO_4 \cdot^-$) for pollutants removal began in 1996 [153]. Compared to HO·, $SO_4 \cdot^-$ has a higher average redox potential (E^0 = 2.5–3.1 V), a wider pH range (2–10) and a longer half-life (30–40 μs) [154]. The decomposition of $SO_4 \cdot^-$ towards PPCPs was mainly achieved via electron transfer [155,156]. By activation through electron transfer, PDS or PMS molecules are converted to $SO_4 \cdot^-$ [149,157]. Generally, persulfates are the main precursors of $SO_4 \cdot^-$ (Figure 9b) [47,149]. Compared with the process without PDS, the decomposition efficiency of organic pollutants in the process with PDS was greatly accelerated [47,158]. In this system, the procedure of persulfate activization was a crucial process in AOPs. Theoretically, most of the organic matters in water environments could be removed by the oxidation of $SO_4 \cdot^-$ [159]. Zhang et al. [160] used carbon nanofiber-loaded Co/Ag bimetallic nanoparticles (Co@CNFS-Ag) as catalysts for the heterogeneous activation of PMS and the efficient oxidation of amoxicillin (AMX). The excellent removal performance of AMX was realized by adjusting the dosage of catalyst, the reaction temperature and the pH condition. The results indicated that the optimal pH of this system was 7, which was environmentally friendly.

Affected by impurities and other interference, the degradation of some PPCPs in the actual water environments by $SO_4 \cdot^-$ may produce intermediates, byproducts and residues that are even more toxic and difficult to degrade. Thus, the utilization of $SO_4 \cdot^-$-based AOPs needs to be further improved to eliminate their negative impacts. In addition, in the quenching identification process of $SO_4 \cdot^-$, the high concentration of added scavengers would cause numerous confounding effects on the persulfate-based process, thus affecting the generation of $SO_4 \cdot^-$. Therefore, adding scavengers may seriously mislead the interpretation of the mechanism of $SO_4 \cdot^-$-based AOPs [161]. Thus, the mechanism of adding scavengers to explain $SO_4 \cdot^-$-based AOPs should be cautious, and some controversial conclusions obtained by adding scavengers may need to be re-examined.

3.3. Superoxide Radical ($O_2 \cdot^-$)

Recently, superoxide radical ($O_2 \cdot^-$) has attracted increasing concern in environmental remediation because of its potential to destroy highly toxic organic chemicals which are carcinogenic in most cases [162]. The redox potential of $O_2 \cdot^-$ is 2.4 V. $O_2 \cdot^-$ can induce the degradation of PPCPs through an initial hydrogen abstraction step, which results in the formation of carbon-based radicals [163,164]. Then, carbon-based radicals combine with O_2 to form peroxide intermediates. Afterwards, the formation of degradation products was realized [165]. $O_2 \cdot^-$ is an important species involved in natural aquatic systems exposed to sunlight [166].

Photocatalysis, an AOP, has been shown to be an effective method for the generation of $O_2 \cdot^-$. Zhao et al. [150] prepared novel carbon quantum dots (CQDs)-modified tubular graphitic carbon nitride (g-C_3N_4) by an adsorption–polymerization method, which showed up to a 100% removal rate towards CBZ under visible light irradiation. It was further confirmed by electron spin resonance (ESR) analysis that the main active species for CBZ degradation were $O_2 \cdot^-$ and photogenerated holes (h^+). The detailed mechanism was shown in Figure 9c. Under the attack of $O_2 \cdot^-$ and h^+, CBZ molecules were mineralized to harmless CO_2 and H_2O. Dong et al. [151] established a visible light-driven PMS activation process dominated by $O_2 \cdot^-$. In the study, the generation of radicals was confirmed via combination with the scavenger test and electron paramagnetic resonance (EPR) detection. During the scavenger test, $HO\cdot$ and $SO_4 \cdot^-$ were captured by methanol, $HO\cdot$ was captured by isopropanol, $O_2 \cdot^-$ was captured by p-BQ, and the results indicated that $HO\cdot$, $SO_4 \cdot^-$ and $O_2 \cdot^-$ were all generated in this system, among which $O_2 \cdot^-$ played a dominant role (Figure 10a). To further verify the generation of free radicals, EPR was employed to detect these free radicals, coupled with 5,5-dimethyl-1-pyrroline (DMPO) as a spin-trapping reagent to capture both $SO_4 \cdot^-$ and $HO\cdot$. The intensity of characteristic peaks for DMPO·-SO_4^- and DMPO·-HO was observed (Figure 10b), verifying the existence of $SO_4 \cdot^-$ and $HO\cdot$. As shown in Figure 10c, after the addition of methanol and DMPO, the characteristic peaks of DMPO·-O_2^- were observed, confirming the generation of $O_2 \cdot^-$. The possible mechanism was exhibited in Figure 9d. Once irradiated with visible light, the charge carriers (i.e., electrons and holes) were generated on the surface of $MoSe_2$ (Equation (28)). Additionally, photo-generated electrons react with O_2 to produce $O_2 \cdot^-$ (Equation (29)). Due to the generation of photoelectrons, IBP, benzophenone-3 (BZP) and CBZ were decomposed in aqueous environments (Equation (30)). These studies on the generation and identification of $O_2 \cdot^-$ could provide mechanisms and theoretical bases for understanding the comprehensive processes of photocatalysis.

$$MoSe_2 + h\nu \rightarrow e^- + h^+ \quad (28)$$

$$O_2 + e^- \rightarrow O_2 \cdot^- \quad (29)$$

$$O_2 \cdot^- + \text{target PPCP} \rightarrow \text{intermediates} + CO_2 + H_2O \quad (30)$$

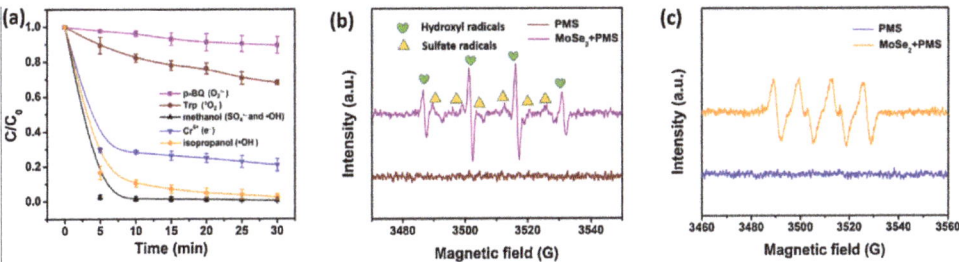

Figure 10. (a) Scavenger test for the degradation of CBZ; (b) EPR spectra for the detection of $HO\cdot$ and $SO_4 \cdot^-$ in the presence of 5,5-dimethyl-1-pyrrolidineN-oxide (DMPO) at room temperature; (c) EPR spectra for the detection of $O_2 \cdot^-$ in the presence of DMPO and methanol, (CH_3OH) at room temperature. Reproduced with permission from [151].

In photocatalysis reactions, $O_2 \cdot^-$ is an important reactive oxygen species. The study of the generation and presence of $O_2 \cdot^-$ could help to promote the understanding of the photocatalysis mechanism. Moreover, it could provide a guideline and theoretical basis for improving photocatalysis efficiency.

3.4. Reactive Chlorine Species (RCS)

The reactive chlorine species (RCS) of the UV/chlorine process is an emerging AOP used for the removal of PPCPs [167]. RCS (including $Cl\cdot$, $Cl_2\cdot^-$ and $ClO\cdot^-$) were found to exhibit excellent removal rates towards many types of PPCPs including chlorine-resistant and UV-resistant PPCPs, i.e., CBZ and CAF [168]. Compared with $HO\cdot$, $Cl\cdot$ has a high redox potential (2.5 V) as well as high selectivity [169]. $Cl\cdot$ can degrade PPCPs by the reactions of hydrogen abstraction, one-electron oxidation and chlorine addition [170]. As known, the UV/chlorine process is a more effective technology to remove PPCPs (i.e., CBZ, sulfamethoxazole and IBP) than the UV/H_2O_2 process for the reason that more effective free radicals are generated in the former process. In addition, the residual chlorine could be used for water disinfection in the former process. Thus, the UV/chlorine process could be considered as a possible alternative to the UV/H_2O_2 process for water treatment plants. Xiang et al. [171] investigated the degradation kinetics and pathways of IBP in the UV/chlorine process. In the same reaction condition, the primary rate constant of in UV/chlorine process was 3.3-fold higher than that of the UV/H_2O_2 process. Guo et al. [172] used a UV/chlorine process to treat various types of different PPCPs. Experimental results showed that $HO\cdot$, $Cl\cdot$, $Cl_2\cdot^-$ and $ClO\cdot^-$ were generated in the process. The concentration of $HO\cdot$ decreased significantly with the increase in pH, while the concentration of $ClO\cdot^-$ decreased gradually, and the concentration of $ClO\cdot^-$ remained essentially constant. The concentration of $ClO\cdot^-$ was 3–4-fold higher than that of $HO\cdot$, $Cl\cdot$ or $Cl_2\cdot^-$, so $ClO\cdot^-$ played a key role in the effective removal of PPCPs. In summary, the UV/chlorine process provides a new idea for the removal of PPCPs from waters.

It is worth noting that although the UV/chlorine process is more effective than the UV/H_2O_2 process, the toxicity of chlorinated products needs to be further evaluated. Meanwhile, when it comes to practical application, the high requirement of equipment and relatively high cost are also crucial issues which need to be further evaluated.

In addition to the above-mentioned free radicals, secondary radicals such as $Br\cdot$, $Br_2\cdot^-$, $ClBr\cdot^-$, and $CO_3\cdot^-$ generated during the UV/chlorine process also contribute to promote the removal rates of PPCPs in waste water [173–178].

In future studies, the AOPs with different types of free radicals could be combined to achieve efficient removal towards PPCPs under facilitated and environmentally friendly conditions, to make it acceptable for large-scale industrial applications finally. Recently, Wang et al. [179] established a novel AOP of bisulfite (BS)/chlorine dioxide (ClO_2) concomitant system, and the removal efficiency of atrazine (ATZ) in water was more than 85% within 3 min. In the above process, the BS was activated by ClO_2. The scavenger experiments and ESR detection results indicated that the dominant radicals were $ClO\cdot^-$ and $SO_4\cdot^-$. Cheng et al. [180] used a combination process of solar irradiation and free available chlorine (FAC) to remove PPCPs from drinking water. It was found that that the in situ generation of $HO\cdot$, RCS and ozone by FAC under solar irradiation contributed greatly to the degradation of PPCPs. PPCPs containing electron-donating groups were degraded more rapidly and preferentially by RCS and/or $HO\cdot$. Paracetamol, IBP and ATZ that contain electron-withdrawing groups were degraded mainly by $HO\cdot$. The combination of solar irradiation and FAC was well-established and inexpensive, which provided a novel idea for the combination of AOPs. In the removal of PPCPs, based on the structure and physicochemical properties of the target PPCPs, studies are expected to develop the combined activation technologies to achieve selective degradation or mineralization of PPCPs.

4. Perspective

This review briefly summarized multiple free radicals generated by AOPs for the removal of PPCPs and look towards the recent progress of various AOPs. In the process of contaminants remediation, AOPs have the advantages of strong oxidation potential, a rapid reaction rate and complete degradation compared to conventional oxidation processes, especially for low-concentration and recalcitrant pollutants. Thus, AOPs are acknowledged as ideal and prospecting technologies in the removal of PPCPs from practical water environments. However, AOPs have disadvantages such as high operation cost and harsh experimental conditions, which make it difficult to apply AOPs at a large scale for industrial application. To achieve efficient degradation of PPCPs, the following suggestions were made for further studies.

(1) To precisely identify the concentration of free radicals, more accurate approaches should be employed. The conventional radical identification methods are challenged because adding scavengers would cause negative effects on AOPs and influence the generation of free radicals. Other methods such as probe-based kinetic models, EPR and laser flash photolysis should be developed and employed to assist in identification of the free radicals in AOPs.

(2) To achieve the practical application of AOPs, convenient and inexpensive approaches should be studied. At present, there are still some issues with AOPs, which need to be further studied. For example, the experimental operations are still complicated and have a high cost. Thus, while researchers focus on the efficiency of water treatment, convenient and inexpensive approaches should be studied to realize large-scale application.

(3) To further study the reaction process and reveal the impacts of AOPs, the mechanism of AOPs in multiple mixture systems should be investigated in depth and the interference of impurity ions on the degradation reaction should be minimized.

(4) To remove PPCPs from complex aqueous mixtures, a pilot-scale plant on the treatment of practical wastewater should be implemented instead of a laboratory scale experiment. Currently, most studies on the removal of PPCPs are focused on water environments containing limited given substances, which is unrealistic. Thus, practical wastewater (i.e., pharmaceutical discharges) should be used in later studies.

(5) To selectively remove PPCPs from aqueous environments, AOPs should be further developed to adjust them to multiple water environments. In the later studies, the removal of PPCPs is expected to be achieved without affecting the existence of trace nutritious natural organic matters (NOMs) in water environments [181].

In addition, eco-friendly AOPs should be studied by adjusting experimental conditions, such as ultrasonic power, radiation dose, current density, temperature, pH, reaction time and chemical dosage. Future studies are advised to focus on the combination of AOPs and other available technologies to selectively and efficiently remove PPCPs from water environments in a green and environmental way.

Author Contributions: Conceptualization, Y.L. and S.Y.; methodology, L.L.; software, J.J.; validation, C.G., Y.L. and T.L.; formal analysis, Q.S.; investigation, L.W.; data curation, J.J.; writing—original draft preparation, J.J.; writing—review and editing, Y.L.; visualization, S.Y.; supervision, Y.L.; project administration, J.J. All authors have read and agreed to the published version of the manuscript.

Funding: This review was funded by the National Natural Science Foundation of China (NO. 42107090), the National Natural Science Foundation of Sichuan Province (2022NSFSC1076), and the Innovative Research Project for Graduate Students of Southwest Minzu University (NO. YB2022193).

Institutional Review Board Statement: Not applicable.

Informed Consent Statement: Not applicable.

Data Availability Statement: Not applicable.

Conflicts of Interest: The authors declare that the work is original research that has not been published previously, and is not under consideration for publication elsewhere. No conflict of interest exist in the submission of this review, and the review is approved by all authors for publication.

References

1. Noguera-Oviedo, K.; Aga, D.S. Lessons learned from more than two decades of research on emerging contaminants in the environment. *J. Hazard. Mater.* **2016**, *316*, 242–251. [CrossRef]
2. Wang, J.; Wang, S. Removal of pharmaceuticals and personal care products (PPCPs) from wastewater: A review. *J. Environ. Manag.* **2016**, *182*, 620–640. [CrossRef]

3. Arpin-Pont, L.; Bueno, M.J.M.; Gomez, E.; Fenet, H. Occurrence of PPCPs in the marine environment: A review. *Environ. Sci. Pollut. Res.* **2016**, *23*, 4978–4991. [CrossRef]
4. Caliman, F.A.; Gavrilescu, M. Pharmaceuticals, Personal Care Products and Endocrine Disrupting Agents in the Environment—A Review. *Clean-Soil Air Water* **2009**, *37*, 277–303.
5. Ngo, T.H.; Van, D.A.; Tran, H.L.; Nakada, N.; Tanaka, H.; Huynh, T.H. Occurrence of pharmaceutical and personal care products in Cau River, Vietnam. *Environ. Sci. Pollut. Res.* **2021**, *28*, 12082–12091. [CrossRef]
6. Spurgeon, D.; Wilkinson, H.; Civil, W.; Hutt, L.; Armenise, E.; Kieboom, N.; Sims, K.; Besien, T. Worst-case ranking of organic chemicals detected in groundwaters an surface waters in England. *Sci. Total Environ.* **2022**, *835*, 155101. [CrossRef]
7. Wang, Y.Q.; Lei, Y.; Liu, X.; Song, L.Y.; Hamid, N.; Zhang, R. Sulfonamide and tetracycline in landfill leachates from seven municipal solid waste (MSW) landfills: Seasonal variation and risk assessment. *Sci. Total Environ.* **2022**, *825*, 153936. [CrossRef]
8. Pisetta, A.M.; Roveri, V.; Guimaraes, L.L.; de Oliveira, T.M.N.; Correia, A.T. First report on the occurrence of pharmaceuticals and cocaine in the coastal waters of Santa Catarina, Brazil, and its related ecological risk assessment. *Environ. Sci. Pollut. Res.* **2022**, *29*, 63099–63111. [CrossRef]
9. Korkmaz, N.E.; Savun-Hekimoglu, B.; Aksu, A.; Burak, S.; Caglar, N.B. Occurrence, sources and environmental risk assessment of pharmaceuticals in the Sea of Marmara, Turkey. *Sci. Total Environ.* **2022**, *819*, 152996. [CrossRef]
10. Anand, U.; Adelodun, B.; Cabreros, C.; Kumar, P.; Suresh, S.; Dey, A.; Ballesteros, F.; Bontempi, E. Occurrence, transformation, bioaccumulation, risk and analysis of pharmaceutical and personal care products from wastewater: A review. *Environ. Chem. Lett.* **2022**. [CrossRef]
11. Liu, J.L.; Wong, M.H. Pharmaceuticals and personal care products (PPCPs): A review on environmental contamination in China. *Environ. Int.* **2013**, *59*, 208–224. [CrossRef]
12. Rahman, M.F.; Yanful, E.K.; Jasim, S.Y. Endocrine disrupting compounds (EDCs) and pharmaceuticals and personal care products (PPCPs) in the aquatic environment: Implications for the drinking water industry and global environmental health. *J. Water Health* **2009**, *7*, 224–243. [CrossRef]
13. Pereira, A.L.; Barros, R.T.D.; Pereira, S.R. Pharmacopollution and Household Waste Medicine (HWM): How reverse logistics is environmentally important to Brazil. *Environ. Sci. Pollut. Res.* **2017**, *24*, 24061–24075. [CrossRef]
14. Insani, W.N.; Qonita, N.A.; Jannah, S.S.; Nuraliyah, N.M.; Supadmi, W.; Gatera, V.A.; Alfian, S.D.; Abdulah, R. Improper disposal practice of unused and expired pharmaceutical products in Indonesian households. *Heliyon* **2020**, *6*, se04551. [CrossRef]
15. Yang, L.; Wang, T.Y.; Zhou, Y.Q.; Shi, B.; Bi, R.; Meng, J. Contamination, source and potential risks of pharmaceuticals and personal products (PPCPs) in Baiyangdian Basin, an intensive human intervention area, China. *Sci. Total Environ.* **2021**, *760*, 144080. [CrossRef]
16. Chen, H.Z.; Chen, W.F.; Guo, H.G.; Lin, H.; Zhang, Y.B. Pharmaceuticals and personal care products in the seawater around a typical subtropical tourist city of China and associated ecological risk. *Environ. Sci. Pollut. Res.* **2021**, *28*, 22716–22728. [CrossRef]
17. Khan, S.A.; Hussain, D.; Abbasi, N.; Khan, T.A. Deciphering the adsorption potential of a functionalized green hydrogel nanocomposite for aspartame from aqueous phase. *Chemosphere* **2022**, *289*, 133232. [CrossRef]
18. Subedi, B.; Lee, S.; Moon, H.B.; Kannan, K. Emission of artificial sweeteners, select pharmaceuticals, and personal care products through sewage sludge from wastewater treatment plants in Korea. *Environ. Int.* **2014**, *68*, 33–40. [CrossRef]
19. Wang, Y.J.; Yin, T.R.; Kelly, B.C.; Gin, K.Y.H. Bioaccumulation behaviour of pharmaceuticals and personal care products in a constructed wetland. *Chemosphere* **2019**, *222*, 275–285. [CrossRef]
20. Sugihara, K. Effect of Environmental Factors on the Ecotoxicity of Pharmaceuticals and Personal Care Products. *Yakugaku Zasshi J. Pharm. Soc. Jpn.* **2018**, *138*, 277–280. [CrossRef]
21. Al-Mashaqbeh, O.; Alsafadi, D.; Dalahmeh, S.; Bartelt-Hunt, S.; Snow, D. Removal of Selected Pharmaceuticals and Personal Care Products in Wastewater Treatment Plant in Jordan. *Water* **2019**, *11*, 2004. [CrossRef]
22. Ismanto, A.; Hadibarata, T.; Kristanti, R.A.; Maslukah, L.; Safinatunnajah, N.; Sathishkumar, P. The abundance of endocrine-disrupting chemicals (EDCs) in downstream of the Bengawan Solo and Brantas rivers located in Indonesia. *Chemosphere* **2022**, *297*, 134151. [CrossRef]
23. Chaves, M.D.S.; Kulzer, J.; de Lima, P.D.P.; Barbosa, S.C.; Primel, E.G. Updated knowledge, partitioning and ecological risk of pharmaceuticals and personal care products in global aquatic environments. *Environ. Sci. Process Impacts* **2022**. [CrossRef]
24. Xie, J.Y.; Liu, Y.F.; Wu, Y.F.; Li, L.R.; Fang, J.; Lu, X.Q. Occurrence, distribution and risk of pharmaceutical and personal care products in the Haihe River sediments, China. *Chemosphere* **2022**, *302*, 134874. [CrossRef]
25. Priya, A.K.; Gnanasekaran, L.; Rajendran, S.; Qin, J.Q.; Vasseghian, Y. Occurrences and removal of pharmaceutical and personal care products from aquatic systems using advanced treatment—A review. *Environ. Res.* **2022**, *204*, 112298. [CrossRef]
26. Mendoza, A.; Acena, J.; Perez, S.; de Alda, M.L.; Barcelo, D.; Gil, A.; Valcarcel, Y. Pharmaceuticals and iodinated contrast media in a hospital wastewater: A case study to analyse their presence and characterise their environmental risk and hazard. *Environ. Res.* **2015**, *140*, 225–241. [CrossRef]
27. Khan, S.A.; Hussain, D.; Khan, T.A. Mechanistic evaluation of metformin drug confiscation from liquid phase on itaconic acid/kaolin hydrogel nanocomposite. *Environ. Sci. Pollut. Res.* **2021**, *28*, 53298–53313. [CrossRef]
28. Slosarczyk, K.; Jakobczyk-Karpierz, S.; Rozkowski, J.; Witkowski, A.J. Occurrence of Pharmaceuticals and Personal Care Products in the Water Environment of Poland: A Review. *Water* **2021**, *13*, 2283. [CrossRef]

29. Caldas, S.S.; Arias, J.L.O.; Rombaldi, C.; Mello, L.L.; Cerqueira, M.B.R.; Martins, A.F.; Primel, E.G. Occurrence of Pesticides and PPCPs in Surface and Drinking Water in Southern Brazil: Data on 4-Year Monitoring. *J. Braz. Chem. Soc.* **2019**, *30*, 71–80. [CrossRef]
30. Liu, S.; Wang, C.; Wang, P.F.; Chen, J.; Wang, X.; Yuan, Q.S. Anthropogenic disturbances on distribution and sources of pharmaceuticals and personal care products throughout the Jinsha River Basin, China. *Environ. Res.* **2021**, *198*, 110449. [CrossRef]
31. Jiang, X.S.; Qu, Y.X.; Zhong, M.M.; Li, W.C.; Huang, J.; Yang, H.W.; Yu, G. Seasonal and spatial variations of pharmaceuticals and personal care products occurrence and human health risk in drinking water—A case study of China. *Sci. Total Environ.* **2019**, *694*, 133711. [CrossRef]
32. Van, D.A.; Ngo, T.H.; Huynh, T.H.; Nakada, N.; Ballesteros, F.; Tanaka, H. Distribution of pharmaceutical and personal care products (PPCPs) in aquatic environment in Hanoi and Metro Manila. *Environ. Monit. Assess.* **2021**, *193*, 847. [CrossRef]
33. Sharma, B.M.; Becanova, J.; Scheringer, M.; Sharma, A.; Bharat, G.K.; Whitehead, P.G.; Klanova, J.; Nizzetto, L. Health and ecological risk assessment of emerging contaminants (pharmaceuticals, personal care products, and artificial sweeteners) in surface and groundwater (drinking water) in the Ganges River Basin, India. *Sci. Total Environ.* **2019**, *646*, 1459–1467. [CrossRef]
34. Zainab, S.M.; Junaid, M.; Rehman, M.Y.A.; Lv, M.; Yue, L.X.; Xu, N.; Malik, R.N. First insight into the occurrence, spatial distribution, sources, and risks assessment of antibiotics in groundwater from major urban-rural settings of Pakistan. *Sci. Total Environ.* **2021**, *791*, 148298. [CrossRef]
35. Benotti, M.J.; Trenholm, R.A.; Vanderford, B.J.; Holady, J.C.; Stanford, B.D.; Snyder, S.A. Pharmaceuticals and Endocrine Disrupting Compounds in US Drinking Water. *Environ. Sci. Technol.* **2009**, *43*, 597–603. [CrossRef]
36. Carballa, M.; Omil, F.; Lema, J.M. Calculation methods to perform mass balances of micropollutants in sewage treatment plants. application to pharmaceutical and personal care products (PPCPs). *Environ. Sci. Technol.* **2007**, *41*, 884–890. [CrossRef]
37. Awfa, D.; Ateia, M.; Fujii, M.; Johnson, M.S.; Yoshimura, C. Photodegradation of pharmaceuticals and personal care products in water treatment using carbonaceous-TiO$_2$ composites: A critical review of recent literature. *Water Res.* **2018**, *142*, 26–45. [CrossRef]
38. Wick, A.; Fink, G.; Joss, A.; Siegrist, H.; Ternes, T.A. Fate of beta blockers and psycho-active drugs in conventional wastewater treatment. *Water Res.* **2009**, *43*, 1060–1074. [CrossRef]
39. Kasprzyk-Hordern, B.; Dinsdale, R.M.; Guwy, A.J. The removal of pharmaceuticals, personal care products, endocrine disruptors and illicit drugs during wastewater treatment and its impact on the quality of receiving waters. *Water Res.* **2009**, *43*, 363–380. [CrossRef]
40. Oluwole, A.O.; Omotola, E.O.; Olatunji, O.S. Pharmaceuticals and personal care products in water and wastewater: A review of treatment processes and use of photocatalyst immobilized on functionalized carbon in AOP degradation. *BMC Chem.* **2020**, *14*, 62. [CrossRef]
41. He, L.; Sun, X.M.; Zhu, F.P.; Ren, S.J.; Wang, S.G. OH-initiated transformation and hydrolysis of aspirin in AOPs system: DFT and experimental studies. *Sci. Total Environ.* **2017**, *592*, 33–40. [CrossRef]
42. Xu, Y.; Liu, T.J.; Zhang, Y.; Ge, F.; Steel, R.M.; Sun, L.Y. Advances in technologies for pharmaceuticals and personal care products removal. *J. Mater. Chem. A* **2017**, *5*, 12001–12014. [CrossRef]
43. Fast, S.A.; Gude, V.G.; Truax, D.D.; Martin, J.; Magbanua, B.S. A Critical Evaluation of Advanced Oxidation Processes for Emerging Contaminants Removal. *Environ. Process.* **2017**, *4*, 283–302. [CrossRef]
44. Somathilake, P.; Dominic, J.A.; Achari, G.; Langford, C.H.; Tay, J.H. Degradation of Carbamazepine by Photo-assisted Ozonation: Influence of Wavelength and Intensity of Radiation. *Ozone Sci. Eng.* **2018**, *40*, 113–121. [CrossRef]
45. Villar-Navarro, E.; Levchuk, I.; Rueda-Marquez, J.J.; Homola, T.; Morinigo, M.A.; Vahala, R.; Manzano, M. Inactivation of simulated aquaculture stream bacteria at low temperature using advanced UVA- and solar-based oxidation methods. *Sol. Energy* **2021**, *227*, 477–489. [CrossRef]
46. Agarkoti, C.; Gogate, P.R.; Pandit, A.B. Comparison of acoustic and hydrodynamic cavitation based hybrid AOPs for COD reduction of commercial effluent from CETP. *J. Environ. Manag.* **2021**, *281*, 111792. [CrossRef]
47. Tian, D.Q.; Zhou, H.Y.; Zhang, H.; Zhou, P.; You, J.J.; Yao, G.; Pan, Z.C.; Liu, Y.; Lai, B. Heterogeneous photocatalyst-driven persulfate activation process under visible light irradiation: From basic catalyst design principles to novel enhancement strategies. *Chem. Eng. J.* **2022**, *428*, 131166. [CrossRef]
48. Zhang, X.; Kamali, M.; Yu, X.B.; Costa, M.E.V.; Appels, L.; Cabooter, D.; Dewil, R. Kinetics and mechanisms of the carbamazepine degradation in aqueous media using novel iodate-assisted photochemical and photocatalytic systems. *Sci. Total Environ.* **2022**, *825*, 153871. [CrossRef]
49. Gmurek, M.; Olak-Kucharczyk, M.; Ledakowicz, S. Photochemical decomposition of endocrine disrupting compounds—A review. *Chem. Eng. J.* **2017**, *310*, 437–456. [CrossRef]
50. Xiong, Y.; Dai, X.L.; Liu, Y.Y.; Du, C.Y.; Yu, G.L.; Xia, Y. Insights into highly effective catalytic persulfate activation on oxygen-functionalized mesoporous carbon for ciprofloxacin degradation. *Environ. Sci. Pollut. Res.* **2022**, *29*, 59013–59026. [CrossRef]
51. Sun, W.J.; Lv, H.X.; Ma, L.; Tan, X.D.; Jin, C.Y.; Wu, H.L.; Chen, L.L.; Liu, M.Y.; Wei, H.Z.; Sun, C.L. Use of catalytic wet air oxidation (CWAO) for pretreatment of high-salinity high-organic wastewater. *J. Environ. Sci.* **2022**, *120*, 105–114. [CrossRef]
52. Mahamuni, N.N.; Adewuyi, Y.G. Advanced oxidation processes (AOPs) involving ultrasound for waste water treatment: A review with emphasis on cost estimation. *Ultrason. Sonochem.* **2010**, *17*, 990–1003. [CrossRef]
53. Ang, W.L.; McHugh, P.J.; Symes, M.D. Sonoelectrochemical processes for the degradation of persistent organic pollutants. *Chem. Eng. J.* **2022**, *444*, 136573. [CrossRef]
54. Gao, H.L.; Wen, Z.N.; Sun, B.C.; Zou, H.K.; Chu, G.W. Intensification of ozone mass transfer for wastewater treatment using a rotating bar reactor. *Chem. Eng. Process.* **2022**, *176*, 108946. [CrossRef]

55. Chavez, A.M.; Gimeno, O.; Rey, A.; Pliego, G.; Oropesa, A.L.; Alvarez, P.M.; Beltran, F.J. Treatment of highly polluted industrial wastewater by means of sequential aerobic biological oxidation-ozone based AOPs. *Chem. Eng. J.* **2019**, *361*, 89–98. [CrossRef]
56. Gholami, M.; Souraki, B.A.; Pendashteh, A. Electro-activated persulfate oxidation (EC/PS) for the treatment of real oilfield produced water: Optimization, developed numerical kinetic model, and comparison with thermal/EC/PS and EC systems. *Process Saf. Environ. Protect.* **2021**, *153*, 384–402. [CrossRef]
57. Nashat, M.; Mossad, M.; El-Etriby, H.K.; Alalm, M.G. Optimization of electrochemical activation of persulfate by BDD electrodes for rapid removal of sulfamethazine. *Chemosphere* **2022**, *286*, 131579. [CrossRef]
58. Khodadadi, T.; Solgi, E.; Mortazavi, S.; Nourmoradi, H. Comparison of advanced oxidation methods of Fenton, UV/Fenton, and O-3/Fenton in treatment of municipal wastewater. *Desalin. Water Treat.* **2020**, *206*, 108–115. [CrossRef]
59. Rueda-Marquez, J.J.; Levchuk, I.; Manzano, M.; Sillanpaa, M. Toxicity Reduction of Industrial and Municipal Wastewater by Advanced Oxidation Processes (Photo-Fenton, UVC/H_2O_2, Electro-Fenton and Galvanic Fenton): A Review. *Catalysts* **2020**, *10*, 612. [CrossRef]
60. Zhu, H.; Ning, S.Y.; Li, Z.Z.Q.; Wang, X.P.; Fujita, T.; Wei, Y.Z.; Yin, X.B. Synthesis of bimetallic NbCo-piperazine catalyst and study on its advanced redox treatment of pharmaceuticals and personal care products by activation of permonosulfate. *Sep. Purif. Technol.* **2022**, *285*, 120345. [CrossRef]
61. Wang, J.L.; Wang, S.Z. Activation of persulfate (PS) and peroxymonosulfate (PMS) and application for the degradation of emerging contaminants. *Chem. Eng. J.* **2018**, *334*, 1502–1517. [CrossRef]
62. Bojanowska-Czajka, A. Application of Radiation Technology in Removing Endocrine Micropollutants from Waters and Wastewaters—A Review. *Appl. Sci.* **2021**, *11*, 12032. [CrossRef]
63. Penalver, J.J.L.; Pacheco, C.V.G.; Polo, M.S.; Utrilla, J.R. Degradation of tetracyclines in different water matrices by advanced oxidation/reduction processes based on gamma radiation. *J. Chem. Technol. Biotechnol.* **2013**, *88*, 1096–1108. [CrossRef]
64. Ghasemian, S.; Nasuhoglu, D.; Omanovic, S.; Yargeau, V. Photoelectrocatalytic degradation of pharmaceutical carbamazepine using Sb-doped Sn-80%-W-20%-oxide electrodes. *Sep. Purif. Technol.* **2017**, *188*, 52–59. [CrossRef]
65. Prado, T.M.; Silva, F.L.; Carrico, A.; Lanza, M.R.D.; Fatibello, O.; Moraes, F.C. Photoelectrocatalytic degradation of caffeine using bismuth vanadate modified with reduced graphene oxide. *Mater. Res. Bull.* **2022**, *145*, 111539. [CrossRef]
66. Lee, J.; von Gunten, U.; Kim, J.H. Persulfate-Based Advanced Oxidation: Critical Assessment of Opportunities and Roadblocks. *Environ. Sci. Technol.* **2020**, *54*, 3064–3081. [CrossRef]
67. Lang, X.J.; Ma, W.H.; Chen, C.C.; Ji, H.W.; Zhao, J.C. Selective Aerobic Oxidation Mediated by TiO_2 Photocatalysis. *Accounts Chem. Res.* **2014**, *47*, 355–363. [CrossRef]
68. Augugliaro, V.; Bellardita, M.; Loddo, V.; Palmisano, G.; Palmisano, L.; Yurdakal, S. Overview on oxidation mechanisms of organic compounds by TiO_2 in heterogeneous photocatalysis. *J. Photochem. Photobiol. C Photochem. Rev.* **2012**, *13*, 224–245. [CrossRef]
69. Zhang, J.; Nosaka, Y. Generation of OH radicals and oxidation mechanism in photocatalysis of WO_3 and $BiVO_4$ powders. *J. Photochem. Photobiol. A Chem.* **2015**, *303*, 53–58. [CrossRef]
70. Yang, J.; Xiao, J.D.; Cao, H.B.; Guo, Z.; Rabeah, J.; Bruckner, A.; Xie, Y.B. The role of ozone and influence of band structure in WO_3 photocatalysis and ozone integrated process for pharmaceutical wastewater treatment. *J. Hazard. Mater.* **2018**, *360*, 481–489. [CrossRef]
71. Savun-Hekimoglu, B.; Eren, Z.; Ince, N.H. Photocatalytic Destruction of Caffeine on Sepiolite-Supported TiO_2 Nanocomposite. *Sustainability* **2020**, *12*, 10314. [CrossRef]
72. Lan, Y.C.; Lu, Y.L.; Ren, Z.F. Mini review on photocatalysis of titanium dioxide nanoparticles and their solar applications. *Nano Energy* **2013**, *2*, 1031–1045. [CrossRef]
73. Ran, Z.L.; Fang, Y.H.; Sun, J.; Ma, C.; Li, S.F. Photocatalytic Oxidative Degradation of Carbamazepine by TiO_2 Irradiated by UV Light Emitting Diode. *Catalysts* **2020**, *10*, 540. [CrossRef]
74. Khraisheh, M.; Kim, J.; Campos, L.; Al-Muhtaseb, A.H.; Al-Hawari, A.; Al Ghouti, M.; Walker, G.M. Removal of pharmaceutical and personal care products (PPCPs) pollutants from water by novel TiO_2-Coconut Shell Powder (TCNSP) composite. *J. Ind. Eng. Chem.* **2014**, *20*, 979–987. [CrossRef]
75. Kumar, A.; Khan, M.; Fang, L.P.; Lo, I.M.C. Visible-light-driven N-TiO_2@SiO_2@Fe_3O_4 magnetic nanophotocatalysts: Synthesis, characterization, and photocatalytic degradation of PPCPs. *J. Hazard. Mater.* **2019**, *370*, 108–116. [CrossRef]
76. Cao, T.T.; Cui, H.; Zhou, D.D.; Ren, X.; Cui, C.W. Degradation mechanism of BPA under VUV irradiation: Efficiency contribution and DFT calculations. *Environ. Sci. Pollut. Res.* **2022**. [CrossRef]
77. Chuang, Y.H.; Chen, S.; Chinn, C.J.; Mitch, W.A. Comparing the UV/Monochloramine and UV/Free Chlorine Advanced Oxidation Processes (AOPs) to the UV/Hydrogen Peroxide AOP under Scenarios Relevant to Potable Reuse. *Environ. Sci. Technol.* **2017**, *51*, 13859–13868. [CrossRef]
78. Wang, J.L.; Wang, S.Z. Reactive species in advanced oxidation processes: Formation, identification and reaction mechanism. *Chem. Eng. J.* **2020**, *401*, 126158. [CrossRef]
79. Fu, H.; Gray, K.A. TiO_2 (Core)/Crumpled Graphene Oxide (Shell) Nanocomposites Show Enhanced Photodegradation of Carbamazepine. *Nanomaterials* **2021**, *11*, 2087. [CrossRef]
80. Wang, Y.; Penas-Garzon, M.; Rodriguez, J.J.; Bedia, J.; Belver, C. Enhanced photodegradation of acetaminophen over Sr@TiO_2/UiO-66-NH_2 heterostructures under solar light irradiation. *Chem. Eng. J.* **2022**, *446*, 137229. [CrossRef]

81. Lozano, I.; Perez-Guzman, C.J.; Mora, A.; Mahlknecht, J.; Aguilar, C.L.; Cervantes-Aviles, P. Pharmaceuticals and personal care products in water streams: Occurrence, detection, and removal by electrochemical advanced oxidation processes. *Sci. Total Environ.* **2022**, *827*, 154348. [CrossRef]
82. Guo, D.; Huang, Z.; Liu, Y.Y.; Zhang, Q.; Yang, Y.L.; Hong, J.M. Incorporation of single-atom copper into nitrogen-doped graphene for acetaminophen electrocatalytic degradation. *Appl. Surf. Sci.* **2022**, *604*, 154561. [CrossRef]
83. Xia, Y.J.; Yan, Y.; Hu, L.B.; Dai, Q.Z.; Ma, X.J.; Lou, J.Q.; Xia, Y. Enhanced Mechanism of Electrochemical Oxidation of Antibiotic Norfloxacin using a $Ti/SnO_2-Sb_2O_3/alpha,beta-Co-PbO_2$ Electrode. *J. Electrochem. Soc.* **2021**, *168*, 106510. [CrossRef]
84. Mishra, V.S.; Mahajani, V.V.; Joshi, J.B. Wet air oxidation. *Ind. Eng. Chem. Res.* **1995**, *34*, 2–48. [CrossRef]
85. Luck, F. Wet air oxidation: Past, present and future. *Catal. Today* **1999**, *53*, 81–91. [CrossRef]
86. Zhu, L.B.; Sheng, D. Study on wet oxidation of antibiotic wastewater. *Oxid. Commun.* **2016**, *39*, 1640–1645.
87. Boucher, V.; Beaudon, M.; Ramirez, P.; Lemoine, P.; Volk, K.; Yargeau, V.; Segura, P.A. Comprehensive evaluation of non-catalytic wet air oxidation as a pretreatment to remove pharmaceuticals from hospital effluents. *Environ. Sci. Wat. Res. Technol.* **2021**, *7*, 1301–1314. [CrossRef]
88. Boffito, D.C.; Crocella, V.; Pirola, C.; Neppolian, B.; Cerrato, G.; Ashokkumar, M.; Bianchi, C.L. Ultrasonic enhancement of the acidity, surface area and free fatty acids esterification catalytic activity of sulphated ZrO_2-TiO_2 systems. *J. Catal.* **2013**, *297*, 17–26. [CrossRef]
89. Joseph, J.M.; Destaillats, H.; Hung, H.M.; Hoffmann, M.R. The sonochemical degradation of azobenzene and related azo dyes: Rate enhancements via Fenton's reactions. *J. Phys. Chem. A* **2000**, *104*, 301–307. [CrossRef]
90. Sierra, R.S.C.; Zuniga-Benitez, H.; Penuela, G.A. Elimination of cephalexin and doxycycline under low frequency ultrasound. *Ultrason. Sonochem.* **2021**, *79*, 105777. [CrossRef]
91. Camargo-Perea, A.L.; Serna-Galvis, E.A.; Lee, J.; Torres-Palma, R.A. Understanding the effects of mineral water matrix on degradation of several pharmaceuticals by ultrasound: Influence of chemical structure and concentration of the pollutants. *Ultrason. Sonochem.* **2021**, *73*, 105500. [CrossRef]
92. Chen, L.; Yin, W.T.; Shao, H.Y.; Tu, M.X.; Ren, Y.F.; Mao, C.K.; Huo, Z.H.; Xu, G. The performance and pathway of benzothiazole degradation by electron beam irradiation. *Chemosphere* **2022**, *303*, 134964. [CrossRef]
93. Trojanowicz, M.; Bojanowska-Czajka, A.; Szreder, T.; Meczynska-Wielgosz, S.; Bobrowski, K.; Fornal, E.; Nichipor, H. Application of ionizing radiation for removal of endocrine disruptor bisphenol A from waters and wastewaters. *Chem. Eng. J.* **2021**, *403*, 126169. [CrossRef]
94. Wang, J.L.; Zhuan, R. Degradation of antibiotics by advanced oxidation processes: An overview. *Sci. Total Environ.* **2020**, *701*, 135023. [CrossRef]
95. Sonmez, G.; Bahadir, T.; Isik, M. Removal of selected pharmaceuticals from tap water by the Fenton process. *Int. J. Environ. Anal. Chem.* **2020**. [CrossRef]
96. Wang, S.Z.; Wang, J.L. Trimethoprim degradation by Fenton and Fe(II)-activated persulfate processes. *Chemosphere* **2018**, *191*, 97–105. [CrossRef]
97. El-Ghenymy, A.; Rodriguez, R.M.; Arias, C.; Centellas, F.; Garrido, J.A.; Cabot, P.L.; Brillas, E. Electro-Fenton and photoelectro-Fenton degradation of the antimicrobial sulfamethazine using a boron-doped diamond anode and an air-diffusion cathode. *J. Electroanal. Chem.* **2013**, *701*, 7–13. [CrossRef]
98. Wang, J.L.; Bai, Z.Y. Fe-based catalysts for heterogeneous catalytic ozonation of emerging contaminants in water and wastewater. *Chem. Eng. J.* **2017**, *312*, 79–98. [CrossRef]
99. Wang, J.L.; Chen, H. Catalytic ozonation for water and wastewater treatment: Recent advances and perspective. *Sci. Total Environ.* **2020**, *704*, 135249. [CrossRef]
100. Yang, Y.L.; Fu, W.Y.; Chen, X.X.; Chen, L.; Hou, C.Y.; Tang, T.H.; Zhang, X.H. Ceramic nanofiber membrane anchoring nanosized Mn_2O_3 catalytic ozonation of sulfamethoxazole in water. *J. Hazard. Mater.* **2022**, *436*, 129168. [CrossRef]
101. Paucar, N.E.; Kim, I.; Tanaka, H.; Sato, C. Ozone treatment process for the removal of pharmaceuticals and personal care products in wastewater. *Ozone Sci. Eng.* **2019**, *41*, 3–16. [CrossRef]
102. Verinda, S.B.; Muniroh, M.; Yulianto, E.; Maharani, N.; Gunawan, G.; Amalia, N.F.; Hobley, J.; Usman, A.; Nur, M. Degradation of ciprofloxacin in aqueous solution using ozone microbubbles: Spectroscopic, kinetics, and antibacterial analysis. *Heliyon* **2022**, *8*, e10137. [CrossRef]
103. Esquerdo, A.A.; Gadea, I.S.; Galvan, P.J.V.; Rico, D.P. Efficacy of atrazine pesticide reduction in aqueous solution using activated carbon, ozone and a combination of both. *Sci. Total Environ.* **2021**, *764*, 144301. [CrossRef]
104. Azuma, T.; Usui, M.; Hayashi, T. Inactivation of Antibiotic-Resistant Bacteria in Wastewater by Ozone-Based Advanced Water Treatment Processes. *Antibiotics* **2022**, *11*, 210. [CrossRef]
105. Johnson, R.L.; Tratnyek, P.G.; Johnson, R.O. Persulfate Persistence under Thermal Activation Conditions. *Environ. Sci. Technol.* **2008**, *42*, 9350–9356. [CrossRef]
106. Li, N.; Wu, S.; Dai, H.X.; Cheng, Z.J.; Peng, W.C.; Yan, B.B.; Chen, G.Y.; Wang, S.B.; Duan, X.G. Thermal activation of persulfates for organic wastewater purification: Heating modes, mechanism and influencing factors. *Chem. Eng. J.* **2022**, *450*, 137976. [CrossRef]
107. Liang, Z.L.; Peng, G.W.; Hu, J.P.; Hou, H.J.; Cai, C.; Yang, X.R.; Chen, S.J.; Liu, L.; Liang, S.; Xiao, K.K.; et al. Mechanochemically assisted persulfate activation for the facile recovery of metals from spent lithium ion batteries. *Waste Manag.* **2022**, *150*, 290–300. [CrossRef]
108. Liu, X.T.; Zhang, X.H.; Shao, K.; Lin, C.Y.; Li, C.B.; Ge, F.Z.; Dong, Y.J. Fe-0-activated persulfate-assisted mechanochemical destruction of expired compound sulfamethoxazole tablets. *RSC Adv.* **2016**, *6*, 20938–20948. [CrossRef]

109. Huang, W.Q.; Xiao, S.; Zhong, H.; Yan, M.; Yang, X. Activation of persulfates by carbonaceous materials: A review. *Chem. Eng. J.* **2021**, *418*, 129297. [CrossRef]
110. Oyekunle, D.T.; Zhou, X.Q.; Shahzad, A.; Chen, Z.Q. Review on carbonaceous materials as persulfate activators: Structure-performance relationship, mechanism and future perspectives on water treatment. *J. Mater. Chem. A* **2021**, *9*, 8012–8050. [CrossRef]
111. Huang, M.Q.; Wang, X.L.; Zhu, C.Y.; Zhu, F.X.; Liu, P.; Wang, D.X.; Fang, G.D.; Chen, N.; Gao, S.X.; Zhou, D.M. Efficient chlorinated alkanes degradation in soil by combining alkali hydrolysis with thermally activated persulfate. *J. Hazard. Mater.* **2022**, *438*, 129571. [CrossRef]
112. Hong, Y.X.; Luo, Z.J.; Zhang, N.; Qu, L.L.; Zheng, M.; Suara, M.A.; Chelme-Ayala, P.; Zhou, X.T.; El-Din, M.G. Decomplexation of Cu(II)-EDTA by synergistic activation of persulfate with alkali and CuO: Kinetics and activation mechanism. *Sci. Total Environ.* **2022**, *817*, 152793. [CrossRef]
113. Song, H.R.; Yan, L.X.; Jiang, J.; Ma, J.; Zhang, Z.X.; Zhang, J.M.; Liu, P.X.; Yang, T. Electrochemical activation of persulfates at BDD anode: Radical or nonradical oxidation? *Water Res.* **2018**, *128*, 393–401. [CrossRef]
114. Zhi, D.; Lin, Y.H.; Jiang, L.; Zhou, Y.Y.; Huang, A.Q.; Yang, J.; Luo, L. Remediation of persistent organic pollutants in aqueous systems by electrochemical activation of persulfates: A review. *J. Environ. Manag.* **2020**, *260*, 110125. [CrossRef]
115. Wang, C.W.; Liang, C.J. Oxidative degradation of TMAH solution with UV persulfate activation. *Chem. Eng. J.* **2014**, *254*, 472–478. [CrossRef]
116. Dubois, V.; Rodrigues, C.S.D.; Alves, A.S.P.; Madeira, L.M. UV/Vis-Based Persulphate Activation for p-Nitrophenol Degradation. *Catalysts* **2021**, *11*, 480. [CrossRef]
117. Anushree, C.; Krishna, D.N.G.; Philip, J. Efficient Dye Degradation via Catalytic Persulfate Activation using Iron Oxide-Manganese Oxide Core-Shell Particle Doped with Transition Metal Ions. *J. Mol. Liq.* **2021**, *337*, 116429. [CrossRef]
118. Huang, M.J.; Han, Y.; Xiang, W.; Zhong, D.L.; Wang, C.; Zhou, T.; Wu, X.H.; Mao, J. In Situ-Formed Phenoxyl Radical on the CuO Surface Triggers Efficient Persulfate Activation for Phenol Degradation. *Environ. Sci. Technol.* **2021**, *55*, 15361–15370. [CrossRef]
119. Cheng, X.; Guo, H.G.; Zhang, Y.L.; Wu, X.; Liu, Y. Non-photochemical production of singlet oxygen via activation of persulfate by carbon nanotubes. *Water Res.* **2017**, *113*, 80–88. [CrossRef]
120. Deng, X.Y.; Zhao, Z.W.; Wang, C.; Chen, R.; Du, J.Y.; Shi, W.X.; Cui, F.Y. Insight into the nonradical mechanism of persulfate activation via visible-light for enhanced degradation of sulfonamides without catalyst. *Appl. Catal. B-Environ.* **2022**, *316*, 121653. [CrossRef]
121. Li, J.; Zou, J.; Zhang, S.; Cai, H.; Huang, Y.; Lin, J.; Li, Q.; Yuan, B.; Ma, J. Sodium tetraborate simultaneously enhances the degradation of acetaminophen and reduces the formation potential of chlorinated by-products with heat-activated peroxymonosulfate oxidation. *Water Res.* **2022**, *224*, 119095. [CrossRef]
122. Weng, M.T.; Cai, M.Q.; Xie, Z.Q.; Dong, C.Y.; Zhang, Y.; Song, Z.J.; Shi, Y.J.; Jin, M.C.; Wang, Q.; Wei, Z.S. Hydrodynamic cavitation-enhanced heterogeneous activation of persulfate for tetracycline degradation: Synergistic effects, degradation mechanism and pathways. *Chem. Eng. J.* **2022**, *431*, 134238. [CrossRef]
123. Miklos, D.B.; Remy, C.; Jekel, M.; Linden, K.G.; Drewes, J.E.; Hubner, U. Evaluation of advanced oxidation processes for water and wastewater treatment—A critical review. *Water Res.* **2018**, *139*, 118–131. [CrossRef]
124. Constantin, M.A.; Chiriac, F.L.; Gheorghe, S.; Constantin, L.A. Degradation of Carbamazepine from Aqueous Solutions via TiO_2-Assisted Photo Catalyze. *Toxics* **2022**, *10*, 168. [CrossRef]
125. Yang, L.W.; Jia, Y.Y.; Peng, Y.Q.; Zhou, P.; Yu, D.A.; Zhao, C.L.; He, J.J.; Zhan, C.A.L.; Lai, B. Visible-light induced activation of persulfate by self-assembled EHPDI/TiO_2 photocatalyst toward efficient degradation of carbamazepine. *Sci. Total Environ.* **2021**, *783*, 146996. [CrossRef]
126. Zhou, S.Q.; Xia, Y.; Li, T.; Yao, T.; Shi, Z.; Zhu, S.M.; Gao, N.Y. Degradation of carbamazepine by UV/chlorine advanced oxidation process and formation of disinfection by-products. *Environ. Sci. Pollut. Res.* **2016**, *23*, 16448–16455. [CrossRef]
127. Feijoo, S.; Kamali, M.; Pham, Q.K.; Assoumani, A.; Lestremau, F.; Cabooter, D.; Dewil, R. Electrochemical Advanced Oxidation of Carbamazepine: Mechanism and optimal operating conditions. *Chem. Eng. J.* **2022**, *446*, 137114. [CrossRef]
128. De la Cruz, N.; Gimenez, J.; Esplugas, S.; Grandjean, D.; de Alencastro, L.F.; Pulgarin, C. Degradation of 32 emergent contaminants by UV and neutral photo-fenton in domestic wastewater effluent previously treated by activated sludge. *Water Res.* **2012**, *46*, 1947–1957. [CrossRef]
129. Yentur, G.; Dukkanci, M. Synergistic effect of sonication on photocatalytic oxidation of pharmaceutical drug carbamazepine. *Ultrason. Sonochem.* **2021**, *78*, 105749. [CrossRef]
130. Zheng, M.; Xu, G.; Zhao, L.; Pei, J.C.; Wu, M.H. Comparison of EB-radiolysis and UV/H_2O_2-degradation of CBZ in pure water and solutions. *Nucl. Sci. Tech.* **2015**, *26*, 020302.
131. Mugunthan, E.; Saidutta, M.B.; Jagadeeshbabu, P.E. Photocatalytic activity of ZnO-WO_3 for diclofenac degradation under visible light irradiation. *J. Photochem. Photobiol. A Chem.* **2019**, *383*, 111993. [CrossRef]
132. Finkbeiner, P.; Franke, M.; Anschuetz, F.; Ignaszak, A.; Stelter, M.; Braeutigam, P. Sonoelectrochemical degradation of the anti-inflammatory drug diclofenac in water. *Chem. Eng. J.* **2015**, *273*, 214–222. [CrossRef]
133. Zhu, J.; Zhang, G.M.; Xian, G.; Zhang, N.; Li, J.W. A High-Efficiency CuO/CeO_2 Catalyst for Diclofenac Degradation in Fenton-Like System. *Front. Chem.* **2019**, *7*, 796. [CrossRef]
134. Gao, G.Y.; Chu, W.; Chen, Z.L.; Shen, J.M. Catalytic ozonation of diclofenac with iron silicate-loaded pumice in aqueous solution. *Water Sci. Technol. Water Supply* **2017**, *17*, 1458–1467. [CrossRef]

135. Xian, G.; Zhang, G.M.; Chang, H.Z.; Zhang, Y.; Zou, Z.G.; Li, X.Y. Heterogeneous activation of persulfate by Co_3O_4-CeO_2 catalyst for diclofenac removal. *J. Environ. Manag.* **2019**, *234*, 265–272. [CrossRef]
136. Makropoulou, T.; Kortidis, I.; Davididou, K.; Motaung, D.E.; Chatzisymeon, E. Photocatalytic facile ZnO nanostructures for the elimination of the antibiotic sulfamethoxazole in water. *J. Water Process. Eng.* **2020**, *36*, 101299. [CrossRef]
137. Wan, J.T.; Jin, C.J.; Liu, B.H.; She, Z.L.; Gao, M.C.; Wang, Z.Y. Electrochemical oxidation of sulfamethoxazole using Ti/SnO_2-Sb/Co-PbO_2 electrode through ANN-PSO. *J. Serb. Chem. Soc.* **2019**, *84*, 713–727. [CrossRef]
138. Al-Hamadani, Y.A.J.; Chu, K.H.; Flora, J.R.V.; Kim, D.H.; Jang, M.; Sohn, J.; Joo, W.; Yoon, Y. Sonocatalytical degradation enhancement for ibuprofen and sulfamethoxazole in the presence of glass beads and single-walled carbon nanotubes. *Ultrason. Sonochem.* **2016**, *32*, 440–448. [CrossRef]
139. Rostami, F.; Ramavandi, B.; Arfaeinia, H.; Nasrzadeh, F.; Hashemi, S. Sulfamethoxazole antibiotic removal from aqueous solution and hospital wastewater using photo-Fenton process. *Desalin. Water Treat.* **2020**, *184*, 388–394. [CrossRef]
140. Wang, S.Z.; Wang, J.L. Synergistic effect of PMS activation by Fe-0@Fe_3O_4 anchored on N, S, O co-doped carbon composite for degradation of sulfamethoxazole. *Chem. Eng. J.* **2022**, *427*, 131960. [CrossRef]
141. Kim, H.Y.; Kim, T.H.; Cha, S.M.; Yu, S. Degradation of sulfamethoxazole by ionizing radiation: Identification and characterization of radiolytic products. *Chem. Eng. J.* **2017**, *313*, 556–566. [CrossRef]
142. Gligorovski, S.; Strekowski, R.; Barbati, S.; Vione, D. Environmental Implications of Hydroxyl Radicals (center dot OH). *Chem. Rev.* **2015**, *115*, 13051–13092. [CrossRef]
143. Sirtori, C.; Aguera, A.; Carra, I.; Perez, J.A.S. Identification and monitoring of thiabendazole transformation products in water during Fenton degradation by LC-QTOF-MS. *Anal. Bioanal. Chem.* **2014**, *406*, 5323–5337. [CrossRef]
144. Babu, S.G.; Ashokkumar, M.; Neppolian, B. The Role of Ultrasound on Advanced Oxidation Processes. *Top. Curr. Chem.* **2016**, *374*, 75. [CrossRef]
145. Cui, T.Y.; Xiao, Z.H.; Wang, Z.B.; Liu, C.; Song, Z.L.; Wang, Y.P.; Zhang, Y.T.; Li, R.Y.; Xu, B.B.; Qi, F.; et al. FeS_2/carbon felt as an efficient electro-Fenton cathode for carbamazepine degradation and detoxification: In-depth discussion of reaction contribution and empirical kinetic model. *Environ. Pollut.* **2021**, *282*, 117023. [CrossRef]
146. Du, X.; Fu, W.; Su, P.; Cai, J.; Zhou, M. Internal-micro-electrolysis-enhanced heterogeneous electro-Fenton process catalyzed by Fe/Fe_3C@PC core-shell hybrid for sulfamethazine degradation. *Chem. Eng. J.* **2020**, *398*, 125681. [CrossRef]
147. Oturan, M.A.; Aaron, J.J. Advanced Oxidation Processes in Water/Wastewater Treatment: Principles and Applications. A Review. *Crit. Rev. Environ. Sci. Technol.* **2014**, *44*, 2577–2641. [CrossRef]
148. Xu, L.J.; Wang, J.L. Fenton-like degradation of 2,4-dichlorophenol using Fe_3O_4 magnetic nanoparticles. *Appl. Catal. B Environ.* **2012**, *123*, 117–126. [CrossRef]
149. He, S.; Chen, Y.; Li, X.; Zeng, L.; Zhu, M. Heterogeneous Photocatalytic Activation of Persulfate for the Removal of Organic Contaminants in Water: A Critical Review. *ACS EST Engg.* **2022**, *2*, 527–546. [CrossRef]
150. Zhao, C.; Liao, Z.; Liu, W.; Liu, F.; Ye, J.; Liang, J.; Li, Y. Carbon quantum dots modified tubular g-C_3N_4 with enhanced photocatalytic activity for carbamazepine elimination: Mechanisms, degradation pathway and DFT calculation. *J. Hazard. Mater.* **2020**, *381*, 120957. [CrossRef]
151. Dong, C.; Wang, Z.; He, J.; Zheng, Z.; Gong, X.; Zhang, J.; Lo, I.M.C. Superoxide radicals dominated visible light driven peroxymonosulfate activation using molybdenum selenide ($MoSe_2$) for boosting catalytic degradation of pharmaceuticals and personal care products. *Appl. Catal. B Environ.* **2021**, *296*, 120223. [CrossRef]
152. Wang, L.; Li, B.Q.; Dionysiou, D.D.; Chen, B.Y.; Yang, J.; Li, J. Overlooked Formation of H_2O_2 during the Hydroxyl Radical-Scavenging Process When Using Alcohols as Scavengers. *Environ. Sci. Technol.* **2022**, *56*, 3386–3396. [CrossRef]
153. Anipsitakis, G.P.; Dionysiou, D. Degradation of organic contaminants in water with sulfate radicals generated by the conjunction of peroxymonosulfate with cobalt. *Environ. Sci. Technol.* **2003**, *37*, 4790. [CrossRef]
154. Hu, P.D.; Long, M.C. Cobalt-catalyzed sulfate radical-based advanced oxidation: A review on heterogeneous catalysts and applications. *Appl. Catal. B Environ.* **2016**, *181*, 103–117. [CrossRef]
155. Ahmed, M.M.; Barbati, S.; Doumenq, P.; Chiron, S. Sulfate radical anion oxidation of diclofenac and sulfamethoxazole for water decontamination. *Chem. Eng. J.* **2012**, *197*, 440–447. [CrossRef]
156. Norman, R.O.C.; Storey, P.M.; West, P.R. Electron spin resonance studies. Part XXV. Reactions of the sulphate radical anion with organic compounds. *J. Chem. Soc. B Phys. Org.* **1970**, 1087–1095. [CrossRef]
157. Yang, Y.; Li, X.; Zhou, C.Y.; Xiong, W.P.; Zeng, G.M.; Huang, D.L.; Zhang, C.; Wang, W.J.; Song, B.A.; Tang, X.; et al. Recent advances in application of graphitic carbon nitride-based catalysts for degrading organic contaminants in water through advanced oxidation processes beyond photocatalysis: A critical review. *Water Res.* **2020**, *184*, 116200. [CrossRef]
158. Gong, J.; Zhang, J.S.; Lin, H.J.; Yuan, J.Y. "Cooking carbon in a solid salt": Synthesis of porous heteroatom-doped carbon foams for enhanced organic pollutant degradation under visible light. *Appl. Mater. Today* **2018**, *12*, 168–176. [CrossRef]
159. Watts, R.J.; Ahmad, M.; Hohner, A.K.; Teel, A.L. Persulfate activation by glucose for in situ chemical oxidation. *Water Res.* **2018**, *133*, 247–254. [CrossRef]
160. Zhang, Y.; Zhang, B.T.; Teng, Y.G.; Zhao, J.J.; Kuang, L.L.; Sun, X.J. Carbon nanofibers supported Co/Ag bimetallic nanoparticles for heterogeneous activation of peroxymonosulfate and efficient oxidation of amoxicillin. *J. Hazard. Mater.* **2020**, *400*, 123290. [CrossRef]
161. Gao, L.W.; Guo, Y.; Zhan, J.H.; Yu, G.; Wang, Y.J. Assessment of the validity of the quenching method for evaluating the role of reactive species in pollutant abatement during the persulfate-based process. *Water Res.* **2022**, *221*, 118730. [CrossRef]

162. Fang, G.D.; Dionysiou, D.D.; Al-Abed, S.R.; Zhou, D.M. Superoxide radical driving the activation of persulfate by magnetite nanoparticles: Implications for the degradation of PCBs. *Appl. Catal. B Environ.* **2013**, *129*, 325–332. [CrossRef]
163. Baum, R.M. Superoxide theory of oxygen-toxicity is center of heated debate. *Chem. Eng. News* **1984**, *62*, 20–26. [CrossRef]
164. Sawyer, D.T.; Valentine, J.S. How super is superoxide. *Accounts Chem. Res.* **1981**, *14*, 393–400. [CrossRef]
165. Frimer, A.A.; Farkashsolomon, T.; Aljadeff, G. Mechanism of the superoxide anion radical (o-2-.) mediated oxidation of diarylmethanes. *J. Org. Chem.* **1986**, *51*, 2093–2098. [CrossRef]
166. Ma, J.; Zhou, H.; Yan, S.; Song, W. Kinetics studies and mechanistic considerations on the reactions of superoxide radical ions with dissolved organic matter. *Water Res.* **2018**, *149*, 56–64. [CrossRef]
167. Sharma, V.K.; Triantis, T.M.; Antoniou, M.G.; He, X.X.; Pelaez, M.; Han, C.S.; Song, W.H.; O'Shea, K.E.; de la Cruz, A.A.; Kaloudis, T.; et al. Destruction of microcystins by conventional and advanced oxidation processes: A review. *Sep. Purif. Technol.* **2012**, *91*, 3–17. [CrossRef]
168. Yang, X.; Sun, J.L.; Fu, W.J.; Shang, C.; Li, Y.; Chen, Y.W.; Gan, W.H.; Fang, J.Y. PPCP degradation by UV/chlorine treatment and its impact on DBP formation potential in real waters. *Water Res.* **2016**, *98*, 309–318. [CrossRef]
169. Hirakawa, T.; Nosaka, Y. Properties of O-2(center dot-) and OH center dot formed in TiO2 aqueous suspensions by photocatalytic reaction and the influence of H_2O_2 and some ions. *Langmuir* **2002**, *18*, 3247–3254. [CrossRef]
170. Minakata, D.; Kamath, D.; Maetzold, S. Mechanistic Insight into the Reactivity of Chlorine-Derived Radicals in the Aqueous-Phase UV Chlorine Advanced Oxidation Process: Quantum Mechanical Calculations. *Environ. Sci. Technol.* **2017**, *51*, 6918–6926. [CrossRef]
171. Xiang, Y.Y.; Fang, J.Y.; Shang, C. Kinetics and pathways of ibuprofen degradation by the UV/chlorine advanced oxidation process. *Water Res.* **2016**, *90*, 301–308. [CrossRef]
172. Guo, K.H.; Wu, Z.H.; Shang, C.; Yao, B.; Hou, S.D.; Yang, X.; Song, W.H.; Fang, J.Y. Radical Chemistry and Structural Relationships of PPCP Degradation by UV/Chlorine Treatment in Simulated Drinking Water. *Environ. Sci. Technol.* **2017**, *51*, 10431–10439. [CrossRef]
173. Wang, Z.Y.; An, N.; Shao, Y.S.; Gao, N.Y.; Du, E.D.; Xu, B. Experimental and simulation investigations of UV/persulfate treatment in presence of bromide: Effects on degradation kinetics, formation of brominated disinfection byproducts and bromate. *Sep. Purif. Technol.* **2020**, *242*, 116767. [CrossRef]
174. Lei, Y.; Lei, X.; Westerhoff, P.; Tong, X.Y.; Ren, J.N.; Zhou, Y.J.; Cheng, S.S.; Ouyang, G.F.; Yang, X. Bromine Radical (Br(center dot)and Br-2(center dot-)) Reactivity with Dissolved OrganicMatter and Brominated Organic Byproduct Formation br. *Environ. Sci. Technol.* **2022**, *56*, 5189–5199. [CrossRef]
175. Liu, Y.Q.; He, X.X.; Duan, X.D.; Fu, Y.S.; Fatta-Kassinos, D.; Dionysiou, D.D. Significant role of UV and carbonate radical on the degradation of oxytetracycline in UV-AOPs: Kinetics and mechanism. *Water Res.* **2016**, *95*, 195–204. [CrossRef]
176. Gao, J.; Duan, X.D.; O'Shea, K.; Dionysiou, D.D. Degradation and transformation of bisphenol A in UV/Sodium percarbonate: Dual role of carbonate radical anion. *Water Res.* **2020**, *171*, 115394. [CrossRef]
177. Zhu, S.M.; Tian, Z.C.; Wang, P.; Zhang, W.Q.; Bu, L.J.; Wu, Y.T.; Dong, B.Z.; Zhou, S.Q. The role of carbonate radicals on the kinetics, radical chemistry, and energy requirement of UV/chlorine and UV/H_2O_2 processes. *Chemosphere* **2021**, *278*, 130499. [CrossRef]
178. Li, Y.; Dong, H.; Xiao, J.; Li, L.; Chu, D.; Hou, X.; Xiang, S.; Dong, Q.; Zhang, H. Advanced oxidation processes for water purification using percarbonate: Insights into oxidation mechanisms, challenges, and enhancing strategies. *J. Hazard. Mater.* **2022**, *442*, 130014. [CrossRef]
179. Wang, Z.Y.; Li, J.; Song, W.; Ma, R.; Yang, J.X.; Zhang, X.L.; Huang, F.; Dong, W.Y. Rapid degradation of atrazine by a novel advanced oxidation process of bisulfite/chlorine dioxide: Efficiency, mechanism, pathway. *Chem. Eng. J.* **2022**, *445*, 136558. [CrossRef]
180. Cheng, S.S.; Zhang, X.R.; Song, W.H.; Pan, Y.H.; Lambropoulou, D.; Zhong, Y.; Du, Y.; Nie, J.X.; Yang, X. Photochemical oxidation of PPCPs using a combination of solar irradiation and free available chlorine. *Sci. Total Environ.* **2019**, *682*, 629–638. [CrossRef]
181. Yang, Z.C.; Qian, J.S.; Shan, C.; Li, H.C.; Yin, Y.Y.; Pan, B.C. Toward Selective Oxidation of Contaminants in Aqueous Systems. *Environ. Sci. Technol.* **2021**, *55*, 14494–14514. [CrossRef]

Article

Green Production of Zero-Valent Iron (ZVI) Using Tea-Leaf Extracts for Fenton Degradation of Mixed Rhodamine B and Methyl Orange Dyes

Diana Rakhmawaty Eddy *, Dian Nursyamsiah, Muhamad Diki Permana, Solihudin, Atiek Rostika Noviyanti and Iman Rahayu

Department of Chemistry, Faculty of Mathematics and Sciences, Universitas Padjadjaran, Jl. Raya Bandung-Sumedang Km. 21 Jatinangor, Sumedang 45363, Indonesia; diannursyamsiah@gmail.com (D.N.); muhamad16046@mail.unpad.ac.id (M.D.P.); solihudin@unpad.ac.id (S.); atiek.noviyanti@unpad.ac.id (A.R.N.); iman.rahayu@unpad.ac.id (I.R.)
* Correspondence: diana.rahmawati@unpad.ac.id; Tel.: +62-81322731173

Abstract: The danger from the content of dyes produced by textile-industry waste can cause environmental degradation when not appropriately treated. However, existing waste-treatment methods have not been effective in degrading dyes in textile waste. Zero-valent iron (ZVI), which has been widely used for wastewater treatment, needs to be developed to acquire effective green production. Tea (*Camellia sinensis*) leaves contain many polyphenolic compounds used as natural reducing agents. Therefore, this study aims to synthesize ZVI using biological reducing agents from tea-leaf extract and apply the Fenton method to degrade the color mixture of rhodamine B and methyl orange. The results show that the highest polyphenols were obtained from tea extract by heating to 90 °C for 80 min. Furthermore, PSA results show that ZVI had a homogeneous size of iron and tea extract at a volume ratio of 1:3. The SEM-EDS results show that all samples had agglomerated particles. The ZVI 1:1 showed the best results, with a 100% decrease in the color intensity of the dye mixture for 60 min of reaction and a degradation percentage of 100% and 66.47% for rhodamine B and methyl orange from LC-MS analysis, respectively. Finally, the decrease in COD value by ZVI was 92.11%, higher than the 47.36% decrease obtained using Fe(II).

Keywords: dye degradation; Fenton method; green synthesis; zero-valent iron

Citation: Eddy, D.R.; Nursyamsiah, D.; Permana, M.D.; Solihudin; Noviyanti, A.R.; Rahayu, I. Green Production of Zero-Valent Iron (ZVI) Using Tea-Leaf Extracts for Fenton Degradation of Mixed Rhodamine B and Methyl Orange Dyes. *Materials* **2022**, *15*, 332. https://doi.org/10.3390/ma15010332

Academic Editor: Ilya V. Mishakov

Received: 15 December 2021
Accepted: 31 December 2021
Published: 3 January 2022

Publisher's Note: MDPI stays neutral with regard to jurisdictional claims in published maps and institutional affiliations.

Copyright: © 2022 by the authors. Licensee MDPI, Basel, Switzerland. This article is an open access article distributed under the terms and conditions of the Creative Commons Attribution (CC BY) license (https://creativecommons.org/licenses/by/4.0/).

1. Introduction

The textile industry is one of the fastest-growing sectors in Indonesia, and according to data from the Indonesian Ministry of Industry, it grew by 15.08% in 2019 [1]. The number of textile industry activities is directly proportional to the amount of wastewater generated. Dyestuff waste produced is toxic because it contains carcinogenic aromatic amines [2,3]. When waste is not appropriately treated, it impacts environmental degradation and disrupts public health. Therefore, effective and environmentally friendly treatment of industrial wastewater, especially dye waste, is needed. The commonly used methods in wastewater treatment to reduce dye content include adsorption, coagulation, and flocculation methods [4–7]. However, using these methods to minimize dyestuffs is often not perfect because it requires a large amount of coagulant. Other methods used in wastewater treatment are the membrane technique and bacterial biodegradation [8,9]. However, the regeneration process is complex, the cost is high, and the sludge generated from treatment residue is significant [10–12]. Therefore, modifications to existing water-treatment methods are needed.

Zero-valent iron (ZVI) was initially used for groundwater remediation. However, along with increasingly complex surface-water problems, ZVI can be applied to surface-water and wastewater treatment to reduce the number of pollutants [13–16]. Zero-valent iron material is made by reducing iron(II) or iron(III) using sodium borohydride [17–19]. Sodium borohydride is considered less environmentally friendly as a reducing agent. Therefore, several studies have developed a synthesis method that is more environmentally friendly. Reducing agents that have been widely used are plant extracts. Subsequently, plants with polyphenol content and high antioxidant capacities, such as strawberries, raspberries, black tea, eucalyptus, mulberry, green tea, pomegranate, and oak leaves, can act as a reducing agent in the production of ZVI [20–26]. Tea leaves (*Camellia sinensis*) are one of the most abundant commodities, with polyphenols as the essential content in the form of flavonoids and catechins ranging from 20 to 30% [27,28]. In 2013, production capacity reached 152,700 tons/year [29]; therefore, tea-leaf extract can be used as an iron-reducing agent.

The performance of ZVI in degrading organic pollutants can be improved when applied to the Fenton method. This method is widely studied because it is considered effective in reducing organic pollutants. It is simple, can be carried out at ambient temperature and pressure, is easy to handle, and is safe for the environment [30]. The Fenton method is an advanced oxidation process (AOPs) system that works by utilizing hydroxyl radicals (•OH) to degrade pollutants [31]. ZVI using this method consists of iron(II) as a catalyst and hydrogen peroxide as an oxidizing agent [32,33]. Several studies have reported that ZVI applied to the Fenton method is effective in reducing the color intensity of bromothymol blue, methylene blue, a mixture of Remazol brilliant blue R (RBB-R) and direct red (DR), Eriochrome blue-black B (EBB), and malachite green [34–37].

The dyes used in the textile dyeing process are classified into several types, including nitro, azo, diphenylmethane, triphenylmethane, xanthene, phthalein, indigo, thioindigo, and anthraquinone dyes [2]. Rhodamine B is one of the xanthene classes often used in the textile dyeing industry because it is cheap and easy to obtain [38]. Subsequently, azo dyes are widely used since about 60–70% of organic dyes produced globally are members of the azo group [39]. One of the azo dyes, one of the most easily found is methyl orange, which is soluble in water. Several studies have shown that the Fenton method can degrade rhodamine B and methyl orange into simpler intermediates [40–44]. However, no study has degraded a mixture of rhodamine B and methyl orange dyes using the Fenton method with ZVI as a catalyst.

This study synthesized the ZVI catalyst leaf extract from tea (*Camellia sinensis*) as a natural reducing agent. Furthermore, the synthesized ZVI was characterized using particle size analysis (PSA), UV-Vis spectrophotometry, Fourier-transform infrared spectroscopy (FTIR), and electron microscopy. The Fenton method was then tested to degrade a rhodamine B-methyl orange dye mixture. In wastewater treatment, the decrease in the value of chemical oxygen demand (COD) should be considered to analyze the effectiveness and quality of the treated water. The COD value can indicate the level of water pollution by organic pollutants [45].

2. Materials and Methods

2.1. Materials

The materials used were iron(II) sulfate heptahydrate ($FeSO_4 \cdot 7H_2O$, 99.5%, Merck 103965, Kenilworth, NJ, USA,), Folin Ciocalteu reagent, which is a mixture of phosphomolybdate and phosphotungstate (2 mol/L, Merck 109001, Kenilworth, NJ, USA), sodium carbonate (Na_2CO_3, 99.9%, Merck 106392, Kenilworth, NJ, USA), gallic acid ($C_7H_6O_5$, 98%, Merck 842649, Kenilworth, NJ, USA), tea leaves (*Camellia sinensis*) from the Citengah tea plantation, Sumedang, Indonesia (Figure S1), concentrated sulfuric acid (H_2SO_4, 95–97%, Smart Lab A-1092 F), hydrogen peroxide (H_2O_2, 30%, Merck 107209, Kenilworth, NJ, USA), digestion solution for COD (prepared from mercury(II) sulfate ($HgSO_4$, 98%, Merck 104480, Kenilworth, NJ, USA) and potassium dichromate ($K_2Cr_2O_7$, 99,9%, Merck 104864, Kenilworth, NJ, USA), sodium hydroxide (NaOH, 99%, Merck 1006498, Kenilworth, NJ, USA), ammonium iron(II) sulfate hexahydrate ((NH_4)$_2$Fe(SO_4)$_2 \cdot 6H_2O$, 99%, Merck 103792, Kenilworth, NJ, USA), ferroin indicator (prepared from 1,10-phenanthroline monohydrate ($C_{12}H_8N_2 \cdot H_2O$, 99,5%, Merck 107225, Kenilworth, NJ, USA) and iron(II) sulfate heptahydrate ($FeSO_4 \cdot 7H_2O$, 99.5%, Merck 103965, Kenilworth, NJ, USA)), rhodamine B ($C_{28}H_{31}ClN_2O_3$, 90%, Merck 107599, Kenilworth, NJ, USA), and methyl orange ($C_{14}H_{14}N_3NaO_3S$, 85%, Merck 114510, Kenilworth, NJ, USA).

2.2. Tea Extract Preparation

Tea leaves (*Camellia sinensis*) were taken from the Citengah tea plantation, Sumedang (Figure S1), and dried until the water was gone [20]. The leaves were then separated from the stems and cut into small pieces to an area of 1 × 1 cm. Furthermore, they were heated at 50 °C for 48 h and at 105 °C for 4 h to determine water content. Finally, the dried tea leaves were ground and then sieved through of 10 mesh.

2.3. Determination of Polyphenol Content in Tea Extract

The tea extract was made by modifying the existing procedures in previous studies [25,46]. First, 3.6 g of tea powder of was added to 100 mL of distilled water as a solvent, then heated at various temperatures of 28–98 °C for 20 min, filtered and concentrated. A total of 10 mg of extract was then dissolved with 50 mL of distilled water. Then, 5 mL of the extract was taken, 0.2 mL of Folin Ciocalteu reagent, 4 mL of 7.5% sodium carbonate was added, and distilled water was added to a volume of 25 mL. The solution was incubated for 45 min at room temperature and measured using a visible spectrophotometer (HACH DR 3900, Loveland, CO, USA) at a wavelength of 698 nm. Then, the optimum heating time was found, with time variations of 20–100 min using the same method. Finally, the total polyphenol content was calculated using Equation (1).

$$TPC = (C \times V)/m \tag{1}$$

where TPC is the total phenolic content (mg GAE/g), C is the concentration of tea leaves (mg/L), V represents the volume of solvent (L), and m is the weight of the tea extract used (g).

2.4. Synthesis and Characterization of Zero-Valent Iron (ZVI)

The synthesis of ZVI was carried out using a modified method from a previous study [25]. First, the tea extract was added with 0.1 mol/L of iron(II) sulfate solution with a volume ratio of iron(II) to tea extract of 1:1, 1:2, and 1:3 while stirring in a sonicator-900W (BEM-900A, Bueno Biotech) in a stream of nitrogen gas. The remaining water was then evaporated, and ZVI was obtained.

The synthesized ZVI was characterized using a UV-Vis spectrophotometer (Thermo Fisher Scientific Genesys 10S, Waltham, MA, USA) to determine the absorption at a 200–600 nm wavelength. ZVI was further described using Fourier-transform infrared spectroscopy (FTIR, Shimadzu IRPrestige-21, Tokyo, Japan) to determine the functional group, while FTIR was used, with a scanning range of 400–4000 cm^{-1}. Scanning electron microscope-energy-dispersive X-ray spectrometry analysis (SEM-EDS, Hitachi SU3500, Tokyo, Japan) was performed using 3 kV at 2000× magnification to determine the shape of the surface morphology. Finally, particle size was determined using a particle-size analyzer (PSA, Horiba SZ-100, Kyoto, Japan).

2.5. ZVI Catalytic Test for Dyes

Samples of a mixture of rhodamine B and methyl orange, 1:1 (each 50 mg/L), were tested for color intensity. First, the sample was adjusted to pH 2, then, 5 mL of 12% hydrogen peroxide and ZVI with various concentrations of 50, 100, and 150 mg/L. Next, the solution was stirred, and the intensity of the color was measured using a UV-Vis spectrophotometer (Thermo Fisher Scientific Genesys 10S, Waltham, MA, USA) at a wavelength of 200–600 nm. The measurement process was carried out for 180 min with an interval of 30 min. Meanwhile, to determine the presence of degradation and intermediate compounds formed, testing was carried out using liquid chromatography-mass spectrometry (LC-MS, Waters Q-tof MS Xevo, Milford, MA, USA).

2.6. Chemical Oxygen Demand (COD) Test

The COD measurement in the sample refers to the measurement of water using the closed reflux method. First, a total of 2.5 mL of sample was put into a reflux tube. Then, 1.5 mL of digestion solution and 3.5 mL of concentrated sulfuric acid were added. The tube was inserted into the COD reactor (Velp ECO 25, Usmate Velate, Italy) and heated at 165 °C for 2 h. After cooling, the sample was titrated using a standard solution of 0.1 mol ek/L ammonium iron(II) sulfate with ferroin as an indicator.

$$COD = \frac{(V_1 - V_2) \times C_{FAC} \times 8 \times 1000}{V_3} \quad (2)$$

where chemical oxygen demand (COD) is represented in mg O_2/g, C_{FAC} is the concentration of ammonium iron(II) sulfate (mol ek/L), V_1 is the blank titration volume (mL), V_2 is the sample titration volume (mL), and V_3 is the sample volume (mL).

3. Results and Discussions

3.1. Preparation and Polyphenol Content of Tea Extract

The extract was prepared using the shoots to leaves without stems, which had a moisture content of 63.43 ± 0.50%. Polyphenols in catechins found in tea leaves were polar compounds extracted with water [47]. The experimental results in Figure 1 show that the higher the heating temperature used, the more concentrated the color of the extract obtained; therefore, more polyphenolic compounds were extracted. The amount of polyphenol content in the extract can be confirmed by the Folin Ciocalteu method. Besides polyphenols, several water-soluble compounds, such as caffeine, amino acids, and sugars in tea, can also be extracted. As a result, the compound components contained in the extract are not in the form of a single compound [48,49].

Figure 1. Tea extract after heating for 20 min and after the addition of Folin Ciocalteu reagent at: (**a**) 28; (**b**) 50; (**c**) 60; (**d**) 70; (**e**) 80; (**f**) 90; and (**g**) 98 °C.

The polyphenol content in tea extract was determined using the Folin Ciocalteu method. Folin Ciocalteu reagent is a complex compound formed from phosphomolybdic acid and heteropoly phosphotungstic acid. The working principle is the oxidation of hydroxyl groups in phenolic compounds by Folin Ciocalteu reagent. Furthermore, the blue color of the solution is the reduced form of the Folin Ciocalteu reagent [50,51]. Gallic acid is used as a standard in the determination of polyphenol content. The maximum wavelength of standard gallic acid and the optimal incubation time are shown in Figure S2 and the calibration curve for gallic acid shown in Figure S3. Polyphenol measurement data showed that the more concentrated the resulting color (Figure 1), the higher the polyphenol content in the extract. Figure 2 shows the redox reaction between polyphenols and the Folin Ciocalteu reagent [51].

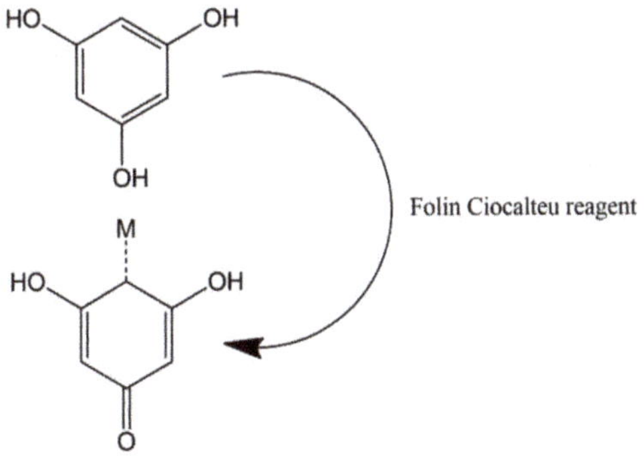

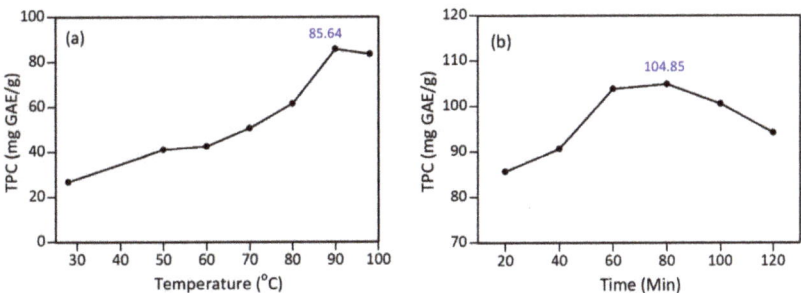

Figure 2. Redox reaction between polyphenols and Folin Ciocalteu reagent.

The test results found that the maximum total phenolic content (TPC) was obtained at a temperature of 90 °C, with a TPC value of 85.64 mg GAE/g (Figure 3a). Furthermore, several time variations were tested using the optimum temperature of 90 °C. The data showed that the polyphenol content of the tea extract heated at 90 °C for 80 min increased by 18.35% to 104.85 mg GAE/g (Figure 3b), compared to heating for 20 min. Therefore, based on these results, the optimum extraction process was conducted at 90 °C for 80 min.

Figure 3. Total polyphenol content (TPC) of tea extract: (**a**) variation of temperature with 20 min heating; and (**b**) variation of heating time at 90 °C.

3.2. Characteristics of ZVI

The particle-size distribution of ZVI was tested using a particle-size analyzer (PSA). Table 1 presents data that shows that ZVI 1:1 and 1:2 are distributed in two sizes, indicated by the presence of two peaks formed, while ZVI 1:3 shows one peak (Figure 4). Therefore, the more tea extract used, the smaller and more uniform the particle size. This is understandable because the polyphenols in tea extract act as reducing and capping agents to prevent the agglomeration process [52]. The presence of hydrophobic poles on the capping agent causes the formation of steric barriers that can control particle growth. It reduces the surface energy of the particles, and aggregation can be avoided [53,54]. Apart from polyphenols, several compounds that are naturally contained in plants, such as citric acid, vitamins, and silica, can also act as natural capping agents [55].

Table 1. Size distribution of ZVI particles.

ZVI	Size Range (nm)		Average Size (nm)	
	First Peak	Second Peak	First Peak	Second Peak
1:1	151–171	945–1207	153	1012
1:2	93–105	740–945	93	733
1:3	279–356	-	301	-

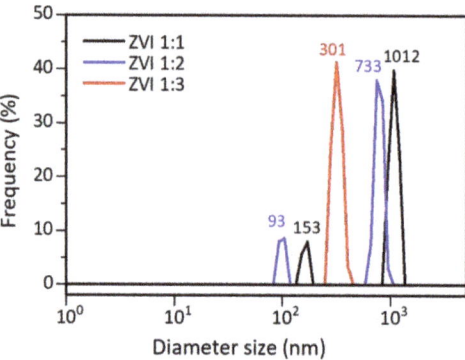

Figure 4. PSA results for ZVI particle size distribution.

UV-Vis spectrum analysis was conducted to analyze the difference in peaks produced by tea extract, iron(II) solution, and ZVI produced from the synthesis process, and the measurement was carried out at a 200–600 nm wavelength [56]. Figure S4 shows the color change of the iron(II) solution on the formation of nZVI. The test results show differences in absorption peaks, indicating that a new product, which is different from the previous constituent materials, was formed (Figure 5). The tea extract has peaks at wavelengths of 230 and 270 nm, indicating the presence of phenolic acid and its derivatives [23]. Meanwhile, the iron(II) solution showed absorption peaks in the 230–260 and 300 nm regions, probably from $[FeHSO_4]^{2+}$ [57]. In the ZVI spectrum, the resulting peak of the $[FeHSO_4]^{2+}$ ion did not appear, indicating that the compound had changed. However, there was a peak at 270 nm at various iron(II) ratios and tea extract but less intense than before. This peak indicates that phenolic compounds and their derivatives are still present in the ZVI material. Around 200 nm, a typical peak of benzoyl in flavonoid compounds is shown from tea extract [58]. Other compounds besides polyphenols identified in ZVI were present since the extract used was not pure isolate. Furthermore, experimental data also showed that the more tea extract used, the higher the absorption peaks at wavelengths of 200 and 270 nm. This occurred because the number of polyphenols and compounds added to the extract increased, while the iron(II) remained. Therefore, there was an excess of unreacted polyphenolic compounds.

FTIR measurements determine the functional groups in tea extract and ZVI. Figure 6 shows the absorption band at wavenumber 3392–3407 cm^{-1} from O–H stretching vibrations of polyphenol compounds [59,60]. Meanwhile, the absorption band at wavenumber 1663–1655 cm^{-1} and 1450–1500 cm^{-1} comes from the C=O vibration and C=C vibration of the aromatic ring [61]. Then, the absorption band at wavenumber 1064–1099 cm^{-1} comes from the C–O bond of the heterocyclic pyranose ring [62].

The sp^3 carbon signal appears in the tea extract spectrum at a wavenumber of 2894 cm^{-1}; this signal can come from lignin dissolved during the extraction process. The C–H stretch signal of aromatic methoxy and methylene groups that are characteristic of lignin are shown at wavenumber 2851–2924 cm^{-1} [63]. However, this signal disappeared in the ZVI spectrum due to the widening of the hydroxyl group band and formation of bonds between iron and hydroxyl groups of polyphenol compounds [64]. In the ZVI spectrum, a signal appears at wavenumber 2354–2363 cm^{-1} from the saturated bond of carbon dioxide [65]. Meanwhile, the absorption band at wavenumber 606–610 cm^{-1} comes from the Fe–O vibration [66].

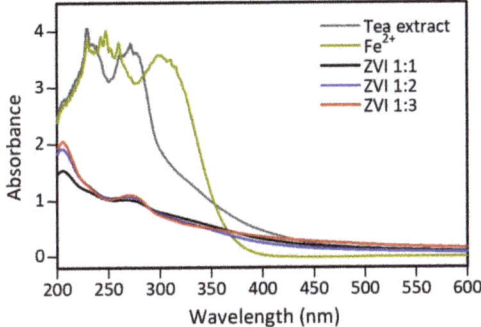

Figure 5. UV-Vis spectra of tea extract, 0.1 M iron(II), and ZVI.

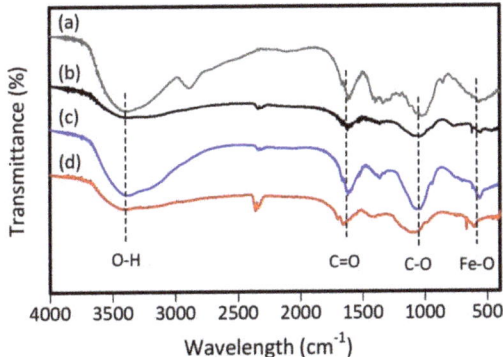

Figure 6. FTIR spectra of (**a**) tea extract, (**b**) ZVI 1:1, (**c**) ZVI 1:2, and (**d**) ZVI 1:3.

The morphology of the ZVI surface was determined using SEM-EDS, and the various shapes and sizes indicated the presence of agglomeration in the three types of ZVI (Figure 7). Agglomeration can occur due to the formation of an organic layer of tea extract and iron oxide on the surface of ZVI [58,59]. This is supported by the results of FTIR measurements, which showed that there are still several functional groups detected on the surface of ZVI. Moreover, the presence of soluble fibers, such as lignin, which was confirmed in the FTIR measurement, can also cause the outer layer of ZVI to be thicker [67].

EDS measurement results show that the main components on the surface of ZVI are carbon, oxygen, sulfur, and iron (Table 2). However, the iron element on the surface is not the primary constituent because there may be an agglomeration of the organic layer that dominates the ZVI surface. Meanwhile, the sulfur in ZVI can come from ferrous sulfate, the precursor of ZVI. Other sodium, nitrogen, and potassium elements may have come from tea extract, which is not a pure isolate [25,68]. Based on the analysis results, the more tea extract used, the more carbon and oxygen measured on the material's surface. This further confirms and supports an organic layer on the ZVI surface. Meanwhile, the estimated iron content was not significantly different.

Figure 7. SEM images of: (**a**) ZVI 1:1; (**b**) ZVI 1:2; and (**c**) ZVI 1:3.

Table 2. Weight percent of EDS measurement results for ZVI 1:1, 1:2, and 1:3.

Element	Weight Percent (%)		
	ZVI 1:1	ZVI 1:2	ZVI 1:3
O	36.97	42.92	44.91
C	19.97	20.96	27.98
S	13.39	14.46	5.40
Fe	10.12	11.30	11.90
Na	3.42	2.69	3.24
K	3.32	4.17	2.25
N	2.97	3.50	4.32

3.3. Color Intensity Test

The dye sample used was a mixture of rhodamine B and methyl orange in a ratio of 1:1 with a concentration of 50 mg/L, respectively. Samples of mixed dyestuffs that have not been catalyzed are determined by their maximum wavelength using UV-Vis. From the test results, each component of rhodamine B, methyl orange, and mixture (methyl orange-rhodamine B) has a maximum wavelength of 555, 464, and 506 nm, respectively (Figure 8). The maximum wavelength of the mixture shifted from each of its constituent components, indicating that there was a change in the chromophore structure due to the interaction between rhodamine B and methyl orange.

Figure 9 shows that ZVI 1:1 obtained the most significant decrease in color intensity, by 49.82%, followed by ZVI 1:2 and ZVI 1:3, by 48.12% and 37.28%, respectively, after 30 min of contact. This happened because ZVI 1:3 has the highest carbon and oxygen content on its surface based on the characterization results. Furthermore, on UV-Vis measurement, the peak of flavonoids and phenolics was most significant at ZVI 1:3 compared to other types of ZVI. Therefore, the presence of these compounds can interfere with the performance of iron(0) in the core of the material. Even though the PSA results of ZVI 1:3 showed the most uniform size, the catalytic process of the Fenton method was influenced by other factors, such as surface composition. Therefore, the particle size of the iron catalyst did not significantly affect the performance [69].

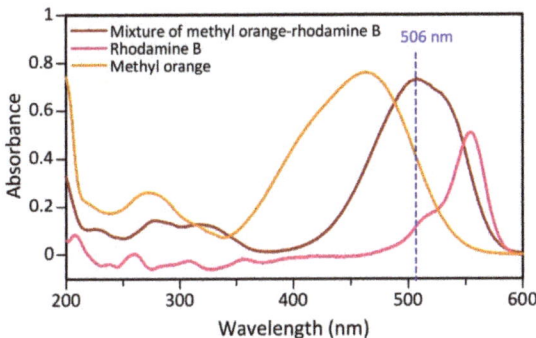

Figure 8. Maximum wavelength determination curve before the catalytic process for the dyes methyl orange, rhodamine B, and mixture (methyl orange-rhodamine B).

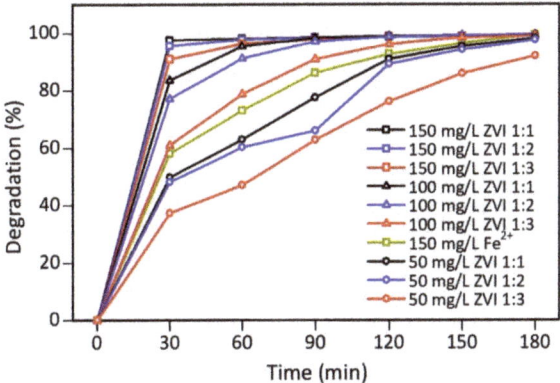

Figure 9. Decreased color intensity of mixture of rhodamine B-methyl orange at 506 nm.

Figure 9 also signifies that the greater the concentration of ZVI used in the catalytic process, the more significant the decrease in the color intensity of the mixture. Furthermore, contact time also affects the decrease in the color intensity of the mixture, where the longer the contact process, the less the color intensity of the mixture. ZVI 1:1, 1:2, and 1:3 catalysts with a concentration of 50 mg/L reduced the dye up to >90% when the contact time reached 120 min. Meanwhile, at a concentration of ZVI 100 mg/L, the color intensity decreased >90% when the Fenton process lasted for 60 min. At ZVI 150 mg/L, the color intensity decreased to 90% only after a contact time of 30 min. Iron(II) sulfate 150 mg/L within 30 min reduced the color intensity of the mixture by 58.07%; therefore, ZVI had 32% more effective ability.

An analysis was also carried out to see any changes in the peak at a wavelength of 200–600 nm during the Fenton process. The results showed that the mixed dyes that underwent a catalytic process using ZVI 1:1, 1:2, and 1:3 with concentrations of 50, 100, and 150 mg/L, respectively, underwent the same changes during the catalytic process from 30–180 min (Figures S5–S8). The peak at the wavelength of 506 nm did not shift and only experienced a decrease in color intensity, while the peak around the wavelength of 200–300 nm experienced a change. This peak is thought to originate from ZVI that has been converted to iron(II) during the Fenton process.

The mechanism of Fenton's reaction with ZVI is illustrated in Figure 10. Fe^0 is oxidized by H_2O_2 to give Fe^{2+} in Equation (3). In addition, Fe^0 can also reduce Fe^{3+} to Fe^{2+} in Equation (4). Then, Fe^{2+} will react with H_2O_2 to form a highly reactive hydroxyl radical

(•OH) (Equation (5)). These hydroxyl radicals then degrade organic pollutants so that they are oxidized, which is written in Equations (6)–(8), which, in the end mineralization, occurs to form CO_2 and H_2O end products.

$$Fe^0 + H_2O_2 \rightarrow Fe^{2+} + 2OH^- \quad (3)$$

$$2Fe^{3+} + Fe^0 \rightarrow 3Fe^{2+} \quad (4)$$

$$Fe^{2+} + H_2O_2 \rightarrow Fe^{3+} + HO^- + HO• \quad (5)$$

$$RH + HO• \rightarrow H_2O + R• \quad (6)$$

$$R• + H_2O_2 \rightarrow ROH + HO• \quad (7)$$

$$R• + O_2 \rightarrow ROO• \quad (8)$$

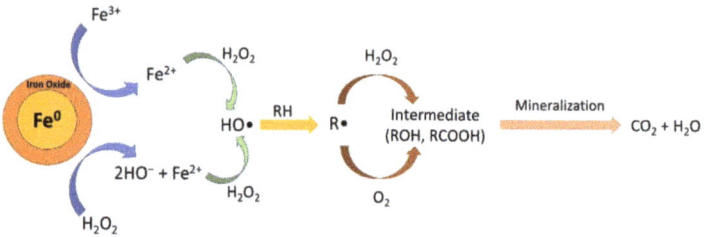

Figure 10. Mechanism of heterogeneous Fenton reactions with ZVI.

3.4. Degradation Pathway and Identification of the Intermediates

Analysis of reaction intermediates and final products is useful for evaluating the efficiency of a catalytic system and can reveal some details of the reaction process. This study used liquid chromatography-mass spectrometry (LC-MS, Waters Q-tof MS Xevo, Milford, MA, USA) to determine the dye-mixture intermediates. The LC-MS chromatogram results showed that the structure of the rhodamine B fragment before the catalytic process was identified to experience chloride-loss ionization with m/z 443. Meanwhile, methyl orange before the catalytic process was identified at m/z 327. A signal appeared with m/z 304 and 306 from methyl orange, which lost sodium ions and underwent rearrangement [70]. The results of a mixture of dyes refer to the retention time of each component before being mixed. The intermediates of rhodamine B and methyl orange are shown in Table 3. The intact structure of rhodamine B did not reappear on the chromatogram after the catalytic process. This indicates that the dye had degraded entirely to another compound with a smaller molecular mass. Meanwhile, the intact structure of methyl orange with m/z 304 and 327 still exists. This proves that the catalytic process to degrade methyl orange requires more difficulty than rhodamine B. The proposed mechanism from the degradation of rhodamine B and methyl orange using ZVI with the Fenton method is shown in Figure 11.

Table 3. Main intermediates from the degradation of rhodamine B and methyl orange detected by LC-MS.

Compound No.	Retention Time (min)	m/z
(1)	12.02	403
(2)	10.32	327
(3)	13.27	167
(4)	0.87	125
(5)	24.13	110
(6)	11.27	279
(7)	12.45	172

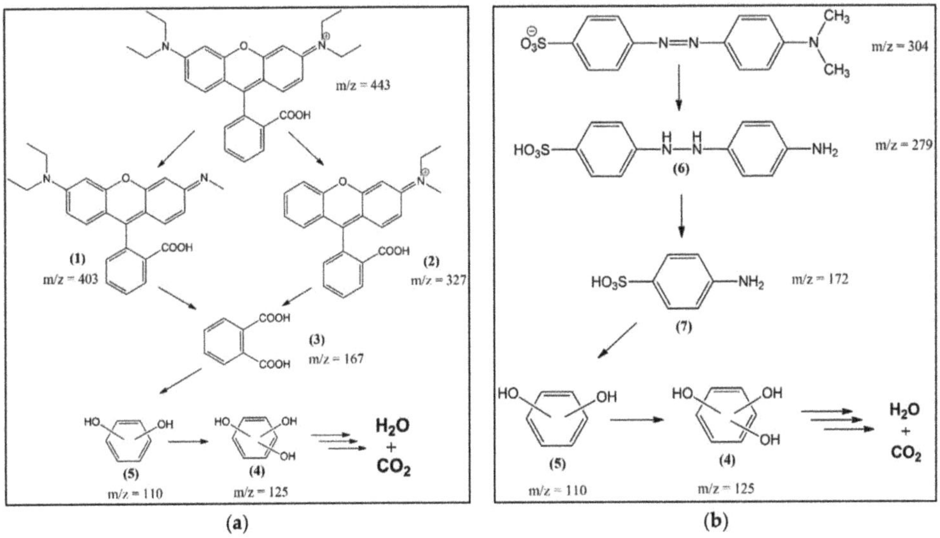

Figure 11. Schematic illustration of the proposed degradation pathway of: (**a**) rhodamine B; and (**b**) methyl orange.

Furthermore, using Origin85 8.5.1 SR_2 software, the area of methyl orange and rhodamine B mixture was measured before and after the catalytic process to determine the degradation level. The peaks selected in the determination of the area were peaks belonging to rhodamine B and methyl orange. Based on the calculation results, rhodamine B and methyl orange areas decreased by 100% and 66.47%, respectively.

3.5. COD Testing

In water treatment, especially wastewater, the chemical oxygen demand (COD) value is one of the essential parameters because it shows the pollution level in water by organic compounds. Therefore, the COD value is directly proportional to the pollution level. Based on the experimental results, the value of the mixture dye before the catalytic process was 92.76 mg O_2/g and decreased by 92.11% to 7.32 mg O_2/g after the catalytic process (Table 4). Meanwhile, iron(II) as a catalyst only reduced the dye by 44.75%.

Table 4. COD value measurement results for the mixture of rhodamine B-methyl orange 150 mg/L before and after the Fenton process.

Sample	COD (mg O_2/g)	Decrease (%)
Mixture before catalytic	92.76	-
ZVI 1:1	7.32	92.11
Iron(II) sulfate	48.83	47.36

4. Conclusions

In this research, ZVI was synthesized using a natural reducing agent from tea-leaf extract and applied in the Fenton method to degrade the color mixture of rhodamine B and methyl orange. The results show that the optimum dose of ZVI to reduce the color intensity of the rhodamine B and methyl orange mixture was ZVI 1:1 150 mg/L. Furthermore, the LC-MS test showed a degradation process for the rhodamine B and methyl orange mixed dyes, where rhodamine B and methyl orange were degraded by 100% and 66.47%, respectively. Meanwhile, the decrease in the COD value obtained in this condition was 92.11%, which is higher than using Fe(II) of 47.36%.

Supplementary Materials: The following supporting information can be downloaded at: https://www.mdpi.com/article/10.3390/ma15010332/s1, Figure S1: Coordinates of tea leaf sampling location; Figure S2: (a) The curve of absorption wavelength of gallic acid; and (b) The curve of optimum incubation time for gallic acid at 698 nm; Figure S3: The curve of the gallic acid calibration standard; Figure S4: (a) Color of 0.1 M iron(II) sulfate solution; (b) Tea extract heated at 90 °C for 80 min; (c) synthesized nZVI before evaporation; (d) nZVI solid 1:1, (e) 1:2, and (f) 1:3; Figure S5: Absorbance of ZVI catalytic test for 50 mg/L ZVI concentration; Figure S6: Absorbance of ZVI catalytic test for 100 mg/L ZVI concentration; Figure S7: Absorbance of ZVI catalytic test for 150 mg/L ZVI concentration; Figure S8: Absorbance of iron(II) sulfate catalytic test for 150 mg/L iron(II) concentration.

Author Contributions: Conceptualization, D.R.E. and S.; methodology, D.N.; software, M.D.P. and D.N.; validation, D.R.E., S., A.R.N., and I.R.; formal analysis, D.N.; investigation, D.N.; resources, D.N.; data curation, D.N.; writing—original draft preparation, D.N.; writing—review and editing, D.R.E. and M.D.P.; visualization, M.D.P.; supervision, D.R.E. and S.; project administration, D.R.E.; funding acquisition, D.R.E. and I.R. All authors have read and agreed to the published version of the manuscript.

Funding: This research was funded by Academic Leadership Grant (ALG) Prof. Iman Rahayu, Universitas Padjadjaran, grant number ID: 1959/UN6.3.1/PT.00/2021, and "The APC was funded by them".

Institutional Review Board Statement: Not applicable.

Informed Consent Statement: Not applicable.

Data Availability Statement: Not applicable.

Conflicts of Interest: The authors declare no conflict of interest.

References

1. Pratiknyo, P. Proyeksi Ketersediaan dan Kebutuhan Air Industri di Kab. Tangerang. In Proceedings of the Nasional Kebumian XI-FTM UPN Veteran Yogyakarta, Yogyakarta, Indonesia, 3–4 November 2016.
2. Singh, K.; Kumar, P.; Srivastava, R. An overview of textile dyes and their removal techniques: Indian perspective. *Pollut. Res.* **2017**, *36*, 790–797.
3. Haley, T.J. Benzidine revisited: A review of the literature and problems associated with the use of benzidine and its congeners. *Clin. Toxicol.* **1975**, *8*, 13–42. [CrossRef] [PubMed]
4. Hassani, A.; Khataee, A.; Karaca, S.; Karaca, M.; Kiranşan, M. Adsorption of two cationic textile dyes from water with modified nanoclay: A comparative study by using central composite design. *J. Environ. Chem. Eng.* **2015**, *3*, 2738–2749. [CrossRef]
5. Agarwal, S.; Tyagi, I.; Gupta, V.K.; Dastkhoon, M.; Ghaedi, M.; Yousefi, F.; Asfaram, A. Ultrasound-assisted adsorption of Sunset Yellow CFC dye onto Cu doped ZnS nanoparticles loaded on activated carbon using response surface methodology based on central composite design. *J. Mol. Liq.* **2016**, *219*, 332–340. [CrossRef]
6. Kasperchik, V.P.; Yaskevich, A.L.; Bil'Dyukevich, A.V. Wastewater treatment for removal of dyes by coagulation and membrane processes. *Pet. Chem.* **2012**, *52*, 545–556. [CrossRef]
7. Rodrigues, C.S.; Madeira, L.M.; Boaventura, R.A. Treatment of textile dye wastewaters using ferrous sulphate in a chemical coagulation/flocculation process. *Environ. Technol.* **2013**, *34*, 719–729. [CrossRef]
8. Cheikh S'Id, E.; Kheribech, A.; Degu, M.; Hatim, Z.; Chourak, R.; M'Bareck, C. Removal of Methylene Blue from Water by Polyacrylonitrile Co Sodium Methallylsulfonate Copolymer (AN69) and Polysulfone (PSf) Synthetic Membranes. *Prog. Color Color. Coat.* **2021**, *14*, 89–100.
9. Etezad, S.M.; Sadeghi-Kiakhani, M. Decolorization of Malachite Green Dye Solution by Bacterial Biodegradation. *Prog. Color Color. Coat.* **2021**, *14*, 79–87.
10. Klavarioti, M.; Mantzavinos, D.; Kassinos, D. Removal of residual pharmaceuticals from aqueous systems by advanced oxidation processes. *Environ. Int.* **2009**, *35*, 402–417. [CrossRef]
11. Seabra, A.B.; Haddad, P.; Duran, N. Biogenic synthesis of nanostructured iron compounds: Applications and perspectives. *IET Nanobiotechnol.* **2013**, *7*, 90–99. [CrossRef]
12. Krishnan, S.; Rawindran, H.; Sinnathambi, C.M.; Lim, J.W. Comparison of various advanced oxidation processes used in remediation of industrial wastewater laden with recalcitrant pollutants. *IOP Conf. Ser. Mater. Sci. Eng.* **2017**, *206*, 012089. [CrossRef]
13. Blowes, D.W.; Ptacek, C.J.; Benner, S.G.; McRae, C.W.; Bennett, T.A.; Puls, R.W. Treatment of inorganic contaminants using permeable reactive barriers. *J. Contam. Hydrol.* **2000**, *45*, 123–137. [CrossRef]
14. Gavaskar, A.; Tatar, L. *Cost and Performance Report Nanoscale Zero-Valent Iron Technologies for Source Remediation*; Naval Facilities Engineering Service Center Port Hueneme: Port Hueneme, CA, USA, 2005.

15. Rashmi, S.H.; Madhu, G.M.; Kittur, A.; Suresh, R. Synthesis, characterization and application of zero valent iron nanoparticles for the removal of toxic metal hexavalent chromium [Cr (VI)] from aqueous solution. *Int. J. Curr. Eng. Technol.* **2013**, *1*, 37–42.
16. Nursyamsiah, D.; Solihudin; Eddy, D.R. Green Synthesis of Zero Valent Iron (ZVI) using Tea Leaves Extract and its Application as Fenton like Catalyst for Textile Dyes Removal. *Asian J. Chem.* **2021**, *33*, 963–968. [CrossRef]
17. Wardiyati, S.; Fisli, A.; Yusuf, S. Sintesis dan Karakterisasi Nano Zero Valent Iron (NZVI) dengan Metode Presipitasi. *J. Kimia Kemasan* **2013**, *35*, 37–44. [CrossRef]
18. Kanel, S.R.; Manning, B.; Charlet, L.; Choi, H. Removal of arsenic (III) from groundwater by nanoscale zero-valent iron. *Environ. Sci. Technol.* **2005**, *39*, 1291–1298. [CrossRef]
19. Nadagouda, M.N.; Castle, A.B.; Murdock, R.C.; Hussain, S.M.; Varma, R.S. In vitro biocompatibility of nanoscale zerovalent iron particles (NZVI) synthesized using tea polyphenols. *Green Chem.* **2010**, *12*, 114–122. [CrossRef]
20. Machado, S.; Pinto, S.L.; Grosso, J.P.; Nouws, H.P.A.; Albergaria, J.T.; Delerue-Matos, C. Application of green zero-valent iron nanoparticles to the remediation of soils contaminated with ibuprofen. *Sci. Total Environ.* **2013**, *445*, 323–329. [CrossRef]
21. Hoag, G.E.; Collins, J.B.; Holcomb, J.L.; Hoag, J.R.; Nadagouda, M.N.; Varma, R.S. Degradation of bromothymol blue by 'greener' nano-scale zero-valent iron synthesized using tea polyphenols. *J. Mater. Chem.* **2009**, *19*, 8671–8677. [CrossRef]
22. Dhuper, S.; Panda, D.; Nayak, P.L. Green synthesis and characterization of zero valent iron nanoparticles from the leaf extract of Mangifera indica. *J. Nanotechnol. Its Appl.* **2012**, *13*, 16–22.
23. Huang, L.; Weng, X.; Chen, Z.; Megharaj, M.; Naidu, R. Synthesis of iron-based nanoparticles using oolong tea extract for the degradation of malachite green. *Spectrochim. Acta Part A Mol. Biomol. Spectrosc.* **2014**, *117*, 801–804. [CrossRef]
24. Murgueitio, E.; Debut, A.; Landivar, J.; Cumbal, L. Synthesis of iron nanoparticles through extracts of native fruits of Ecuador, as capuli (*Prunus serotina*) and mortiño (*Vaccinium floribundum*). *Biol. Med.* **2016**, *8*, 1. [CrossRef]
25. Kecić, V.; Kerkez, Đ.; Prica, M.; Lužanin, O.; Bečelić-Tomin, M.; Pilipović, D.T.; Dalmacija, B. Optimization of azo printing dye removal with oak leaves-nZVI/H_2O_2 system using statistically designed experiment. *J. Clean Prod.* **2018**, *202*, 65–80. [CrossRef]
26. Desalegn, B.; Megharaj, M.; Chen, Z.; Naidu, R. Green synthesis of zero valent iron nanoparticle using mango peel extract and surface characterization using XPS and GC-MS. *Heliyon* **2019**, *5*, e01750. [CrossRef]
27. Sumpio, B.E.; Cordova, A.C.; Berke-Schlessel, D.W.; Qin, F.; Chen, Q.H. Green tea, the "Asian paradox," and cardiovascular disease. *Chen. J. Am. Coll. Surg.* **2006**, *202*, 813–825. [CrossRef]
28. Anjarsari, I.R.D. Indonesia tea catechin: Prospect and benefits. *J. Kultivasi* **2016**, *15*, 99–106.
29. Chang, K. World Tea Production and Trade Current and Future Development. 2015. Available online: http://www.fao.org/3/i4480e/i4480e.pdf (accessed on 20 May 2020).
30. Neyens, E.; Baeyens, J.A. A review of classic Fenton's peroxidation as an advanced oxidation technique. *J. Hazard Mater.* **2003**, *98*, 33–50. [CrossRef]
31. Haddad, M.E.; Regti, A.; Laamari, M.R.; Mamouni, R.; Saffaj, N. Use of Fenton reagent as advanced oxidative process for removing textile dyes from aqueous solutions. *J. Mater. Environ. Sci.* **2014**, *5*, 667–674.
32. Walling, C. Fenton's reagent revisited. *Acc. Chem. Res.* **1975**, *8*, 125–131. [CrossRef]
33. Goi, A. Advanced Oxidation Processes for Water Purification and Soil Remediation. Ph.D. Thesis, Tallinn University of Technology, Tallinn, Estonia, 2005.
34. Shahwan, T.; Sirriah, S.A.; Nairat, M.; Boyacı, E.; Eroğlu, A.E.; Scott, T.B.; Hallam, K.R. Green synthesis of iron nanoparticles and their application as a Fenton-like catalyst for the degradation of aqueous cationic and anionic dyes. *Chem. Eng. J.* **2011**, *172*, 258–266. [CrossRef]
35. Truskewycz, A.; Shukla, R.; Ball, A.S. Iron nanoparticles synthesized using green tea extracts for the fenton-like degradation of concentrated dye mixtures at elevated temperatures. *J. Environ. Chem. Eng.* **2016**, *4*, 4409–4417. [CrossRef]
36. Ghanim, D.; Al-Kindi, G.Y.; Hassan, A.K. Green synthesis of iron nanoparticles using black tea leaves extract as adsorbent for removing eriochrome blue-black B dye. *Eng. Technol. J.* **2020**, *38*, 1558–1569. [CrossRef]
37. Xiao, C.; Li, H.; Zhao, Y.; Zhang, X.; Wang, X. Green synthesis of iron nanoparticle by tea extract (polyphenols) and its selective removal of cationic dyes. *J. Environ. Manag.* **2020**, *275*, 111262. [CrossRef] [PubMed]
38. Sibarani, J.; Purba, D.L.; Suprihatin, I.E.; Manurung, M. Fotodegradasi Rhodamin B menggunakan ZnO/UV/Reagen Fenton. *Cakra Kimia* **2016**, *4*, 84–93.
39. Bafana, A.; Devi, S.S.; Chakrabarti, T. Azo dyes: Past, present and the future. *Environ. Rev.* **2011**, *19*, 350–370. [CrossRef]
40. Hou, M.F.; Liao, L.; Zhang, W.D.; Tang, X.Y.; Wan, H.F.; Yin, G.C. Degradation of rhodamine B by Fe (0)-based Fenton process with H_2O_2. *Chemosphere* **2011**, *83*, 1279–1283. [CrossRef]
41. Youssef, N.A.; Shaban, S.A.; Ibrahim, F.A.; Mahmoud, A.S. Degradation of methyl orange using Fenton catalytic reaction. *Egypt J. Pet.* **2016**, *25*, 317–321. [CrossRef]
42. Yoon, S.; Bae, S. Novel synthesis of nanoscale zerovalent iron from coal fly ash and its application in oxidative degradation of methyl orange by Fenton reaction. *J. Hazard. Mater.* **2019**, *365*, 751–758. [CrossRef]
43. Liang, L.; Cheng, L.; Zhang, Y.; Wang, Q.; Wu, Q.; Xue, Y.; Meng, X. Efficiency and mechanisms of rhodamine B degradation in Fenton-like systems based on zero-valent iron. *RSC Adv.* **2020**, *10*, 28509–28515. [CrossRef]
44. Devi, L.G.; Kumar, S.G.; Reddy, K.M.; Munikrishnappa, C. Photo degradation of Methyl Orange an azo dye by Advanced Fenton Process using zero valent metallic iron: Influence of various reaction parameters and its degradation mechanism. *J. Hazard. Mater.* **2009**, *164*, 459–467. [CrossRef]

45. Effendi, H.; Romanto; Wardiatno, Y. Water quality status of Ciambulawung River, Banten Province, based on pollution index and NSF-WQI. *Procedia Environ. Sci.* **2015**, *24*, 228–237. [CrossRef]
46. Alara, O.R.; Abdurahman, N.H.; Olalere, O.A. Ethanolic extraction of bioactive compounds from Vernonia amygdalina leaf using response surface methodology as an optimization tool. *J. Food Meas. Charact.* **2018**, *12*, 1107–1122. [CrossRef]
47. Barchan, A.; Bakkali, M.; Arakrak, A.; Pagán, R.; Laglaoui, A. The effects of solvents polarity on the phenolic contents and antioxidant activity of three Mentha species extracts. *Int. J. Curr. Microbiol. Appl. Sci.* **2014**, *3*, 399–412.
48. Unachukwu, U.J.; Ahmed, S.; Kavalier, A.; Lyles, J.T.; Kennelly, E.J. White and green teas (*Camellia sinensis* var. *sinensis*): Variation in phenolic, methylxanthine, and antioxidant profiles. *J. Food Sci.* **2010**, *75*, C541–C548. [CrossRef]
49. Peng, L.; Song, X.; Shi, X.; Li, J.; Ye, C. An improved HPLC method for simultaneous determination of phenolic compounds, purine alkaloids and theanine in Camellia species. *J. Food Compos. Anal.* **2008**, *21*, 559–563. [CrossRef]
50. Khadijah, K.; Jayali, A.M.; Umar, S.; Sasmita, I. Penentuan total fenolik dan aktivitas antioksidan ekstrak etanolik daun samama (*Anthocephalus macrophylus*) asal Ternate, Maluku Utara. *J. Kimia Mulawarman* **2017**, *15*, 11–18.
51. Ford, L.; Theodoridou, K.; Sheldrake, G.N.; Walsh, P.J. A critical review of analytical methods used for the chemical characterisation and quantification of phlorotannin compounds in brown seaweeds. *Phytochem. Anal.* **2019**, *30*, 587–599. [CrossRef]
52. Puspitasari, L.; Arief, S.; Zulhadjri, Z. Ekstrak Daun Andalas sebagai Capping Agent dalam Green Hydrothermal Synthesis Nanopartikel Mangan Ferrit dan Aplikasinya sebagai Antibakteri. *Chim. Nat. Acta* **2019**, *7*, 20–26. [CrossRef]
53. Phan, C.M.; Nguyen, H.M. Role of capping agent in wet synthesis of nanoparticles. *J. Phys. Chem. A* **2017**, *121*, 3213–3219. [CrossRef]
54. Yunita, Y.; Nurlina, N.; Syahbanu, I. Sintesis Nanopartikel Zink Oksida (ZnO) dengan Penambahan Ekstrak Klorofil sebagai Capping Agent. *Positron* **2020**, *10*, 123–130. [CrossRef]
55. Gulati, S.; Sachdeva, M.; Bhasin, K.K. Capping agents in nanoparticle synthesis: Surfactant and solvent system. *AIP Conf. Proc.* **2018**, *1953*, 030214.
56. Yuvakkumar, R.; Elango, V.; Rajendran, V.; Kannan, N. Preparation and characterization of zero valent iron nanoparticles. *Dig. J. Nanomater. Biostruct.* **2011**, *6*, 1771–1776.
57. Buckingham, M.A.; Marken, F.; Aldous, L. The thermoelectrochemistry of the aqueous iron (ii)/iron (iii) redox couple: Significance of the anion and pH in thermogalvanic thermal-to-electrical energy conversion. *Sustain. Energy Fuels* **2018**, *2*, 2717–2726. [CrossRef]
58. Joseph, J.; Nagashri, K. Novel copper-based therapeutic agent for anti-inflammatory: Synthesis, characterization, and biochemical activities of copper (II) complexes of hydroxyflavone Schiff bases. *Appl. Biochem. Biotechnol.* **2012**, *167*, 1446–1458. [CrossRef]
59. Kumar, K.M.; Mandal, B.K.; Kumar, K.S.; Reddy, P.S.; Sreedhar, B. Biobased green method to synthesise palladium and iron nanoparticles using Terminalia chebula aqueous extract. *Spectrochim. Acta Part A Mol. Biomol. Spectrosc.* **2013**, *102*, 128–133. [CrossRef]
60. Wang, T.; Lin, J.; Chen, Z.; Megharaj, M.; Naidu, R. Green synthesized iron nanoparticles by green tea and eucalyptus leaves extracts used for removal of nitrate in aqueous solution. *J. Clean Prod.* **2014**, *83*, 413–419. [CrossRef]
61. Geng, S.; Shan, S.; Ma, H.; Liu, B. Antioxidant activity and α-glucosidase inhibitory activities of the polycondensate of catechin with glyoxylic acid. *PLoS ONE* **2016**, *11*, e0150412. [CrossRef]
62. Moreno-Vásquez, M.J.; Valenzuela-Buitimea, E.L.; Plascencia-Jatomea, M.; Encinas-Encinas, J.C.; Rodríguez-Félix, F.; Sánchez-Valdes, S.; Rosas-Burgos, E.C.; Ocaño-Higuera, V.M.; Graciano-Verdugo, A.Z. Functionalization of chitosan by a free radical reaction: Characterization, antioxidant and antibacterial potential. *Carbohydr. Polym.* **2017**, *155*, 117–127. [CrossRef]
63. Mohamed, S.E.; Khalifa, M.G.; Sayed, S.A.; Kamel, A.M.; Shalabi, M.E.H. Removal of Lignin from Pulp Waste Water's Black Liquor via By-Pass Cement Dust. *Eurasian Chem.-Technol. J.* **2009**, *11*, 51–59.
64. D'Souza, L.; Devi, P.; Divya Shridhar, M.P.; Naik, C.G. Use of Fourier Transform Infrared (FTIR) spectroscopy to study cadmium-induced changes in Padina tetrastromatica (Hauck). *Anal. Chem. Insights* **2008**, *3*, 135–143. [CrossRef]
65. Dong, H.; Zhao, F.; He, Q.; Xie, Y.; Zeng, Y.; Zhang, L.; Tang, L.; Zeng, G. Physicochemical transformation of carboxymethyl cellulose-coated zero-valent iron nanoparticles (nZVI) in simulated groundwater under anaerobic conditions. *Sep. Purif. Technol.* **2017**, *175*, 376–383. [CrossRef]
66. Mishra, D.; Arora, R.; Lahiri, S.; Amritphale, S.S.; Chandra, N. Synthesis and characterization of iron oxide nanoparticles by solvothermal method. *Prot. Met. Phys. Chem. Surfaces* **2014**, *50*, 628–631. [CrossRef]
67. Fitria, W.; Fatriasari, E.; Hermiati, N.S. Pengaruh lokasi tempat tumbuh terhadap kandungan kimia kayu sengon (*Paraserianthes falcataria*) sebagai bahan baku pulp. *J. Ilmu Teknol. Has. Hutan* **2010**, *3*, 45–50.
68. Muhammad, F.; Xia, M.; Li, S.; Yu, X.; Mao, Y.; Muhammad, F.; Huang, X.; Jiao, B.; Yu, L.; Li, D. The reduction of chromite ore processing residues by green tea synthesized nano zerovalent iron and its solidification/stabilization in composite geopolymer. *J. Clean Prod.* **2019**, *234*, 381–391. [CrossRef]
69. Wang, W.; Jin, Z.H.; Li, T.L.; Zhang, H.; Gao, S. Preparation of spherical iron nanoclusters in ethanol–water solution for nitrate removal. *Chemosphere* **2006**, *65*, 1396–1404. [CrossRef]
70. Aziztyana, A.P.; Wardhani, S.; Prananto, Y.P.; Purwonugroho, D.; Darjito. Optimisation of Methyl Orange Photodegradation Using TiO$_2$-Zeolite Photocatalyst and H$_2$O$_2$ in Acid Condition. *IOP Conf. Ser. Mater. Sci. Eng.* **2019**, *546*, 042047. [CrossRef]

MDPI
St. Alban-Anlage 66
4052 Basel
Switzerland
Tel. +41 61 683 77 34
Fax +41 61 302 89 18
www.mdpi.com

Materials Editorial Office
E-mail: materials@mdpi.com
www.mdpi.com/journal/materials